Lecture Notes in Networks and Systems

Volume 10

Series editor

Janusz Kacprzyk, Polish Academy of Sciences, Warsaw, Poland
e-mail: kacprzyk@ibspan.waw.pl

The series "Lecture Notes in Networks and Systems" publishes the latest developments in Networks and Systems—quickly, informally and with high quality. Original research reported in proceedings and post-proceedings represents the core of LNNS.

Volumes published in LNNS embrace all aspects and subfields of, as well as new challenges in, Networks and Systems.

The series contains proceedings and edited volumes in systems and networks, spanning the areas of Cyber-Physical Systems, Autonomous Systems, Sensor Networks, Control Systems, Energy Systems, Automotive Systems, Biological Systems, Vehicular Networking and Connected Vehicles, Aerospace Systems, Automation, Manufacturing, Smart Grids, Nonlinear Systems, Power Systems, Robotics, Social Systems, Economic Systems and other. Of particular value to both the contributors and the readership are the short publication timeframe and the world-wide distribution and exposure which enable both a wide and rapid dissemination of research output.

The series covers the theory, applications, and perspectives on the state of the art and future developments relevant to systems and networks, decision making, control, complex processes and related areas, as embedded in the fields of interdisciplinary and applied sciences, engineering, computer science, physics, economics, social, and life sciences, as well as the paradigms and methodologies behind them.

More information about this series at http://www.springer.com/series/15179

Durgesh Kumar Mishra · Malaya Kumar Nayak
Amit Joshi
Editors

Information and Communication Technology for Sustainable Development

Proceedings of ICT4SD 2016, Volume 2

Springer

Editors
Durgesh Kumar Mishra
Microsoft Innovation Centre
Sri Aurobindo Institute of Technology
Indore, Madhya Pradesh
India

Amit Joshi
Department of Information Technology
Sabar Institute of Technology
Ahmedabad, Gujarat
India

Malaya Kumar Nayak
Dagenham
UK

ISSN 2367-3370 ISSN 2367-3389 (electronic)
Lecture Notes in Networks and Systems
ISBN 978-981-10-9998-4 ISBN 978-981-10-3920-1 (eBook)
https://doi.org/10.1007/978-981-10-3920-1

Softcover reprint of the hardcover 1st edition 2017

Printed on acid-free paper

This Springer imprint is published by Springer Nature
The registered company is Springer Nature Singapore Pte Ltd.
The registered company address is: 152 Beach Road, #21-01/04 Gateway East, Singapore 189721, Singapore

Preface

The volume contains the best selected papers presented at the ICT4SD 2016: International Conference on Information and Communication Technology for Sustainable Development. The conference was held during 1 and 2 July 2016 in Goa, India and organized communally by the Associated Chambers of Commerce and Industry of India, Computer Society of India Division IV and Global Research Foundation. It targets state of the art as well as emerging topics pertaining to ICT and effective strategies for its implementation for engineering and intelligent applications. The objective of this international conference is to provide opportunities for the researchers, academicians, industry persons and students to interact and exchange ideas, experience and expertise in the current trend and strategies for information and communication technologies. Besides this, participants are also enlightened about vast avenues, and current and emerging technological developments in the field of ICT in this era and its applications are thoroughly explored and discussed. The conference attracted a large number of high-quality submissions and stimulated the cutting-edge research discussions among many academically pioneering researchers, scientists, industrial engineers and students from all around the world. The presenter proposed new technologies, shared their experiences and discussed future solutions for designing infrastructure for ICT. The conference provided common platform for academic pioneering researchers, scientists, engineers and students to share their views and achievements. The overall focus was on innovative issues at international level by bringing together the experts from different countries. Research submissions in various advanced technology areas were received, and after a rigorous peer-review process with the help of program committee members and external reviewer, 105 papers were accepted with an acceptance ratio of 0.43. The conference featured many distinguished personalities like Prof. H.R. Vishwakarma, Prof. Shyam Akashe, Mr. Nitin from Goa Chamber of Commerce, Dr. Durgesh Kumar Mishra and Mr. Aninda Bose from Springer India Pvt. Limited, and also a wide panel of start-up entrepreneurs were present. Separate invited talks were organized in industrial and academia tracks on both days. The conference also hosted tutorials and workshops for the benefit of participants. We are indebted to CSI Goa Professional Chapter, CSI Division IV, and Goa Chamber of Commerce for their immense support to make this conference possible in such a grand scale. A total of 12 sessions were organized as a part of ICT4SD 2016 including nine technical, one plenary, one inaugural and one valedictory session. A total of 73 papers were presented in nine technical sessions with high discussion insights. Our sincere thanks to all sponsors, press, print and electronic media for their excellent coverage of this conference.

Indore, India
Dagenham, UK
Ahmedabad, India
July 2016

Durgesh Kumar Mishra
Malaya Kumar Nayak
Amit Joshi

Organisation Committee

Advisory Committee

Dr. Dharm Singh, Namibia University of Science and Technology, Namibia
Dr. Aynur Unal, Standford University, USA
Mr. P.N. Jain, Add. Sec., R&D, Government of Gujarat, India
Prof. J. Andrew Clark, Computer Science, University of York, UK
Dr. Anirban Basu, Vice President, CSI
Prof. Mustafizur Rahman, Endeavour Research Fellow, Australia
Dr. Malay Nayak, Director-IT, London
Mr. Chandrashekhar Sahasrabudhe, ACM, India
Dr. Pawan Lingras, Saint Mary's University, Canada
Prof. (Dr.) P. Thrimurthy, Past President, CSI
Dr. Shayam Akashe, ITM, Gwalior, MP, India
Dr. Bhushan Trivedi, India
Prof. S.K. Sharma, Pacific University, Udaipur, India
Prof. H.R. Vishwakarma, VIT, Vellore, India
Dr. Tarun Shrimali, SGI, Udaipur, India
Mr. Mignesh Parekh, Ahmedabad, India
Mr. Sandeep Sharma, Joint CEO, SCOPE
Dr. J.P. Bhamu, Bikaner, India
Dr. Chandana Unnithan, Victoria University, Australia
Prof. Deva Ram Godara, Bikaner, India
Dr. Y.C. Bhatt, Chairman, CSI Udaipur Chapter
Dr. B.R. Ranwah, Past Chairman, CSI Udaipur Chapter
Dr. Arpan Kumar Kar, IIT Delhi, India

Organising Committee

Organising Chairs
Ms. Bhagyesh Soneji, Chairperson ASSOCHAM Western Region

Co-Chair
Dr. S.C. Satapathy, ANITS, Visakhapatnam

Members
Shri. Bharat Patel, COO, Yudiz Solutions
Dr. Basant Tiwari, Bhopal
Dr. Rajveer Shekhawat, Manipal University, Jaipur
Dr. Nilesh Modi, Chairman, ACM Ahmedabad Chapter
Dr. Harshal Arolkar, Assoc. Prof., GLS Ahmedabad
Dr. G.N. Jani, Ahmedabad, India
Dr. Vimal Pandya, Ahmedabad, India
Mr. Vinod Thummar, SITG, Gujarat, India
Mr. Nilesh Vaghela, Electromech, Ahmedabad, India
Dr. Chirag Thaker, GEC, Bhavnagar, Gujarat, India
Mr. Maulik Patel, SITG, Gujarat, India
Mr. Nilesh Vaghela, Electromech Corp., Ahmedabad, India
Dr. Savita Gandhi, GU, Ahmedabad, India
Mr. Nayan Patel, SITG, Gujarat, India
Dr. Jyoti Parikh, Professor, GU, Ahmedabad, India
Dr. Vipin Tyagi, Jaypee University, Guna, India
Prof. Sanjay Shah, GEC, Gandhinagar, India
Dr. Chirag Thaker, GEC, Bhavnagar, Gujarat, India
Mr. Mihir Chauhan, VICT, Gujarat, India
Mr. Chetan Patel, Gandhinagar, India

Program Committee

Program Chair
Dr. Durgesh Kumar Mishra, Chairman, Div IV, CSI

Members
Dr. Priyanka Sharma, RSU, Ahmedabad
Dr. Nitika Vats Doohan, Indore
Dr. Mukesh Sharma, SFSU, Jaipur
Dr. Manuj Joshi, SGI, Udaipur, India
Dr. Bharat Singh Deora, JRNRV University, Udaipur
Prof. D.A. Parikh, Head, CE, LDCE, Ahmedabad, India
Prof. L.C. Bishnoi, GPC, Kota, India

Mr. Alpesh Patel, SITG, Gujarat
Dr. Nisheeth Joshi, Banasthali University, Rajasthan, India
Dr. Vishal Gaur, Bikaner, India
Dr. Aditya Patel, Ahmedabad University, Gujarat, India
Mr. Ajay Choudhary, IIT Roorkee, India
Dr. Dinesh Goyal, Gyan Vihar, Jaipur, India
Dr. Devesh Shrivastava, Manipal University, Jaipur
Dr. Muneesh Trivedi, ABES, Gaziabad, India
Prof. R.S. Rao, New Delhi, India
Dr. Dilip Kumar Sharma, Mathura, India
Prof. R.K. Banyal, RTU, Kota, India
Mr. Jeril Kuriakose, Manipal University, Jaipur, India
Dr. M. Sundaresan, Chairman, CSI Coimbatore Chapter
Prof. Jayshree Upadhyay, HOD-CE, VCIT, Gujarat
Dr. Sandeep Vasant, Ahmedabad University, Gujarat, India

Contents

Editors and Contributors

About the Editors

Dr. Durgesh Kumar Mishra has received his M. Tech degree in Computer Science from DAVV, Indore, in 1994 and Ph.D. degree in Computer Engineering in 2008. Presently, he has been working as a Professor (CSE) and Director, Microsoft Innovation Center at Sri Aurobindo Institute of Technology, Indore, MP, India. He is also a visiting faculty at IIT-Indore, India. He has 24 years of teaching and 10 years of research experience. He has completed his Ph.D. under the guidance of late Dr. M. Chandwani on "Secure Multi-Party Computation for Preserving Privacy". Dr. Mishra has published more than 90 papers in refereed international/national journals and conferences including IEEE and ACM conferences. He has organized many conferences such as WOCN, CONSEG and CSIBIG in the capacity of conference General Chair and Editor of conference proceedings. He is a Senior Member of IEEE and held many positions such as Chairman, IEEE MP Subsection (2011–2012) and Chairman IEEE Computer Society Bombay Chapter (2009–2010). Dr. Mishra has also served the largest technical and professional association of India, the Computer Society of India (CSI) by holding positions as Chairman, CSI Indore Chapter, State Student Coordinator—Region III MP, Member–Student Research Board and Core Member–CSI IT Excellence Award Committee. He has been recently elected as Chairman CSI Division IV Communication at National Level (2014–2016). Recently, he has been awarded with "Paper Presenter at International Level" by Computer Society of India. He is also the Chairman of Division IV Computer Society of India and Chairman of ACM Chapter covering Rajasthan and MP State.

Dr. Malaya Kumar Nayak is a technology and business solution expert and an academia visionary with over 17 years of experience in leading design, development, and implementation of high-performance technology solutions. He has proven abilities to bring the benefits of IT to solve business issues while delivering applications, infrastructure, costs, and risks. Dr. Nayak has provided strategic direction to senior leadership on technology. He is skilled at building teams of top 1% performers as well as displaying versatility for the strategic, tactical, and management aspects of technology. He has managed a team of professionals noted for integrity and competency. He is a member of Program Committee of Ninth International Conference on Wireless and Optical Communications Networks (WOCN 2012); Program Committee of 3rd International Conference on Reliability, Infocom Technologies and Optimization (ICRITO 2014); Editorial Advisory Board of International Book titled "Issues of Information Communication Technology (ICT) in Education"; Editorial Advisory Board of International Book titled "Global Outlook on Education"; Editorial Board of International Journal titled "Advances in Management"; Editorial Advisory Board of International Journal titled "Delving: Journal of Technology and Engineering Sciences (JTES)"; Editorial Advisory Board of International Journal of Advanced Engineering Technology

(IJAET) and Journal of Engineering Research and Studies (JERS); and Editorial Advisory Board of ISTAR: International Journal of Information and Computing Technology. Dr. Nayak has also served as reviewer in international reputed conferences such as CSI Sixth International Conference on Software Engineering (CONSEG 2012); Tenth International Conference on Wireless and Optical Communications Networks (WOCN 2013); Eleventh International Conference on Wireless and Optical Communications Networks (WOCN 2014); and CSI BIG 2014 International Conference on IT in Business, Industry and Government.

Mr. Amit Joshi is a young entrepreneur and researcher who has completed his graduation (B. Tech.) in information technology and M.Tech. in computer science and engineering and pursuing his research in the areas of cloud computing and cryptography. He has an experience of around 6 years in academic and industry in prestigious organizations of Udaipur and Ahmedabad. Currently, he is working as an Assistant Professor in Department of Information Technology at Sabar Institute in Gujarat. He is an active member of ACM, CSI, AMIE, IACSIT Singapore, IDES, ACEEE, NPA, and many other professional societies. He also holds the post of Honorary Secretary of CSI Udaipur Chapter and Secretary of ACM Udaipur Chapter. He has presented and published more than 30 papers in national and international journals/conferences of IEEE and ACM. He has edited three books on diversified subjects including Advances in Open Source Mobile Technologies, ICT for Integrated Rural Development, and ICT for Competitive Strategies. He has also organized more than 15 national and international conference and workshops including International Conference ICTCS 2014 at Udaipur through ACM–ICPS. For his contribution toward the society, he has been given Appreciation Award by the Institution of Engineers (India), ULC, on the celebration of engineers, September 15, 2014, and by SIG-WNs Computer Society of India on the Occasion of ACCE 2012 on February 11, 2012.

Contributors

Parakh Agarwal Amity University Uttar Pradesh, Noida, UP, India

Shikha Agarwal Department of Computer Science, Central University of South Bihar, Patna, India

Akash Agrawal Department of Computer Engineering and Information Technology, Veermata Jijabai Technological Institute, Mumbai, India

Anshul Agrawal ITM University, Gwalior, Madhya Pradesh, India

Sanjay Ahuja Amity University Uttar Pradesh, Noida, UP, India

Chakshu Ahuja Department of Computer Science and Engineering, National Institute of Technology, Tiruchirappalli, Tiruchirappalli, India

Shyam Akashe ITM University, Gwalior, Madhya Pradesh, India

Nathaniel Albuquerque Department of Electronics and Communication Engineering, ASET, Amity University, Noida, Uttar Pradesh, India

Khwairakpam Amitab Department of Information Technology, School of Technology, North Eastern Hill University, Shillong, India

V. Amruth Maharaja Institute of Technology, Mysore, India

Sanchita Arora School of Computer Science and Engineering, Galgotias University, Greater Noida, India

Shefali Arora Department of Computer Science and Engineering, Thapar University, Patiala, India

Kanika Bahl Department of Computer Science and Engineering, Lovely Professional University, Phagwara, India

Amlan Basu Department of Electronics and Communication Engineering, ITM University, Gwalior, Madhya Pradesh, India

Neeraj Bhargava Department of Computer Science, School of Engineering & System Sciences, M.D.S. University Ajmer, Ajmer, India

Ritu Bhargava Department of Computer Science, Aryabhatt International College, Ajmer, India

Debashree Bhattacharjee Department of Information Technology, School of Technology, North Eastern Hill University, Shillong, India

Gajanan K. Birajdar Department of Electronics & Telecommunication, Pillai HOC College of Engineering & Technology, Raigad, Maharashtra, India

Nikhil Bisht Amity University Uttar Pradesh, Noida, UP, India

Sushma Niket Borade Department of Computer Science and Information Technology, Dr. Babasaheb Ambedkar Marathwada University, Aurangabad, India

Ajay P. Chainani Thadomal Shahani Engineering College, Mumbai, Maharashtra, India

Deepali Chalak Department of E&TC, Maharashtra Institute of Technology, Pune, India

Inderveer Chana Department of Computer Science and Engineering, Thapar University, Patiala, India

Ekta Chauhan ITM, Tonk, Gwalior, India

Vijay Chauhan Gtu Pg School, Ahmedabad, Gujarat, India

Mahesh Chavan KIT College of Engineering, Kolhapur, India

Shamkumar Chavan Department of Technology, Shivaji University, Kolhapur, India

Shweta Chawla SC Cyber Solutions, Pune, India

Megha Chhabra Amity University Uttar Pradesh, Noida, UP, India

Santosh S. Chikne Thadomal Shahani Engineering College, Mumbai, Maharashtra, India

Pritam Dash Amity University Uttar Pradesh, Noida, UP, India

Priyanka Deshmane Department of Computer Engineering, Dr. D.Y. Patil Institute of Engineering and Technology (DYPIET), Pimpri, Pune, India

Ratnadeep R. Deshmukh Department of Computer Science and Information Technology, Dr. Babasaheb Ambedkar Marathwada University, Aurangabad, India

Poonam Dhaka University of Namibia, Windhoek, Namibia

Sujit Dharmapatre Department of E&TC, Maharashtra Institute of Technology, Pune, India

Meghana A. Divakar M S Ramaiah Institute of Technology, Bangalore, India

Neelesh Dixt ITM University, Gwalior, Madhya Pradesh, India

Nikunj D. Doshi Thadomal Shahani Engineering College, Mumbai, Maharashtra, India

Pooja Dubey Gtu Pg School, Ahmedabad, Gujarat, India

Deepak Gadde Sinhgad Institute of Technology, Lonavala, Pune, India

Rahul S. Gaikwad Department of Computer Enginering and Information Technology, VJTI, Mumbai, India

Vinita P. Gangraj Department of Electronics & Telecommunication, Pillai HOC College of Engineering & Technology, Raigad, Maharashtra, India

Anchal Garg Department of Computer Science & Engineering, Amity School of Engineering & Technology, Amity University, Uttar Pradesh, Noida, India

Roopali Garg Department of IT, UIET, Panjab University, Chandigarh, India

Nidhi Gaur Department of Electronics and Communication Engineering, ASET, Amity University, Noida, Uttar Pradesh, India

Jayashri C. Gholap Sinhgad Institute of Technology, Lonavala, Pune, India

Shivani Goel Department of Computer Science and Engineering, Thapar University, Patiala, India

Ayush Goyal Amity University Uttar Pradesh, Noida, UP, India

Somya Goyal PDM College of Engineering, Bahadurgarh, India

Sugandha Gupta Department of Computer Science and Engineering, Thapar University, Patiala, India

Heena Handa Department of Computer Science & Engineering, Amity School of Engineering & Technology, Amity University, Uttar Pradesh, Noida, India

Vinayak Hegde Deparment of Computer Science, Amrita Vishwa Vidyapeetham Mysuru Campus, Amrita University, Mysuru, Karnataka, India

Saroj Hiranwal Computer science, Rajasthan Institute of Engineering and Technology, Rajasthan Technical University, Jaipur, India

C.R. Jadhav Department of Computer Engineering, D.Y. Patil Institute of Engineering & Technology, Pimpri, Pune, India

Hitesh Jangir Mody University of Science and Technology, CET, Lakshmangarh, Rajasthan, India

Kuntesh Jani Government Engineering College, Gandhinagar, Gujarat, India

Manfred Janik University of Namibia, Windhoek, Namibia

Dharm Singh Jat Namibia University of Science and Technology, Windhoek, Namibia

R. Jenkin Suji Department of Electronics and Communication Engineering, ITM University, Gwalior, Madhya Pradesh, India

Jenifer Mariam Johnson Department of Electrical Engineering, National Institute of Technology, Raipur, CG, India

Prashant Johri School of Computer Science and Engineering, Galgotias University, Greater Noida, India

Mansi Joshi School of Computer Studies, Ahmedabad University, Ahmedabad, India

Sanaj Singh Kahlon Amity University Uttar Pradesh, Noida, UP, India

Monika Kalra Mody University of Science and Technology, Lakshmangarh, Sikar, Rajasthan, India

Debdatta Kandar Department of Information Technology, School of Technology, North Eastern Hill University, Shillong, India

Asim Z. Karel Thadomal Shahani Engineering College, Mumbai, Maharashtra, India

Anita Gautam Khandizod Department of Computer Science and Information Technology, Dr. Babasaheb Ambedkar Marathwada University, Aurangabad, India

Iani de Kock University of Namibia, Windhoek, Namibia

Siddhartha Kosti Indian Institute of Technology Kanpur, Kanpur, India

Akhouri Pramod Krishna Department of Remote Sensing, Birla Institute of Technology, Mesra, Ranchi, India

Divya Kumar Computer Science & Engineering Department, Motilal Nehru National Institute of Technology Allahabad, Allahabad, UP, India

Mithun Kumar School of Computer Science and Engineering, Galgotias University, Greater Noida, India

Jeril Kuriakose St. John College of Engineering and Technology, Palghar, India; Manipal University Jaipur, Jaipur, India

Niranjan Lal Mody University of Science and Technology, Lakshmangarh, Sikar, Rajasthan, India; Suresh Gyan Vihar University, Jaipur, Rajasthan, India

Madhulika Department of Computer Science & Engineering, Amity School of Engineering & Technology, Amity University, Uttar Pradesh, Noida, India

Hilma Mbandeka University of Namibia, Windhoek, Namibia

Anu Mehra Department of Electronics and Communication Engineering, ASET, Amity University, Noida, Uttar Pradesh, India

Phalit Mehta School of Information Technology, Centre for Development of Advanced Computing, Noida, India

Alok Bhushan Mukherjee Department of Remote Sensing, Birla Institute of Technology, Mesra, Ranchi, India

Atty Mwafufya University of Namibia, Windhoek, Namibia

Prafull Chandra Narooka Department of Computer Science, MJRP University, Jaipur, India

Nemi Chandra Rathore Central University of South Bihar, Patna, India

Nidhi Department of IT, UIET, PU, Chandigarh, India

Priyanka Paliwal Computer Science & Engineering Department, Motilal Nehru National Institute of Technology Allahabad, Allahabad, UP, India

Shivam Pandey Mody University of Science and Technology, CET, Lakshmangarh, Rajasthan, India

Mayuri Pandya Maharaja Krishnakumarsinhji, Bhavnagar University, Bhavnagar, India

Amrita Parashar Amity University, Gwalior, Madhya Pradesh, India

Anubha Parashar Vaish College of Engineering, Rohtak, India

Apoorva Parashar Maharshi Dayanand University, Rohtak, India

Vivek Parashar Amity University, Gwalior, Madhya Pradesh, India

Nilanchal Patel Department of Remote Sensing, Birla Institute of Technology, Mesra, Ranchi, India

Abha Pathak Department of Computer Engineering, Dr. D.Y. Patil Institute of Engineering and Technology (DYPIET), Pimpri, Pune, India

Priya Pathak MPCT, Banasthali, Gwalior, India

Pramod Patil Department of Computer Engineering, Dr. D.Y. Patil Institute of Engineering and Technology (DYPIET), Pimpri, Pune, India

Kamni Patkar ITM University, Gwalior, Madhya Pradesh, India

Karishma Patkar ITM University, Gwalior, Madhya Pradesh, India

Priyanka Powar Department of Computer Engineering, D.Y. Patil Institute of Engineering & Technology, Pimpri, Pune, India

Kritika Prakash Department of Electronics and Communication Engineering, ASET, Amity University, Noida, Uttar Pradesh, India

Samimul Qamar King Khalid University, Abha, Kingdom of Saudi Arabia

Agrawal Avni Rajeev Computer science, Rajasthan Institute of Engineering and Technology, Rajasthan Technical University, Jaipur, India

Shiwani Rana Department of IT, UIET, Panjab University, Chandigarh, India

Prabhat Ranjan Department of Computer Science, Central University of South Bihar, Patna, India

Sugandha Rathi SRCEM, Gwalior, Rajasthan, India

Ravi Sahran Central University of Rajasthan, Ajmer, Rajasthan, India

Vijay K. Sambhe Department of Computer Engineering and Information Technology, Veermata Jijabai Technological Institute, Mumbai, India

Kriti Saroha School of Information Technology, Centre for Development of Advanced Computing, Noida, India

Darpan Shah Government Engineering College, Gandhinagar, Gujarat, India

Dipti B. Shah G.H. Patel Post Graduate Department of Computer Science, S. P. University, Gujarat, India

S. Shanthi Therese Information Technology, Thadomal Shahani Engineering College, Mumbai, India

Anu Sharma MMICT & BM, Maharishi Markandeshwer University, Haryana, India

Bharati Sharma Department of Electronics and Communication Engineering, ITM University, Gwalior, Madhya Pradesh, India

Bhavna Sharma Apex Institute of Engineering & Technology, Jaipur, India

Nitin Sharma Department of IT, UIET, PU, Chandigarh, India

Rohitt Sharma Department of Computer Science and Engineering, Lovely Professional University, Phagwara, India

Vijay Kumar Sharma Computer science, Rajasthan Institute of Engineering and Technology, Rajasthan Technical University, Jaipur, India

Aarti Singh MMICT & BM, Maharishi Markandeshwer University, Haryana, India

Ayush Singh SCIT, Manipal University Jaipur, Jaipur, Rajasthan, India

Manjeet Singh Central University of Rajasthan, Ajmer, Rajasthan, India

E. Sivasankar Department of Computer Science and Engineering, National Institute of Technology, Tiruchirappalli, Tiruchirappalli, India

Devesh Kumar Srivastava SCIT, Manipal University Jaipur, Jaipur, Rajasthan, India

Aishwarya Suresh Deparment of Computer Science, Amrita Vishwa Vidyapeetham Mysuru Campus, Amrita University, Mysuru, Karnataka, India

H.S. Sushma Rao Deparment of Computer Science, Amrita Vishwa Vidyapeetham Mysuru Campus, Amrita University, Mysuru, Karnataka, India

Jinal H. Tailor S.P. University, Gujarat, India

Prakash Singh Tanwar Department of Computer Science, MJRP University, Jaipur, India

Manthan Thakker Information Technology, Thadomal Shahani Engineering College, Mumbai, India

Shanthi S. Therese Thadomal Shahani Engineering College, Mumbai, Maharashtra, India

Solley Thomas Carmel College of Arts Science & Commerce for Women, Nuvem, Salcete, Goa, India

Heena Timani School of Computer Studies, Ahmedabad University, Ahmedabad, India

Vineeta Tiwari CDAC-Acts, Pune, India

Rashmeet Toor Department of Computer Science and Engineering, Thapar University, Patiala, India

Ankita Tripathi Mody University of Science and Technology, CET, Lakshmangarh, Rajasthan, India

Sandeep S Udmale Department of Computer Engineering and Information Technology, Veermata Jijabai Technological Institute, Mumbai, India

Prachi Ved Information Technology, Thadomal Shahani Engineering College, Mumbai, India

Shivangi Vyas Information Technology, Thadomal Shahani Engineering College, Mumbai, India

Anamika Yadav Department of Electrical Engineering, National Institute of Technology, Raipur, CG, India

Ravindar Yadav Manipal University Jaipur, Jaipur, India

Green IT and Environmental Sustainability Issues

Priyanka Paliwal and Divya Kumar

Abstract The climate change catastrophe, global warming, and environmental sustainability are some of the major challenges our globe is facing today. There is an urgent requirement of the global rules and policies to ensure environment sustainability. The Green IT is a genuine effort initiated from computer science world toward sustained environment. It is a worldwide rung which lays the foundations for the proficient use of computers and other similar computing resources so as to produce minimal carbon emissions. Because of its direct impact on environment, Green IT has also emerged as a research topic of prime importance. In this paper, we have studied and quantified these varied impacts. As well as we have proposed an action plan to diminish these awful impacts by changing the designs and development strategies of various silicon valley products. The stated action plan has also been verified on a sample test case study in this manuscript.

Keywords Environment sustainability · Green information technology
Carbon emission · Energy efficiency

1 Introduction

Sustainability is the ability of any system to remain productive indefinitely. It defines the capability of a system to remain endure. The objective of sustained development can be defined as meeting the demands of the present generation without compromising the capacity of future generations to meet their own needs [1]. The three pillars of sustainability [2] are: (a) economic, (b) social, and (c) environment as shown in Fig. 1. Environmental sustainability has emerged as a fundamental interdisciplinary

P. Paliwal (✉) · D. Kumar
Computer Science & Engineering Department, Motilal Nehru National Institute of Technology Allahabad, Allahabad, UP, India
e-mail: priyanka07mnnit@gmail.com

D. Kumar
e-mail: divyak@mnnit.ac.in

D.K. Mishra et al. (eds.), *Information and Communication Technology for Sustainable Development*, Lecture Notes in Networks and Systems 10,
https://doi.org/10.1007/978-981-10-3920-1_1

Fig. 1 Three pillars of sustainability

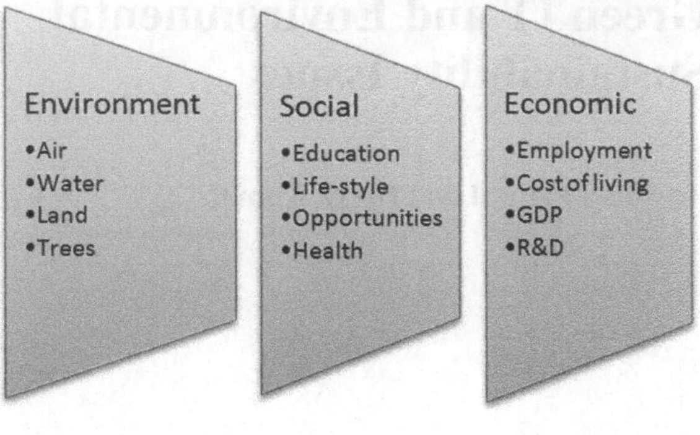

research area since past decade. We as human being have learnt well that optimum resource utilization for any production is very necessary but in such a way so as to cause insignificant negative impacts on environment [3]. The IT (Information Technology) industry has always been acknowledged as a capital and money-generating industry. The total revenue generated amounts to nearly $407.3 billion in the year 2013 [4]. Besides finance, this sector has also shown tremendous growth in nurturing innovations and designs which are prolific and beneficial for human beings in numerous aspects. Thus, this industry is touching the lives of almost every common man in many ways by making things easier and efficient [5]. However, IT sector has another important impact on environmental sustainability and that is the accountability for 2% of share in the global CO_2 production [6–10]. Because of this downbeat impact, the term Green IT/Green ICT (Information and Communication Technology) has gained impetus since past decade [11, 12]. Green IT refers to the direct application of science and technology to the world of information technology with an objective of reducing the resource consumption and harmful emissions during the entire ICT life cycle ranging from designing and development to deployment, use and disposal [8, 13]. It has been clearly identified that ICT may help in sustaining the environment in two ways. One is to alleviate the ecological impact of Information Technology; this involves optimizing their own activities like production and disposal of computers, i.e., minimizing the direct negative impact of IT and its processes on environment. The other is to maximize the positive impacts of computer on various vital sectors. This could be achieved by making software industry competent enough in serving other industries to reduce their carbon foot prints. Green IT is thus an indispensable research dimension. This research area is promising and goal oriented. To measure the potential of Green IT in supporting environmental sustainability is the challenge for this paper. The rest of this manuscript is organized as follows: First, we have discussed and quantified the major source of emissions in the computer industry then we have proposed an action plan for IT industries to reduce their CO_2 emissions. Finally, we have concluded the manuscript with the challenges that any IT industry may face in applying this plan into action.

2 ICT Advantages

Computers and similar computing devices have always been actively used as computing sources from the year 1970. Since their onset, these devices have changed the face of computing era because of their manifold development and intensification [14]. ICT has a sound potential for offering a great quality of service in manufacturing and production processes. It also offers unquestionable support for management and information dissemination. ICT has not only helped the different organizations but also individuals to attain a whole new zenith. It could also be used for raising environmental concerns in various ways leading to the advancement in all the pillars of sustainability. ICT is the main contributor in the social and industrial activities as it provides various obligatory services both for home-entertainment and office-work. Some of the few examples include distance-education, video-conferencing, robotics, e-governance, e-commerce, e-billing, telemedicine, social-networking, Internet, and other numerous R&D programs [15].

3 CO_2 Emission Sources in Industry

In spite of having various advantages as discussed in the previous section the computer science industry is pinked for the annual production of approximately 850 million tons of CO_2, the main greenhouse gas. This amounts to the 2% of the global carbon emission. What is more risky and unsafe is the estimate that this may increase to 6–8% by year 2020 [3, 4, 16–19]. The major elements of ICT industry that are responsible for this emission chiefly because of power consumption are: Personal Computers and Monitors, Servers/Data Centers, Networks (both mobile and fixed-line), Other Computer Peripherals like printers, mouse, storage, etc. The share of emission from each of these elements is shown in the following Fig. 2. Energy, predominantly as electricity, is used during the whole service lifecycle of ICT products. The mathematical steps for deriving the volume of CO_2 produced per kilowatt hour generated energy/electricity is shown in Algorithm 1. For

Fig. 2 Carbon emission percentage from various ICT elements

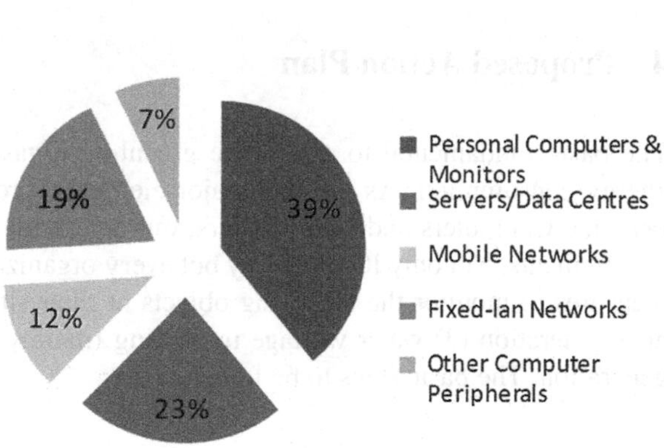

example, suppose a 19 in. TFT monitor consumes 50 W of energy. Its kWh rating will be 0.050 kWh. The annual CO_2 emission from this monitor will be 383.25 pounds (0.050 × 10 × 365 × 2.1) if the monitor is used for 10 h daily. For more detailed calculation of the carbon emission from various real world entities we redirect the readers to [20]. In addition to this the prodigious e-waste generated from the disposal/dumps of ICT products like outdated monitors, mobile phones, silicon-chips, etc., is full of toxicants like lead and mercury [21]. The amount of e-waste is swiftly rising in the current time. Recycling of this e-waste further requires space and energy which creates air pollution and extra green house gas emission as by-products.

In addition to this, the prodigious e-waste generated from the disposal/dumps of ICT products which includes outdated monitors, mobile phones, silicon-chips, etc., is full of toxicants like lead and mercury [21]. The amount of e-waste is swiftly rising in the current time. Recycling of this e-waste further requires space and energy which creates air pollution and extra green house gas emission as by-products. The problem of e-waste management is getting so much grave and legislative that the Government of India (and other countries also) have framed various guidelines for the disposal of the e-waste [22].

Algorithm 1 Calculation of Volume of CO_2

Inputs: *kWh* or *Wh*, rating of the appliance.
U, appliance usage hours/day.
Temporary Variable: *Z*, volume of CO_2 produced.
Output: *ACE*, Annual CO_2 emissions
Step 1: If rating is in *Wh* convert it into *kWh*

$$kWh = Wh/1000$$

Step 2: Set $Z = 2.1$ pounds according to [20, 23].
Step 3: Set $ACE = kWh * U * 365 * Z$
Step 4: Return *ACE*

4 Proposed Action Plan

The basic fundamental to shrink the greenhouse gas emission is to decrease the energy consumption. As the two major elements, producing carbon emission are Personal Computers and Data Centers, our action plan will mainly focus on these two elements. Not only ICT industry but every organization must utilize metrics and measures to monitor the following objects at their sites: (1) energy utilization (2) heat generation (3) water wastage in cooling (mainly for data centers) (4) e-waste generation. The basic steps to be followed are:

1. Finalize, the ICT-related investments and develop a calculated plan to quantify each of the aforementioned objects.
2. Take the Gartner's [4] approach for reduce, reuse and recycle:
   (a) Trim down power distribution losses, e.g., using renewable resources.
   (b) Lessen cooling loads, e.g., improvements in air flow movements.
   (c) Decrease IT loads, e.g., using virtualization and efficient servers and designing efficient software.
   (d) Decrease mobile computing loads to enhance battery life cycle.
   (e) Print only what is necessary. Use recycled paper.
   (f) Disassemble the outdated products and try to reuse the spare parts. The products those are no longer utilizable for our organization might be used in some different organization.
3. Learn that eco-friendly operations and sustained environment is the responsibility of all of us. Everyone in the organization should be committed for energy efficiency.

4.1 *Remedies for Reducing CO_2 Emission from Personal Computers*

1. Use eco-labeling of all IT products.
2. Design software to make most use of multi-core processors.
3. Set your computer in power-saving mode so as to dim the screen lights when computer remains inactive for some predefined duration.
4. Use thin clients with least processing powers.

4.2 *Remedies for Reducing CO_2 Emission from Data Centers*

1. To increase the server utilization use virtualization so as to merge servers to form a single-host server which could answer multiple different clients with different requirements.
2. Retire old servers which are no longer in use.
3. Use storage data management strategies like de-duplication and compression for energy efficient data structure operations like update and retrieval.
4. Use variable speed fans which run according to the processor utilization.
5. Properly design the hot and cold airflows.

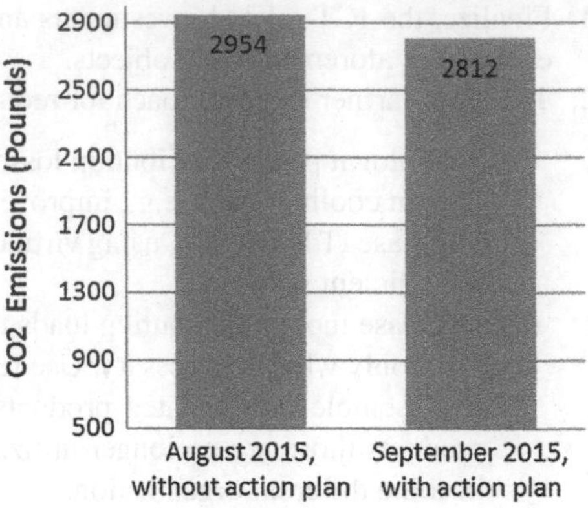

Fig. 3 Decline in CO_2 levels with defined action plan

4.3 Remedies for Reducing CO_2 Emission from Other ICT Elements

1. Switch off the computing devices and other electrical appliances when not in use.
2. Take time for regular maintenance of all IT products.
3. Use multi-functional devices such as attached printer and scanner or attached monitor and touch pad.

5 Results

Our sample experiment lasts for two consecutive months for same place. In the month of August 2015, we did not follow any approach while in the month of September 2015, we have followed some of the points discussed in the previous section using Gartner's approach with $4.1 - a$, $4.1 - c$, $4.1 - d$, $4.2 - a$, $4.3 - a$, $4.3 - b$, $4.3 - c$. The results are highly encouraging and reflects an acute decrease of 4.8 in the produced CO_2 levels when the defined action plan is followed. The measured information is shown in graph in Fig. 3. This decrease is mainly due to the reduction of used energy.

6 Conclusion

This paper has taken into account a bird-eye look at the energy consumption of the ICT industry around the globe. Although the ICT industry has innumerous advantages but it does have a negative effect on the environment, predominantly due to the use of personal computers and data centers which consumes energy and CO_2 is emitted during its production. Green IT focuses on reducing energy and resource consumption in an entire ICT cycle. We have discussed about different methods or remedies by which the content of CO_2 emission can be reduced by the optimized use of ICT products. The various techniques like Gartner's approach, using virtualization to increase server utilization and using storage management strategies etc. can help to reduce the CO_2 emission. All in all, Green IT is an effort of emancipating the ICT industry of its negative impact on the atmosphere.

References

1. Gro Harlem Brundtland. Our common future. *World Commission on Environment and Development, Brussels*, 1987.
2. Bob Giddings, Bill Hopwood, and Geoff O'brien. Environment, economy and society: fitting them together into sustainable development. *Sustainable development*, 10(4):187–196, 2002.
3. Lorenz Hilty, Wolfgang Lohmann, and Elaine Huang. Sustainability and ICT-an overview of the field. *Politeia*, 27(104):13–28, 2011.
4. Online Article. Gartner says: Worldwide software market grew 4.8 percent in 2013. http://www.gartner.com/newsroom/id/2696317, 2013.
5. Juha Taina. Good, bad, and beautiful software-in search of green software quality factors. *Green ICT: Trends and Challenges, Cepis Upgrade*, 12(4):22–27, 2011.
6. Richard Hodges and W White. Go green in ICT. *Green Tech News*, 2008.
7. Tracy A Jenkin, Jane Webster, and Lindsay McShane. An agenda for green information technology and systems research. *Information and Organization*, 21(1):17–40, 2011.
8. Simon Mingay. Green IT: the new industry shock wave. *Gartner RAS Research Note G*, 153703:2007, 2007.
9. San Murugesan. Harnessing green IT: Principles and practices. *IT professional*, 10(1):24–33, 2008.
10. Barbara Pernici, Danilo Ardagna, and Cinzia Cappiello. Business process design: Towards service-based green information systems. In *E-Government ICT Professionalism and Competences Service Science*, pages 195–203. Springer, 2008.
11. Kuan-Siew Khor, Ramayah Thurasamy, Noor Hazlina Ahmad, Hasliza Abdul Halim, and Lo May-Chiun. Bridging the gap of green IT/IS and sustainable consumption. *Global Business Review*, 16(4):571–593, 2015.
12. Wietske van Osch and Michel Avital. From green IT to sustainable innovation. In *AMCIS*, page 490, 2010.
13. Online Article. 10 key elements of a green IT strategy. https://www.gartner.com/doc/559114/-key-elements-green-it, 2007.
14. David J Staley. *Computers, visualization, and history: How new technology will transform our understanding of the past*. Routledge, 2015.
15. Juan-Carlos, Giovanna Sissa, and Lasse Natvig. Green ICT: The information society's commitment for environmental sustainability. *Cepis Upgrade*, 12(4):2–5, 2011.
16. News Paper Article. Internet emits 830 million tonnes of carbon dioxide. The Hindu, 6 January 2013.
17. Erol Gelenbe and Yves Caseau. The impact of information technology on energy consumption and carbon emissions. *Ubiquity*, June 2015.
18. Andy Lewis. Energy, carbon, waste and water: the ICT angle. Sustainability Perspectives, 2014.
19. Ward Van Heddeghem, Sofie Lambert, Bart Lannoo, Didier Colle, Mario Pickavet, and Piet Demeester. Trends in worldwide ICT electricity consumption from 2007 to 2012. *Computer Communications*, 50:64–76, 2014.
20. Online Article. Carbon footprint calculator. http://www.carbonindependent.org/, November 2015.

21. Jennifer Mankoff, Robin Kravets, and Eli Blevis. Some computer science issues in creating a sustainable world. *iEEE Computer*, 41(8):102–105, 2008.
22. Gazette. E-wastes (management and handling) rules:2011. Ministry of Environment, Forest and Climate Change, Government of India, 2011.
23. Online Article. How much carbon dioxide is produced per kilowatt hour when generating electricity with fossil fuels? Independent Statictics & Analysis, U.S. Energy Information Administartion, March 2014.

Design and Simulation of Microstrip Array Antenna with Defected Ground Structure Using ISM Band for Bio Medical Application

Deepali Chalak and Sujit Dharmapatre

Abstract For detecting the malignant cancerous tissues in the breast, various techniques are used like Mammography, MRI, Ultra Sound, etc. But such types of technology are miserable while examining. The Microwave Breast Imaging technique has generated scattered signals after transmission of Microwave signals. This causes a loss of information. Therefore 1 × 4 array antenna is design, to detect the tumor at early stage and this array antenna is more specific to detect the malignant tissues. The array antenna has 1 × 4 array, with Defected Ground Structure which will increase the efficiency of this model. This model is used in ISM (Industrial, Scientific, and Medical) band having frequency 2.4 GHz. This band is unlicensed frequency band.

Keywords Microstrip antenna · Defected ground structure · ISM band
Microwave breast imaging technique

1 Introduction

Cancer is the leading cause of death in the world. Breast cancer is mostly found in women. To prevent breast cancer various treatments are available, but there is a need to early detection of malignant tissues to reduce the mortality rate of cancer diseases. The technology is used for detecting the malignant tissues of Breast cancer are X-ray mammography, MRI, Ultra sound imaging techniques. X-ray

---

D. Chalak (✉) · S. Dharmapatre
Department of Electronics & Communication Engineering,
Maharashtra Institute of Technology, Pune, India
e-mail: deepalichalak@gmail.com

S. Dharmapatre
e-mail: d_sujit@yahoo.com

D.K. Mishra et al. (eds.), *Information and Communication Technology for Sustainable Development*, Lecture Notes in Networks and Systems 10,
https://doi.org/10.1007/978-981-10-3920-1_2

mammography is widely used technique. This technique provides high percentage of successful detection rate. These are some limitations like, it cannot detect small tissues and due to X-rays ionizing radiations are penetrating on human body. MRI technique uses the magnet to create detail images of internal structure with radioactive techniques.

Ultrasound techniques use emission of sound wave and pick up the echoes and predict the condition of body tissues. The echoes can be seen as black and white images on computer screen as well as the technique also uses ionized radiation. To overcome these drawbacks imagining technique is used.

2 Microwave Breast Imaging Technique

In imaging technique, antennas used for detecting the malignant tissues. These antennas are placed near the skin and send micro waves on the body for detection purpose. In case of Microwave Breast Cancer, while scanning the breast, microwave signals apply on the breast. It will generate a scattering signal. The scattering signals are received by antenna. The information will be extracted from the signal. Scattering of signal depends on the various factors like strength of signal, electrical properties of tissues. This technique will provide the position of the tumor and volume of the tumor. Malignant tissues can be detected in various ways. If water content level is high the tissues are cancerous tissue. This technique categorized into 3 types. Passive method differentiates the malignant tissues and healthy tissues by increasing the temperature of tumor. The region will inspect the malignant tissues of higher temperature and then tumor is detected. The second category is Hybrid Method which will examine the conductivity in malignant tissues of higher conductivity.

Microwave energy will be absorbed by the tumor and tumor size gets expanded. Thus tumor is detected. Third method is Active method which generates full map image by controlling and reconstructing the back scattering signals. Malignant tissues have more scattering signal as compared to normal tissues. This way detecting the malignant tissues. Microstrip Patch Antenna is used to overcome the disadvantages of these Microwave breast imaging techniques (Fig. 1).

Fig. 1 Microwave breast imaging techniques

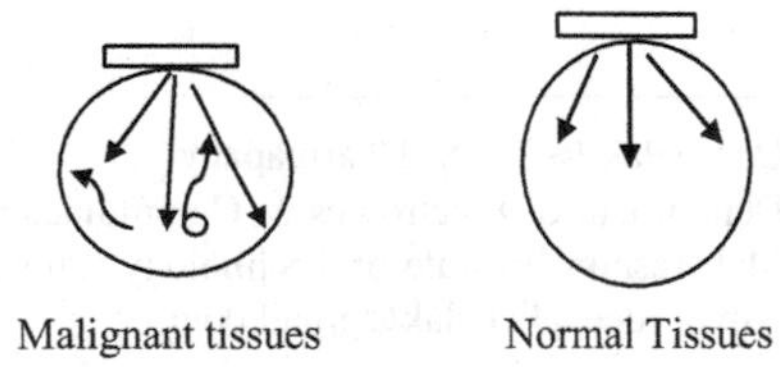

3 Microstrip Antenna Design

Figure 2 shows, single Microstrip Patch Antenna with probe feed input feeding technique Probe feed type is fabricated on same substrate so that total structure will remain planer and elimination of spurious feed network radiation.

3.1 *Designing of Microstrip Patch Antenna for 2.4 GHz ISM Band Steps*

Step I Substrate Dielectric constant: 4.4

Step II Frequency of operation: 2.4 GHz

Step III Calculation of Width of Patch (W),
Where, C = 3 × 10^8, ε_r = 4.4, *fo* = 2.4 GHz, Width of Patch (W) = 38.22

$$W = \frac{c}{2f_{o\sqrt{\frac{\varepsilon_r+1}{2}}}} \tag{1}$$

Step IV Calculation of the Effective Dielectric Constant. It is based on the height of the substrate, dielectric constant of the substrate, and the calculated width of the patch antenna. Effective Dielectric Constant = 3.99

$$\varepsilon_{eff} = \frac{\varepsilon_r + 1}{2} + \frac{\varepsilon_r - 1}{2}[1 + 12\frac{h}{w}]^{-\frac{1}{2}} \tag{2}$$

Step V Calculation of the Effective length, Effective length = 30.25 mm

$$L_{eff} = \frac{c}{2f_o\sqrt{\in_{eff}}} \tag{3}$$

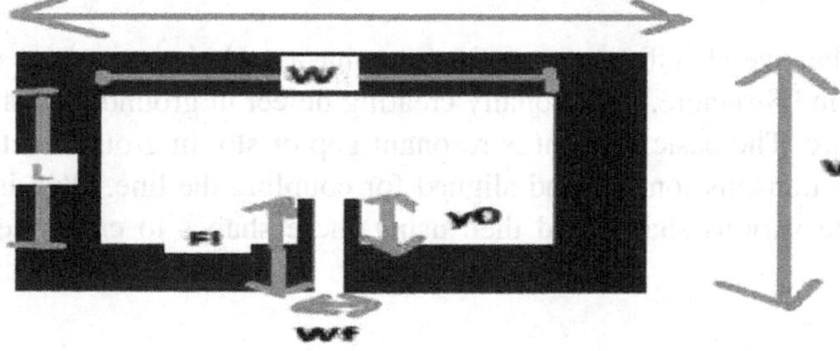

Fig. 2 Design microstrip antenna

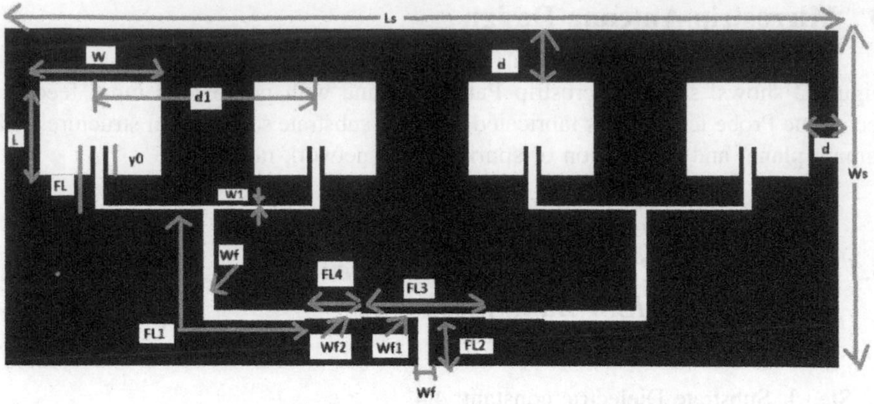

Fig. 3 Shows the microstrip antenna with 4 × 1 array

Step VI Calculation of the length extension (ΔL) = 0.70 mm

$$\Delta L = 0.412\, h \frac{\left(\in_{eff} + 0.3\right)\left(\frac{w}{h} + 0.264\right)}{\left(\in_{eff} - 0.258\right)\left(\frac{W}{h} + 0.8\right)} \tag{4}$$

Step VII Calculation of actual length of the patch, (L) = 28.4 mm

$$\mathrm{L} = \mathrm{Leff} - 2\Delta\mathrm{L} \tag{5}$$

Step VIII Calculation of the ground plane dimensions (Lg and Wg) (Fig. 3)

$$\mathrm{Lg} = 2 * 6\mathrm{h} + \mathrm{L} = 12 \times 1.6 + 29.44 = 58\ \mathrm{mm} \tag{6}$$

$$\mathrm{Wg} = 2 * 6\mathrm{h} + \mathrm{W} = 12 \times 1.6 + 38.03 = 66\ \mathrm{mm} \tag{7}$$

4 Defected Ground Structure

To increase the overall efficiency of the antenna a technique is used called as Defected Ground Structure. Intentionally creating defect in ground means defected ground structure. The basic element is resonant gap or slot in ground metal placed directly on the transmission line and aligned for coupling the line. DGS is realized by creating the various shapes and then using these shapes to create defect in a

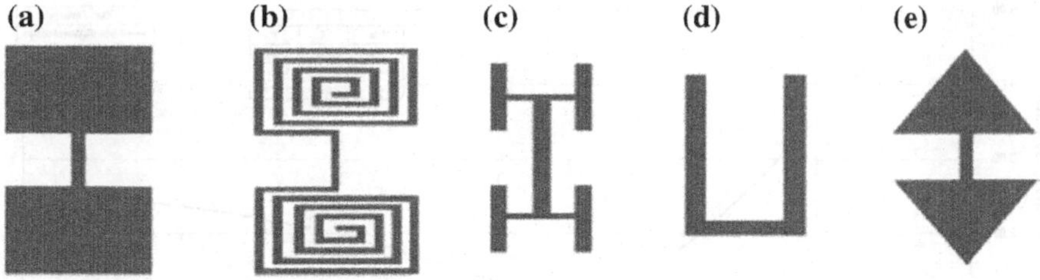

Fig. 4 Shapes of DGS

ground structure. Thus, it will disturb the current distribution depending upon the shape defect. These disturbances at the shielded current will influence the input impedance and the current flow of antenna. DGS is also useful for controlling excitation and electromagnetic waves. The DGS geometries simple shapes are shown in Fig. 4.

The Defected Ground Structure (DGS) can be used in two ways: Single and Periodic. In 2-D DGS, the single shape is used in repetition pattern. The periodic DGS uses the parameters like shape, distance between two DGS, and distribution of DGS. DGS will change the current distribution in the ground plane of microstrip line which will increase in the inductance and capacitance. The repetition of single cell in a periodic pattern, gives much better result and deeper characteristics. DGS which are formed by the periodic repetition of unit cells is referred as periodic DGS. DGS behaves like L-C resonator. Circuit is coupled to microstrip line. When the RF signal is transmitted through a DGS integrated microstrip line, strong coupling occurs between the line and DGS (Figs. 5, 6, 7, 8 and 9).

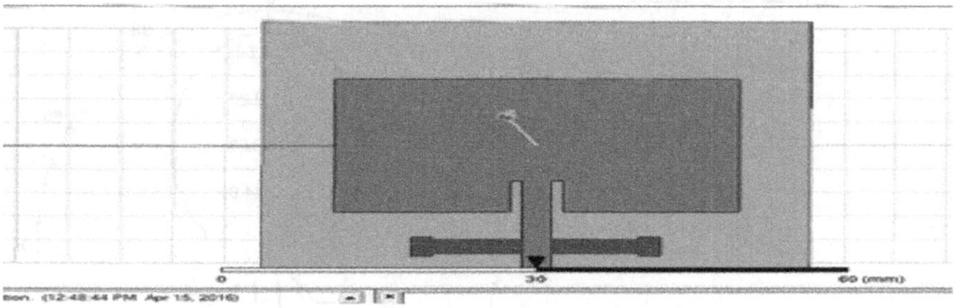

Fig. 5 Shows the DGS with microstrip antenna

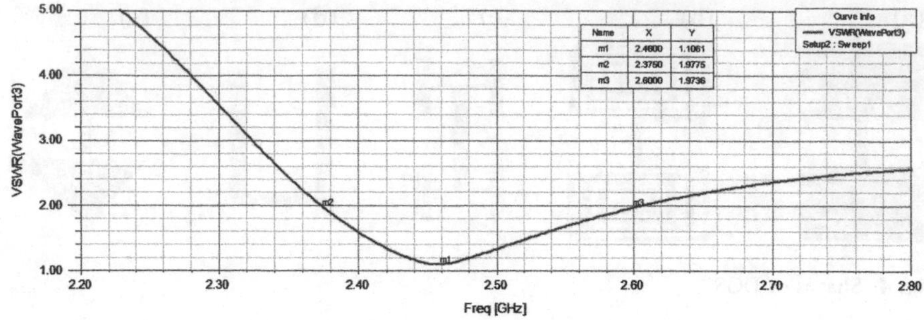

Fig. 6 VSWR for ISM band frequency is which is 2.4 GHz is 1.10

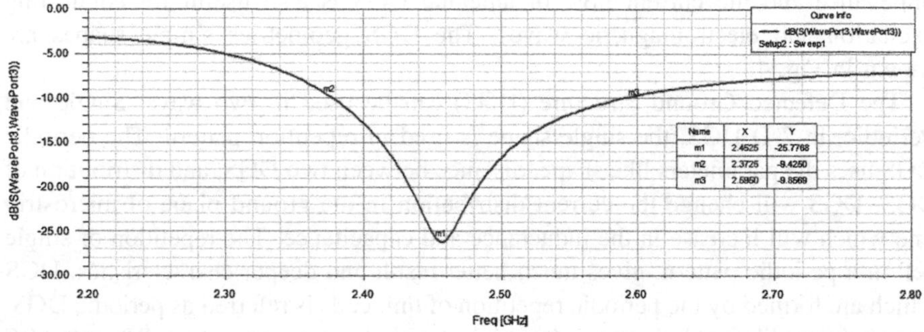

Fig. 7 Return loss is −25.77

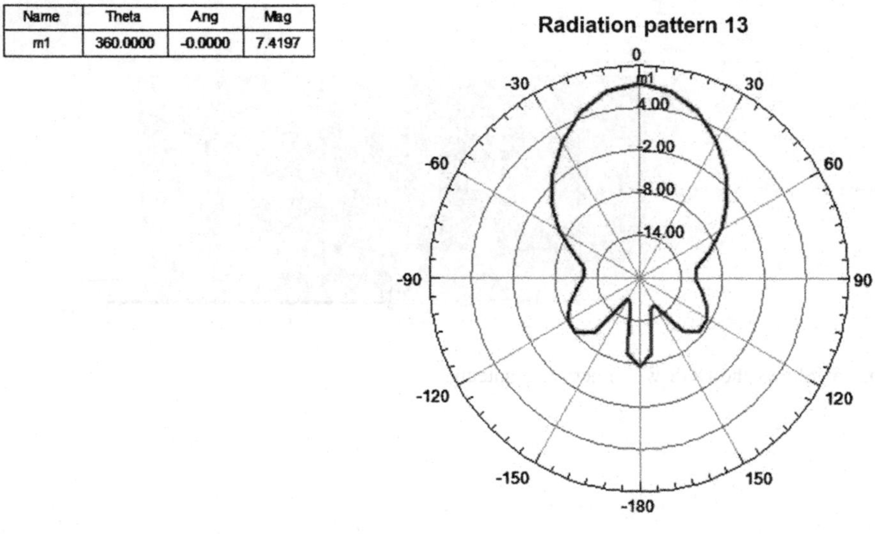

Fig. 8 Radiation pattern of 1 × 8 array with defected ground structure of microstrip patch antenna

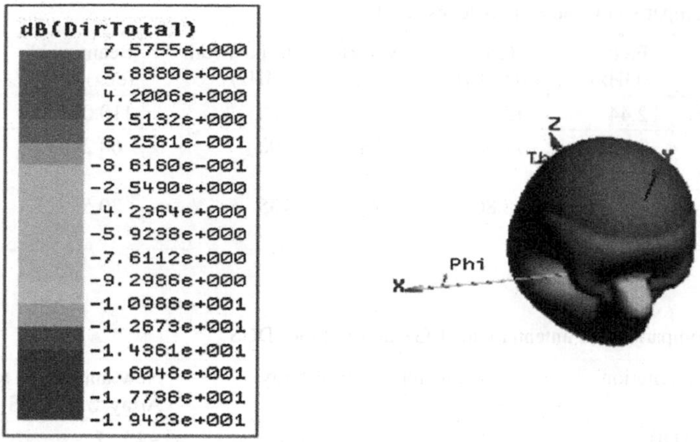

Fig. 9 Directivity of antenna 7.57 (DBI)

5 Simulation Results

A. VSWR
B. Return Loss
C. Radiation Pattern
D. Directivity

Performance of Microstrip Patch Antenna from Single Patch antenna to 4 × 1 Patch array antennas. As the number of patches of antenna increases bandwidth of antenna increases as well as the beam width increases. The gain of antenna also increases (Tables 1 and 2).

Table 3 shows the results of Microstrip Patch array antenna with DGS and without DGS. At the frequency of 2.4 GHz, return loss increases with using DGS.

Table 4 shows the results of various shapes of defected ground structure. Depending upon the geometric shapes of DGS, results of antenna change. In case of Dumbbell Circular-shaped antenna, return loss decreases as compared to dumbbell shape slot DGS. There is a change in VSWR. Bandwidth of the antenna also

Table 1 Input parameters for designing antenna

Input parameters	Value
Width of patch	38.22
Length of patch	30.25
Height of substrate	1.6 mm
Patch material	Copper
Length of ground plane (Lg)	58
Width of ground plane (Wg)	66

Table 2 Comparison of antenna patches results

Type of MSA	Freq (GHz)	Return loss (dB)	VSWR	Band width (MHz)	Beam width (deg)	Gain (dB)
Single patch	2.44	−15.28	1.41	80	110.0	3.50
2 × 1 Patch array	2.46	−17.98	1.28	100	60.2	7.62
4 × 1 Patch array	2.46	−18.86	1.21	135	20.5	10.00

Table 3 Comparison of antenna with DGS and without DGS

Performance criterion	New antenna (1 × 4 array)	New antenna 1 × 4 Array with DGS
Return loss (DB)	−18.86	−25.77
VSWR	1.21	1.10
Bandwidth (MHz)	5.62	9.37
Gain	13.55	7.50

Table 4 Comparison of various shapes of DGS

Type of MSA	Return loss (dB)	VSWR	Bandwidth (MHz)	Gain (dB)
Dumbbell circle shape slot DGS	−11.09	1.77	158	7.90
Dumbbell shape slot DGS	-25.77	1.10	225	7.50

changes in case of dumbbell circular shape slot DGS In dumbbell circular shaped antenna bandwidth is 6.58% and of dumbbell shape slot DGS it is 9.37%. There is little bit change in gain of antenna.

6 Conclusion

Newly miniaturized antenna structure has been successfully designed and simulated by using HFSS software. The performance criteria include Return loss, VSWR, Radiation pattern in the proposed design. The 1 × 4 array antenna with defected ground structure has low profile, low weight and having thin substrate working in ISM Band. The Future work will be focused on increasing the gain by maintaining or increasing its bandwidth.

References

1. Rabia Çalkana,*, S. Sinan Gültekina, Dilek Uzera, Özgür Dündarb, 'A Microstrip Patch Antenna Design for Breast Cancer Detection', Elsevier, Procedia—Social and Behavioral Sciences 195 (2015) 2905–2911
2. S. Banu, A. Vishwapriya, R. Yogamathi, Periyar Maniammai University, Vallam, Thanjavur, India, 'Performance Analysis of circular patch antenna for breast cancer Detection', IEEE – 31661
3. S. Adnan1, A.F. Mirza1, R.A. Abd-Alhameed1, M. Al Khambashi1,2, Q. Yousuf1, R. Asif1 and C.H. See1, 3 and P.S Excell4, 'Microwave Antennas for Near Field Imaging', © 2014 IEEE
4. Amal Afyfl, Larbi Bellarbi, Abdelhamid Errachid, M. Adel Sennouni, 'Flexible Microstrip CPW Slated Antenna for Breast Cancer Detection', © 2015 IEEE
5. Pankaj Kumar Singh, Subodh Kumar Tripathi, Rahul Sharma, Ashok Kumar, 'Design and Simulation of Microstrip Antenna for Cancer Diagnosis', International Journal of Scientific and Engineering Research, Volume 4, Issue 11, November-2013 1821 ISSN 2229-5518.
6. Sudhir Shrestha, Mangilal Agarwal, Joshua Reid, and Kody Varahramyan Integrated Nanosystems Development Institute (INDI) Department of Electrical and Computer Engineering, 'Microstrip Antennas for Direct Human Skin Placement for Biomedical Applications', Progress In Electromagnetics Research Symposium Proceedings, Cambridge, USA, July 5,8, 2010.
7. Maneesha Nalam, Nishu Rani, Anand Mohan, 'Biomedical Application of Microstrip Patch Antenna', International Journal of Innovative Science and Modern Engineering (IJISME), ISSN: 2319-6386, Volume-2, Issue-6, May 2014
8. John P. Stang and William T. Joines, 'A Tapered Micro strip Patch Antenna Array for Microwave Breast Imaging', @2008 IEEE
9. Gagandeep Kaur, Geetanjali Singla, 'Design of Wideband Micro strip Patch Antenna Using Defected Ground Structure for Wireless Applications'. International Journal of Advanced Research in Computer Science and Software Engineering, 2013
10. Chandra Shekhar Gautam Indian Institute of Information Technology, Allahabad, India rs62@iiita.ac.in Avanish Bhadauria MEMS and Micro-sensors Group, CSIR-Central Electronics Engineering Research Institute, 2D Defected Ground Structures in Microstrip Line, IEEE 2015
11. L. H. Weng, Y. C. Guo, X. W. Shi, and X. Q. Chen, 'AN OVERVIEW ON DEFECTED GROUND STRUCTURE', Progress In Electromagnetics Research B, Vol. 7, 173–189, 2008
12. Dharani K R Sampath P Soukath Ali K Pavithra 0 Renuga Devi M, 'Design and Simulation of Microstrip Phased Array Antenna for Biomedical Applications'

Design and Implementation of a 32-Bit Incrementer-Based Program Counter

Nathaniel Albuquerque, Kritika Prakash, Anu Mehra and Nidhi Gaur

Abstract The paper presents the design and implementation of a 32-bit program counter that has been used in DLX-RISC processor which uses Tomasulo Algorithm for out of order execution and a 32-bit program counter based on incrementer logic that was self designed on Virtex-7 FPGA board. The results for power, delay, and area were compared in order to obtain optimal results for the program counter. The power delay product (PDP) of the program counter design based on incrementer logic was found to be 94.4% less than that of the program counter used in DLX-RISC processor. Thereby, the improvised program counter enhances the overall performance of any processor it is used in as the power and delay have been substantially reduced in the proposed design. The designs are simulated and synthesized on Xilinx Vivado 2015.4 using VHDL and are implemented on Virtex-7 FPGA.

Keywords Program counter · Incrementer-based program counter
DLX-RISC processor · Tomasulo algorithm · Out of order execution
Virtex-7

N. Albuquerque (✉) · K. Prakash · A. Mehra · N. Gaur
Department of Electronics and Communication Engineering, ASET,
Amity University, Noida, Uttar Pradesh, India
e-mail: nathanielalbuquerque@gmail.com

K. Prakash
e-mail: prakashkritika2@gmail.com

A. Mehra
e-mail: amehra@amity.edu

N. Gaur
e-mail: ngaur@amity.edu

D.K. Mishra et al. (eds.), *Information and Communication Technology for Sustainable Development*, Lecture Notes in Networks and Systems 10,
https://doi.org/10.1007/978-981-10-3920-1_3

1 Introduction

Earlier computers were manufactured using mechanical devices. Later, they were built using electromagnetic relays. In the mid 1940s, electronic computers were built using vacuum tubes. In the early 1960s, transistors were used and in later part of 1960s standard integrated circuits were introduced. By 1970s, supercomputers had been built and by 1990s, high speed microprocessors were developed. The performance of a microprocessor is based on its internal architecture and the clock speed [1]. The RISC processor allows building simple architectures with high clock speed. The aim of architecture improvement is to enhance the performance of the processor in terms of power, speed, and size. High performance of processors can be achieved by applying various power and delay reduction techniques.

There are various data paths used for the execution of an instruction in a microprocessor. In case of DLX-RISC processors, instructions are fetched, decoded, executed, and finally written back into the register memory. The speed of instruction fetch depends on the speed of program counter which points to the address of the next instruction that is to be executed. In order to improve the overall performance of the processor, the performance of individual components of the processor need to be improved.

1.1 Contribution

Here the design and implementation of 32-bit incrementer-based program counter on 28 nm FPGA is focused upon. The program counter used in 32-bit DLX-RISC processor was also implemented on the 28 nm technology. The implementation was done using VHDL and simulated on ModelSim. The codes were then run on Xilinx Vivado for power and delay estimation on the Virtex-7 FPGA which works on 28 nm technology. The results are compared and analyzed for power, delay, and area.

2 System Description

2.1 Program Counter

The program counter is a special purpose register that points to the address of the next instruction that is to be executed. It keeps a track of the position of instructions that are being executed by the processor. The programmable logic array (PLA) automatically updates the value of the program counter to the next value. Thus, the program counter is automatically updated as each instruction is executed. In every cycle, the value at the pointer is read by the decoder of the processor and

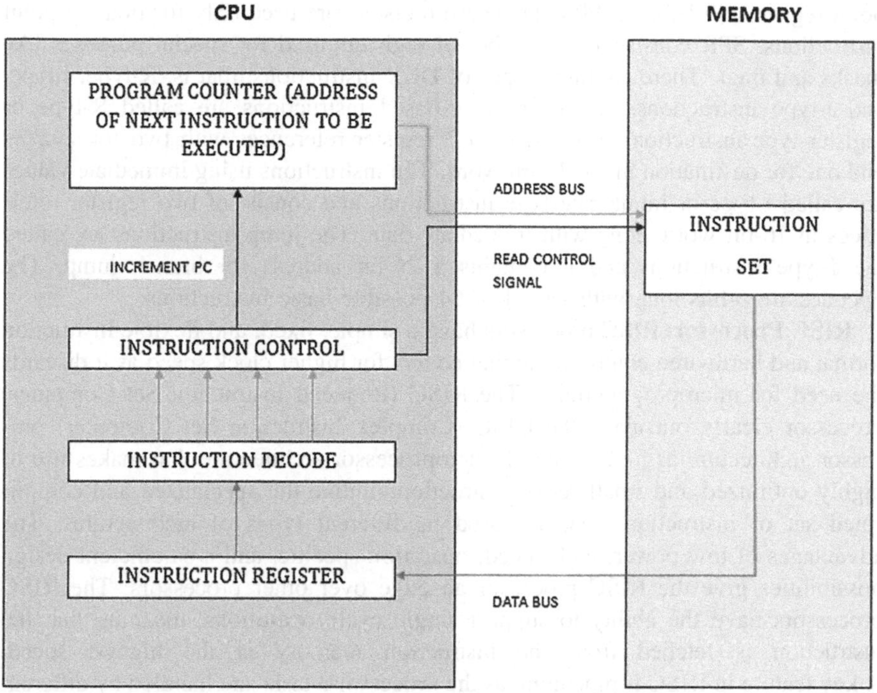

Fig. 1 Block diagram of a CPU comprising of PC and decode environment

then the value is changed to that of next instruction as shown in Fig. 1. After fetching an instruction, the program counter increments the address value by one. This address is the location of the next instruction to be executed. The 'jump' or 'branch' are special instructions in which the control is moved to an instruction located at a different location in the application program, rather than the next instruction. In branch instructions, a new address value is loaded for the program counter from the instruction. This new value is updated in the program counter and subsequent instructions are loaded from this location onwards.

2.2 *Program Counter Used in DLX-RISC Processor*

DLX Architecture: DLX architecture was first designed by Patterson and Hennessy [2]. It is a load and store architecture that allows the integer and floating point instructions to be executed correctly. Information is stored in three types of registers namely, the general purpose register (GPR), the floating point register (FPR), and the special purpose register (SPR). GPR consists of integer registers (R0, …, R31) with R0 always having all its bits set to 0. This register is used for memory addressing and integer operations. FPR consists of single precision floating point registers (FPR0, …, FPR31). These registers are used only for floating point instructions. SPR consists of a number of registers used for special purposes like masks and flags. There are three types of DLX instructions, that is, R-type, I-type, and J-type instructions. All the register-based instructions are called R-type or register type instructions. It consists of 3 register references with two for sources and one for destination in the 32-bit word. The instructions using immediate values are called I-type or immediate type instructions and consist of two register references in 16-bit word along with immediate data. The jump instructions are called the J-type instructions and it contains a 26-bit address for branch jump. The opcodes are 6-bits long with a total of 64 possible basic instructions.

RISC Processor: RISC processors have a simple, fixed, and flexible instruction format and hardwired control logic that covers for higher clock speed as it discards the need for microprogramming. The RISC (Reduced Instruction Set Computer) processor clearly outsmarts the CISC (Complex Instruction Set Computer) processor architecture as it is a type of microprocessor architecture that makes use of highly optimized and small set of instructions unlike the specialized and complicated set of instructions that are used in different types of architectures. The advantages of low power, high speed, operation-specific, and area-efficient design possibilities give the RISC processor an edge over other processors. The RISC processors have the ability to support single cycle operations, meaning that the instruction is fetched from the instruction memory at the highest speed. A key feature in RISC is pipelining as the processor's tasks are handled by different stages or units at the same time [3]. Thus, more instructions can be executed and written back in less amount of time given that the units/stages function simultaneously. Moreover, RISC processors are less in cost in terms of design and manufacturing. The architecture consists of a 5 stage pipeline as shown in Fig. 2. The primary stage is the Instruction Fetch '(IF)' which performs the fetching of instructions which it retrieves from its memory registers. The second stage is that of the Decode/Issue '(D/I)' environment where the instruction word that has been fetched and is decoded and passed into a reservation station. The third stage is the Execution Unit '(EX)' which consists of the function units that execute the instruction [4]. Depending on the nature of the task to be executed, some instructions are executed quickly while others may take a significant time to compute. Hence, the time taken for each instruction to execute varies. In the penultimate stage, known as Completion, the result of the instruction is stored in reorder buffer. The fifth stage is the Write Back '(WB)' stage which performs the 'Write Back' of the result into the destination register.

2.3 *Tomasulo Algorithm*

Tomasulo Algorithm works on the principle that the hardware determines which instructions are to be executed as opposed to the in order, static scheduled logic. Thereby, the algorithm decides on which order should the instructions be executed

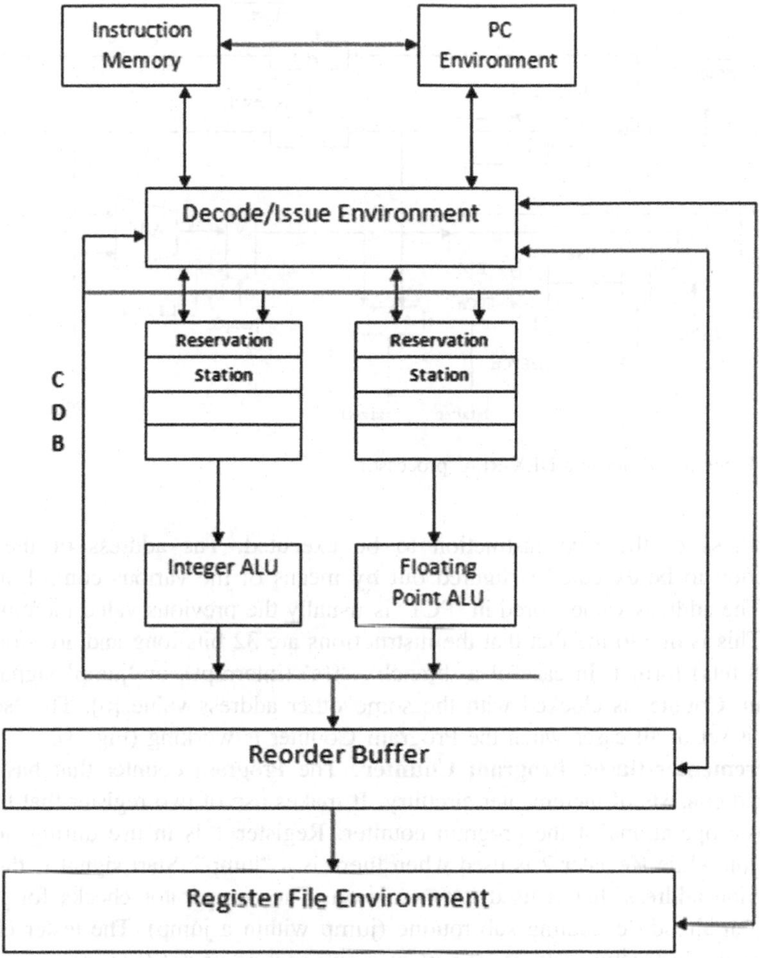

Fig. 2 Overview of the data path for Tomasulo scheduling

giving rise to the out of order execution. In out of order execution, the processor executes instructions in an order governed by the availability of the operands as well as considering the machine cycles for the execution to complete instead of being executed by the original order in which the program was written [5].

2.4 PC Environment

The Program Counter in the DLX-RISC processor consists of 'oPC1' and 'PC1'. 'oPC1' is a register that hold the address of the current instruction and 'PC1' stores

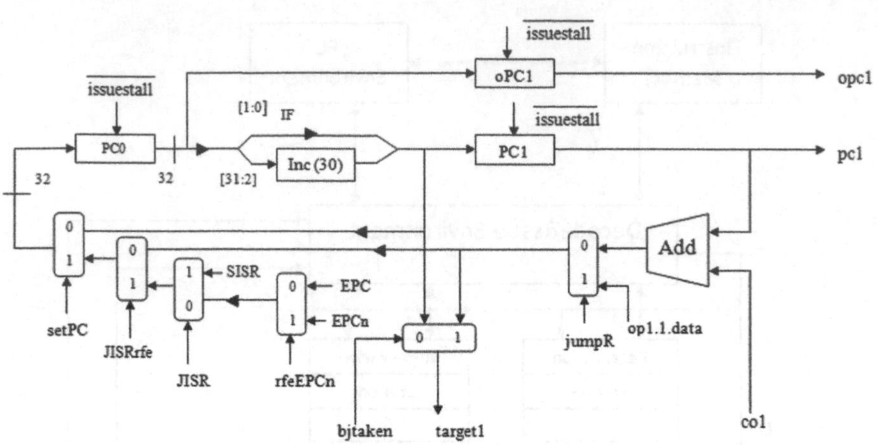

Fig. 3 Program counter in a DLX-RISC processor

the address of the next instruction to be executed. The address of the next instruction to be executed is figured out by means of the various control signals used. The address value stored in 'PC1' is usually the previous value incremented by 4. This is due to the fact that the instructions are 32 bits long and are stored in byte (8 bits) format. In case of a 'branch', 'rfe' (interrupt), or 'jump' signal, the Program Counter is clocked with the some other address value [6]. The 'setPC' signal is set in all cases when the Program Counter is working (Fig. 3).

Incrementer-Based Program Counter: The Program counter that has been designed consists of incrementer circuitry. It makes use of two register that handle the basic operations of the program counter. Register 1 is in use during normal execution while Register 2 is used when there is a "jump". Start signal is the first instruction address that is used for execution. The comparator checks for jumps within an already executing sub routine (jump within a jump). The tester checks whether the instruction address is part of a sub routine or not (Fig. 4).

2.5 *Field Programmable Gate Arrays (FPGA)*

FPGAs have the capability to be reprogrammed in the field wherever the errors in design need to be corrected and upgraded or modified [7]. They provide flexibility to synthesize designs and are low in cost. Hardware description languages like VHDL and Verilog allow FPGAs to be more apt for the different types of designs where errors and component failures can be limited.

Virtex-7 FPGA: Virtex-7 is a FPGA which is optimized for high system performance and integration at 28 nm technology. It provides best-in-class performance for circuit designs. The device used is xc7vx485tffg1761-2. The I/O pin count of the device is 1761. It consists of 1030 blocks of RAM, 2800 DSPs, 607200

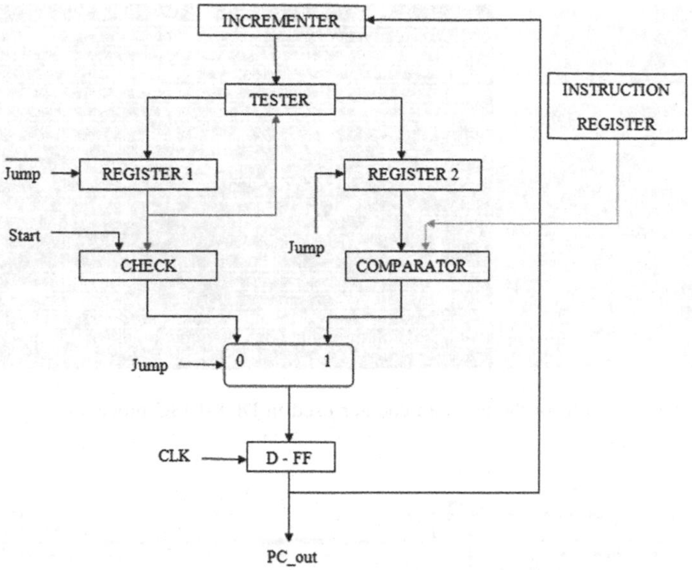

Fig. 4 Incrementer-based program counter

flip flops, 700 Input/Output Blocks and 303600 LUTs. The maximum and minimum operating voltages are 1.03 V and 0.97 V, respectively, and the maximum and minimum operating temperatures are 85 °C and 0 °C, respectively.

3 Experimental Results and Discussion

The program counter applied in DLX-RISC processor and incrementer-based program counter (PC) were simulated as shown in Fig. 5 and Fig. 9, respectively. They were synthesized and implemented on Virtex-7 FPGA board. The power, area and delay were analyzed for the same. The overall on-chip power (including dynamic and static power) on Virtex-7 for PC based on DLX-RISC processor and incrementer-based PC is depicted in Fig. 7 and Fig. 11, respectively. The number of LUTs, flip flops, global buffers and IO pins for PC based on DLX-RISC processor and incrementer-based PC are depicted in Fig. 8 and Fig. 12, respectively. The elaborated designs of the Program Counters obtained from the netlists are shown in Figs. 6 and 10. A comparative study was done based on the values obtained for the program counters.

From Tables 1 and 2 it can be concluded that the novel incrementer-based PC has its power reduced by 13.1% and delay reduced by 93.5% as compared to the PC that was used in DLX-RISC processor. The Power Delay Product (PDP) of the novel incrementer-based PC was found to be 94.4% less than that of PC used in

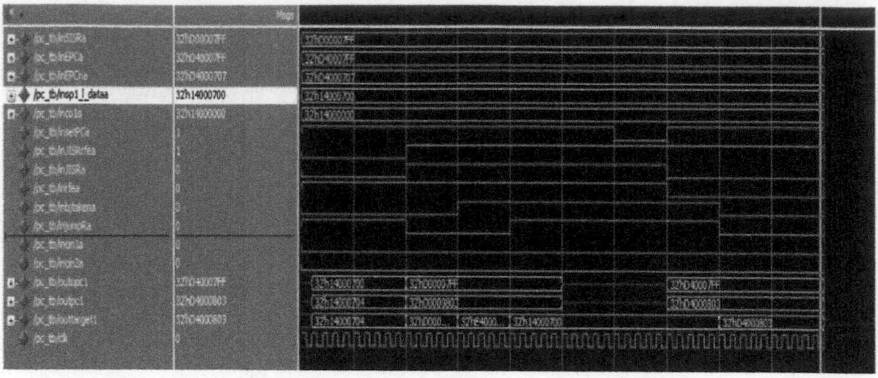

Fig. 5 Simulation result for the program counter used in DLX-RISC processor

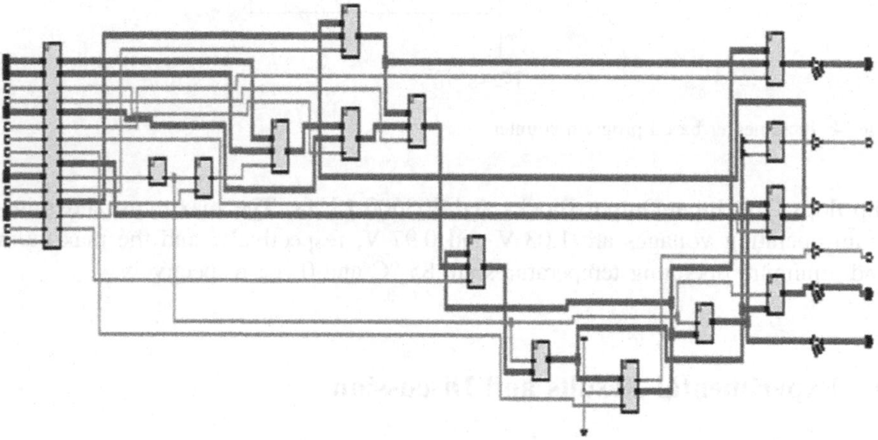

Fig. 6 Elaborated design of the program counter used in DLX-RISC processor

Fig. 7 On-chip power of the program counter used in DLX-RISC processor

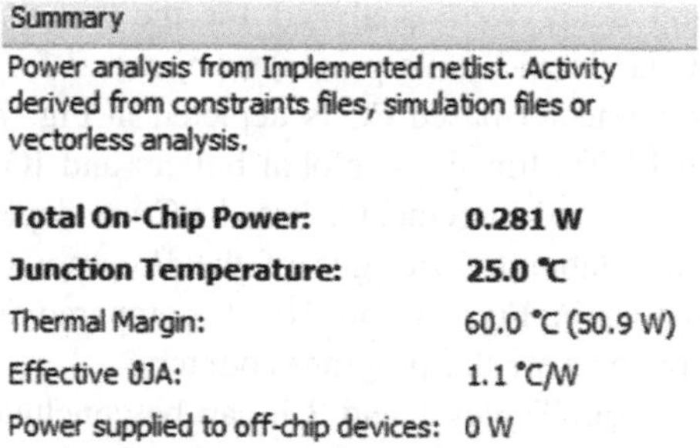
Summary
Power analysis from Implemented netlist. Activity derived from constraints files, simulation files or vectorless analysis.
Total On-Chip Power: 0.281 W
Junction Temperature: 25.0 °C
Thermal Margin: 60.0 °C (50.9 W)
Effective ϑJA: 1.1 °C/W
Power supplied to off-chip devices: 0 W

Resource	Utilization	Available	Utilization %
LUT	186	303600	0.06
FF	168	607200	0.03
IO	268	700	38.29
BUFG	1	32	3.12

Fig. 8 Utilization of program counter used in DLX-RISC processor

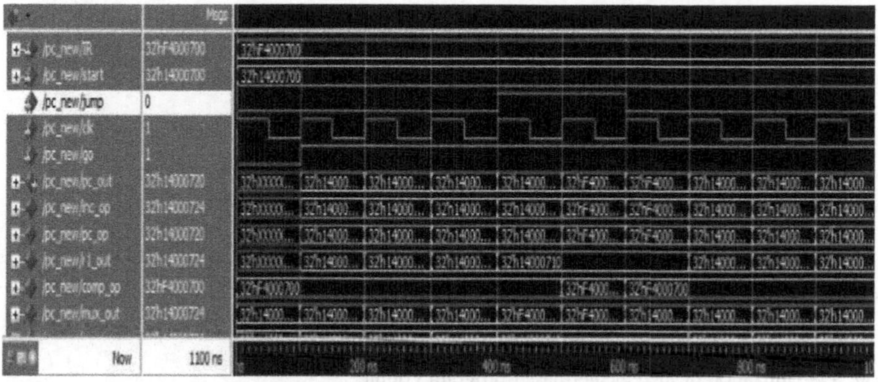

Fig. 9 Simulation result for incrementer-based program counter

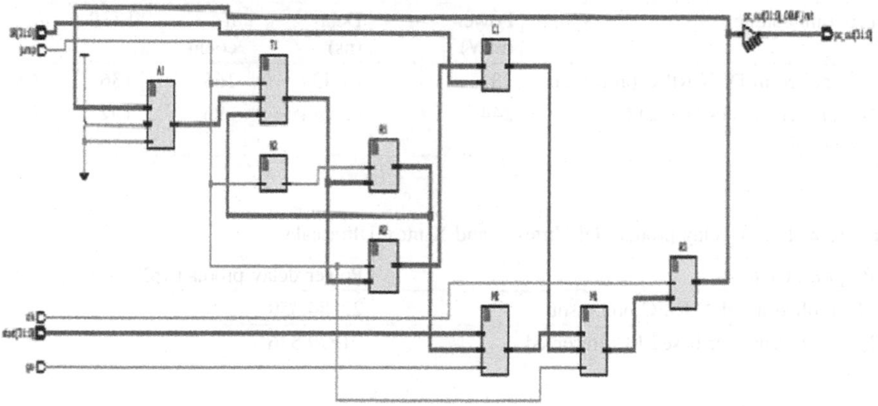

Fig. 10 Elaborated design of incrementer-based program counter

DLX-RISC processor. Hence, there is a significant improvement in the overall efficiency of the program counter with a marginal increase in LUT of approximately 3%.

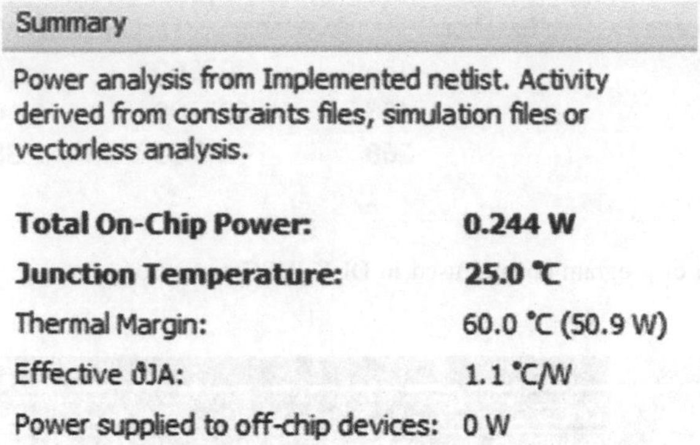
Summary
Power analysis from Implemented netlist. Activity derived from constraints files, simulation files or vectorless analysis.
Total On-Chip Power: 0.244 W
Junction Temperature: 25.0 °C
Thermal Margin: 60.0 °C (50.9 W)
Effective ϑJA: 1.1 °C/W
Power supplied to off-chip devices: 0 W

Fig. 11 On-chip power of the incrementer-based program counter

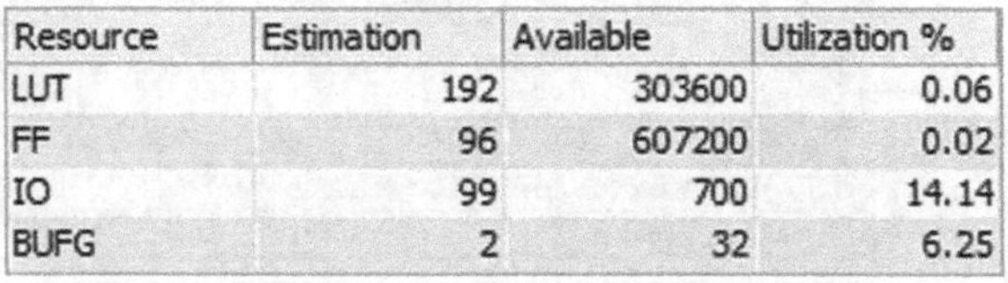

Resource	Estimation	Available	Utilization %
LUT	192	303600	0.06
FF	96	607200	0.02
IO	99	700	14.14
BUFG	2	32	6.25

Fig. 12 Utilization of incrementer-based program counter

Table 1 Comparison of Pc based on incrementer circuit and Pc Used In DLX-RISC processor

Program counter	Power (mW)	Delay (ns)	ff count	LUT	i/o
PC applied in DLX-RISC processor	281	81.439	168	186	268
Novel incrementer-based PC presented	244	5.244	96	192	99

Table 2 Power delay product Of Virtex-7 and Kintex Ultrascale

Program counter	Power delay product (pJ)
PC applied in DLX-RISC processor	22884.359
Novel incrementer-based PC presented	1279.536

4 Conclusion

Individual blocks of the processor should be optimally designed in order to achieve low power and high speeds. The performance of the processor can be improved by reducing the power consumed and delay of the PC. In this paper, the program counter that had been used in DLX-RISC processor and a proposed program counter using lesser hardware were implemented and compared. The proposed incrementer-based PC was comparatively better in performance as the overall PDP was 94.4% lesser as compared to the PC used in DLX-RISC processor. By just changing one part of the processor, namely the Program Counter, the overall performance of the processor was improved tremendously.

References

1. V. Korneev and A. Kiselev, "Modern Microprocessors," in Computer Engineering Series, 3rd ed. New Delhi, India: Dreamtech Press, 2005, ch. 1, sec. 4, pp. 107–150.
2. David A. Patterson and John L. Hennessy, "Instruction level parallelism and its exploitation," in Computer Architecture, 4th ed. San Fransisco, USA: Morgan Kaufmann, 2007, ch. 2, sec. 4, pp. 092–128.
3. J. Poornima, G.V. Ganesh, M. Jyothi, M. Sahithi, A. Jhansi Rani B and Raghu Kanth, "Design and Implementation of Pipelined 32-bit Advanced RISC Processor for Various D.S.P Applications," in Internation Journal of Advanced Research in Computer Science and Software Engineering Conf., 2012, pp. 3208–3213.
4. Jinde Vijay Kumar, Boya Nagaraju, Chinthakunta Swapna and Thogata Ramanjappa, "Design and Development of FPGA Based Low Power Pipelined 64-Bit RISC Processor with Double Precision Floating Point Unit," in International Conference on Communication and Signal Processing Conf., 2014, pp. 1054–1058.
5. Daniel Kröning, "Design and Evaluation of a RISC Processor with a Tomasulo Scheduler," M.S. thesis, Dept. of Comp. Sc., Saarland Univ., Saarbrücken, Germany, 1999.
6. Divya M, Ritesh Belgudri, V S Kanchana Bhaskaran "Design and Implementation of Program Counter using Finite State Machine and Incrementer Based Logic," in Internation Conference on Advances in Computing, Communication and Informatics, 2014, pp. 582–587.
7. Soumya Murthy and Usha Verma, "FPGA based Implementation of 32 bit RISC processor Core using DLX-RISC Architecture," in International Conference on Computer Control and Architecture, 2015, pp. 964–968.

Application of Remote Sensing Technology, GIS and AHP-TOPSIS Model to Quantify Urban Landscape Vulnerability to Land Use Transformation

Alok Bhushan Mukherjee, Akhouri Pramod Krishna and Nilanchal Patel

Abstract This study demonstrated the efficacy of remote sensing technology, GIS and AHP-TOPSIS model to quantify vulnerability of different segments of urban landscape to land use transformation. Six different factors such as Accommodating Index (AI), Mean Heterogeneity Index (MHI), Landscape Shape Index (LSI), Division Index (DI), Cohesion Index (CI), and Distance (D) were identified as inputs to the AHP-TOPSIS model. The Landsat 8 satellite data was classified using supervised classification to determine the aforementioned factors. The influencing factors may have varying intensity in triggering land use conversion. Therefore, the relative importance of the aforementioned factors was quantified using AHP. Furthermore the influence of factors can be either positive or negative on the phenomenon. Thus, Technique of Order of Preference for Similarity to Ideal Solution (TOPSIS) was employed in the present study. The results succeeded in handling the varying characteristics of the variables, and are very close to the actual field-scenario.

Keyword Remote sensing technology · GIS · Vulnerability
Urban landscape · AHP · TOPSIS · Supervised classification

1 Introduction

Urbanization signifies processes which are dynamic in nature. Characterization of its behavior requires multidimensional inputs such as land use change analysis, urban sprawl pattern identification and computation of landscape metrics [12].

A.B. Mukherjee (✉) · A.P. Krishna · N. Patel
Department of Remote Sensing, Birla Institute of Technology, Mesra, Ranchi, India
e-mail: alokbhushan@bitmesra.ac.in

A.P. Krishna
e-mail: apkrishna@bitmesra.ac.in

N. Patel
e-mail: npatel@bitmesra.ac.in

D.K. Mishra et al. (eds.), *Information and Communication Technology for Sustainable Development*, Lecture Notes in Networks and Systems 10,
https://doi.org/10.1007/978-981-10-3920-1_4

Land use influences terrestrial functions and is significant in assessment of sustainability challenges such as urbanization, climate change, and changes in ecosystem function. In a land system, different factors can trigger changes in land use/land cover and finally, defines the landscape pattern. The characteristics of factors may vary from physical to socio-economic. Spatial–temporal analysis of landscape transformations can unveil complex relationship between anthropogenic activities and land use modifications [6, 7, 11]. Causes and consequences of urbanization were investigated by [3, 4, 5, 6, 14]. Furthermore, [1, 9, 13] have performed studies pertaining to land use science to understand different dimensions of the land system.

There are different factors which can influence, or trigger land use transformation process in an urban landscape. Primarily, demographic and socio-economic factors trigger transitions in land use and land cover of a city. However, the role of other potential significant factors that can instigate land use transformation in an urban landscape should not be ignored. For example, segments of an urban landscape which are vulnerable to transformation are dependent on its geometry and pattern. The way a segment of urban landscape is surrounded by other segments of a city may have huge impact in bringing about structural changes in the urban morphology. That suggests, there could be various factors that act as a catalytic force for landscape transformation. On the other hand, it also needs to be considered that there is likelihood of presence of factors that discourage the phenomenon, i.e., landscape transformation. Therefore, a phenomenon which is influenced by both positive and negative factors needs a model that can accommodate nonlinear inputs. Nonlinear inputs in the present context correspond to the domain of inputs that contain inputs having opposite characteristics. Therefore, models like Technique of Order of Preference for Similarity to Ideal Solution (TOPSIS) should be encouraged in these kinds of applications where event is influenced by factors possessing opposite characteristics.

This work demonstrated the efficacy of AHP-TOPSIS model in accommodating the factors of nonlinear characteristics. Efficacy of the proposed model in quantifying the vulnerability of different segments of the urban landscape to land use transformation was tested in a reference area. Ranchi, capital of the Jharkhand state, India was chosen as the reference study area. The reference study area is in evolving stage of urbanization since its formation as the capital of Jharkhand state. Hence, it became a prominent commercial centre of the state. That encourages swarming of people from other parts of the state. Consequently, there have been abrupt land use transformations to accommodate the growing demand from different sections of the city.

The main objective of the proposed model is to assess the vulnerability of different segments of urban landscape to land use conversions. It begins with identifying the influential factors that trigger land use conversions. Then the relative importance of these factors in escalating land use transformation process was determined using AHP. Finally, TOPSIS model was employed to determine the vulnerability of different segments of the urban landscape to land use conversion.

2 Methods

The research methodology of the present study comprises of four stages, i.e., identification and computation of different significant factors, quantification of relative importance of such significant factors using AHP, quantification of the vulnerability of different landscape portions to land use transformation using AHP-TOPSIS model, and normalization of bias of the result determined from AHP-TOPSIS model. The research methodology flowchart is shown in Fig. 1.

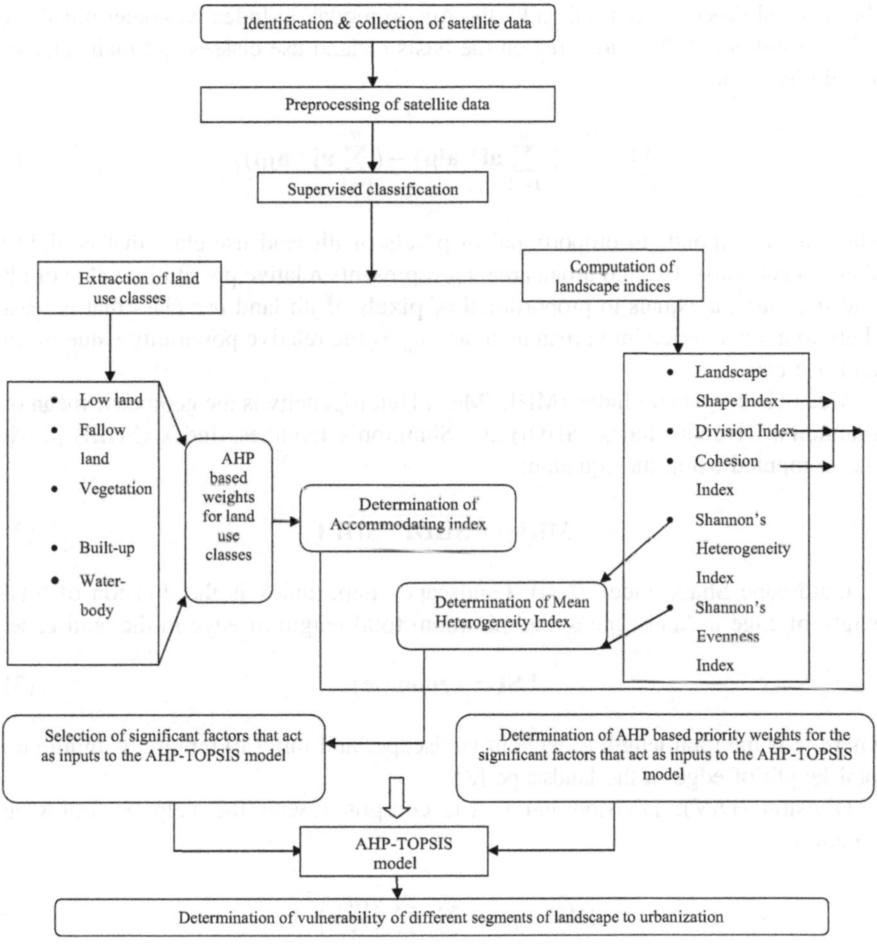

Fig. 1 Research methodology flowchart

2.1 Identification and Computation of Significant Factors

Six factors were identified as potential variables, i.e., Accommodating Index, Mean Heterogeneity Index, Landscape Shape Index, Division, Cohesion, and Distance for quantifying vulnerability of different landscape portions to land use conversion. The aforementioned factors were computed for different buffers of the study area.

Accommodating Index (AI): Accommodating Index was determined from land use classes of the study area. First, different land use classes of the study area, i.e., Low Land, Fallow Land, Vegetation, Built-up, and Water Body were classified using satellite data by supervised classification technique. Thereafter, their relative possibility of transformation into urban areas was quantified using Analytical Hierarchical Process (AHP). Finally, the Accommodating Index was determined for different buffers of the study area on the basis of land use classes and their relative possibility values.

$$\mathbf{AI} = 1 + \{(\sum_{\mathbf{i=1}}^{\mathbf{n}} \mathbf{ai} * \mathbf{aip}) - (\sum_{\mathbf{j=1}}^{\mathbf{n}} \mathbf{aj} * \mathbf{ajp})\} \quad (1)$$

where a_i corresponds to proportional of pixels of ith land use class that is highly likely to get converted into urban area, a_{ip} represents relative possibility value of ith land use class, a_j equals to proportional of pixels of jth land use class that is least likely to get converted into urban area, and a_{jp} is the relative possibility value of jth land use class.

Mean Heterogeneity Index (MHI): Mean Heterogeneity is the geometric mean of Shannon's Diversity Index (SHDI) and Shannon's Evenness Index (SHEI) [8]. It was computed using the equation:

$$\mathbf{MHI} = \sqrt[2]{\mathbf{SHDI} * \mathbf{SHEI}} \quad (2)$$

Landscape Shape Index (LSI): Landscape shape index is the division of total length of edge in landscape to the minimum total length of edge in the landscape.

$$\mathbf{LSI} = \mathrm{le}/\min(\mathrm{le}) \quad (3)$$

where le is the total length of edge in landscape, and min (le) represents minimum total length of edge in the landscape [8].

Division (DIV): Division index was computed with the help of following equation:

$$\mathbf{DIV} = \{\mathbf{1} - \sum_{i=1}^{m} \sum_{j=1}^{n} \left(\frac{ar_{\mathrm{ij}}}{Ar}\right)^2\} \quad (4)$$

where ar_{ij} is the area of ijth patch, and Ar corresponds to total area [8].

Cohesion (COH): Cohesion index was computed with the following equation:

$$\left[1-\left\{\sum_{i=1}^{m}\sum_{j=1}^{n} probij / \sum_{i=1}^{m}\sum_{j=1}^{n} probij\sqrt{arij}\right\}\right] * \left[1-1/\sqrt{TNC}\right]^{-1} * 100 \quad (5)$$

where $prob_{ij}$ represents perimeter of patch ij, ar_{ij} equals to area of patch ij, and TNC corresponds to total number of cells in the landscape [8].

Distance: The factor, i.e., Distance is the length of buffers from a chosen reference point. It was used in the model as a proxy variable. It acts as a negatively influencing variable that discourages the possibility of landscape conversion into an urbanized area.

2.2 Analytic Hierarchy Process (AHP)

The Analytic Hierarchy Process (AHP) is a decision-making method. It was developed by Saaty in 1980 [2]. The main objective of the Analytic Hierarchy Process (AHP) is to obtain priority weights for the alternatives. There are different steps required to implement Analytic Hierarchy Process (AHP) such as Hierarchical decomposition of research problem, Construction of pair wise comparison matrix, Normalization of pair wise comparison matrix, Computation of priority vectors, and Consistency analysis. The relative importance of the influential variables, i.e., Accommodating Index, Mean Heterogeneity Index, Landscape Shape Index, Division, Cohesion, and Distance in assessment of landscape transformation was determined using AHP.

2.3 Application of AHP-TOPSIS Model to Assess Landscape Vulnerability to Urbanization

TOPSIS (Technique for order preference by similarity to ideal solution) [10] is a multi-criteria method. It is used to determine the optimal solution from a finite set of alternatives. Two different ideal solutions are determined with the help of TOPSIS method, i.e., Positive Ideal Solution and Negative Ideal Solution. The underlying principle of TOPSIS method is to identify the alternative which is at the shortest distance from positive ideal solution and at the farthest distance from negative ideal solution.

Following steps need to be followed to perform TOPSIS operation:

Development of a decision matrix: A decision matrix needs to be developed to assess the relative importance of alternatives. The structure of the decision matrix should be as follows:

$$\begin{array}{c|ccccc} & \mathbf{X1} & \mathbf{X2} & \mathbf{X3} & .\ . & \mathbf{Xn} \\ \mathbf{A1} & X11 & X12 & X13 & .\ . & X1n \\ \mathbf{A2} & X21 & X22 & X23 & .\ . & X2n \\ \mathbf{A3} & X31 & X32 & X33 & .\ . & X3n \\ . & . & . & . & .\ . & . \\ . & . & . & . & .\ . & . \\ \mathbf{Am} & Xm1 & Xm2 & Xm3 & .\ . & Xmn \end{array} \quad (6)$$

where A_i is the ith alternative, and X_{ij} equals to value of ith alternative with respect to the jth alternative.

Determination of normalized decision matrix (n_{ij}): The decision matrix needs to be normalized with the following formula:

$$n_{ij} = \frac{Xij}{\sqrt{\sum_{i=1}^{m} Xij^2}} \quad (7)$$

where i = 1, …, m and j = 1, …, n.

Determination of normalized weight in the decision matrix (v_{ij}): The weighted normalization matrix is obtained by multiplying the weight of the criteria with the computed normalized value of the matrix. Following formula is used for obtaining weighted normalized matrix:

$$v_{ij} = n_{ij} * w_j \quad (8)$$

where w_j represents weight of the jth criterion.

Determination of positive ideal solution (PIS^+) *and negative ideal solution* (NIS^-): The positive and negative ideal solutions are determined with the help of following formulas:

$$PIS^+ = \{(\max v_{ij} | j \in J), (\min v_{ij} | j \in J')\} = \{v_1^+, \ldots, v_n^+\} \quad (9)$$

$$NIS^- = \{(\min v_{ij} | j \in J), (\max v_{ij} | j \in J')\} = \{v_1^-, \ldots, v_n^-\} \quad (10)$$

where J is benefit criteria, and J′ equals to the cost criteria.

Determination of separation measure: The distance between each alternative from positive ideal solution and negative ideal solution can be determined with the help of following equations:

$$d^+ = \left\{ \sqrt{\sum_{j=1}^{n} (Vij - PIS+)^2} \right\} \quad (11)$$

$$d^- = \left\{ \sqrt{\sum_{j=1}^{n} (Vij - NIS-)^2} \right\} \quad (12)$$

where i = 1, 2, …, m.

Quantification of relative closeness of different alternatives to the ideal solution (R_i): Following equation can be used to assess the relative closeness of alternatives to the ideal solution;

$$R_i = \{d^- / (d^+ + d^-)\} \quad (13)$$

where i = 1, …, m.

In the proposed research investigation, influential variables such as Accommodating Index, Mean Heterogeneity Index, Landscape Shape Index, Division, and Cohesion were incorporated into the proposed model as positive variables. On the other hand, the variable, i.e., Distance was considered as negative variable in the present research investigation.

2.4 Normalization of Bias of the Result Obtained from AHP-TOPSIS Model (Norm (B))

A bias is generated in the result of AHP-TOPSIS model because of negative variable, i.e., Distance. The bias needs to be reduced from the result to get more accurate result. It was reduced by using the following formulae:

$$\text{Norm (B)} = \sqrt{R * \{AI/\max(AI)\}} \quad (14)$$

where R is the result of the AHP-TOPSIS model, AI corresponds to accommodating index, and max (AI) represents maximum value of AI of all the buffers in the study area.

3 Results and Discussion

Test study area is comprised of a multiple ring buffer consisting ten different buffers from a reference point to assess the vulnerability of different buffers to urbanization. The uniqueness of influencing factors, i.e., Accommodating Index, Mean Heterogeneity Index, Landscape Shape Index, Division Index, Cohesion Index, and Distance were incorporated into the proposed model to simulate the impact of these aforementioned factors in assessing the possibility of land use conversion into an urbanized class. The results of the proposed research investigation (Table 1, Figs. 2 and 3) suggest that the outermost buffer is the least vulnerable to land use conversion. On the other hand, it is not the innermost buffer but the third buffer which is the most vulnerable to land use conversion. Furthermore it is worth noting that the vulnerability of first buffer to land use conversion is more than second buffer.

Table 1 Degree of vulnerability of different buffers to urbanization

Buffers	Degree of vulnerability to urbanization based on AHP-TOPSIS model	Rank
1	0.651	8
2	0.639	10
3	0.700	4
4	0.725	1
5	0.724	2
6	0.716	3
7	0.709	5
8	0.684	6
9	0.661	7
10	0.643	9

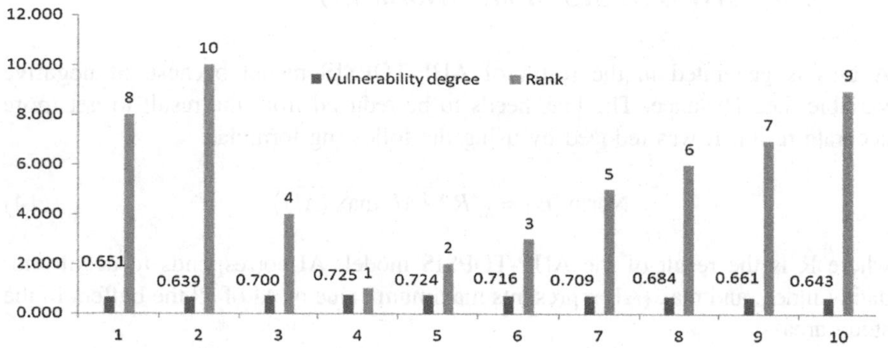

Fig. 2 Graphical representation for the degree of vulnerability of different buffers to urbanization

That means, first there is a dip in the possibility of land use conversion when transiting from first buffer to second buffer. Then there is a rise in the susceptibility of middle buffers to land use conversion. Finally, the outer buffers witness a consistent decline in the possibility of land use conversion into an urbanized class. The aforementioned results demonstrate the capability of the proposed model to simulate the influence of unique characteristics of different influencing factors. Had the proposed model failed to consider the unique characteristics of the influencing variables, there would have been a linear result instead of a nonlinear graph representing vulnerability of different buffers to land use conversion.

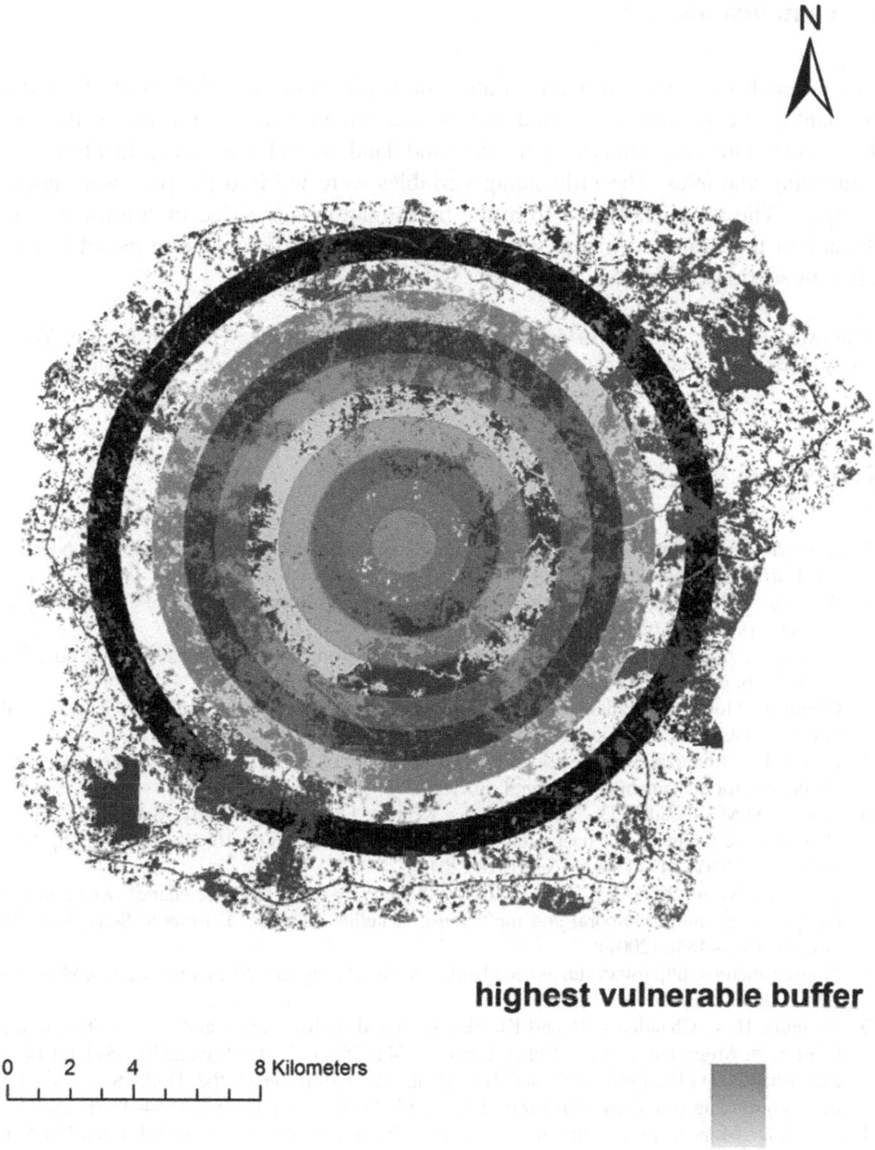

Fig. 3 Representation of the buffer which is possibly the highest vulnerable to urbanization on GIS platform

4 Conclusions

This research investigation demonstrated the applicability of AHP-TOPSIS model to quantify the possibility of land use conversion in different buffers of the test study area. Different landscape indices and land use classes were identified as influencing variables. The influencing variables were fed into the proposed model as inputs. The results obtained from the investigation succeeded in confirming the efficacy of the proposed model to represent the land use conversion possibility in different section of the study area.

Acknowledgements The authors are grateful to the Vice Chancellor, Birla Institute of Technology, Mesra, and Ranchi for providing necessary facilities to perform the present study.

References

1. Akineymi, F.O.: An assessment of land-use change in the Cocoa Belt of south-west Nigeria. Int. J. Remote Sens. (2013)
2. Al-Harbi, K.M.A.S. (2001). Application of the AHP in project management. Int. J. Project Manage, 19, 19–27 (2001)
3. Cabral, P., Augusto, G., Tewolde, M, Araya, Y.: Entropy in Urban Systems. Entropy, 15, 5223–5236, doi:10.3390/e15125223 (2013)
4. Cheng, J.: Modelling spatial and temporal growth. PhD dissertation, Faculty of Geographical Sciences, Utrecht University, The Netherland (2003)
5. Cohen, B.: Urbanization in developing countries: Current trends, future projections, and key challenges for sustainability. Technol. Soc. 28, 63–80 (2006)
6. Caldas, M.M., Goodin, D., Sherwood, S., Krauer, J.M.C. and Wisely, S.M.: Land-cover change in the Paraguayan Chaco: 2000–2011. J Land Use Sci. Vol. 10, No. 1, 1–18, http://dx.doi.org/10.1080/1747423X.2013.807314 (2015)
7. Deng, J.S., Wang, K., Deng, Y.H., Qi, G.J. (2007). PCA-based land-use change detection and analysis using multi temporal and multi sensor satellite data. Int. J. Remote Sens. Vol. 29, No. 16, 4823–4838 (2007)
8. Fragstatsmetrics. http://www.umass.edu/landeco/research/fragstats/documents/Metrics/Metrics%20TOC.htm (accessed Jan 10, 2015)
9. Gartaula, H.N., Chaudhary, P. and Khadka, K.: Land Redistribution and Reutilization in the Context of Migration in Rural Nepal. Land, 3, 541–556. doi:10.3390/land3030541 (2014)
10. Jahanshahloo, G.R., Lotfi, F.H. and Izadikhah, M.: Extension of the TOPSIS method for decision-making problems with fuzzy data. Appl. Math. Comput. 181, 1544–1551 (2006)
11. Miller, D., Munroe, D.K.: Current and future challenges in land use science. J. Land Use Sci. (2014)
12. Sudhira, H.S., Ramachandra, T.V., Jagadish, K. S.: Urban sprawl: metrics, dynamics and modelling using GIS. Int. J. Appl. Earth Obs. Geoinf. 5 29–39, (2004)
13. Veldkamp, A., Lambin, E.F.: Predicting land use change. Agr. Ecosyst. Environ. 85, 1–6, (2001)
14. Zeng, C., He, S., Cui, J.: A multi-level and multi-dimensional measuring on urban sprawl: a case study in Wuhan metropolitan area, central China. Sustainability.6, 3571–3598. doi:10.3390/su6063571, (2014)

Enhanced Feature Selection Algorithm for Effective Bug Triage Software

Jayashri C. Gholap and N.P. Karlekar

Abstract For developing any software application or product it is necessary to find the bug in the product while developing the product. At every phase of testing the bug report is generated, most of the time is wasted for fixing the bug. Software industries waste 45% of cost in fixing the bug. For fixing the bug one of the essential techniques is bug triage. Bug triage is a process for fixing the bugs whose main object is to appropriately allocate a developer to a novel bug for further handling. Initially manual work is needed for every time generating the bug report. After that content categorization methods are functional to behavior regular bug triage. The existing system faces the problem of data reduction in the fixing of bugs automatically. Therefore, there is a need of method which decreases the range also improves the excellence of bug information. Traditional system used CH method for feature selection which is not give accurate result. Therefore, in this paper proposed the method of feature selection by using the Kruskal method. By combining the instance collection and the feature collection algorithms to concurrently decrease the data scale also enhance accuracy of the bug reports in the bug triage. By using Kruskal method remove noisy words in a data set. This method can improve the correctness loss by instance collection.

Keywords Bug · Bug triage · Kruskal method · Feature selection
Instance selection

J.C. Gholap (✉) · N.P. Karlekar
Sinhgad Institute of Technology, Lonavala, Pune, India
e-mail: jayagholap29@gmail.com

N.P. Karlekar
e-mail: npk.sit@sinhgad.edu

D.K. Mishra et al. (eds.), *Information and Communication Technology for Sustainable Development*, Lecture Notes in Networks and Systems 10,
https://doi.org/10.1007/978-981-10-3920-1_5

1 Introduction

In IT industries for developing the software, software repositories are used like the big range of database for accumulating the output of the developing the software. Starting step in bug repository is to oversee bugs in the software. Fixing the bug is an essential and time-consuming procedure in development process. In the development of open-source projects are characteristically incorporated by an open bug repository in which bug can be reported by both software developers and users or flaws or issues in the software, recommend conceivable improvements, also remark on accessible bug reports. The benefit of an open bug repository is that this report may be authorizing more bugs to be recognized and is also explained, enhancing the nature of the product item. Bug triage is most crucial step for fixing the bug, is to allocate another bug to a significant developer for further handling. For open-source programming projects, large number of bugs is produced day by day from which the process of triaging process is exceptionally troublesome and it was more demanding. Fundamental goal of bug triage is to appoint a developer for fixing the bug. Once a developer is appointed to another bug report, he will fix the bug or attempt to correct it. The main goal of this work is to decrease the large range of the training set and to evacuate the noisy and repetitive reports of bug for bug triage [1, 2]. Data reduction is the procedure of reducing the bug data by using two methods which are, instance collection and feature collection which intends toward to get low range as well as quality data [3]. Now, we discuss the existing work done for this concept, limitation of the existing work and proposed system.

1.1 Existing System

The existing system represents the issues of reduction of data for bug triage. This issues aspire to expand the data set of bug triage in different two aspects, which are first one for concurrently diminishing the scales of the bug measurement and the word measurement and to enhance the correctness of bug triage. This system is an approach which handles the issues of reduction of information. The above issues are solved by the application of instance collection and feature collection in bug repositories. This system builds a method of binary classifier for predicting the order of applying instance collection and feature collection.

Limitations of Existing System: In existing system, the sequence of relating instance collection and feature collection has not been examined in connected domains.

1.2 Proposed Scheme

This scheme is proposed to develop an effective model for doing data reduction on bug data set for reducing the range of the information and also improve the excellence of the data by falling the time and cost of bug reducing techniques. The proposed improved feature selection method by using Kruskal model for handling the issues for reduction of information. The main output of this system is:

- For removing the noisy words from the dataset, feature collection is used.
- Instance collection can remove uninformative bug reports.
- The accuracy of the bug triage is improved by eliminating the redundant words;
- Instance collection can recover the accuracy loss.

Rest of the paper is discussed as: Section. 2 discusses the related work for removing the bug triage; Sect. 3 discussed the description of the proposed scheme. Finally, comparison result among the proposed and existing system is done. At last the paper is concluded and describes the future scope.

2 Literature Review

The author for decreasing the range of information on the measurement of bug and the measurement of word this paper merge instance collection method with feature collection. The author removes out attributes from the historical bug data set and constructs a prescient model for novel bug information set for deciding the order of applying instance collection and feature collection [4].

The author proposes a markov based scheme for evaluating the quantity of bugs which will be developed in further advancement. The author proposed a strategy for evaluating the aggregate sum of time requisite to repair them on the basis of the experimental distribution of bug fixing time resultant from information this can be discussed for the given number of defect. Also the author is able to likewise develop a grouping model for predicting moderate or quick fix for the given number of input [5].

The author discussed the method in which the training place reduces with both feature collection and instance collection procedures for triaging bug. They consolidate feature collection for example instance collection to enhance the exactness of bug triage. The feature collection algorithm X2-test, instance collection algorithm Iterative Case Filter, and their grouping are contemplated in this paper [6].

The author introduced a retrieval function to quantify the comparison among two reports of bug. This system completely uses data accessible in a bug report together with not just the comparability of textual matter in outline also depiction fields, additionally similarity of non-textual fields, for content, item, segment, adaptation, and so forth. For more precise estimation of textual comparison, author extended BM25F an efficient comparison equation in data recovery group, specifically for copy report retrieval [7].

The author targets on two directions to tackle this issue: Initially, they reformulate the issue as a development issue of both the precision and rate. Second, they accept a content boosted collaborative filtering (CBCF), combining a current CBR with a collaborative filtering recommender (CF), this method improves the proposal nature of either advances alone. Then again, dissimilar general proposal situations, bug fix record is particularly sparse. Because way of bug fixes, one bug is settled by one and only developer, which makes it demanding to follow after the above two directions. To address this problem, they expand a subject model to sparseness the inadequacy and improve the nature of CBCF [8].

The authors depict iterative outline mining which yields patterns which are rehashed regularly within a program trace, or over various follows, or both. Continuous iterative patterns reflect successive program behaviors that sensible evaluate to software details. To decrease the quantity of prototypes and enhance the proficiency of the algorithm, author has additionally presented mining congested iterative patterns, which are maximal patterns with any great pattern having the same sustain. In this paper, for theoretically extending investigation on iterative pattern mining; they present mining iterative generators, i.e., minimal pattern with any subpattern having the same support. Iterative generators can be paired with closed patterns to deliver an arrangement of rules communicating forward, backward, and in the middle of sequential restriction among events in one general demonstration [9].

System is the effect of distributed software advancement and organizational structure. Frequently vast projects are created in a distributed fashion around the world. Is the adequacy of the bug fixing process affected by hierarchical and geographic barriers? The author addresses this inquiry with information from two releases of Microsoft Windows. Further, they recognize the effect of the reputation of the bug opener and the architect fixing the bug on the likelihood of a reopen. They likewise observe that how bugs are discovered, noticeably affects bug revives [10].

3 Implementation Details

3.1 System Overview

Figure 1 describes the working of the proposed scheme in detail: The above diagram shows the Bug Triage block which shows the architecture of work done by the existing scheme on bug triage. In this system, initially a phase of data reduction in added after that classifier is trained a with a bug data set. Big data reduction block unites the method of instance collection and feature collection to diminish the range of bug information. In the process of bug data reduction, the issues are how to

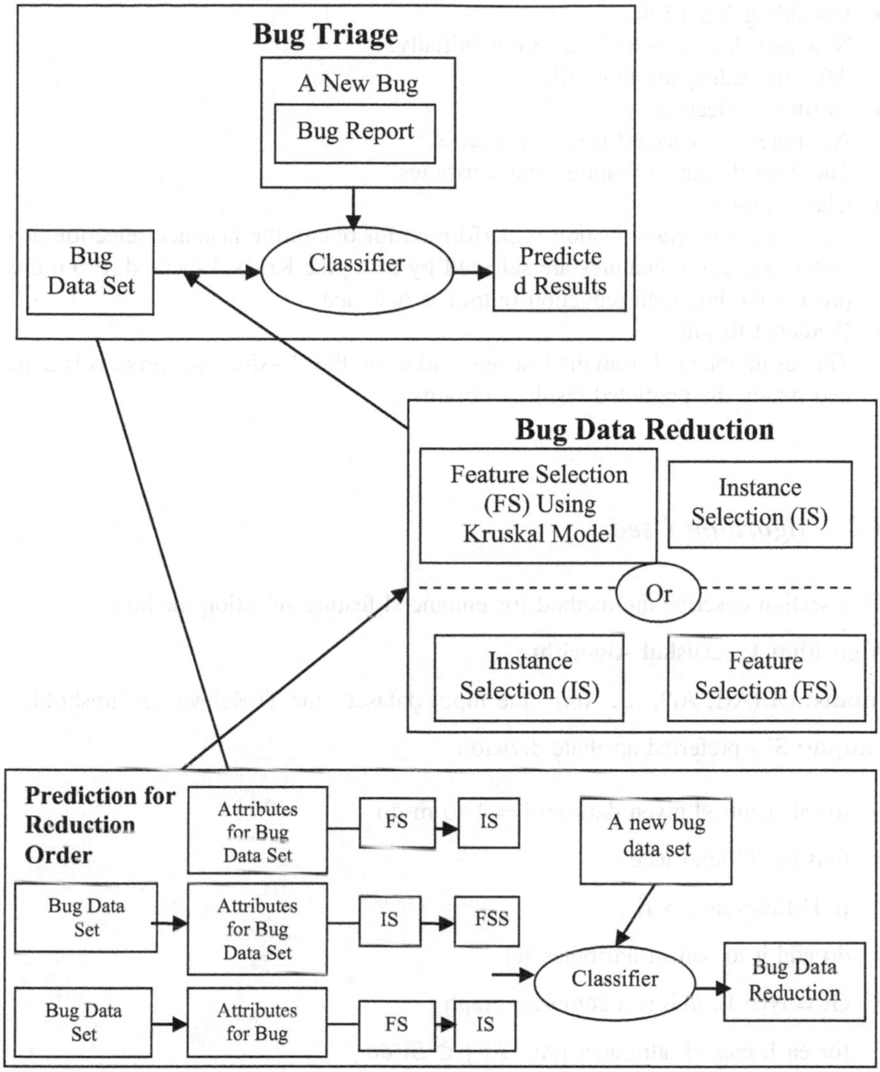

Fig. 1 System architecture

conclude the order of two reduction methods. Prediction for reduction order on the basis of the attributes of historical bug data sets, system introduced a binary categorization method to forecast decrease orders. Also, proposed the improved feature collection technique by using Kruskal model for tackling the issues of data reduction.

- Uploading Input File
  New bug dataset is taken as input initially.
  After uploading the input file.
- Attribute Selection
  Attributes are selected from the dataset.
  The dataset contain features and attributes.
- Classification
  The process of classification is performed for obtain the instance selection and feature selection. Features are selected by using the Kruskal method. From this process the bug data reduction output is obtained.
- Predicted Result
  The result obtained from the last step and again the classification process is done and finally the predicted result is obtained.

3.2 Algorithm Used

This section describe the method for enhanced feature selection method.

Algorithm 1: Kruskal Algorithm

Inputs: DB(Ai1, Ai2, ..., Ain) - the input dataset - the Tl-Relevance threshold.

Output: Si - preferred attribute division

1. for element od given data set i = 1 to m do
2. find the Tl-relevance
3. if Tl-Relevance > ri
4. do add it to pair of attributes set
5. Gt = NULL; //Gt is a complete graph
6. for each pair of attributes {Ai′, Ai′} $\subset$ Si do
7. F-Correlation = SUi (Ai′, Ai′) add the edge to the
8. tree // per h weight of matching tree
9. minSpanTree = KRUSKALS(G); //KRUSKALS

Algorithm 2: Algorithm for generating the minimum spanning tree

Forest = minSpanTree for each edge $\in$ Forest do

if SUi(Ai′, Ai′) < SUi(Ai′,) $\wedge$ SUi(Ai′, Ai′) < SUi(Ai′,)

Then

Forest = Forest − //remove the edge

Si = r

for each tree ∈ Forest do

find the strongest attribute set

Si = Si ∪ {Ai};

Return Si

3.3 Experimental Setup

The system is built using Java framework (version jdk 8) on Windows platform. The Net beans (version 8.1) is used as a development tool. The system does not require any specific hardware to run; any standard machine is capable of running the application.

4 Result and Discussion

4.1 Dataset Discussion

Bug dataset and bug report are used as an input dataset in the proposed scheme.

4.2 Results

Table 1 shows the accuracy of the proposed scheme and the existing scheme. It shows that the accuracy of the proposed system is more than the accuracy of the existing system.

Figure 2 shows the comparison of the proposed system over the existing system.

Table 1 Accuracy table

List size	Existing scheme (%)	Proposed scheme (%)
1	61	68
2	40	51
3	73	76
4	83	84
5	57	62

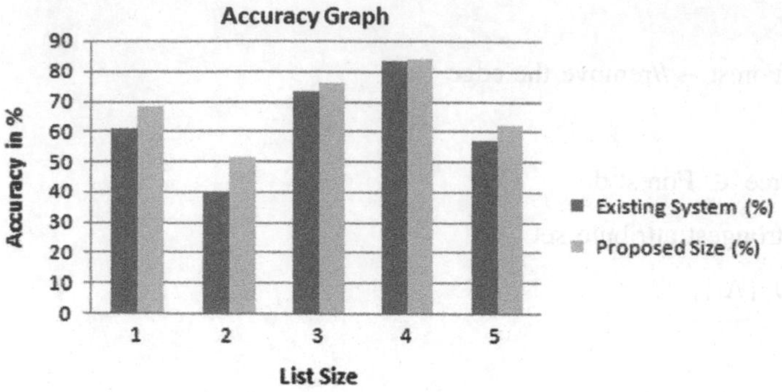

Fig. 2 Accuracy graph

5 Conclusion

Bug triage is an expensive step for maintaining the labor cost and time cost of the software. In this scheme, unique feature collection with instance collection to reduce the range of bug datasets as well as improve the quality data was used. Kruskal model is used for feature collection instead of CH feature collection which extracts attributes of each bug data set and train an extrapolative model based on historical data sets. Provides an approach to leveraging techniques on data processing to reduced form and software development and maintenance using high-quality bug data. The experimental result demonstrates that the correctness of the proposed scheme is more than the accuracy of the existing scheme.

References

1. V. Cerveron and F. J. Ferri: Another move toward the minimum consistent subset: A tabu search approach to the condensed nearest neighbor rule. IEEE Trans. Syst., Man, Cybern. Part B, Cybern. pp. 408–413 (2001).
2. S. Breu, R. Premraj, J. Sillito, and T. Zimmermann: Information needs in bug reports: Improving cooperation between developers and users. In: Proc. ACM Conf. Comput. Supported Cooperative Work, pp. 301–310 (2010).
3. J. W. Park, M. W. Lee, J. Kim, S. W. Hwang, and S. Kim: Costriage: A cost-aware triage algorithm for bug reporting systems. In: Proc. 25th Conf. Artif. Intell., pp. 139–144 (2011).
4. Jifeng Xuan, He Jiang, Member, Yan Hu, Zhilei Ren, Weiqin Zou, Zhongxuan Luo, and Xindong Wu: Towards Effective Bug Triage with Software Data Reduction Techniques. In: IEEE transactions on knowledge and data engineering (2015).
5. H. Zhang, L. Gong, and S. Versteeg: Predicting bug-fixing time: An empirical study of commercial software projects. In: Proc. 35th Int. Conf. Softw. Eng., pp. 1042–1051 (2013).
6. W. Zou, Y. Hu, J. Xuan, and H. Jiang: Towards training set reduction for bug triage. In: Proc. 35th Annu. IEEE Int. Comput. Soft. Appl. Conf., pp. 576–581(2011).
7. C. Sun, D. Lo, S. C. Khoo, and J. Jiang: Towards more accurate retrieval of duplicate bug reports. In: Proc. 26th IEEE/ACM Int. Conf. Automated Softw. Eng., pp. 253–262 (2011).

8. J. W. Park, M. W. Lee, J. Kim, S. W. Hwang, and S. Kim: Costriage: A cost-aware triage algorithm for bug reporting systems. In: Proc. 25th Conf. Artif. Intell., pp. 139–144 (2011).
9. D. Lo, J. Li, L. Wong, and S. C. Khoo: Mining iterative generators and representative rules for software specification discovery. IEEE Trans. Knowl. Data Eng., pp. 282–296, (2011).
10. T. Zimmermann, N. Nagappan, P. J. Guo, and B. Murphy: Characterizing and predicting which bugs get reopened. In: Proc. 34th Int. Conf. Softw. Eng., pp. 1074–1083 (2012).

Reliable Data Delivery on the Basis of Trust Evaluation in WSN

Deepak Gadde and M.S. Chaudhari

Abstract Different applications have come out as in the field of Wireless Sensor Network (WSN) by years of research. At present, Lossy Networks (LLNs) are in the center of area of studies. LLNs are made up of Wireless Personal LANs, low-power Line Communication Networks, and Wireless Sensor Networks. In such LLNs, for sending protected data, routing IPv6 routing protocol is used for minimum-power as well as not reliable networks (RPL) controlled by Internet Engineering Task Force (IETF) routing protocol. A route created by rank-based RPL protocol for the sink based in the rank value and link quality of its neighbors. But, it has few site backs for example, high packet loss rates, maximized latency, and very low security. Packet losses as well as latency get maximized as the length of path (in hops) increases as with every hope a wrong parent link is chosen. For increasing the RPL system and to solving issue stated above, a Trust Management System is proposed. Every node is having a trust value. Trust value will increase or decrease depending to behavior of node and trusted path is selected for delivering the data. For increasing energy efficiency, at the time of data transferring "compress then send" method is used, this results in minimum utilization of energy reduced data size. By making use of cryptography, we gained data security which is the key concern. By analyzing the test outcomes conducted on JUNG simulator shows, our proposed system have increased the packet delivery ratio by having trust management system while transferring the data, increases energy efficiency by utilizing the data compression, network security is improved by utilizing encryption decryption method as compared to present system.

Keywords Shortest path discovery · Data compression techniques
LLNs · Routing protocol for RPL

D. Gadde (✉) · M.S. Chaudhari
Sinhgad Institute of Technology, Lonavala, Pune, India
e-mail: dgadde1234@gmail.com

M.S. Chaudhari
e-mail: mschaudhari20@gmail.com

D.K. Mishra et al. (eds.), *Information and Communication Technology for Sustainable Development*, Lecture Notes in Networks and Systems 10,
https://doi.org/10.1007/978-981-10-3920-1_6

1 Introduction

After the Internet Engineering Task Force (IETE) formed, they have given Internet norms. It norms has Internet convention suite (TCP/IP). It is an association having open standard.

Currently, in research regions Low Power and Lossy Networks (LLNs) are in boom. LLNs include Wireless Personal Area Networks, low-Power Line Interaction Systems, and Wireless Sensor Networks. In LLNs, for security of data while transferring it iPv6 is used for low power and lossy framework (RPL) is controlled by Internet Engineering Task Force (IETF).

RPL is a Distance Vector IPv6 oversees convention for LLNs. Searching of data which is not required as well as removing shows its bits by lossy compression. The procedure of representing the bits of a data document is known as information compression. For security the system will encrypts data before sending. System makes use of Elliptic Curve Cryptography (ECC) algorithm for encryption. Compared to non-ECC cryptography, ECC uses small size key for providing security.

The dependability of each node with other nodes opinion, representation of it is a trust model. Each node has a trust value with each reaming nodes. The modification of trust value has several parts. Aging is the first part. Author gives different weights to the latest and current trust values is based on the real needs and different sinks. If author gives the current trust value a higher weight, it gradually holds the node in normal state. If author gives high weight to past trust values, it may prohibit the nodes deceptions. Faulty nodes work well in a brief to encourage trust value.

2 Literature Review

In this paper [1], author analyzes and provides results of the analysis. The protocol features and design selection that creates unreliability problems in the analysis. The important attribute of their RPL deployment is an adaptable cross-layering design with simple optimal routing, increased link estimation capacities as well as effective management of neighbor tables.

The author [2] given a simple technique, i.e., meta-data is used as an efficient way to form clusters. Energy efficient way presented for a novelty multi-hop and fault-tolerance routing protocol capable of transporting information from sensor nodes to its cluster-head and with the order reversed. A tedious task is misbehavior of nodes in the routing process and the results are heavy efficiency demotion in network outcome, packet delivery ratio, packet loss raising, etc.

To overcome this issue [3], analysis of the existing misbehavior detection schemes and proposed multi-hop acknowledgment scheme as novel solution used to detect misbehaving nodes.

In this paper [4], to prevent the black hole and gray hole attacks on the delay and energy proficient routing protocol in sensor-actor networks, proposed an effective trust-based secured coordination system. This mechanism analyzes each sensor's trust level 1 hop sensors on the basis of the experience, recommendation, and knowledge.

This paper [5] presents to overcome routing issues in the IPv6 Routing Protocol for Low power and Lossy Networks (RPL), which merged in LLNs. When data packets have to be sent it implements measures to reduce energy utilization such as dynamic sending rate of control messages and addressing topology inconsistencies. The protocol helps to use of IPv6 and supports, not only in the upward direction, but also traffic flowing from a gateway node to all other network participants.

3 Problem Definition

Forwarding of data packets with no packet loss is the aim in low-power sensor network. As we take low power and lossy network in account, for considering lifetime when getting higher packet delivery ratios. For searching the neighbors having lowest distances, the rank-based RPL protocol is used and for checking the quality of the links between nodes periodical link estimations are used. The process consumes time and memory; because of changing nature of the sensor nodes the algorithms are not able to give trusted routes. Security is also the issue in existing system. A new trust-based RPL protocol is implemented which removes the complex periodical estimations and replaced it with the most realistic trust computation which can give shortest lossless paths. Alongside the trustworthiness security is achieved by ECC algorithm. Notwithstanding this data compression technique is utilized to compress the information so that lightweight information is exchanged and spare more system energy. This proposed framework is likewise capable to identify and prevent diverse attackers that endeavor to break the dependability of the framework.

4 Implementation Details

4.1 System Overview

Every node has a routing table which has information with link quality. It is used by nodes to connect to other nodes. In LLNs RPL is used protocol for data routing. RPL is a gradient-based routing which implements a destination oriented (DO) DAG rooted at an information collector or sink node. The gradient is also named rank and based on description of the node's personal position corresponds with another nodes on the DODAG root. So, sharing a packet to the DODAG root approximately include in taking the neighbor node with the less rank. A routing

Fig. 1 System architecture

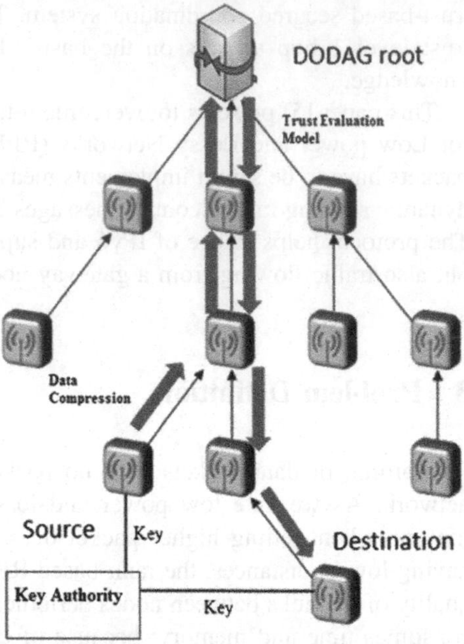

objective function (OF) describes RPL nodes calculations with their rank values and select their parents. It chooses neighbor nodes having less rank.

By making use of RPL protocol, all paths from sources to destination nodes are computed. From these paths, the shortest path is chosen based on lowest hop count. After this system evaluates the trust value of all nodes and computes the trust factor of all paths. Then the shortest path with maximum trust value is chosen for actual data routing from source to destination node. All data is encrypted using ECC algorithm to provide the data security. System performs data compression after this encryption. This step gives lightweight data which will take less energy than actual data forwarding (Fig. 1).

The main goal of proposed system is to improve the packet delivery ratio as well as the secure and energy efficient data forwarding. That's why, in our system data compression and trusts evaluation is done, which will minimize data forwarding time and also increase lossless data transmission along with that this system takes care of the security requirement of data.

- Data Compression:
  It is a method to compress the data which minimize the actual size of the data and outcome in saving the network energy. While sending data from source to destination we are doing data compression. Because of this nodes utilize less energy compared to energy sending while original data. The framework also takes care of lossless data compressions which blocks data loss while decompression operation.

- Trust Calculation models:
  In place of complex link estimation methods, this framework develops a trust model which computes a trust value for each node by its last behavior and which will help us in sending information using the secure and optimal path. The most trusted path is chosen to send the data between sources to destination by computing the average trust of all the nodes in the path.
- Secure Data Sending:
  However, system develops the trust model to choose the trusted paths for data forwarding; data secrecy is the biggest threat by outside attackers or outliers. To solve this issue, proposed system creates and assigns the pair wise keys by making use of ECC algorithm in the nodes which are responsible to transmit and receive the information. The data is encrypted using private key at the sender side and this encrypted data is send via trusted path to destination node, after the key generation and distribution. Data send can only be decrypted by destination node by using valid key. With its asymmetric behavior and less energy consumption, ECC algorithm gives more security which is needed to this system.
- Attack prevention model:
  There are various types of attacks such as man in the middle attack and in addition Dos attacks which is able to listen or modify the information which threatens the security in the network system. In this way, to prevent these attacks on the system framework gives node authentication using ECC algorithm. Only node having valid keys can send and get the information in the system.

By doing Bad Mouthing attack or ON-OFF attacks outlier is able to break in our trusted system. In such type of attack, the attacker give bad approval to misguide the source form choosing the trusted nodes. To provide protection from such attack, system maintains other recommendation grade list. If a node has miner recommendation grade than its recommendation have minor or no effect on the actual recommended trust value. In the ON-OFF attack, the attacker behaves randomly to keep up its trust value so that he cannot be found. To keep out from such type of attacks, forgetting factor is implemented and modified depending on the cases, which reflects the latest trust of the node.

4.2 Algorithm

Algorithm for detection crater on MOC datasets.

1: Create Network graph G (V, E) Where V = Vertices, E = Edges.
2: Select the source S and Destination D from the network nodes.
3: Send the lightweight probe packet from destination.
4: Update the rank and link estimation of each node.
5: Generate all available paths to destination.
6: Select the shortest path from source to destination.

7: Generate and distribute valid key pairs to required nodes.
8: Compress the data at source by replacing the common data with patterns.
9: Encrypt the data.
10: Send the encrypted data through shortest route.
11: Check for the packet loss at each node.
12: Calculate the trust value of each node.
13: Check for the valid key
If (key matched)
{
Decrypt the data
If (Packet loss occurred)
{Resend the data with max.
Trusted paths
}
Else
Decompress the data at destination.
}
Else
{
Deny the access for the data
}

4.3 Mathematical Model

$$E_{Tx}(k,d) = E_{Tx-elec}(k) + E_{Tx-amp}(k,d)$$
$$E_{Tx}(k,d) = E_{elec} * k + \in_{amp} * k * d^2 \quad (1)$$

d: Distance for neighboring sensor node
2_{amp}: Energy required for the transmitter amplifier
E_{elec}: Energy consumed for driving the transmitter or receiver circuitry
K: bits per message

Above formula is used to calculate the energy required to transfer the data. So,
Node's updated Energy = Available Energy − Energy Consumed;
Error Rate = Number of Packets Send − Number of Packets Received;
Latency (in ms) = Packet Received Time − Packet Send Time;
Trust Calculation: Available Trust (T_{ava}) = Trust value of that node.
Average Trust (T_{cal}) = Calculated based On (Error Rate, previous Successful Transaction, latency);
Direct trust (t_{dx}): Trust assigned directly after transaction.
Recommended Trust (t_{rx}) = recommended trust from neighbors.
Indirect Trust (t_{ix}) = t_{rx} + T_{cal}.

Forgetting factor (β) = $0 < \beta < 1$;
So;
Node's Present Trust value = $T_{ava} + t_{dx} + t_{ix} + \beta$;
//(If Successful Transaction), where $\beta < 0{:}5$
Node's Present Trust value = $T_{ava} - t_{dx} - t_{ix} - \beta$;
//(If Unsuccessful Transaction), where $\beta > 0{:}75$

For ECC Algorithm:

- Key generation
  W = r * c
  Where,
  W = public key
  r = random number from (1 to n − 1)
  c = point on curve
- Encryption
  C1 = K * p
  C2 = M + k * W
  C1 & c2 = cipher text
  K = random number between {1 − (n − 1)}
  C = point on curve
  M = original message
  W = public key
- Decryption
  M = c2 – d * c1

5 Result and Discussion

The graph in Fig. 2 shows that the existing system having maximum amount of packet drops than the proposed system, because the proposed system uses trust score calculation in every iterations of all nodes. Only node with the highest trust is

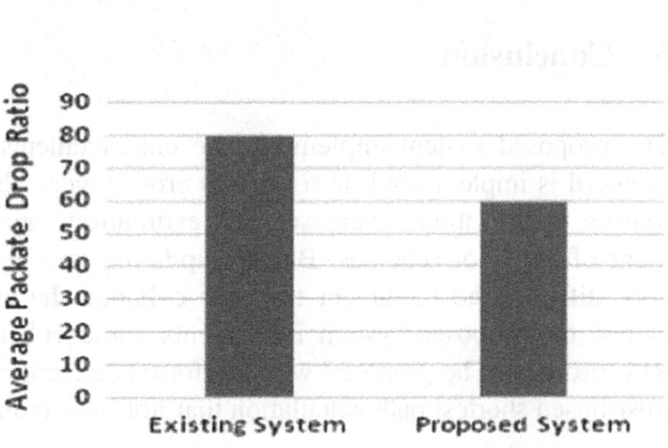

Fig. 2 Average packet drop ratio graph comparison

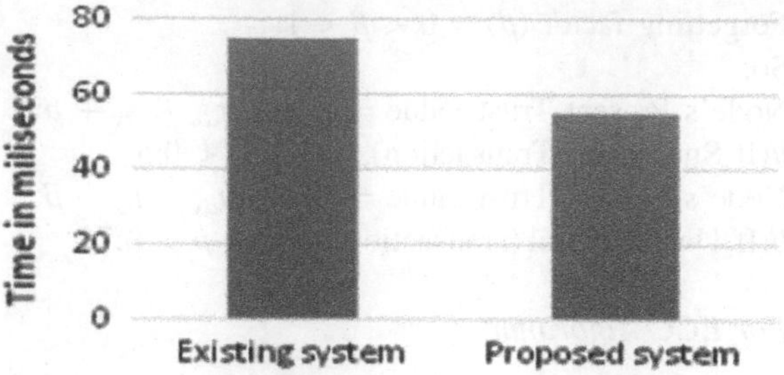

Fig. 3 Time graph comparison

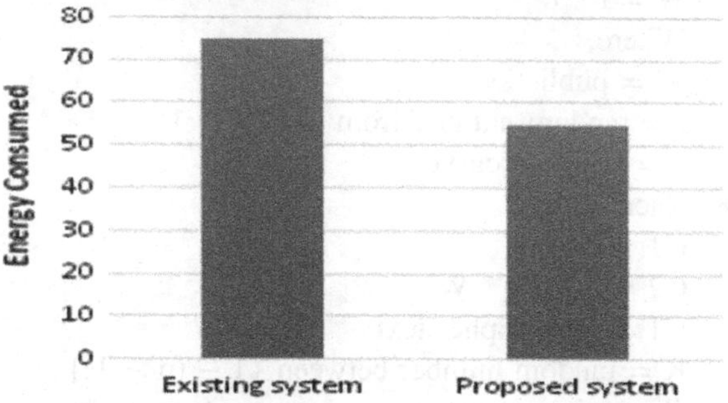

Fig. 4 Energy consumed graph comparison

considered for further data transmission. Therefore, the probability of packet loss at trusted node will be reduced.

Figure 3 shows that the proposed system takes minimum time to forward the data from source to destination node that the existing system as we pass data through trusted nodes only there will be no chance for data to travel through long path so the chances of packet loss is reduced.

Figure 4 shows that the energy consumed by proposed system is less than existing system. We are performing data compression while sending the data from the source. By this, the author is sending the lightweight data which consume less energy than actual data sending.

6 Conclusion

The proposed system implements the enhancements in RPL protocol. The RPL protocol is implemented on a flexible cross-layer, which is responsible for simple routing optimizations, increased link estimation capacities, and effective management of neighbor schemas. But the updating and maintenance of such tables are very difficult and inefficient task. To enhance the efficiency of RPL protocol in LLNs, the proposed system implements some enhanced features combined with RPL protocol. The proposed work performs data encryption, data compression, and trust-based shortest path calculation that not only reduce data sending time but also improves lossless data transmission. For data encryption system uses ECC algorithm, which makes data secure during transmission. Data compression is decreases bits by finding not needed data and deleting it. This technique minimizes resource utilization like data storage capacity or transmission capacity. Finally, system calculates the trust value of each node by which data is transmitted over most trusted and shortest path to the destination.

References

1. Emilio Ancillotti, Raffaele Bruno: Reliable Data Delivery With the IETF Routing Protocol for Low-Power and Lossy Networks. IEEE transactions on industrial informatic (2014).
2. E. Canete, M. Diaz, L. Llopis, and B. Rubio: HERO: A hierarchical, efficient and reliable routing protocol for wireless sensor and actor networks. Comput. Commun. pp. 1392–1409 (2012).
3. Usha. Sakthivel and Radha. S: Routing layer Node Misbehavior Detection in Mobile Ad hoc Networks using Nack Scheme. International Conference on Trends in Electrical, Electronics and Power Engineering (ICTEEP'2012) (2012) Singapore.
4. Jagadeesh Kakarla, Banshidhar Majhi: A Trust Based Secured Coordination Mechanism for WSAN. International Journal of Sensor Networks, 978-1-4799-1823-2/15 (2015).
5. TsvetkoTsvetkov: RPL: IPv6 Routing Protocol for Low Power and Lossy Networks. Seminar SN SS2011, doi:10.2313/NET-2011-07-1 09, Network Architectures and Services (2011).

K-Mean Clustering Algorithm Approach for Data Mining of Heterogeneous Data

Monika Kalra, Niranjan Lal and Samimul Qamar

Abstract The increasing rate of heterogeneous data gives us new terminology for data analysis and data extraction. With a view toward analysis of heterogeneous sources of data, we consider the challenging task of developing exploratory analytical techniques to explore clustering techniques on heterogeneous data consist of heterogeneous domains such as categorical, numerical, and binary or combination of all these data. In our paper, we proposed a framework for analyzing and data mining of heterogeneous data from a multiple heterogeneous data sources. Clustering algorithms recognize only homogeneous attributes value. However, data in the every field occurs in heterogeneous forms, which if we convert data heterogeneous to homogeneous form can loss of information. In this paper, we applied the K-Mean clustering algorithm on real life heterogeneous datasets and analyses the result in the form of clusters.

Keywords Heterogeneous data · Clustering · K-Mean

1 Introduction

With rapid growth of heterogeneous information in every field brings the need of data mining technique. Data Mining is the powerful technique used for the purpose of extracting the useful knowledge from huge database by analyzing the data from

M. Kalra (✉) · N. Lal
Mody University of Science and Technology, Lakshmangarh, Sikar, Rajasthan, India
e-mail: monikakalra11091425@gmail.com

N. Lal
e-mail: niranjan_verma51@yahoo.com

N. Lal
Suresh Gyan Vihar University, Jaipur, Rajasthan, India

S. Qamar
King Khalid University, Abha, Kingdom of Saudi Arabia
e-mail: jsqamar@gmail.com

D.K. Mishra et al. (eds.), *Information and Communication Technology for Sustainable Development*, Lecture Notes in Networks and Systems 10,
https://doi.org/10.1007/978-981-10-3920-1_7

different prospective. It extracts the previously unknown, predictive information, hidden pattern from the data available in database. The increasing rate of heterogeneous data is required intelligent technique and tools, so that we extract useful knowledge from heterogeneous data [1]. Knowledge-driven results perform primary role in today's generation. Either we take in Business, Financial institutes, Government departments, or any other development organizations, etc., all are collecting excessive amounts of heterogeneous data in their own fields. In more general terms, with the availability of multiple heterogeneous information sources, it is a great challenging problem occurs to perform integrated exploratory analyses with the goal of extracting more information than what is possible from only a particular single heterogeneous source [2].

To summarize the contribution of the paper as follows:

1. Retrieve the result individually from all the data sources into one format.
2. Analysis of all the heterogeneous sources including text corpus, social media, image, and homogeneous data.
3. Applying the K-Mean clustering algorithm individually on each heterogeneous data source for extracting the hidden knowledge.
4. Applying the clustering algorithm K-Mean on heterogeneous large dataset.

2 Literature Survey

Azra Shamin et al. [3], proposed a framework for bio data analysis data mining technique on bio data as well as their proprietary data, bio database is often distributed in nature. This system take input from the user, preprocess the query and load it into local bio database. System will search the knowledge from database and send it back to the user, if the data related to user query exists.

Rumi Ghosh and Sitaram Asur [4], author represent a schema to aggregate multiple heterogeneous documents to extract data from them. Therefore, in this paper, author proposed a novel topic modeling schema, probabilistic source LDA which is planned to handle heterogeneous sources. Probabilistic source LDA can compute latent topics for each source maintain topic–topic correspondence between sources and yet retain the distinct identity of each individual source. Therefore, it helps to mine and organize correlated information from many different sources.

Prakash R. Andhale and S.M. Rokade [5] present the characteristics of HACE theorem which provide the features of heterogeneous data and proposes a model for processing on heterogeneous data from the view data mining. This information extraction model involves the information extraction, data analysis, and provides the security and privacy mechanism to the data.

Amir Ahmad and Lipika De [6], performs clustering algorithm such as K-Mean for both numerical, categorical data. The author represents a better characterization of clusters to conquered the constraints of clustering K-Mean algorithm for numerical data.

3 Data Mining of Heterogeneous Data

Heterogeneous data source has grown rapidly in recent years. Data explosion and 91% of the appellant in the survey say they are aware of heterogeneous data files are used in every field [7]. Today knowledge becomes the biggest asset of all companies so that maximum of the knowledge is recorded in heterogeneous format. Heterogeneous data is the most important part of the Big Data explosion. It is not possible to convert heterogeneous data into homogeneous data. The effectiveness of information retrieval is reduced by the presence of heterogeneous data in the various data sources. Heterogeneous data is the combination of homogeneous, structured, semistructured, and unstructured data. It is generated from digital images, mp3, video, audio, online transaction, social media, E-commerce websites, data from different domains including business, industries, medical, etc. [8].

In today's world due to presence of big data or heterogeneous data, the volume of data is increasing and the data has no specific format. For analysis purposes, a proper structured format data is given.

Consider a description of dataset as shown in Table 1, we analyze the time of one of the numerical attributes is 251 > 164. Discretizing of such numeric values will allocate 164 and 256 the same categorical value. This will create lots of confusion and also loss of useful information. Status is another attribute which is also categorical. There exist no proper existence between the values of status attribute Married, Single and Divorced. It is very difficult task to conversion of categorical values into numeric value. Here, we also consider the binomial value which has only two possible values yes and no. However, binary attributes will follow a Bernoulli distribution. Thus, every attribute having numerical, categorical, and binary attribute has separate characteristics and each attribute will treated separately [6]. A methodology to clustering of attributes of heterogeneous data is present in this paper. This type of task is still in progress to measure effectiveness and validity of cluster. The goal of this paper is to apply clustering algorithm for heterogeneous dataset and measure the point variability of each point in the dataset.

Table 1 Dataset description

Id	Time	Designation	Status	Age
1	251	Employed	Married	35
2	164	Business	Married	38
3	253	Entrepreneur	Single	24
4	254	Management	Divorced	50

4 K-Mean on Heterogeneous Data

Data clustering is the process of aggregation of items in a manner such that items in the same group are more identical to each other than in other groups. In other words, main objective of clustering is to divide the items into uniform and different group for an output. Mostly, clustering methods can made for only handling of numerical data. We also manipulate or change the operations on our data according to our requirement. Clustering algorithms depends on multiple purposes as the type of data available for clustering. There exists a several numbers of techniques for clustering such as hierarchical, center-based partitioning, density, and graph-based clustering.

4.1 *Similarity-Based Cluster*

This type of cluster contains data items which are similar or related to each other in some way. Generally, partitioning clustering methods are K-mediods and K-Mean. Here we are using only K-Mean for clustering. The basic requirement of K-Mean clustering is that taking the no. of cluster as 'k' from the user initially. Representation of cluster as mean value is based on similarity of the data items in a cluster. The mean or center point of cluster is known as 'centroid'. Centroid is a value which can be find out through the mean of related points. K-Mean algorithms have $\sqrt{a^2+b^2}$ several advantages such as simplicity, high speed to access database on very large scale.

4.2 *Standard K-Mean Algorithm*

Step 1: Initial step will be done by the user is to define the number of clusters and the centroid for each cluster.
Step 2: To find out the distance between each data, item is predicted by computing the minimum distance between centroid and the data items.
Step 3: Next step is to generate the centroid again, means recomputing of the centroid.
Step 4: Convergence Condition: Severals conditions are defined as below:

(a) Stopping the process when a given number of iterations by the user is attained.
(b) Stopping the process when there is no changing of data items between the clusters.
(c) Stopping the process when a threshold value is obtained.

Step 5: If all the conditions present above are not fulfilled, then go back to step 2 again and perform the whole process again, until the above present conditions will not fulfilled.

4.3 Idea of Getting Initial Centroids

Input: number of clusters k, number of objects n.

Output: k clusters that specify the least error.

Process of the algorithm is like:

There are n data items in set A and wants to divide the set A into k clusters. Euclidean distance formula is used to find out the distance between data items, e.g., distance between one vector P = (p1, y1) and the other vector Y = (p2, y2) is describing as set A into k clusters.

$$D(P, Y) = |p2 - p1| + |y2 - y1| \tag{1}$$

Algorithm:

Input: A set of data item P = {p1, p2, …, pn}. A Set of initial Centroids C = {c1, c2, ck}
Output: List of Output which contains pairs of (Ci, pj) where $1 \le i \le n$ and $1 \le j \le k$

Procedure N1 ← {p1, p2, …, pm} current_centroids ← C
distance P, Y = $(p_i - y_i)^2 d_i = 1$ where pi (or yi) is the coordinate of p (or y) in dimension i **for all** pi€ N1 such that $1 \le i \le m$ do Centroid ← null, minDist ← ∞
for all centroid c€current_centroids do dist ← distance (pi, c)

4.4 Similarity Measure Between Two Numerical Values

For finding the numerical distance between two data points and the centroids, Euclidean distance will be considered. Euclidean distance between two numeric attribute values of X and Y is given as [6].

$$D(X, Y) = |x(n+1) - x(n)| + |y(n+1) - y(n) \tag{2}$$

Consider two points X(2, 5), Y(2, 10) and the Euclidean distance between two points are:

$$|2 - 2| + |5 - 10| = 0 + 5 = 5$$

4.5 Similarity Measure Between Two Binary Values

For binary attribute, Hamming distance will be considered to find out the binary distance between two Boolean attribute. The Binary distance between two Boolean attribute values a and b is defined as [6]

$$\delta(a,b) = 0 \text{ if } x = y \text{ and } \delta(a,b) = 1 \text{ if } x \neq y. \quad (3)$$

For finding out the distance between a pair of binary data items, Hamming distance will be considered. A binary attribute is a type of categorical attribute. Then it will always follow a Bernoulli distribution. Finding the distance between two binary values will always be either 1 or 0. However, the existence of distinct categorical values in a categorical attribute will change but the existence of number of distinct values in a binary attribute will always remain 2(0 or 1) [6].

4.6 Similarity Between Two Categorical Attribute

Here we are dealing with heterogeneous data, data having no predefined format. First, we have to convert the heterogeneous categorical data in the term document matrix form. After that we apply the Euclidian distance to find out the distance between values of attributes [6].

5 Implementation and Results

R is a generally used programming language for statistics and various data operation. As we know statistical problems have turn into ordinary place today, thousands of parallel R packages have been developed to solve these problems.

In the proposed framework as shown in Fig. 1, first we have to read files from the folder, after that data of various heterogeneous sources is analyzed. After that result are visualized in more details we are described below. After that we stored the results. After combining all the results we will store in one format.

In this paper, as shown in Fig. 2, first we convert the heterogeneous data into same format for analyzing data, then we apply K-Mean clustering algorithm using Euclidean distance on the heterogeneous data. We can also apply Euclidean distance for heterogeneous data. Before we studied Euclidean distance is possible only for numerical values. It is also possible for heterogeneous data. It only depends how we store data and convert in a format which is suitable for clustering of heterogeneous data. According to survey almost 91% of the data is heterogeneous. So, we require a technique or tool which can handle heterogeneity of data. Table 2 shows the description of Different type of data for analysis.

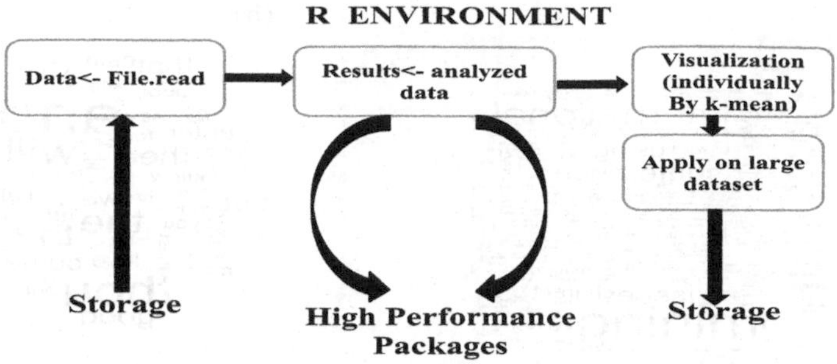

Fig. 1 Data Flow in the framework

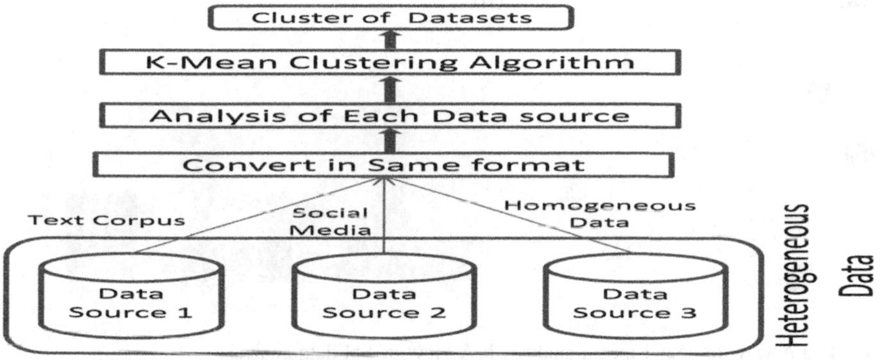

Fig. 2 A view of proposed framework

Table 2 Description of data

Data	Data Type	Contents
Text analysis	Unstructured data	No. of files = 4
Facebook analysis	Unstructured data	No. of friends = 6
Sql data analysis	Homogeneous data	No. of records = 22

5.1 Text Mining of Heterogeneous Data

We received the result when we perform analysis of text files, pdf, docs, or unstructured data as shown in Fig. 3a. And the complete works of William Shakespeare written by William Shakespeare. We perform analysis on large-scale dataset and analysz the results as shown in Fig. 3b.

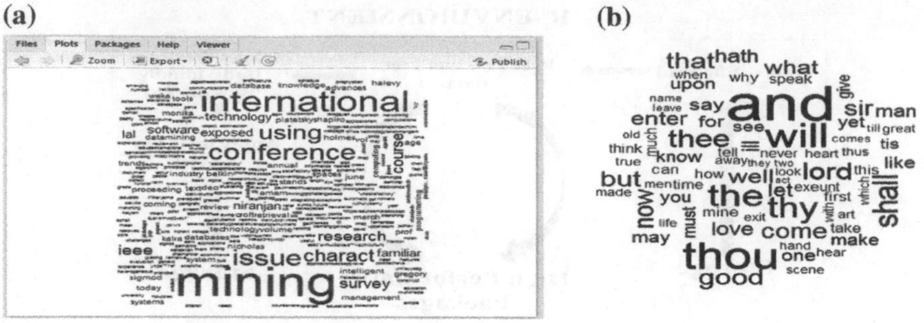

Fig. 3 **a** A view of text analysis **b** A view of William Shakespeare book

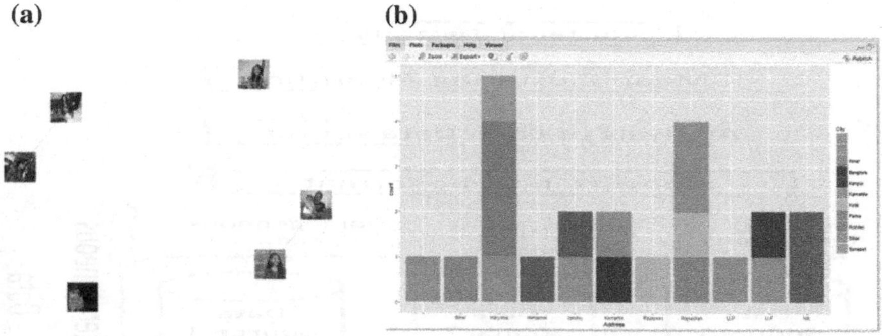

Fig. 4 **a** A view of social media analysis **b** A view of Sql data analysis

5.2 *Social Media Mining and*

Here, we considered the social media analysis like facebook. We received the result of analysis of facebook data. Clustering of data based on mutual friends, edges between the nodes shows the similarity between mutual friends as shown in Fig. 4a, and SQL data analysis is analysis of homogeneous data. SQL data is in the form of specific format. It is store in the form of tables. We analyze the data in histogram format as shown in Fig. 4b.

6 K-Mean Individually on Dataset

6.1 *K-Mean on Text Data and Social Media Data*

Large document corpus may afford a lot of useful information to people. Corpus consist of number of documents like pdf, ppt, and text files. If the document is in the form of ppt or pdf or any other form first we have to convert it into text format.

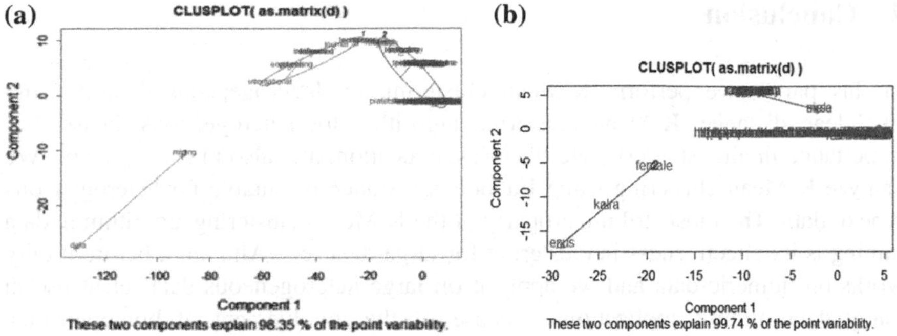

Fig. 5 A view using K-Means **a** Text clustering **b** Social media clustering

Then we perform preprocessing step to remove noise from the data. After that we convert the data in csv format. After converting the whole data in csv format we analyze the data by word cloud or histogram. After analyzing the data we apply the clustering algorithm on heterogeneous data as shown in Fig. 5a, and clustering of facebook data using K-Mean as sown in Fig. 5b.

6.2 SQL Data K-Mean Clustering

In sql data analysis, we have notice that data will be in specific format like data is in the form of tables. The variability of occurence of data points in the cluster will be less as shown in Fig. 6a and clustering of book using K-Mean as shown in Fig. 6b. The Shakespeare book has no predefined format. We apply the clustering for large-scale dataset and analyze the cluster. After we get the solution of our goal, we can apply the K-Mean clustering algorithm using distance metric for heterogeneous data and extract the useful data from huge dataset.

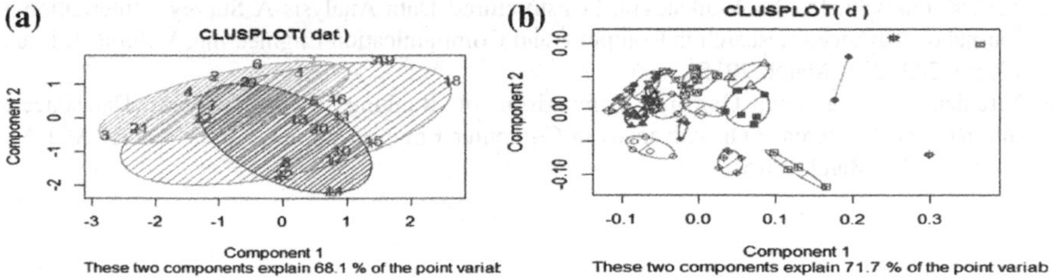

Fig. 6 A view using K-Means clustering of **a** SQL data **b** Shakespeare data

7 Conclusion

In this paper, we perform K-Mean clustering on heterogeneous dataset using Euclidean distance. K-Mean clustering algorithm for heterogeneous dataset has importance in almost every field business, education, and also in health sector. We analyze K-Mean clustering using Euclidean distance is suitable for heterogeneous type of data. The most striking property of the K-Means clustering algorithm in data mining is its effectiveness in clustering for large data sets. Although, before it only works on numeric data and we apply it on large heterogeneous data set, it use in many data mining applications because of the involvement of homogeneous, unstructured data, semistructured, or the combination of all data. Our main contribution in this paper, we have designed the framework which can work on heterogeneous data using existing clustering technique.

References

1. Rohit Yadav, Kratika Varshney, Kapil Arora, "Extracting Data Mining Applications through Depth Analyzed Data", International Journal of Emerging Research in Management &Technology, Volume 4, Issue-9, pp 115–119, September 2015.
2. Prabhat Kumar, Berkin Ozisikyilmaz et.al, "High Performance Data Mining Using R on Heterogeneous Platforms", IEEE International Parallel & Distributed Processing Symposium, pp. 1715–1724, 2011.
3. Azra Shamim, Vimala Balakrishnan, Madiha Kazmi, and Zunaira Sattar, "Intelligent Data Mining in Autonomous Heterogeneous Distributed and Dynamic Data Sources", 2nd International Conference on Innovations in Engineering and Technology (ICCET'2014) Sept. 19–20, 2014.
4. Rumi Ghosh, Sitaram Asur, "Mining Information from Heterogeneous Sources: A Topic Modeling Approach" ACM 978-1-4503-2321, 2013.
5. Prakash R. Andhale1, S.M. Rokade2, "A Decisive Mining for Heterogeneous Data", International Journal of Advanced Research in Computer and Communication Engineering Vol. 4, Issue 12, pp. 43–437, December 2015.
6. Amir Ahmad, Lipika De, "A K-Mean clustering algorithm for mixed numeric and categorical data" Data & Knowledge Engineering Elsevier, pp. 503–527, 2007.
7. K.V. Kanimozhi, Dr. M. Venkatesan, "Unstructured Data Analysis-A Survey", International Journal of Advanced Research in Computer and Communication Engineering, Volume 4, Issue 12, pp. 223–225, March 2015.
8. Niranjan Lal, Samimul Qamar, "Comparison of Ranking Algorithm with Dataspace", International Conference On Advances in Computer Engineering and Application(ICACEA), pp. 565–572, March 2015.

User Authentication System Using Multimodal Biometrics and MapReduce

Meghana A. Divakar and Megha P. Arakeri

Abstract Establishing the identity of a person with the use of individual biometric features has become the need for the present technologically advancing world. Due to rise in data thefts and identity hijacking, there is a critical need for providing user security using biometric authentication techniques. Biometrics is the science of recognizing a person by evaluating the distinguished physiological and biological traits of the person. A unimodal biometric system is known to have many disadvantages with regard to accuracy, reliability, and security. Multimodal biometric systems combine more than one biometric trait to identify a person in order to increase the security of the application. The proposed multimodal biometric system combines three biometric traits for individual authentication namely Face, Fingerprint, and Voice. MapReduce is the technique used for analyzing and processing big data sets that cannot fit into memory.

Keywords Multimodal biometrics · Machine learning · Face recognition
Fingerprint recognition · Voice matching · MapReduce

1 Introduction

User authentication is essential to provide security that restricts access to system and data resources. Person identification has become one of the core tasks for providing security in today's electronically wired information society. Traditionally, for person identification, systems use knowledge-based- or token-based systems. Token-based system use keys, ID card, etc., whereas knowledge-based systems use passwords that only the user should know. Biometric system is used for recognition of authorized user based on a feature set or vector which is extracted

M.A. Divakar (✉) · M.P. Arakeri
M S Ramaiah Institute of Technology, Bangalore, India
e-mail: meghanadiva91@gmail.com

M.P. Arakeri
e-mail: meghaarakeri@gmail.com

D.K. Mishra et al. (eds.), *Information and Communication Technology for Sustainable Development*, Lecture Notes in Networks and Systems 10,
https://doi.org/10.1007/978-981-10-3920-1_8

from a user's distinguishing biometric trait like face, finger, speech iris, etc. Biometrics is a field of science where an individual is recognized based on the biological or physiological traits.

Currently, there are many different biometric indicators which are widely used namely face, fingerprint, iris, facial thermogram, hand vein, voice, signature, hand geometry, and retinal pattern. These biometric traits have their respective advantages and disadvantages in terms of accuracy, applicability, and user acceptance. It is the requirements of where the system will be deployed which will determine the choice of the needed biometric trait for authentication.

However, unimodal biometric system is not able to satisfy reliability, speed, and acceptability constraints of authentication in real-time applications. This is due to noise in data collected, data quality, spoof attacks, restricted degree of freedom, lack of distinctiveness, nonuniversality, and other factors. Therefore, multimodal biometric systems are used to provide increased security and faster performance.

The proposed user authentication system uses multimodal biometrics to integrate face, fingerprint and voice to achieve person identification. The selection of these three specific biometric traits is based on the fact that they have been used routinely in law enforcement community. Many present biometric systems currently rely on face, fingerprint or voice matching for successful authentication. Also these three biometric traits enhance and complement one another in their strengths and advantages. Fingerprint is used for high verification accuracy; it is considered to be unique for every individual. Face and voice are commonly used by all in daily life. The system is targeted for authentication to verify the identity of a user in a multiple user database system.

MapReduce is a parallel computing programming technique used for processing big data and to perform statistical analysis to big data. It can be used to process large datasets and perform complex calculations on large amount of data.

2 Multimodal Biometric System

In general, multimodal biometric systems operate in enrolment phase and authentication phase. Where biometric traits are captured in enrolment phase and matched in authentication phase. The images for face and fingerprint are captured and a voice sample is recorded for a user and required features are extracted for each biometric trait by applying suitable algorithms. The features obtained are stored in the database and a dataset is created for each biometric trait, i.e., face dataset, fingerprint dataset, and voice dataset for all the users who are enrolled. If a large dataset (big data set) consisting of more number of users' data is present, then mapreduce is performed for the extraction of features of users and stored it in datasets.

When a user needs to be authenticated, images of the user face and fingerprint are captured along with a voice sample and stored as test file samples. Features of the biometric traits for the test images are then extracted and matched against the

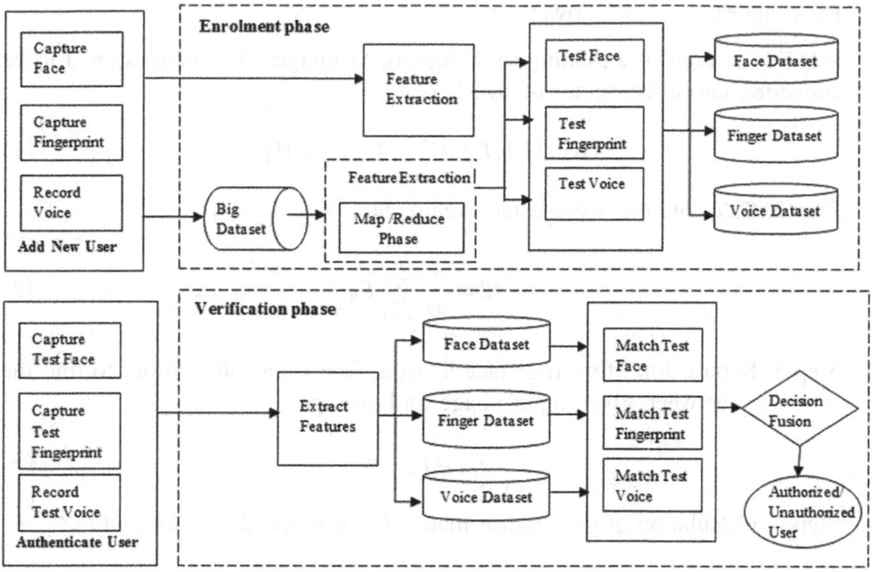

Fig. 1 Design of user authentication system using multimodal biometrics

dataset, and a decision is processed as to whether the user is present in the dataset. Since the database size is large due to multiple users, mapreduce can be used to process large datasets and store features of the users. Authentication is done by match of user data against each dataset present in the database, a one-to-one match is performed until the user data is found. Based on the results obtained for each biometric trait, decision fusion is applied to authenticate a given user is an authorized user or unauthorized user. The design of user authentication system using multimodal biometrics is shown in Fig. 1.

2.1 Face Recognition Module

In face recognition, feature extraction and classification of face images are two primary steps. In this system, face recognition is performed by combining PCA (Principle Component Analysis) and SVM (Support Vector Machine). PCA is used for extraction of features and SVM is used for classification of features.

I. *Feature Extraction using PCA*
PCA (Principal Component Analysis) algorithm is an effective technique used for selection and extraction of facial features. The main goal of PCA algorithm is dimensionality reduction.

PCA algorithm is as follows:

Step 1. Creation of a training set S having M images. The images need to be converted into a facevector of size N.

$$S = \{\Gamma 1, \Gamma 2, \Gamma 3, \Gamma 4, \ldots, \Gamma M\} \quad (1)$$

Step 2. Calculate the average face vector Ψ.

$$\Psi = \frac{1}{M} \sum_{n=1}^{M} \Gamma_n \quad (2)$$

Step 3. Subtraction of average face Ψ from face vector Φ in order to find the difference between given input image and average

$$\Phi_i = \Gamma_i - \psi \quad (3)$$

Step 4. Calculation of Covariance matrix C for reduced dimensionality

$$C = \frac{1}{M} \sum_{n=1}^{M} \Phi_n \Phi_n^T \quad (4)$$

Step 5. Eigenvectors must be calculated from the obtained covariance matrix.
Step 6. Representing each face image as a combination of M training images in order to form the eigenfaces u_l

$$u_l = \sum_{k=1}^{M} v_{lk} \Phi_k \quad (5)$$

Step 7. Select K eigenfaces where K < M and K represents the most relevant eigenfaces. Conversion of higher dimensional K eigenvectors into the original dimensionality of face.
Step 8. Represent each facial image as a linear combination of K eigenvectors, i.e., summation of weights from K Eigenfaces and Average face. The weight vectors obtained are stored in the training set [1]

$$\omega_k = u_k^T(\Gamma - \psi) \quad \Omega^T = [\omega_1, \omega_2, \ldots, \omega_M] \quad (6)$$

where u_k is the kth eigenvector and ω_k is the kth weight vector.

II. *SVM Classification*
SVM is a classification technique used to find a linear decision surface by structural risk minimization. This linear decision surface is obtained by combining weights of elements in a training set. The elements present are known as support vectors and define the boundary between any two classes. The idea behind SVM classification technique is to have a classifier trained on the input

data. It is a supervised learning algorithm, where the training vectors are trained and mapped into a space with clearly defined gaps in between them. This is done by using standard kernel functionalities. For testing purpose the input vectors are mapped onto the same space for predicting the class of the face. SVMs are used for pattern recognition between two classes based on the decision surface. By determining the support vectors which have maximum distance between nearby points in a training set. Using this, the test image is assigned to a particular class and thereby the image is identified and better accuracy is achieved using a combination of both PCA and SVM.

2.2 Fingerprint Recognition Module

Minutiae points are significant for fingerprint recognition because no two fingerprints are considered to be identical. Most long-established minutiae which can be used in fingerprint recognizing are centered on bifurcation and ridge ending. Minutiae algorithms basically comprise of the phases Minutiae Extraction and Minutiae Matching.

i. *Minutia Extraction*

The algorithm is described below

Step 1. Fingerprint image enhancement is performed in order to eliminate the noise from the image. This is done by using Histogram.

Step 2. Fingerprint image binarization is done to assign binary values of 0 and 1 to the image. The fingerprint image is changed into 1 bit image where zero (0) is assigned for a ridge and one (1) is assigned to a furrow.

Step 3. Extraction of ROI (Region of Interest) is performed after binarization. The subsequent assignment is the extraction of the ROI from the binary image. It is completed using OPEN and CLOSE functions.

Step 4. Ridge Thinning is the next step performed, which is the elimination of recurring pixels from the ridges. Then ridges present are transformed to 1-pixel width.

Step 5. Minutiae Marking is done after the process of ridge thinning. It is performed by means of considering a 3 × 3-pixel window. For every pixel, it is neighboring 8 pixels are analyzed to decide if the considered central pixel is a ridge ending or a ridge branch.

Step 6. False Minutiae removal is performed to eliminate the false points. The space between each adjacent minutia point is calculated and accordingly the minutiae points are kept or dismissed as faulty.

ii. *Minutiae Matching*

1. Alignment stage: When matching two fingerprints, decide upon any individual minutia from each fingerprint; and calculate the similarity between the two ridges and its corresponding minutia. If the similarity between them

is greater than the minimum acceptable value, then the two fingerprint minutia are changed to a new set of values such that x-axis coincides with the referenced factor and the origin is at the reference point.

2. Match stage: After obtaining two sets of converted minutia features, the matched minutia pairs are counted using the elastic match algorithm and two minutiae which have practically similar position and identical direction are matched.

2.3 Voice Recognition Module

The method of recognizing a speaker by analyzing information present in a recorded voice is known as Voice recognition. The data is obtained by way of MFCC, i.e., mel-frequency cepstrum coefficients. MFCC is the outcome acquired by a cosine transformation of the real logarithm of the short-term energy spectrum depicted on a mel-frequency scale. These coefficients have demonstrated to be very efficient in recognition. MFCC is calculated as follows:

i. *Mel-frequency wrapping*

For every human tone having a genuine frequency (*f*), which is measured in Hertz, alternatively another subjective pitch is measured using another scale referred to as the 'mel' scale. Used mel-frequency scale has a linear frequency spacing beneath 1000 Hz and also a logarithmic spacing above 1000 Hz. For reference, the pitch of a 1 kHz tone, which is 40 dB above the hearing threshold has been defined as 1000 mels. Mels for any given frequency (*f*) can be calculated, in *Hz*, using the following formula.

$$Mel(f) = 2595 * \log_{10}(1 + f/700) \quad (7)$$

For the simulation of a spectrum, filter bank is used, preferably one filter is used for every one component (mel-frequency). The filter bank used for mel scale is a series of one triangular band cross filters, which are designed for simulation of the band pass filter which occurs in the auditory system. The result is a series of band pass filters which have consistent bandwidth along with space on a mel-frequency scale.

ii. *Cepstrum*

In this step, the log mel spectrum is converted back to time in order to get MFCC. For the given frame analysis, the cepstral representation of processed speech spectrum gives the spectral properties of voice signal. Considering that the mel spectrum coefficients are real numbers, by using the discrete cosine transform (DCT) it needs to be modified to time domain. Following which the log mel spectrum is changed back to time. Then, the obtained output is referred to as MFCC or Mel-Frequency Cepstrum Coefficients. The DCT is used for conversion of mel coefficients back to time domain.

2.4 Decision Fusion

In user authentication system using multimodal biometrics, recognition results of the three biometric traits (face, fingerprint, voice) are combined using decision-level fusion and is used to authenticate an authorized or unauthorized user. Some of the common decision fusion techniques are 'AND' or 'OR' rules, simple rule of sum and majority voting.

The implemented system uses fuzzy logic principles and above-mentioned fusion techniques to integrate the results of the three biometric recognition systems. Fuzzy logic is a form of logic in which the truth values can vary based on degrees of truth as compared to Boolean logic of 'true or false'. It allows for intermediate values between false and true. The output is then evaluated using the fuzzy logic conditions below:

- If a user's three biometrics features match; face, fingerprint and voice recognition is successful, then the final decision is '*Authorized User*'.
- If a user's two biometric features match; the possibilities being face and fingerprint recognition, face and voice recognition, and fingerprint and voice recognition then the final decision is '*Authorized User*'. Consideration is given to users who cannot provide a good fingerprint image due to cut in the finger, or background is noisy when recording a voice or the background is cluttered when capturing face image.
- If user's three biometric features do not match; face, fingerprint and voice recognition is unsuccessful, then the final decision is '*Unauthorized User*'.

2.5 MapReduce

MapReduce is a technique of parallel programming which is used for filtering big datasets and to perform statistical analysis on big data. MapReduce explicitly uses datastore for the processing of big data into small chunks of data that can fit into memory. The Map phase and Reduce phase is performed by executing *map* function and *reduce* function in sequence. Many combinations of map function and reduce function can be used to process large amount of data, so mapreduce framework is very powerful as well as flexible to handle the processing of big data.

Mapreduce is used to read a block of data from the created datastore, which consists of input data, and then map function is called to process that data block. The map function performs the specified calculation, or can also just organize the data and then calls add function or add multi-functions which is used to add the key-value pairs to an intermediate or temporary data storage object which is called a KeyValueStore. Then, mapreduce performs grouping of all values stored in the KeyValueStore by referencing the values with a unique key. Then, the reduce function is called for each unique key which has been added by the preceding map function. All the values are then passed to reduce function by using a ValueIterator object, which is used for iterating over the values. Then, after aggregating the intermediate results, the reduce function will add the final key-value pairs obtained to the output. The order of the keys present in the output will be the same order in which the reduce function has added them to the output or final KeyValueStore object.

3 Experimental Results

A sample database of face, fingerprints and voice samples of 50 users was recorded. The fingerprint images were acquired using Secugen fingerprint scanner in.*bmp* format. The face images were captured using Laptop web camera in.*jpg* format. The voice samples were collected by using a microphone in.*wav* format.

The Face database was then divided into Training and Testing Samples of 100 face images each. Then Support Vector Machine Classifier was used to obtain a SVM Classifier with accuracy 87%. All the fingerprint images obtained are processed to create a Fingerprint Database. In fingerprint, minutiae are extracted and the corresponding bifurcations and terminations are stored for each image in a text (*.txt*) file. Similarly, a Voice Database is created when recording the voice samples from the users. The voice samples are processed in order to obtain mel-frequency cestrum coefficients (MFCC). The respective coefficients are saved for each user in a Sound Database file. The accuracy obtained for each is shown in Table 1.

In biometric, the performance of a system is specified as per terms of FAR (false acceptance rate) and FRR (false rejection rate). FAR is that the quantitative relation of fraud been recognized as genuine and FRR is the quantitative relation of genuine being accepted as fraud. The performance of the multimodal system is shown in Fig. 2.

The authentication of a user as 'Authorized User' or 'Unauthorized User' happens in a serial manner. For a given user, each biometric trait is verified in order of face followed by fingerprint and voice. For face recognition, using the Face Database and SVM Classifier, test image of a person is input to authenticate if the person was present in the Database. For fingerprint recognition, the user's Test fingerprint image is captured and from the termination and bifurcation obtained fingerprint matching is done. If the corresponding fingerprint image is shown as matched, it is displayed if not, no image found is displayed. For voice recognition,

Table 1 Table showing accuracy for individual biometric modules

Feature	Algorithm	Accuracy (%)	TPR (%)	FPR (%)
Face	PCA	81.3	7.08	11.7
	SVM + PCA	87.7	4.22	8.19
Fingerprint	Minutiae matching	96	1.88	4.23
Voice	MFCC	78.6	8.55	12.6

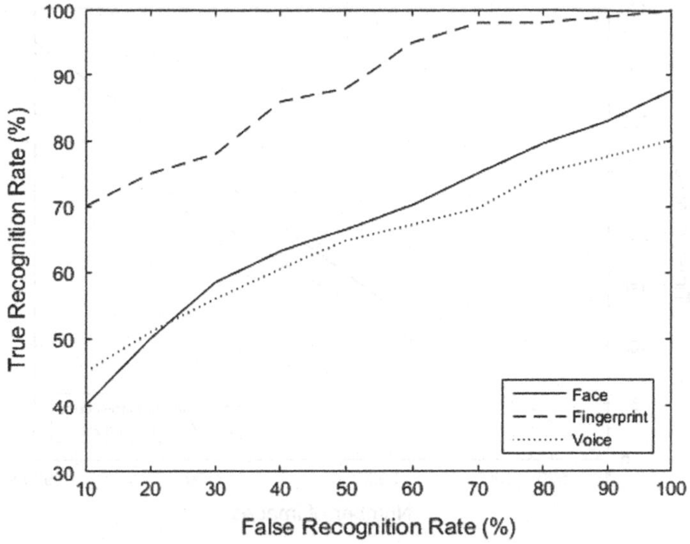

Fig. 2 Performance of the user authentication system

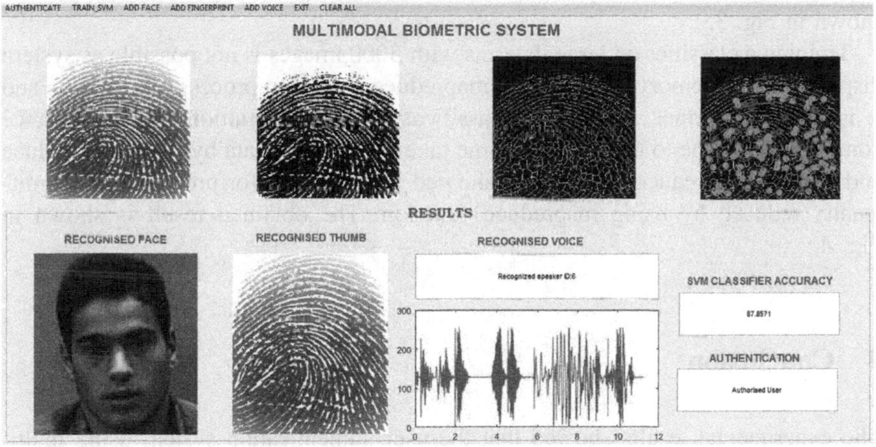

Fig. 3 Interface of implemented user authentication system

the user's voice is recorded and if it is matches the voice sample in database, then corresponding User ID of the sample and plot of the voice sample is displayed. Based on the recognition values of the biometric traits, the final result is displayed. If any two of the three biometric traits are recognized, then the user is an 'Authorized User', as shown in Fig. 3. If only one biometric trait is recognized and other two fail, then the user is an 'Unauthorized User'. The interface of the

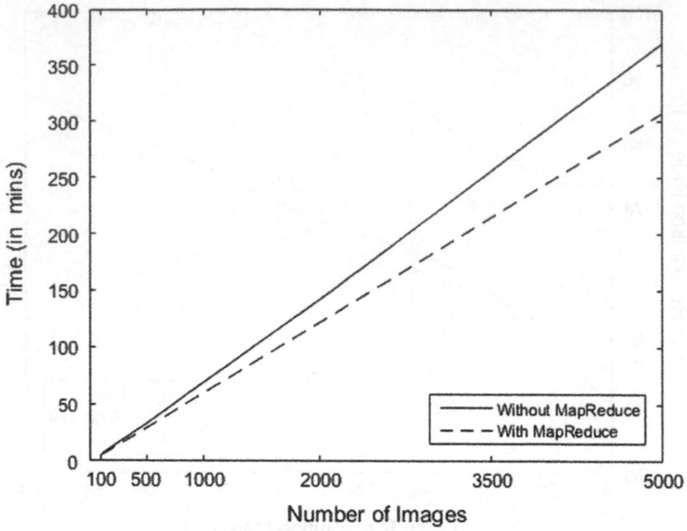

Fig. 4 Time taken with map/reduce and without map/reduce

implemented multimodal system with the final result of an 'Authorized User' is shown in Fig. 3.

Training a classifier on large datasets with 5000 images is not possible as system displays out of memory error. Hence mapreduce is used to process big datasets and it is used for further analysis in classification and recognition. A performance comparison is done to calculate the time taken to process data by using mapreduce and without mapreduce, and results showed that the time for processing is significantly reduced by using mapreduce function. The obtained result is shown in Fig. 4.

4 Conclusion

The experimental results showed that the user authentication system using multimodal biometrics combining face, fingerprint and voice is more accurate than traditional systems. The proposed system provides robust security and minimal error. By using MapReduce to process big datasets, performance is significantly improved and time taken is reduced. In future, other biometric traits can be added to increase the security of the system. Different fusion methods can be tested to comprehend their influence on the performance of the system.

References

1. Kumar, Ch Srinivasa, and P. R. Mallikarjuna. "Design of an automatic speaker recognition system using MFCC, Vector Quantization and LBG algorithm." International Journal on Computer Science and Engineering 3.8 (2011): 2942–2954 (2011)
2. Prabu, U., et al. "Efficient personal identification using multimodal biometrics." Circuit, Power and Computing Technologies (ICCPCT), 2015 International Conference on. IEEE, 2015 (2015)
3. AS Raju and V Udayashankara. Biometric person authentication: A review. In Contemporary Computing and Informatics (IC3I), 2014 International Conference on, pages 575–580. IEEE, 2014 (2014)
4. Mandeep Kaur, Rajeev Vashisht, and Nirvair Neeru. Fingerprint recognition techniques and its applications. International Conference on Advances in Engineering Technology Research (ICAETR - 2014), 2014 (2014)
5. Rozeha Rashid, Nur Hija Mahalin, Mohd Adib Sarijari, Ahmad Abdul Aziz, et al. Security system using biometric technology: Design and implementation of voice recognition system (vrs). In Computer and Communication Engineering, 2008. ICCCE 2008. International Conference on, pages 898–902. IEEE, 2008 (2008)
6. Didier Roy and Anupam Shukla. Speaker recognition using multimodal biometric system. In Oriental COCOSDA held jointly with 2013 Conference on Asian Spoken Language Research and Evaluation (O-COCOSDA/CASLRE), 2013 International Conference, pages 1–7. IEEE, 2013 (2013)
7. Khan, Asif Iqbal, and Mohd Arif Wani. "Strategy to extract reliable minutia points for fingerprint recognition", 2014 IEEE International Advance Computing Conference (IACC), 2014 (2014)
8. Ross, Arun, and Anil K. Jain. "Multimodal biometrics: An overview." Signal Processing Conference, 2004 12th European. IEEE, 2004. (2004)
9. Hong, Lin, and Anil K. Jain. "Multimodal biometrics." Biometrics. Springer US, 1996. 327–344. (1996)
10. Jain, Anil K., Arun Ross, and Salil Prabhakar. "An introduction to biometric recognition." Circuits and Systems for Video Technology, IEEE Transactions on 14.1 (2004): 4–20. (2004)
11. Dave, Pushpak, and Jatin Agarwal. "Study and analysis of face recognition system using Principal Component Analysis (PCA)." Electrical, Electronics, Signals, Communication and Optimization (EESCO), 2015 International Conference on. IEEE, 2015 (2015)
12. Frolov, Igor, and Rauf Sadykhov. "Face recognition system using SVM-based classifier." Intelligent Data Acquisition and Advanced Computing Systems: Technology and Applications, 2009. IDAACS 2009. IEEE International Workshop on. IEEE, 2009. (2009)
13. Wang, Chengliang, et al. "Face recognition based on principle component analysis and support vector machine." Intelligent Systems and Applications (ISA), 2011 3rd International Workshop on. IEEE, 2011. (2011)
14. Lim, Jin Fei, and Renee Ka Yin Chin. "Enhancing Fingerprint Recognition Using Minutiae-Based and Image-Based Matching Techniques." Artificial Intelligence, Modelling and Simulation (AIMS), 2013 1st International Conference on. IEEE, 2013. (2013)
15. Nijhawan, Geeta, and M. K. Soni. "Speaker Recognition Using MFCC and Vector Quantisation." Int. J. on Recent Trends in Engineering and Technology 11.1 (2014).
16. Jain, Anil K., Lin Hong, and Yatin Kulkarni. "A multimodal biometric system using fingerprint, face and speech." Proceedings of 2nd Int'l Conference on Audio-and Video-based Biometric Person Authentication, Washington DC. 1999. (1999)

17. Alkandari, Abdulrahman, and Soha Jaber Aljaber. "Principle Component Analysis algorithm (PCA) for image recognition", 2015 Second International Conference on Computing Technology and Information Management (ICCTIM), 2015. (2015)
18. Debnath, Saswati, B. Soni, U. Baruah, and D.K. Sah. "Text-dependent speaker verification system: A review", 2015 IEEE 9th International Conference on Intelligent Systems and Control (ISCO), 2015. (2015)

Fake Profile Identification on Facebook Through SocialMedia App

Priyanka Kumari and Nemi Chandra Rathore

Abstract In today's life almost everyone is in association with the online social networks. These sites have made drastic changes in the way we pursue our social life. But with the rapid growth of social networks, many problems like fake profiles, online impersonation have also grown. Current announces indicate that OSNs are overspread with abundance of fake user's profiles, which may menace the user's security and privacy. In this paper, we propose a model to identify potential fake users on the basis of their activities and profile information. To show the effectiveness of our model, we have developed a Facebook canvas application called "SocialMedia" as a proof of concept. We also conducted an online evaluation of our application among Facebook users to show usability of such apps. The results of evaluation reported that the app successfully identified possible fake friend with accuracy 87.5%.

Keywords Online social networks (OSNs) · Fake profile · SocialMedia app Trust weight (TW)

1 Introduction

Social networking sites mimic real-life interaction and behavior of users. OSNs, such as Facebook [1], Google+ [2], LinkedIn [3], Twitter [4], have hundreds of millions of active users that publish huge amount of private information. Therefore, online social network users are oblivious to the innumerable security risks [5] like identity theft, privacy violations, and sexual harassment, just to name a few. Many times attackers use a fake profile as a tool to steal sensitive information of target users. According to the report [6], Online Social Networks are invaded with mil-

P. Kumari (✉) · N.C. Rathore
Central University of South Bihar, Patna, India
e-mail: priyankakumari073@hotmail.com

N.C. Rathore
e-mail: nemichandra@cub.ac.in

D.K. Mishra et al. (eds.), *Information and Communication Technology for Sustainable Development*, Lecture Notes in Networks and Systems 10,
https://doi.org/10.1007/978-981-10-3920-1_9

lions of fake profiles that are made to harvest the personal information and breach the security issues of the OSNs users. These fake profiles can jeopardize the life of the users.

Fire et al. [5] describes a fake profile as an automatic or semi-automatic profile that mimics human behaviors in OSNs. In other words, fake profiles are profiles of persons who claim to be someone they are not. Such profiles are intended to perform some malicious and undesirable activities, causing problems to the social network users. The issues like privacy theft, online bullying, potential for misuse, trolling, slandering, etc., are mostly done using fake profiles. Attackers may create fake account on the name of the victim and post vulgar pictures and posts to deceive everyone to believe that the person is inferior and thus defaming the person. Social networking platform are used by spammers to send advertisement messages to other users by creating fake profiles. Some attacker creates fake account for advertising and campaigning, hacking or cyber-bullying to name a few. All of these affairs threaten OSNs users and can harm them badly both in their cyber and real life. Despite such potential risks, none of the social network operators have equipped themselves with any sophisticated features that may alarm the users of these fake profiles. So it becomes vital to hamper these ill-intentions with a sound and full-proof safety measure. This paper presents a model to identify the fake profiles present in the user's friend list.

The organization of remainder of this paper is as follows. We provide an overview of various related solutions in Sect. 2. In Sect. 3, we have described our proposed model for the identification of fake friends. Section 4, describes implementation and evaluation work with architecture of SocialMedia app in detail. In Sect. 5, we discuss the obtained results from the app and also give statistics obtained from the questionnaire feedback from the user. Finally, in Sect. 6, we conclude this paper while providing future research directions.

2 Related Work

Due to the increase in security and privacy violation on OSNs many solutions are proposed by different researchers to protect the OSNs users from these risks. Fong et al. [7] proposed to use different decision tree classification algorithms to find fake profiles. The limitations of this solution are large number of attributes and amount of time taken for training. Conti et al. [8] have given a framework for detecting fake profile attack where the victim has no prior online profile. The approach is based on the growth rate of the social network graph, and on social network interactions of typical users with their friends. The solution is not efficient in terms of time and space. Wei et al. proposed anti-sybil detection schemes called SybilDefender [9] which used community detection approach. If a Sybil node is detected, the Sybil community is found based on the behavior that Sybil nodes generally use to connect with other Sybil nodes only. Fire et al. [10] have developed SPP software to identify real and fake user. It has two versions, in its initial version it calculates the connection strength between the user with his each friend, using different features and based on the score, it recommends user's friend as fake. In later version supervised learning techniques were used to construct fake profile identification classifiers. The initial version of SPP software fails to detect cyber predators and later version faces dataset imbalance issue. Also, they have used large number of attributes which results time and space inefficiency.

Ahmed et al. [11] in their work offered a hybrid approach that identified coordinated spam or malware attacks controlled by Sybil accounts on OSNs. They conclude that the nodes belonging to Sybil communities have higher closeness centrality values in comparison to normal users. This solution also faces time and space complexity. Yang et al. in his initial work [12] studied the link creation behavior of Sybils on Renren [13]. They proposed a threshold-based detector to detect the Sybils. In their later work [14] they used click-stream data of Sybils and compared session level characteristics of Sybils and normal users using Markov chain model. These models helped to recognize the differences between normal and abnormal behavior on Renren.

After taking a glance at literature survey, we found several drawbacks like high dimensionality, space and time complexity, data set imbalance, etc., present in existing solutions, so this work is meant to identify the possible fake friends present in the user's friendlist by using less number of attributes and complexity.

3 Proposed Mechanism

To better identify the fake friend on Facebook, we have taken few attributes that are listed below, that can characterize these fake accounts more efficiently. To support our work, we have also done survey on the OSNs user's and recorded their feedback that are discussed in Sect. 5.

1. **Number of mutual friends**: Number of mutual friends is the number of common friend between the user and his/her friend. Number of mutual friends is higher in case of real user but reverse in fake user. So it is a strong attribute to distinguish between fake and real.
2. **Is family**: It checks whether the user's friend on Facebook, is user's family member or not. If the user's friend is a family member then there will be negligible chance of fake profile of that family member.
3. **Age group**: According to the barracuda labs social network analysis [15], it is found that the age specified in an online profile by a fake user most probably belongs to 17–25 age groups.
4. **Average likes**: Average likes is the average number of likes done on the user posts and photos by their friends. It is assumed that the number of likes on genuine person's photos and posts are more than the fake person.
5. **Average comments**: An average comment is the average number of comments done on the user posts and photos by their friends. It is assumed that the number

of comments on genuine person's photos and posts are more than the fake person.

After getting the attributes value from users account we used the following heuristic to define the "Trust Weight" function between a user and its friend:

$$\text{Trust Weight}(u, v) = TW(u, v) := \text{No. of mutual friends}(u, v) + \text{Average likes} + \text{Average comments} + \text{Age group} + 100.(\text{Is Family}(u, v))$$

The flow of the proposed framework is shown in the Fig. 1. User login to the app on their Facebook account and give permission to the app to access their profile information. App extracts the value of the required attributes and calculates the trust weight between the user and his Facebook friend. Now the application sorts the friends according to their TW so that the friend having lowest score is at the top of the list and who got highest score is at the bottom. According to a recent report [6], it was found that Facebook estimates that 8.7% accounts do not belong to real profiles. Following the report, the app returns a Facebook page as shown in Fig. 2 containing name of 10% of total friends from the top of the list.

It is assumed that friends having lower TW value are highly likelihood of fake because trust weight will be high between the user and his real friend and low with the possibly fake friends. Now the user accepts or declines the name of friends suggested by the app as possible fake friends.

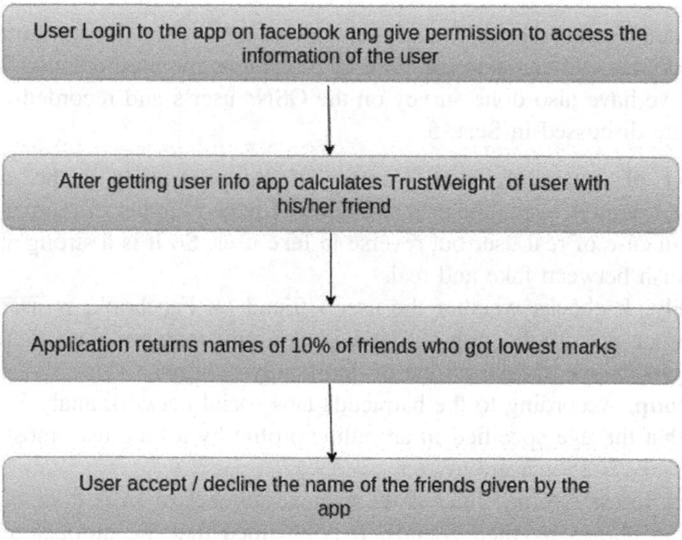

Fig. 1 Flowchart of proposed method

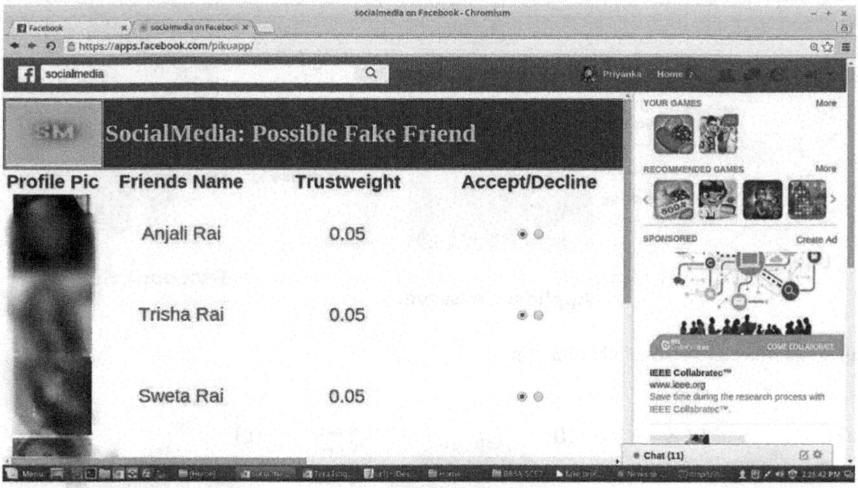

Fig. 2 Facebook page containing names of possible fake friend

4 Implementation and Evaluation

Fake profile identification is a real-world task. The need of real-world data is necessary to understand the habitual behavior of fake user. In our work, we have considered Facebook to collect data for our experiment. To gather the real-time data from user profiles that fulfill our requirements, we have developed an app named "SocialMedia".

4.1 SocialMedia App

SocialMedia is a Facebook canvas application used to collect the data from user Facebook account. The application is developed using PHP v5, HTML5, CSS, and MySQLi v5.1 languages. The URL for the app is https://apps.facebook.com/pikuapp.

In Fig. 3 architecture of data collection through the SocialMedia app is shown. User logins to Facebook using our application and request to the application server for user information, which contains SocialMedia database is shown in Fig. 4. After receiving the request from the user, the application server makes Facebook API call to Facebook server, in the reply of that Facebook server sends requested data from user Facebook profile, which is then saved in the application database present on the application server. The data that is stored in SocialMedia database is transferred to the requesting user.

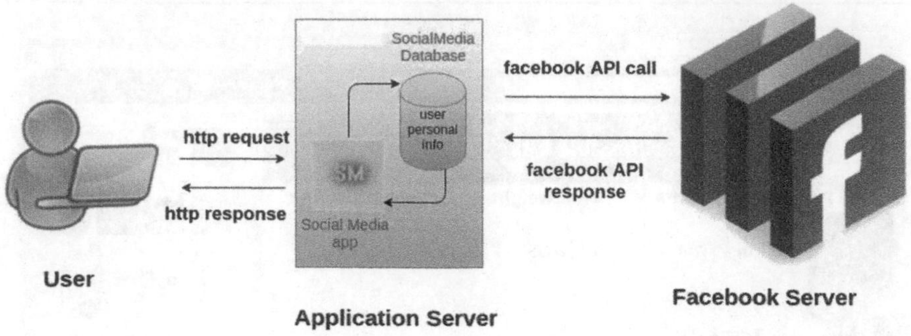

Fig. 3 Architecture of SocialMedia app

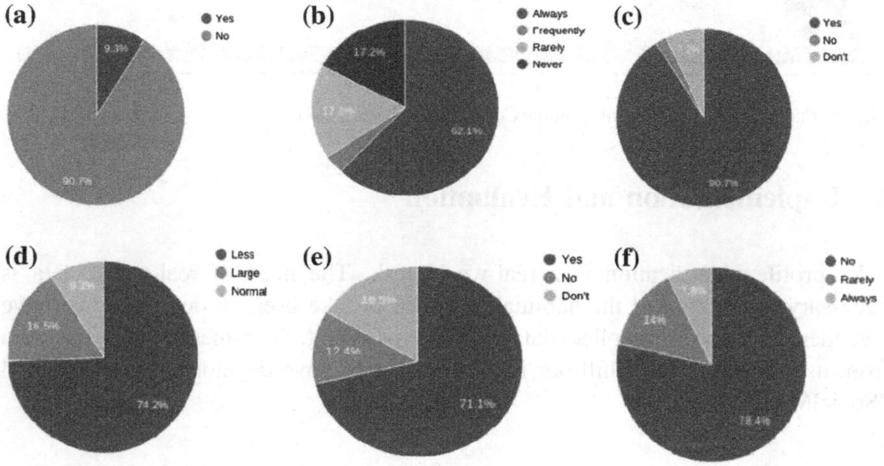

Fig. 4 Questionnaire feedback of OSN user

After collecting and storing the dataset into the database, we have analyzed the data and calculated the efficiency of our model. The results are described in the next section.

5 Result and Discussion

To know the outlook of the OSN users upon the behavior of normal and fake user on the Facebook, we prepared a questionnaire and got response from 100 student of Central University of South Bihar, who are regular users of Facebook; out of which 54.7% are male and 45.3% are female. Most of the participants belong to below 30 age group. The questionnaire contains several questions that show the conduct of the OSN user. The feedback from the user strengthens the behavior that we have taken in our approach (refer Sect. 3) to identify the possible fake profile. In the Fig. 4 statistics of different activities are given.

Figure 4a shows that 90.7% of users do not like or comment on unknown users' posts and photos, and in Fig. 4f, 78.4% user thinks that genuine users do not like or comment on the fake user's stuffs. This behavior reinforces that likes or comments on possible fake user will be less than the genuine users. Figure 4b shows that 62.1% users add new friends based on the mutual number of friends and Fig. 4d tells that 74.2% user thinks that mutual number of possible fake friends is less. This conduct shows that the mutual number of fake friend is very small. Figure 4c indicates, 90.7% people believe that their family member cannot be a fake friend. This strengthens our assumption considered in our model. Finally in Fig. 4e, 71.1% users accept that fake user belongs to 17–25 age groups as we have considered in this work. We also asked users in the questionnaire that whether such type of Facebook app is needed to preserve the user's privacy, out of which 90.7% of users were found, agreed. After seeing questionnaire result and user feedback, we can conclude that the characteristics taken by our approach is significant and useful in identifying OSN user behavior and detecting possible fake friend on Facebook.

Now, if it comes to the application result then, we can say that we have successfully identified the possible fake friends through SocialMedia app that were present in the user's friend list. For the experiment purpose, we gathered required data of all the users that logged into the app. Based on the trust weights, the app returns the name of possible fake friends to the user. The app has achieved its goal in detecting the possible fake friends with accuracy of 87.5%. It lacked achieving 100% accuracy because some of the user's are less active on Facebook so the TW score will be less for them compared to other genuine friend therefore, the app may treat them as possible fake friend.

Our model has used very less number of parameters to characterize fake users, compared to the existing solution proposed by different authors. Hence we can say that it is space and time efficient. Also the approach is easy to implement and user friendly.

6 Conclusion and Future Work

Despite the existence of several mechanisms to address the issue of fake profile identification, the need for a sophisticated methodology was felt unavoidable. In the lieu of this much sought need for the overwhelmingly increasing world of social networking, our model targets to a safer and more secure social networking in future. To ensure this safety and security of private information of OSNs users on social networking sites a model based on ranking of trust weight among the various users has been proposed. The feasibility and effectiveness of our model is demonstrated through Facebook app named as *SocialMedia*. Our model promises to tackle the problem of time complexity, space complexity, and abundance of attribute, etc., very efficiently.

Due to the limited access to user data on Facebook, some of the attributes of OSN user have not been taken into account. Therefore, there are further chances to exploit some other attributes like user activities, type of contents posted, etc., to improve upon our model.

References

1. Facebook, available: http://www.facebook.com/.
2. Google+, available: http://www.plus.google.com/.
3. LinkedIn, available: http://www.linkedin.com/.
4. Twitter, available: http://www.twitter.com/.
5. M. Fire, R. Goldschmidt, and Y. Elovici. Online social networks: Threat and solutions. Communications Surveys Tutorials, IEEE, 16(4):2019–2036 Fourthquarter 2014.
6. I. Facebook or 15(d) quarterly of the report pursuant to securities exchange Act section of 13 1934, url https://www.sec.gov/archives/edgar/data/1326801/000132680115000006/fb-12312014 x10k.htm.
7. S Fong, Yan Zhuang, and Jiaying He. Not every friend on a social network Can be trusted: Classifying imposters using decision trees. In Future Generation Communication Technology (FGCT), 2012 International Conference on, pages 58–63. IEEE, 2012.
8. M. Conti, R. Poovendran and M. Secchiero. Fakebook: Detecting fake profiles in on-line social networks. In Proceedings of the 2012 International Conferenceon Advances in Social Networks Analysis And Mining (ASONAM 2012), ASONAM '12, pages 1071–1078, Washington, DC, USA, 2012. IEEE Computer Society.
9. Wei Wei, F. Xu, C. C. Tan, and Qun Li. Sybildefender: Defend gainst A sybil attacks in large social networks. In INFOCOM, 2012 Proceedings IEEE, pages 1951–1959, March 2012.
10. M. Fire, Dima Kagan, Aviad Elyashar, and Yuval Elovici. Friend or foe? Fake profile identification in online social networks. Social Network Analysis and Mining, 4(1), 2014.
11. F. Ahmed and M. Abulaish Identification of Sybil communities Generating context – aware spam on online social networks. In Yoshiharu shikawa, Jianzhong Li, Wei Wang, Rui Zhang, and Wenjie Zhang, editors, Web Technologies and Applications, volume 7808 of Lecture Notes in ComputerScience, pages 268–279. Springer Berlin Heidelberg, 2013.
12. Zhi Yang, Christo Wilson, Xiao Wang, Tingting Gao, Ben Y. Zhao, and Yafei Dai. Uncovering social network sybils in the wild. In Proceedings of the 2011 ACM SIGCOMM Conference on Internet Measurement Con., IMC '11, pages 259–268, New York, NY, USA, 2011. ACM.
13. Renren, available: http://www.renren.com/en/.
14. Zhi Yang, Christo Wilson, Xiao Wang, Tingting Gao, Ben Y. Zhao, and Yafei Dai. Uncovering social network sybils In the wild. ACM Trans. KnowlDiscov. Data, 8(1):2:1–2:29, February 2014.
15. Barracuda labs social network analysis on real people vs fake profiles, url = https://barracudalabs.com/research-resources/sample-page/.

Disease Inference from Health-Related Questions via Fuzzy Expert System

Ajay P. Chainani, Santosh S. Chikne, Nikunj D. Doshi, Asim Z. Karel and Shanthi S. Therese

Abstract Automatic disease inference is vital to shorten the gap for patients seeking online remedies for health-related issues. Some of the persistent problems in offline medium are hectic schedules of doctors engrossed in their workload which abstain them to supervise on all health-related aspects of patients seeking advice and also community-based services which may be trivial in nature because of factors such as vocabulary gaps, incomplete information, and lack of available preprocessed samples limiting disease inference. Thus, we motivate users with proposed expert system by answering the underlying challenges. It is an iterative process working on multiple symptoms and compiles overall symptoms and causes required for inference of diseases. First, symptoms are mapped from extracted raw features. Second, fuzzy inference is made from weight-based training and catalyst factors. Thus, fuzzy-based expert systems will be boon for online health patrons who seek health-related and diagnosing information.

Keywords Fuzzy set theory · Fuzzy logic · Fuzzy expert system
Symptoms · Diseases · Disease inference · Data mining · Catalyst
Clustering

A.P. Chainani (✉) · S.S. Chikne · N.D. Doshi · A.Z. Karel · S.S. Therese
Thadomal Shahani Engineering College, Bandra (W), Mumbai 400050,
Maharashtra, India
e-mail: chainaniajay@yahoo.in

S.S. Chikne
e-mail: itssantoshchikane@gmail.com

N.D. Doshi
e-mail: nikunjdoshi67@gmail.com

A.Z. Karel
e-mail: asim.karel@gmail.com

S.S. Therese
e-mail: shanthitherese123@gmail.com

D.K. Mishra et al. (eds.), *Information and Communication Technology for Sustainable Development*, Lecture Notes in Networks and Systems 10,
https://doi.org/10.1007/978-981-10-3920-1_10

1 Introduction

The aging of society, augmenting costs of healthcare and proliferating computer technologies are together driving more people to spend more time online to explore health-related information. As per the factual analyses from particular surveys, nearly three-fifth of total population of developed countries in the world have considered online Internet medium as a pioneering tool for diagnose purpose. Extracts from other studies imparts on an average annually more than 80 h are dedicated by adult population on Internet for seeking and solving health-related issues. On contrary, it is clarified that visiting doctors by them are just 5–6 times approximately yearly for minimal and same set of issues. With the advent of the Internet paying way to new horizons, the medical expert systems have burgeoned in an enormous manner. Also, the Android and IOS app based healthcare systems have also added inter-operability and hand on results swiftly. But due to the technology glitches and lack of availability of such features is a concern which cannot be fixed with ease. Previously, there were online tools for automatic disease inference but they failed in providing necessary information. But that is not enough since various constraints behold the necessary execution and guiding of proper expertise to the consumers. Usually, the busy schedule and hectic workload delimit the doctors to process the uncertainty that may limit the efficiency and optimality. Yet, doctor's decisions can prove to be in contrast with underlying disease inference because of personal information pertaining to an individual. They are based on their demographics, personal life style and choices, sustaining long-term health-related issues. With these, an innovative solution that could be an answer to automatic health ecosystem is the advent of Fuzzy Inference Based Expert System [1] for disease inference. Thus this provides complete command over pertaining issues and this research is novelty for automatic diagnose of diseases. The fuzzy expert system is queried by the user for the disease they might be suffering with and to ease the functionality it provides symptoms [2]. It also deals with the obnoxious terms and uncertainty prevailing over the user queries and gives the outcome or output in fuzzy form. This is done so, by assigning ranks to the pretrained weights of symptoms and adding catalysts factor which may or may not be optional for certain diseases.

2 Literature Survey

In this section, some information about the existing system using similar technologies is discussed and techniques used in this system have been more focused. Many online tools are available online in different platforms for predictive analysis and mining of information from data. A common example of such existing software is Weka which provides solution to all data mining queries. But given the fact that algorithms used for fuzzy set operations have crisp values, it cannot be scaled every time. Thus, in our project we used Enterprise Java Beans deployed on Netbeans software to make our project an enterprise project. The purpose of using Netbeans is to use its existing functionalities like Java Servlets, Server and on demand web pages and java files to seamlessly execute it. The built-in integration features like database and glassfish server makes the execution swiftly and along with ease. Also enterprise java beans are state full which means they store the data set values of every phase of execution till they are not removed making it available for further processing. The core of the project lies in the data mining techniques and soft computing techniques like Fuzzy Set Theory or Fuzzy Logic along with the mathematical models. Execution is based on the user query extracted using the aforementioned techniques and generating the crisp output to the users. Following core concepts and technologies are conveyed as how they are used in the system implementation.

2.1 Clustering

Clustering is basically used for analyzing data sets where class label mostly used in classification and prediction is not consulted. Since the beginning is not known of training data, class label is absent. In general, clustering is used to group existing similar kind of data from the data set. It is done so by comparing the similarity between intra-class and minimal inter-class of the objects. For example, in our system clustering is widely used for mapping the symbols for inference of the diseases with its symptoms. Considering the fact that the disease [3, 4] 'Diabetes' will have symptoms like 'excessive thirst', 'frequent urination' etc., so clustering will map this similar symptom together closely with the disease 'Diabetes'. Also, main advantage of clustering is outliers can be easily detected. Adding to the example, any symptoms like 'vomiting', 'headache' etc., are considered as outliers. Thus, clustering provides optimal values from the data sets and reducing workload.

2.2 Fuzzy Logic

Fuzzy logic uses veracity meaning truthfulness of values from zero to one which depicts the membership degree for certain value of category given to represent fuzzy sets. In traditional crisp data sets there can be only two classes where the value can belong either in the set or in its complement. One set which is fuzzy and all elements belonging to that set can belong to multiple fuzzy sets [5–7]. Rule-based classification of fuzzy set theory is helpful in mining of data. Certain operations can be easily combined for fuzzy measurements. Thus, exact truth values need to be known and thus after performing the operations they can be added. Fuzzy logic helps in easily identifying the untruth values and rectifying them. For example, if a user of the system inputs a wrong query due to lack of knowledge or vocabulary gap, case mismatch or inconsistent data then fuzzy logic corrects that. If a user enters a wrong spelling of a symptom, then it is rectified using fuzzy logic.

After the fuzzification is done, many procedures are available for translation of crisp value also known as defuzzified value which system returns.

3 Proposed System

First step in every data mining system is to segregate the data as per the constraints and inferences. By usage of the information collected from the copious sources, the information is transformed and the associated data is mined and use for inference. This work signifies and uses health-related questions and inferences are based on fuzzy inference system. In the second step, the health-related questions are mined using fuzzy inference algorithms [8] and different strata are classified accordingly. The system consists of the UI which consists on interaction engine and help which provides the users to have hands on to the frequently asked questions and also a track of recently asked questions is kept for better interaction and getting updated to latest diseases and their symptoms. The diseases will be tracked as per the user's information provided by them. Users will give the query in the form of questions like currently what is the problem they are facing, is there any particular symptom they have been came to know, any kind of pain which may or may not be acute, etc. Taking all these questions into consideration, the decision-making process comes into the scenario and the decision engine will infer the questions of the users. All the questions queried by the users are tagged to the system for actual sparse deep learning. Since that requires more computations due to processing over three nodes, i.e., input layer node, hidden layer node, and output layer node which itself is a part of Artificial Neural Network [9, 10] along with the set of several associations rules mined to get inference of diseases through classification and prediction fuzzy logic along with clustering has undue advantage while working in tandem [11, 12]. Thus, the system will use techniques such as clustering and algorithms on fuzzy logic [13] to mine the knowledge from the user's questions. The system will use techniques and algorithms to mine the knowledge from the user's questions. The system has a predefined dataset for the relevant symptoms and diseases which will be vital cog in commencing a decision computed by the system with the knowledge gained from the inferences. All the information will be responded to the user through interactivity in UI.

3.1 Design Approach

3.1.1 Disease Inference

In these, we have reviewed a system which mines the data via fuzzy sets. By using extensive computation through formulae and algorithms applied on the data sets a relative inference will be concluded. The databases will be source of data sets where

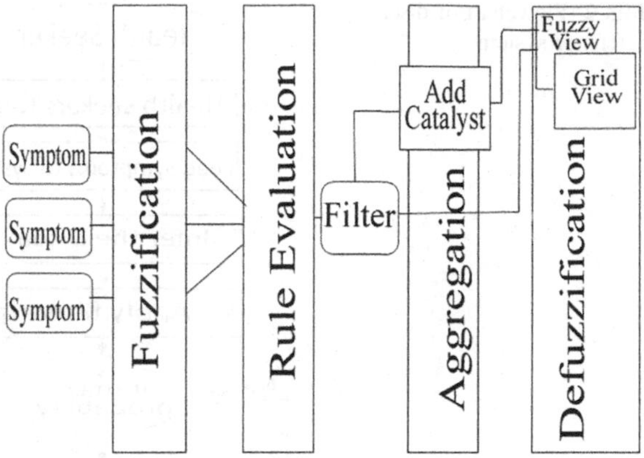

Fig. 1 Design of the system

different repositories will be required for diseases, symptoms, patients, users, allergies, hereditary diseases, etc. Here the relevant information will be extracted as close as to meet the proximity of the actual subject (Fig. 1).

3.2 *Methodology and Analysis*

The system consists of the UI which consists on interaction engine the diseases will be tracked as per the user's information provided by them. Users will give the query in the form of questions like currently what is the problem they are facing, is there any particular symptom they have been came to know, any kind of pain which may or may not be acute, etc. Taking the questions into consideration the decision-making process comes into the scenario and the decision engine will infer the questions of the users. The system will use techniques and algorithms to mine the knowledge from the user's questions.

3.2.1 Design Approach of System

See (Fig. 2).

3.2.2 Implementation of the System

After studying vivid diagnosis strategies, a standard structure is made and mathematical formula is proposed to be implemented for the fuzzy expert system. Proceeding with the implementation of expert system is commenced, the system is tested for checking the output whether it is producing correct and desired result or not. In case of alterations or corrections in the results it is verified and

Fig. 2 Flowchart of disease inference system

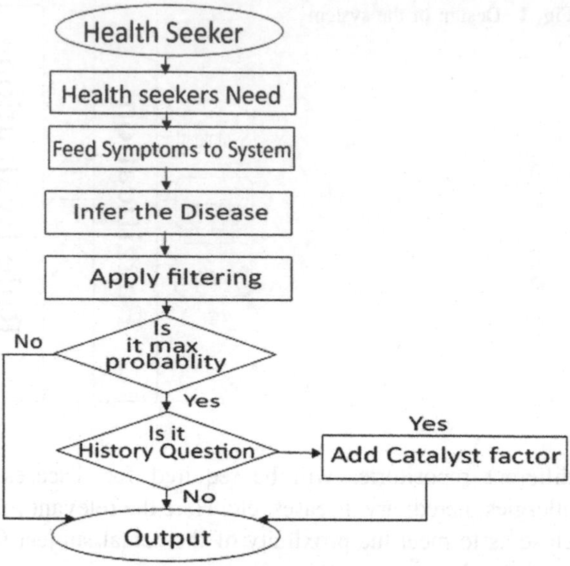

acknowledged by the doctor. Once the output is generated, it is checked for the nearness in the output. The output is measured through the equation or formula and several factors are dependent in the formula. Also the output can vary due to individual differences. Thus catalyst factor is added which is based on person's demographics, personal background, medical history, etc.

Formula for Calculating Output:

Probability for disease

$$\frac{(\sum Rk * Wk - \sum (Rmajor)k(Wmajor)k) - Minth * 100\% + \sum Cxf}{Maxth - Minth}, \quad (1)$$

where

k	counter of different symptoms
Rk	is kth symptom selected or not
Wk	weight of kth symptom
Maxthd	maximum threshold of disease
Minthd	minimum threshold of disease
ΣRkWk	total weight of a disease based on the selected symptom
Σ(Rmajor)k(Wmajor)k	total weight of unselected major symptoms of a disease
Cxf	catalyst factor based on patient's history/personal Information

Example: Asthma Symptoms—Cough (0.9), Chest pain (0.2), Wheezing Noise (0.9), Short Breath (0.6).

4 Results

4.1 Home Page of Disease Inference System

This is the home page for the respective project where the user and admin login are mentioned. The users have to register themselves with the system and then for their respective queries they can query the system. Also the system provides provisioning of asking generic queries based on their causes and symptoms and the user can get the assistance of default questions generated by the system (Fig. 3).

4.2 Options Available with User

System interaction happens when user queries are extracted with the help of tags. The keywords from the query are extracted for example: name of causes, symptoms, etc., for any particular disease or multiple diseases. Also, the system has predefined sets of questions or frequently asked questions for assisting naïve users for optimality (Fig. 4).

4.3 Probabilistic Output

The output generated by the system is based on various factors like the symptoms of the particular disease, causing agents for particular disease, etc. Processing of these done through the signs and tags and training is given at each layer and thus diseases having multiple common symptoms can also be calculated with the given

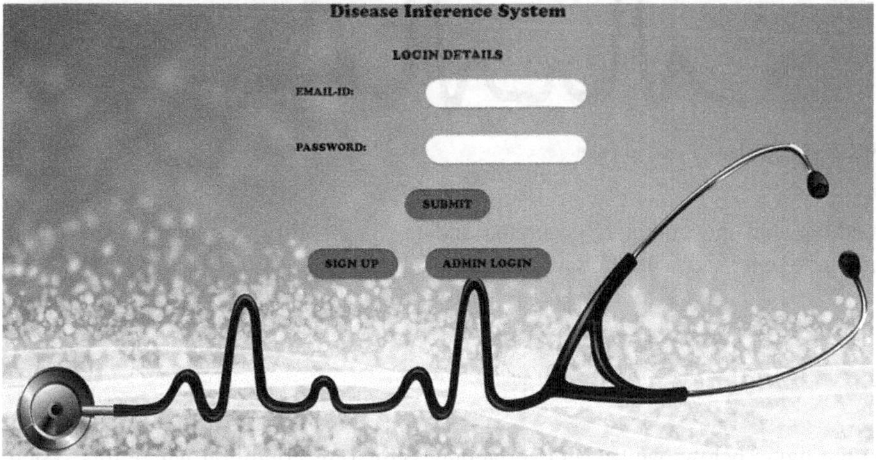

Fig. 3 Screenshot of home page of the system

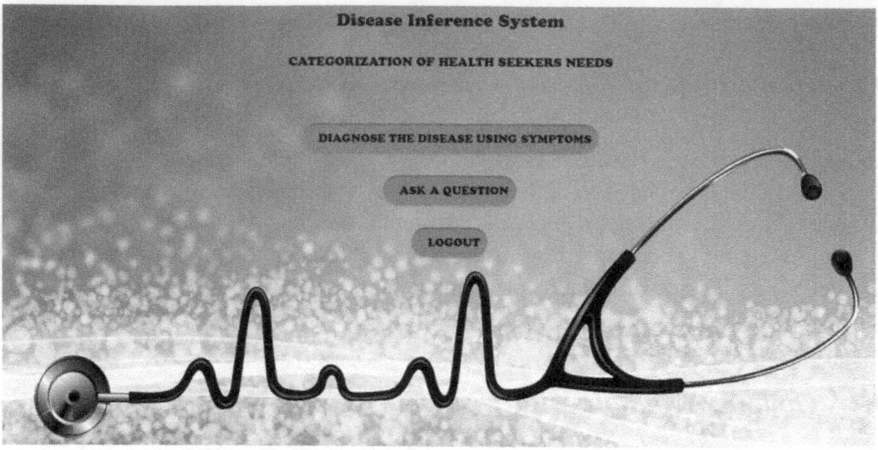

Fig. 4 Screenshot of options available with user

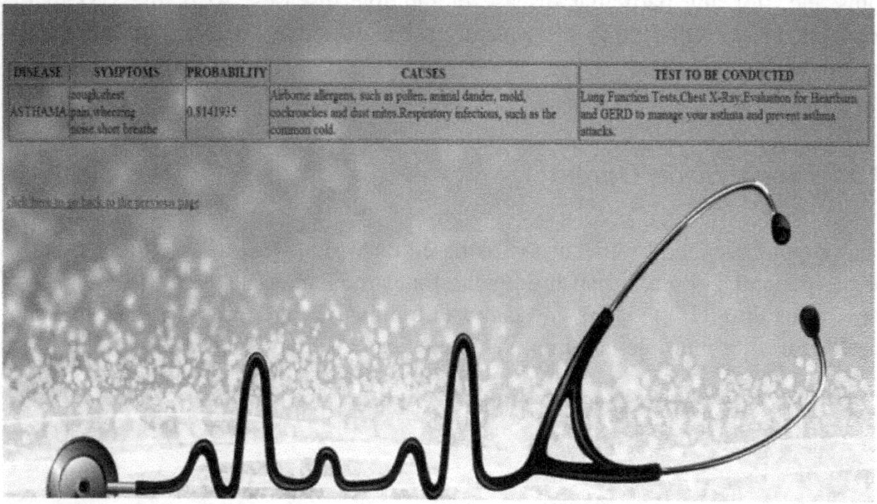

Fig. 5 Screenshot of output page

methods and the output is generated and the probability is generated each time query is generated with aforementioned symptoms and causes (Fig. 5).

4.4 Handling Overlapping

In the figure below, we see an overlapping symptoms condition where this system handles the all overlapping condition and calculates the probability for respective disease. After that it provides output of disease with the highest probability. As we can see in the figure given below, when user enters symptom as fever, the system looks for this symptom in the given dataset and returns the list of symptoms which contains fever. Once the overlapping symptoms have been identified, the next task system does is calculating the probability of having that disease with respect to the given symptom by the user. After the probability for the overlapping disease has been calculated, the system returns the respective disease as output whose probability is highest among the calculated ones (Fig. 6).

```
run:
enter the symptoms
fever
******List of Symptoms where we find overlapping******
Symptoms match are-->burning sensation,hot weather,fever,seating problem
Symptoms match are-->weakness,fever,headaches,cough
Symptoms match are-->poor appetite,headaches,body pain,fever,diarrhea
Symptoms match are-->Dry cough,Fever,Chest pain,Sore throat,Rapid breathing


INDEX-->0 5 7 11

Disease Name-->FABRY DISEASE
 Symptoms-->burning sensation,hot weather,fever,seating problem
Probability-->13.333334

Disease Name-->Viral Fever
 Symptoms-->weakness,fever,headaches,cough
Probability-->21.42857

Disease Name-->TYPHOID FEVER
 Symptoms-->poor appetite,headaches,body pain,fever,diarrhea
Probability-->13.333334

Disease Name-->Pneumonia
 Symptoms-->Dry cough,Fever,Chest pain,Sore throat,Rapid breathing
Probability-->11.764711

******Disease with maximum Probability******

Disease Name->Viral Fever
Symptoms-->weakness,fever,headaches,cough
Probabilty-->21.42857
BUILD SUCCESSFUL (total time: 11 seconds)
```

Fig. 6 Working of overlapping condition

```
run:
enter the symptoms
Pain or burning when you urinate,back pain,pain in sidebelow the ribs
******List of Symptoms where we find overlapping******
Symptoms match are-->Pain or burning when you urinate,back pain,pain in sidebelow the ribs

INDEX-->4

Disease Name-->URINARY TRACT INFECTION
 Symptoms-->Pain or burning when you urinate,back pain,pain in sidebelow the ribs
Probability-->46.15385

******Disease with maximum Probability******

Disease Name->URINARY TRACT INFECTION
Symptoms-->Pain or burning when you urinate,back pain,pain in sidebelow the ribs
Probabilty-->46.15385
BUILD SUCCESSFUL (total time: 38 seconds)
```

Fig. 7 Working of nonoverlapping condition

4.5 Handling Nonoverlapping Condition

In the figure below, we see a nonoverlapping symptoms condition where this system calculates the probability for the respective disease and returns it as an output to the user. Here, there is no intervention of multiple diseases having common symptoms (Fig. 7).

5 Conclusion

The proposed system has been envisioned for the purpose of reducing the complexity that occurs in the traditional system as well as to enhance the accuracy of the result. The aim is to identify the disease via the symptoms from the datasets that are stored. The database is crucial to the execution of the system and symptoms should be stored carefully with respect to disease. The health-related questions are the source of the input to the system. The final phase is the identification of the disease from the questions while inferring there causes and symptoms. We will be developing the Disease Inferring Expert System. Since it is a complicated and research-oriented project, we will try to develop the maximum possibility using Enterprise software's and similar platforms available.

6 Further Work

6.1 Work on Complicated Diseases

The current proposed system is underlined to work for limited set of diseases due to the lack of availability of expertise on various diseases that are complicated in nature and decoding the same is itself a challenging task. For example, Diseases like Cancer, Heart Diseases, Zika Virus, and Neurological Disorders, etc. There are many complications in mining and inference this disease but expertise knowledge at apex level may suffice that.

6.2 Inclusion of Interaction via Bots

Bots are the self-automated tools or algorithms that can emulate instructions as requested. They respond to the requests just by inference set of instructions and information. For example, in our existing system bots will be used on messaging platforms. The clear vision for this purpose is through messaging apps with which our expert system can also work. Users just need to query their problems and the automated bots will fetch their queries and then it will be passed to the Expert System which will infer the disease and then it will be again passed to the bots where finally users get their queries answer in the form of messages. It is a mammoth task but could be gigantic makeshift in the realm of humanity where Internet is the master in the era of e-portals.

6.3 Customized User Support

As of now, the regular users or the users who constantly seek help through the system, for them a customized prescription or a dashboard were keeping track of their disease history seems to be mandatory thus enhancing the productivity of the system and better for the benefits of the user more commonly the customers. "Catalyst Factor" as explained earlier which is more user oriented, where it keeps vital or unique information about user's health, plays an important factor for user customization-based inference by the system.

References

1. Shi, Y. and Eberhart, R. and Chen, Y., (1999) "Implementation of evolutionary fuzzy systems", IEEE Transactions on Fuzzy Systems, Vol. 7, No. 2, pp 109–11.
2. "Human Disease Diagnosis Using a Fuzzy Expert System" Journal of Computing, Volume 2, Issue 6, June 2010, ISSN 2151-9617.
3. Symptoms of different diseases available on: http://www.webmd.com/
4. Symptoms of different diseases available on: http://www.mediplus.com/
5. Zadeh, LA, (1983) "The role of fuzzy logic in the management of uncertainty in expert systems", Fuzzy sets and Systems On Elsevier.,Vol. 11, No. 1–3, pp 197–198.
6. Timothy J. Ross "Fuzzy Logic With Engineering Applications" Wiley.
7. J.-S. R. Jang "Neuro-Fuzzy and Soft Computing" PHI 2003.

8. M. Negnevitsky, (2005) "Artificial intelligence: A guide to intelligent systems", Addison Wesley Longman.
9. Jacek M. Zuarada——Introduction to Artificial Neural System ||; West Publishing Company. 1992.
10. S. Rajasekaran and G. A. Vijayalakshmi Pai "Neural Networks, Fuzzy Logic and Genetic Algorithms" PHI Learning.
11. A. Sudha, P. Gayathri, N. Jaisankar——Utilization of Data mining Approaches for Prediction of Life Threatening Diseases Survivability ||; International Journal of Computer Applications (0975 – 8887) Volume 41– No. 17, March 2012.
12. Carlos Ordonez,——Comparing association rules and decision trees for disease prediction ||; ACM 2006.
13. S. N. Sivanandam, S. N. Deepa "Principles of Soft Computing" Second Edition, Wiley Publication.
14. Han, J., Kamber, M,——Data Mining Concepts and Techniques||; Morgan Kaufmann Publishers, 2006 (Second Edition).

Spectral Biometric Verification System for Person Identification

Anita Gautam Khandizod, Ratnadeep R. Deshmukh and Sushma Niket Borade

Abstract Automatic person identification is possible through many biometric techniques which provide easy solution like identification and verification. But there may be chances of spoofing attack against biometric system. Biometric devices can also be spoofed artificially by plastic palmprint, copy medium to provide a false biometric signature, etc., so existing biometric technology can be enhanced with a spectroscopy method. In this paper, ASD FieldSpec 4 Spectroradiometer is used to overcome this problem, the palmprint spectral signatures of every person are unique in nature. Preprocessing technique including smoothing was done on the palmprint spectra to remove the noise. Statistical analysis were done on preprocessed spectra, FAR (False acceptance Rate), and FRR (False Rejection Rate) values against different threshold values were obtained and equal error rate was acquired. EER of the system is approximately 12% and the verification threshold 0.12.

Keywords Spectra ASD FieldSpec 4 · Hyperspectral palmprint
False acceptance · False rejection

1 Introduction

Numerous types of algorithm and system have been proposed in palmprint recognition, although a great success has been achieved, but spoofing mechanism and accuracy are limited [1]. Current biometric and palmprint techniques can

A.G. Khandizod (✉) · R.R. Deshmukh · S.N. Borade
Department of Computer Science and Information Technology,
Dr. Babasaheb Ambedkar Marathwada University, Aurangabad, India
e-mail: anukhandizod@gmail.com

R.R. Deshmukh
e-mail: rrdeshmukh.csit@bamu.ac.in

S.N. Borade
e-mail: sushma.borade@gmail.com

D.K. Mishra et al. (eds.), *Information and Communication Technology for Sustainable Development*, Lecture Notes in Networks and Systems 10,
https://doi.org/10.1007/978-981-10-3920-1_11

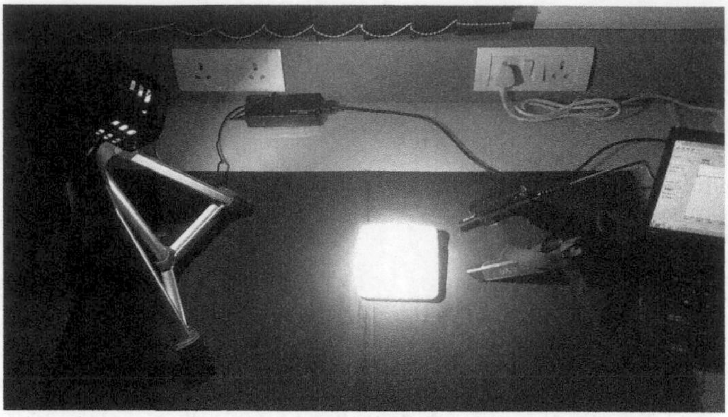

Fig. 1 Laboratory experimental setup for spectral palmprint

provide an image of a person's palmprint. Impression of the palmprint can be left on the surface by sweat or secretion from glands present in palmprint region or can be used transferring ink to copy the palmprint features; many times palmprint sensors accepted artificial palmprint therefore the accuracy is limited. This problem is solved by ASD Fieldspec4 spectroradiometer, in order to increase the accuracy and prevent spoofing attack. ASD Spectroradiometer method is used to enhance existing palmprint recognition system. ASD (Analytical Spectral Devices) Spectroradiometer is the study of interaction between the physiochemical characteristics and spectral signature characteristics of object [2]. ASD spectroradiometer have been used in detection, identification, verification, and quantification of object. ASD Fieldspec4 spectroradiometer device are portable, rugged and robust, yet with no compromise in performance and acquires continuous spectra from wavelength range of 350–2500 nm; this device consists of three separate detectors: UV/VNIR range (300–1000), for the SWIR1 range (1,000–1,800 nm) and the SWIR2 range (1,800–2,500 nm). Sampling interval for the Fieldspec4 is 1.4 nm for the region 350–1000 nm and 2 nm for the region 1000–2500 nm (Fig. 1).

2 Comparative Analysis with Existing Technique

Davar Pishva et al. (2011), [3], proposed a novel method which is quite feasible to use fingerprint and iris spectra biometric to preventing spoofing of existing biometric technology. Marion Leclerc, Benjamin Bowen, Trent Northen (2015), [4] Suggested that a novel fingerprint based indentification approach by using nanostructure initiator mass spectrometry (NIMS), mass spectrum pattern compare to a known pattern, there by identifying the subject. There is no work carried out on hyperspectral palmprint spectral signature, this approach is new and can be applied on other biometric characteristics such as face, hand, etc.

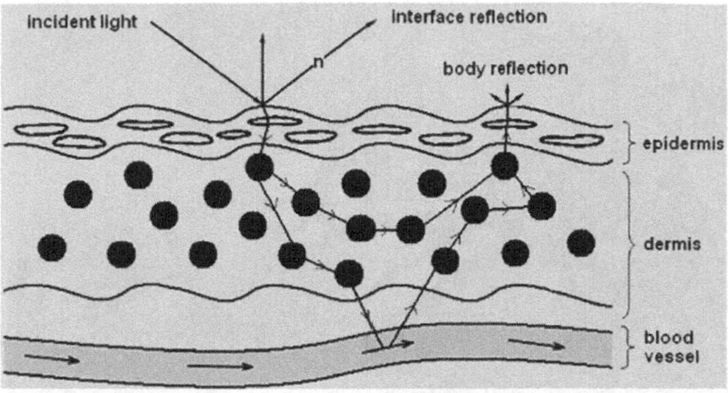

Fig. 2 Principle of skin reflectance

3 Skin Reflectance

Skin is largest organ of the human body, having complex biological structure made of different layers like epidermis, dermis, and hypodermis. Light reflected from the skin, there are two reflection components: a specular reflection component and diffuse reflection component. The specular reflection occurs at the surface [5], in which light from a single incoming direction (incident ray) is reflected into a single outgoing direction (reflected ray) and make the same angle with respect to the normal surface, thus the angle of incidence equals the angle of reflection (Qi = Qr).

As shown in Fig. 2, the incident light is not entirely reflected at the surface, some incident light penetrate into the skin; these light travels through the skin and hitting physiological particles. This carries information about the skin color of person's and his/her biological spectral signature. Blood has hemoglobin, bilirubin, Beta-carotene, melanin in the epidermis absorbs blue light at ~470 nm; hemoglobin absorbs green and red light at ~525, ~640 nm, and papillary dermis absorbs near infrared light at 850 nm.

4 Laboratory Spectral Palmprint Data Collection

The spectral signature of palmprint database from 50 individuals was built in Geospatial technology research laboratory, Dr. Babasaheb Ambedkar Marathwada University. The age distribution was from 20 to 40 years old. Database was collected by 1 and 8 degrees of FOV. In each degree, the subject was asked to provide around ten spectral signatures of each of his/her left and right palms, so the total database contains 2000 reflectance spectra (350–2500 nm) of palmprint (Fig. 3).

As can be observed in Fig. 4a shows the three reflectance spectra of palmprint look identical because it comes from an object of the same palmprint pattern,

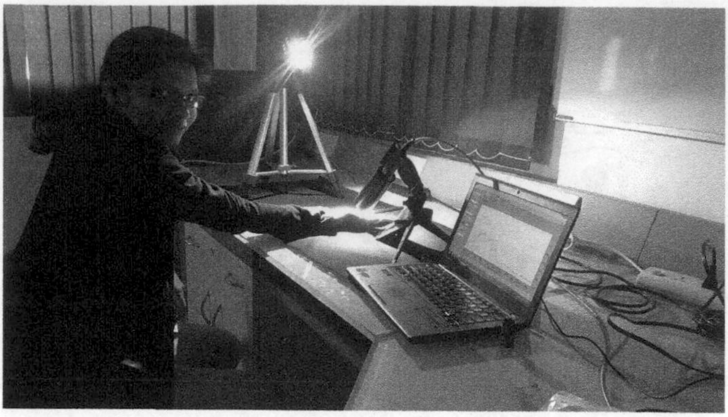

Fig. 3 Laboratory experimental measurement setup of palmprint Spectra

Fig. 4b show reflectance spectra of five different persons, having variation among different individuals as there are variation in palmprint pattern, physiological characteristics. Thus, reflectance spectra of the palmprint check the authentication of the person during verification process of biometric.

5 Smoothing

Smoothing technique is applied on collected database, smoothing helps to remove noise from spectral signature without reducing the number of variables [6]. In smoothing Moving Average, Savitzky-Golay, Median Filter smoothing techniques are used. Figure 5 shows the raw reflectance spectra and filtered reflectance spectra, the moving average smoothing technique gives good result compared to other smoothing technique.

6 Statistical Analysis for the Identification of Palmprint Reflectance Spectra

In the present study, descriptive statistics like mean, standard deviation, they describe the distribution and relationship among variables. A spectral signature sample from person 1, P1 is tested against four sample of person 1, numbered P1-1, P1-2, P1-3, P1-4 and four other samples from person 12 (P12), person 17 (P17), person 21 (P21), person 29 (P29). Table 1 shows matching percentage of sample P1 against other samples (Fig. 6).

The matching percentages for sample P1 against four samples from the same person are very close and high, and matching percentages for sample P1 against

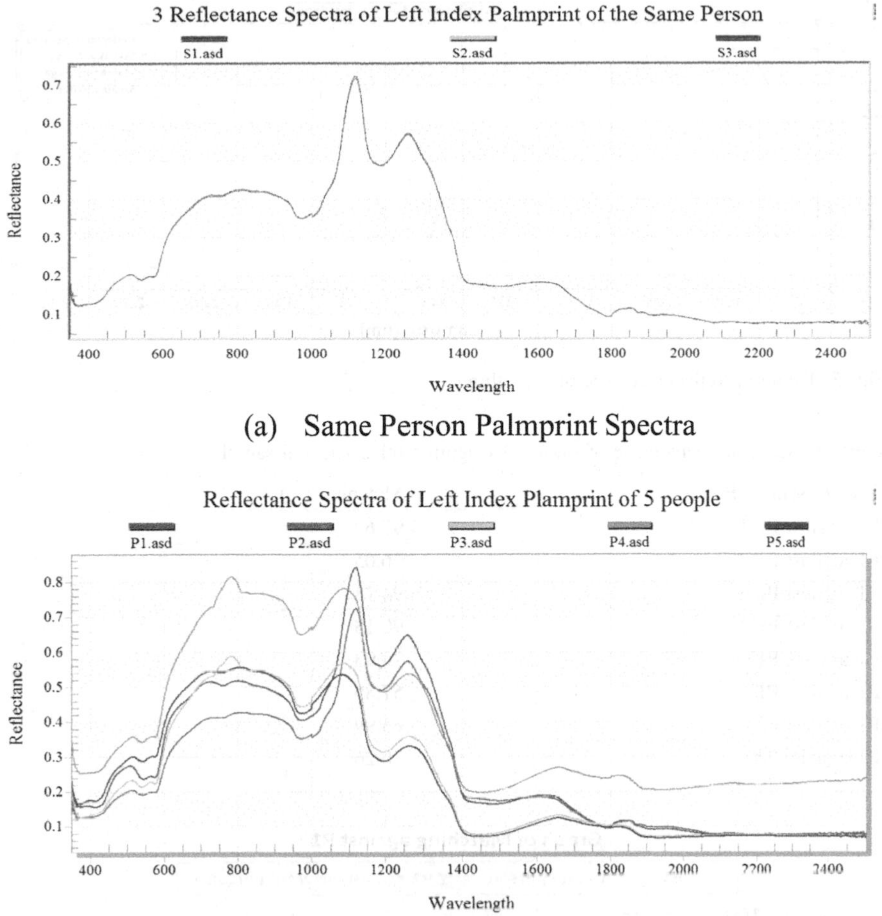

(a) Same Person Palmprint Spectra

(b) Variation among different Persons Palmprint Reflectance

Fig. 4 An individual's palmprint spectra and spectral variation among different individuals

four samples from the different person are low, therefor threshold can be set to determine whether two samples are from the same spectral signature of palmprint or different.

7 Euclidian Distance

It compares the relationship between actual ratings, The formula to Euclidean distance between two points a(x1, y1) and b(x2, y2) is as follows:

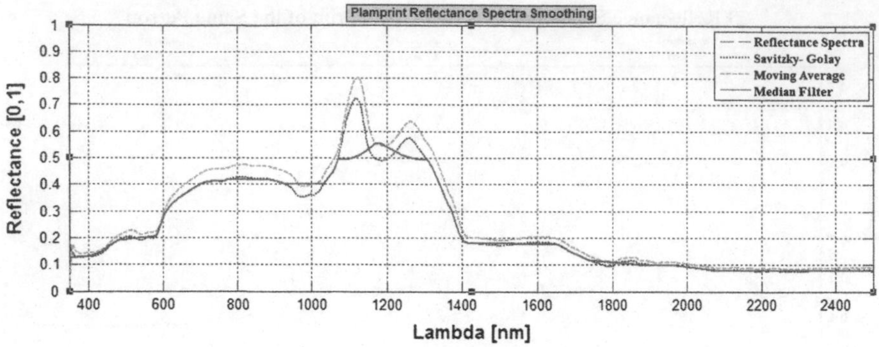

Fig. 5 Palmprint reflectance spectra smoothing

Table 1 Matching percentage of sample P1 against other spectral sample

Spectra sample Pair	Matching (%)
P1 against P1-1	97.69
P1 against P1-2	96.05
P1 against P1-3	94.73
P1 against P1-4	90.13
P1 against P12	79.93
P1 against P17	57.89
P1 against P21	83.99
P1 against P29	70.26

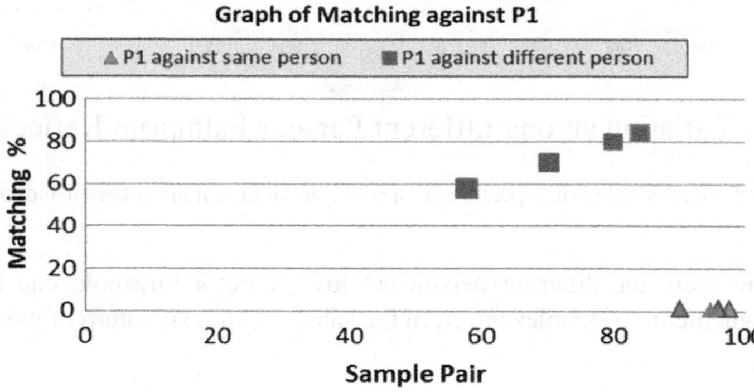

Fig. 6 Scatter graph of matching against sample P1

$$\sqrt{(x_1 - x_2)^2 + (y_1 - y_2)^2} \tag{6}$$

Euclidean distance is used for recognition purpose, but before this classifier cloud be used it was necessary to calculate the mean for each spectral reflectance of

Table 2 Distance matrix of palmprint spectra

	1	2	3	4	5	6	7	8
1	0.00	0.05	0.04	0.06	0.02	0.00	0.07	0.02
2	0.05	0.00	0.01	0.01	0.02	0.05	0.11	0.06
3	0.04	0.01	0.00	0.02	0.02	0.04	0.11	0.06
4	0.05	0.00	0.01	0.00	0.03	0.05	0.12	0.07
5	0.03	0.02	0.02	0.04	0.00	0.02	0.09	0.04
6	0.02	0.03	0.02	0.04	0.04	0.00	0.09	0.04
7	0.07	0.12	0.11	0.13	0.09	0.07	0.00	0.05
8	0.02	0.07	0.06	0.08	0.04	0.02	0.05	0.00

palmprint processed with derivative spectra. Euclidean distance was applied to the obtained mean of derivative spectra.

Table 2 shows the distance matrix of palmprint spectra, diagonal elements of the matrix representing the similarity between the spectral samples between the persons; while non-diagonal elements are having value greater than diagonal elements which represent the difference between the corresponding spectral samples of persons.

8 Threshold Decision

Threshold value is chosen range between maximum and minimum value of the Euclidian distance of each spectral signature of palmprint from other spectral signature. Choosing the threshold value is a very important factor to increase the performance of spectral palmprint recognition system [7]. The acceptance and rejection of biometric palmprint spectral signature data is dependent on the threshold match score falling above or below the threshold value.

8.1 False Accept Rate (FAR) and False Reject Rate (FRR)

In FAR, nonauthorized person incorrectly authorizes the system, due to incorrectly matching the biometric input with a template while in case of FRR authorized person is not match with their own existing biometric template FAR is the portion of imposter score > threshold [8]. FAR is the portion of imposter score > threshold.

Table 3 shows the threshold value against FAR and FRR. The threshold value is decided within the range of the minimum and maximum value of distance matrix minimum distance: 0.02 and maximum distance: 0.24, from Table 4 threshold value is increased then FAR will decrease but FRR will increases, for given

Table 3 Threshold value against FAR and FRR

Threshold	FAR	FRR
0	90	0
0.02	80	0
0.04	73	0
0.06	70	0
0.08	60	0
0.1	55	0
0.12	43	0
0.14	40	2.2
0.16	32	2.2
0.18	20	7.7
0.2	15	9.9
0.22	10	15.5
0.24	0	24.4

palmprint spectral biometric system both the errors cannot be decreased simultaneously by varying threshold value.

8.2 *Equal Error Rate (EER)*

Equal error rate (EER) is also called as Crossover Error Rate (CER). Equal error rate is the location on ROC (Receiver operating characteristic) curve where the FAR and FRR overlaps and equal [9], lower the equal error rate value, the higher the accuracy of the biometric system. As shown in Fig. 7, EER of the system is approximately 12% and the threshold taken for verification is 0.12.

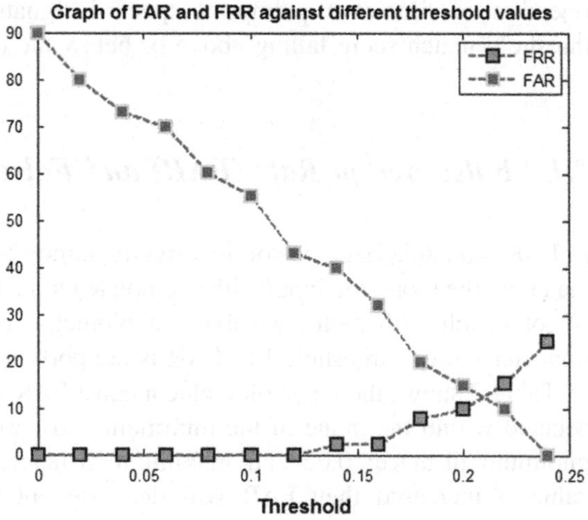

Fig. 7 Graph of RAR and FRR against different threshold values

9 Conclusion

Existing biometrics technologies enhances by using palmprint spectroscopic method in order to prevent spoofing, palmprint spectral signature vary from person to person, and makes spoofing a very difficult task. In this paper we developed the database of palmprint spectra signature, to remove noise three types of smoothing techniques are used like Moving Average, Savitzky-Golay, and Median Filter, statistical analysis could be used for discriminating palmprint spectral signature those belonging to the same or different persons. For recognition purpose, Euclidian distance was used, and takes the decision with the help of threshold value, false acceptance rate decreases and false rejection rate increases, equal error rate of the system is quite low which is about 12%.

References

1. Anita G. Khandizod., R. R. Deshmukh.: Analysis and Feature Extraction using Wavelet based Image Fusion for Multispectral Palmprint Recognition. International J Enhanced Research in Management & Computer Applications, ISSN: 2319-7471, Vol. 3 Issue 3, pp: 57–64, (2014)
2. Francesca Garfagnoli., Gianluca Martelloni., Andrea Ciampalini., Luca Innocenti., Sandro Moretti.: Two GUIs-based analysis tool for spectroradiometer data pre-processing. Springer-Verlag Berlin Heidelberg, (2013)
3. Davar Pishva.: Use of Spectral Biometrics for Aliveness Detection, Advanced Biometric Technologies, Dr. Girija Chetty (Ed.), ISBN: 978-953-307-487-0, InTech, DOI:10.5772/17100, (2011)
4. Marion Leclerc., Benjamin Bowen., Trent Northen .: Nanostructureinitiator mass spectrometry biometrics, US9125596 B2,(2015)
5. T. Weyrich., W. Matusik., H. Pfister., J. Lee., A. Ngan., H.W. Jensen., M. Gross.: A Measurement-Based Skin Reflectance Model for Face Rendering and Editing. MITSUBISHI ELECTRIC RESEARCH LABORATORIES, (2005)
6. Shalini Singha., Dibyendu Duttaa., Upasana Singha., Jaswant Raj Sharmab. Vinay Kumar Dadhwalb.: HYDAT-A HYPERSPECTRAL DATA PROCESSING TOOL FOR FIELD SPECTRORADIOMETER DATA, The International Archives of the Photogrammetry, Remote Sensing and Spatial Information Sciences, Volume XL-8, ISPRS Technical Commission VIII Symposium, (2014)
7. Otsu., Nobuyuki.: A Threshold Selection Method from Gray-Level Histograms, IEEE Trans. On Systems, Man., and Cybernetics, vol. smc-9, no. 1, (1979)
8. G. Seshikala., Umakanth Kulkarni., M.N. Goroprasad.: Biometric Parameters & Palm Print Recognition," International J Computer Applications (0975 – 8887) Volume 46– No.21, (2012)
9. Sheetal Chaudhary., Rajender Nath.:A New Multimodal Biometric Recognition System Integrating Iris, Face and Voice, International J Advanced Research in Computer Science and Software Engineering, Volume 5, Issue 4, ISSN: 2277 128, (2015)

A Proposed Pharmacogenetic Solution for IT-Based Health Care

Rashmeet Toor and Inderveer Chana

Abstract Health care is a vast domain and has a large-scale effect on population. It has been facing critical issues of safety, quality, and high costs. Technical innovations in health care since the last decade have led to emergence of various computational, storage and analysis tools and techniques which are high quality, easily accessible, and cost-effective. In this paper, we have summarized the emerging trends of IT in medical domain. Further, we have proposed a pharmacogenetic solution for health care which can act as an aid to customized medicine.

Keywords Health care · Data mining · Pharmacogenetics · Personalized medicine

1 Introduction

The developments of Information Technology have immensely impacted health-based outcomes. Innovative solutions in the past few years have propelled the medical field forwards. Improvements in the bandwidth of Internet services over the years led to the creation of trillions of images, videos, and speech files. With the accumulation of such large volumes of biomedical data, the invention of varied technologies came into existence leading to better information retrieval, extraction, analysis, and storage.

Before the technology-led healthcare revolution, medical records of patients were in the form of handwritten notes. These were difficult to store and access. The emergence of Electronic Health Records (EHR) served the purpose of an electronic repository of not only patient's medical history, but also other valuable information

R. Toor (✉) · I. Chana
Department of Computer Science and Engineering, Thapar University, Patiala 147004, India
e-mail: rashmeett@gmail.com

I. Chana
e-mail: inderveer@thapar.edu

D.K. Mishra et al. (eds.), *Information and Communication Technology for Sustainable Development*, Lecture Notes in Networks and Systems 10,
https://doi.org/10.1007/978-981-10-3920-1_12

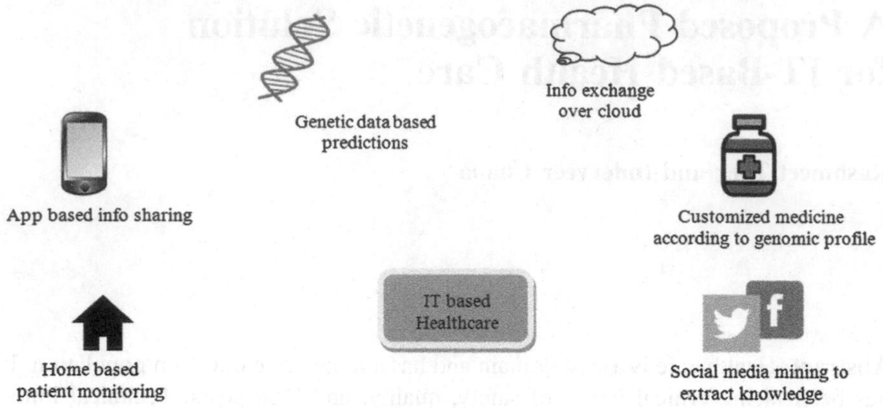

Fig. 1 IT-based future medical solutions

about drugs, prescriptions, medical tests, and billing information. This repository acts as a central place for doctors, patients and pharmacists thereby enhancing the decision making process. Microsoft HealthVault is an example of personal EHR where people can store and share health information [19]. EHRs were followed by invention of Clinical Decision Support Systems (CDSS) which automated decision-making for doctors. Timely and cost-efficient solutions were proposed by the system with the help of an extensive knowledge repository. Gradually, machine-learning techniques assisted decision-making which transformed it into an expert system. Although these applications were the pillars of transformation in health care, they were not much successful due to various reasons. With further advances in technology, various new trends were observed. Figure 1 describes the various IT-based medical solutions which are currently in progress.

The promising potential of Cloud computing and its convergence with technologies like wireless networks and sensors led to innovation of cloud services. Similarly, one of the new domains in health care is data mining which is the base for medical prediction models. In this paper, we first elaborate the current technologies and then discuss the proposed solution for IT-based healthcare.

The rest of the paper is as follows: Sect. 2 elaborates the trends in healthcare technologies since the last decade. Section 3 summarizes the issues and solutions that need to be worked upon followed by Sect. 4 which describes our proposed solution for an IT-based medical application. Section 5 concludes the paper by suggesting possible future directions.

2 Emerging Technologies

Recently, the wave of IT-led reform has swept some areas of the medical domain. Text mining is one of the emerging technologies, which when used in combination with other technologies will prove beneficial for healthcare industry. Signals of adverse drug reactions can be identified by extracting knowledge from social media, medical records, or search engine logs. This can be done using machine learning and text mining [9, 20]. EHRs can be mined to extract valuable information. Apart from text mining, analyzing networks like disease–gene or gene–gene using similarity-based measures is widely used to extract candidate genes for a disease [5]. Cytoscape is open-source software which is used for network analysis and visualizations [17, 18]. Machine learning is extensively used for medical solutions. An application was developed for monitoring pulmonary patients using a smart phone with underlying algorithm of Support Vector Machine [7]. Artificial intelligence enhanced decision support systems [13]. In the past few years, data mining and cloud computing technologies have elevated the healthcare industry to a new level. The data has now become more reliable, accessible and cost-effective. In this section, we discuss these two major technologies which play a vital role in IT revolution.

2.1 Data Mining in Healthcare

Mining large amount of data to extract valuable relations is the notion of data mining. It has been used vastly for medical tasks like unknown gene finding, treatment optimization, identifying similarity of genes and proteins and it is still progressing. One of the recent works done in genomic domain is [3]. Due to technologies like cDNA microarrays, it is now easy to measure the expression levels of thousands of genes simultaneously. In this study, high-throughput gene expression data was analyzed along with clinical data to create a prediction model for lung cancer. Patients were categorized into two classes that is, low risk and high risk using Artificial Neural Network algorithm.

Due to ease of performing genetic tests, analysis of SNPs/genes has become uncomplicated. We can now determine patterns for a particular disease or relationships between genes and physical characteristics. This in turn would lead to better drugs development or foundation for customized medicine. Due to massive amount of genetic data, high costs are required to directly analyze it. So, before the processing, relevant genes/SNPs are selected using mining [16]. In [11], researchers propose approaches to parallelize tasks for such massive data. Data mining and machine-learning approaches have helped in innovating prediction systems for various diseases as well [2, 8].

2.2 Cloud Computing in Health Care

The intersection of Cloud Computing with medical applications has lowered the costs and increased the speed of deployment over the internet. Storage of massive data and communication across remote areas has been made seamless through cloud technologies. Many recent applications are using cloud platforms like [15], where it provides a reliable and easily accessible healthcare monitoring system. The data from various remote patients is stored on a cloud-based medical repository. Health care could reach global level than only personal level because of the advent of cloud. Patients of chronic diseases can share the health information online over the cloud for different purposes [12]. A hybrid cloud using OpenStack and Amazon Web Services (AWS) was proposed for such application.

In [4], an application was developed on android for tracking of daily life activities of the elderly people by health professionals or family members can keep an eye on their daily activities from remote location. The elderly person himself can be helped through this app so that they can perform multiple activities. The application communicates via web services on the cloud. Sensors are deployed in the smart home to sense the environment continuously.

Another related study is [14] where monitoring of a patient is possible at home as well as multiple patients in hospitals or public health care units. Similar application of Electronic Medical System for emergency healthcare services is discussed in [10].

3 Issues in Current Technologies

Healthcare industry has seen tremendous growth in past few years and will continue to do so in coming years. For successful execution of various technologies, few issues which are currently prevalent in healthcare industry need to

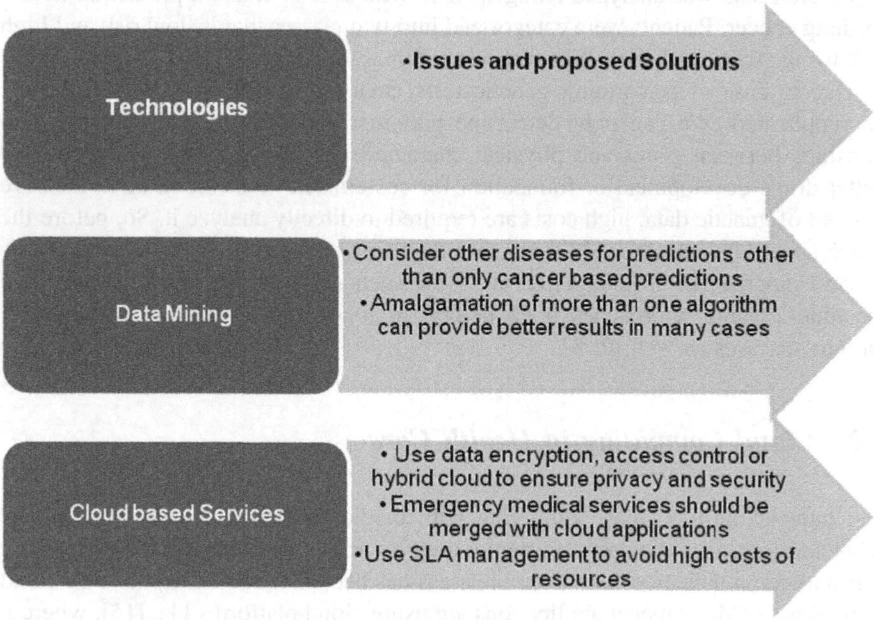

Fig. 2 Issues and their solutions leading to improved healthcare

be eliminated. Figure 2 summarizes the issues and proposed solutions in each technology. If we work on these issues, they can collectively culminate into a better future healthcare.

4 Proposed Work

Mining of data for gene–gene, gene–protein, gene–disease associations has been performed to extract unknown associations directing us to the notion of personalized medicine. Similarly, mining of gene–drug/chemical associations can be performed to extract relationships between drugs for individuals. With the advancements in technology, performing genetic tests is easier than ever before. Various services are available where we just need to provide our DNA samples and they provide us with a report of SNPs present like 23andMe, Promethease, FamilyTreeDNA, Ancestory.com, and many more. The use of this genomic data for predicting possible combination of drugs/chemicals for specific patients is a new domain. This domain is known as pharmacogenetic as it inculcates the knowledge of correlation of drugs and genes. It requires the relation of genes with drugs and a data mining algorithm to find the probable drug combinations. We have proposed a prototype where a druggist can enter the list of SNPs/genes present in a patient and the application returns possible drug combinations for that patient. These relations can also be used for replacing a drug with another in case of known allergies or in drug repositioning. Table 1 describes the comparison of the proposed work with other related work. This study is focused on gene and chemical associations whereas other studies revolve around clinical or disease–gene data. In our proposed work, we have collected data from the Comparative Taxicogenomics Database. For our application, we have limited our data to Type 2 Diabetes Mellitus (T2DM). It consists of interactions among chemicals and genes associated with T2DM. Since,

Table 1 Comparative analysis of proposed method with other related works

Author	Algorithm/Method	Type of data	Purpose
Ahmed et al. [2]	K-means clustering, AprioiriTID	Clinical	Lung cancer risk detection
Freudenberg et al. [5]	Similarity measures	Clinical and disease-gene	Identify disease relevant genes
Jung et al. [8]	LDA classification	Gene expression	Prediction of cancer therapy response
Chen et al. [3]	ANN	Clinical and gene expression	Cancer survival prediction
Proposed method	Filtered associator with Apriori	Gene-chemical associations	Predict drugs according to patients

	A	B	C	D	E	F	G	H	I	J	K	L	M	N	O	P	Q
1	gene	bispl	aflat	ben	buty	potassi	quer	roten	thiram	arse	diet	tetrac	cisp	irin	pent	resve	rosig
2	CDKAL1	?	T	T	T	T	T	T	T	?	?	?	?	?	?	?	?
3	CDKN2B	?	?	?	?	?	?	?	?	T	T	T	?	?	?	?	?
4	CDKN2A	?	?	?	?	?	?	?	?	T	T	T	?	?	?	T	?
5	FTO	T	?	T	T	?	?	?	?	?	?	?	T	T	T	?	?
6	HHEX	T	?	T	?	?	T	?	?	?	?	T	?	?	?	?	?
7	IGFBP2	?	?	?	?	?	?	?	?	?	?	T	?	?	?	T	T
8	KCNJ11	T	?	?	?	?	?	?	?	?	?	?	?	?	?	?	?
9	PPARG	T	?	T	?	?	?	T	?	T	T	T	T	?	?	T	T
10	TCF7L2	T	?	?	?	?	T	?	?	?	?	T	T	T	?	T	?
11	PDE4B	?	?	?	?	?	?	?	?	T	?	T	?	?	?	?	?
12	RBMS1	?	?	?	?	?	?	?	?	?	?	?	?	?	?	T	?
13	SREBF2	?	?	?	?	?	?	?	?	?	?	T	?	?	?	?	?
14	SLC12A3	T	?	?	?	?	?	?	?	?	?	?	?	?	?	?	?
15	ELMO1	T	?	?	?	?	?	?	?	?	?	T	?	?	?	?	?
16	CCDC33	T	?	?	?	?	?	?	?	?	?	?	?	?	?	?	?
17	ZFAND6	?	?	?	?	?	?	?	?	T	?	T	?	?	?	?	?

Fig. 3 Screenshot of formatted data used by Apriori algorithm

we were keen to find out associations among chemicals, we have proposed the use of one of the association algorithms, which is Apriori algorithm [1, 6]. In order to be able to apply this algorithm, we have refined the data. 60 genes were confirmed to be the factors for T2DM from various literature sources like SNPedia. A chemical either increases or decreases the expression of gene. Increase of the expression of a gene implies more chances of occurrence of disease. So we have changed the format of data by specifying a value of true to those drugs which decrease the expression of disease, thereby decreasing the risk of disease. We have ignored the false values and kept them unknown as we do not require it in our results. We have applied the filtered associator where Apriori algorithm was used as the associator and a filter of removeType was used for proper results. So, the druggist just needs to enter a list of genes and the application retrieves the possible associations between drugs. Figure 3 depicts the screenshot of formatted data for our application and Fig. 4 shows screenshots of application where the user is selecting genes.

We are considering only one disease, but this application can be enhanced further by adding more diseases. For successful and efficient working of such an application, it can be moved to a cloud environment where the resources will be provided on the fly.

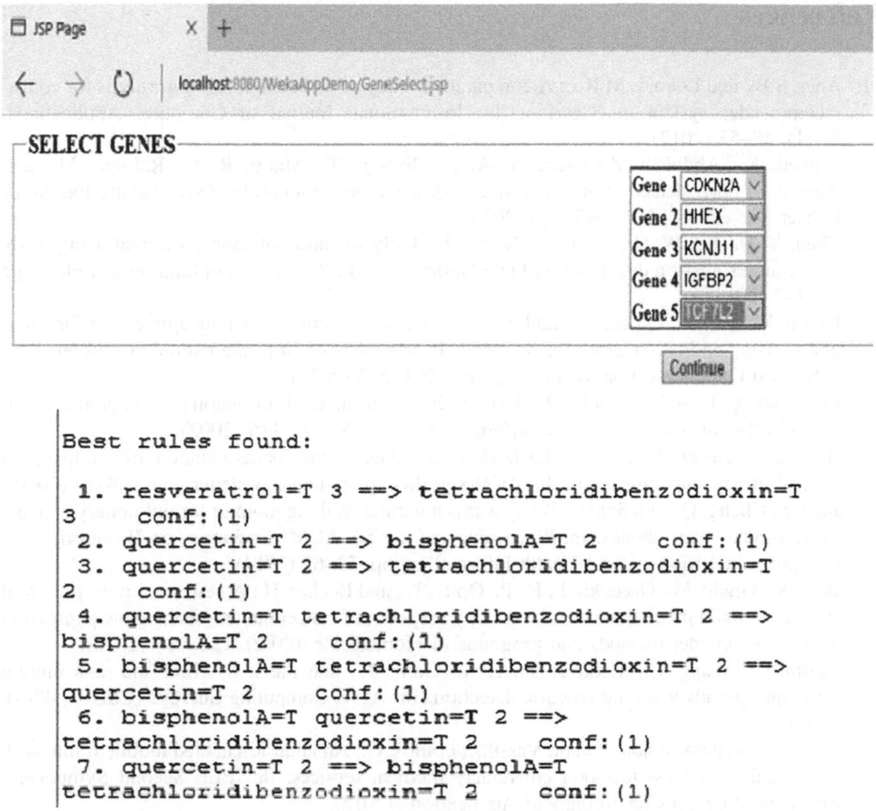

Fig. 4 Screenshots of prototype application

5 Conclusion and Future Work

In this paper, we have described some major technologies which are shaping the future of biomedical world. We have supported the fact that new ideas will be needed to eliminate prevalent issues in these technologies so as to overcome the reluctance to accept them. Finally, we have proposed a pharmacogenetic solution to predict possible drugs for an individual. This solution might act as an aid to drug repositioning or replacement as well. In future, we can enhance this application and make it more reliable and efficient by providing it a cloud platform.

We have also realized two new domains in health care which are pharmacogenetics and diet-based treatments. Our proposed work is oriented towards pharmacogenetics as it utilizes drug and genes data for diabetes patients. In coming future, diet-related data can also be used to deduce relations between our diet and diseases, leading to diet-based treatments. Changes in diet and exercise can be used to predict nutrition suitable to an individual. So, personalized medicine and nutrition are expected to be the next breakthrough technologies in coming years.

References

1. Aher, S.B., and Lobo, L.M.R.J.: A comparative study of association rule algorithms for course recommender system in e-learning. In: International Journal of Computer Applications, 39 (1), 48–52 (2012).
2. Ahmed, K., Abdullah-Al-Emran, A. A. E., Jesmin, T., Mukti, R. F., Rahman M., and Ahmed F.: Early detection of lung cancer risk using data mining. In: Asian Pacific Journal of Cancer Prevention 14(1), 595–598 (2013).
3. Chen, Y. C., Ke, W. C., and Chiu, H. W.: Risk classification of cancer survival using ANN with gene expression data from multiple laboratories. In: Computers in biology and medicine 48, 1–7 (2014).
4. Fahim, M., Fatima, I., Lee, S., and Lee, Y.K.: Daily life activity tracking application for smart homes using android smartphone. In: IEEE Proceedings of 14th International Conference on Advanced Communication Technology, pp. 241–245 (2012).
5. Freudenberg, J., and Propping, P.: A similarity-based method for genome-wide prediction of disease-relevant human genes. Bioinformatics, 18(2), S110–S115 (2002).
6. Hipp, J., Güntzer, U. and Nakhaeizadeh, G.: Algorithms for association rule mining—a general survey and comparison. In: ACM sigkdd explorations newsletter, 2(1), 58–64 (2000).
7. Juen, J., Cheng, Q., and Schatz, B.: Towards a natural walking monitor for pulmonary patients using simple smart phones. In: Proceedings of the 5th ACM Conference on Bioinformatics, Computational Biology, and Health Informatics, pp. 53–62 (2014).
8. Jung, K., Grade, M., Gaedcke, J., Jo, P., Opitz, L., and Becker, H.: A new sensitivity-preferred strategy to build prediction rules for therapy response of cancer patients using gene expression data. In: Computer methods and programs in biomedicine 100(2), 132–139 (2010).
9. Karimi, S., Wang, C., Metke-Jimenez, A., Gaire, R., and Paris, C.: Text and data mining techniques in adverse drug reaction detection. In: ACM Computing Surveys (CSUR), 47(4), 56 (2015).
10. Koufi, V., Malamateniou, F., and Vassilacopoulos, G.: An android-enabled mobile framework for ubiquitous access to cloud emergency medical services. In: IEEE Second Symposium on Network Cloud Computing and Applications (2012).
11. Kutlu M., and Agrawal, G.: Cluster-based SNP calling on large-scale genome sequencing data. In: 14th IEEE/ACM International Symposium on Cluster, Cloud and Grid Computing, pp. 455–464 (2014).
12. Ma, J., Peng, C., and Chen, Q.: Health Information Exchange for Home-Based Chronic Disease Self-Management–A Hybrid Cloud Approach. In: 5th IEEE International Conference Digital Home, pp. 246–251 (2014).
13. Meireles, A., Figueiredo, L., Lopes, L.S., and Almeida, A.: Portable decision support system for heart failure detection and medical diagnosis. In: Proceedings of the 18th ACM International Database Engineering & Applications Symposium, pp. 257–260 (2014).
14. Mukherjee, S., Dolui, K., and Datta, S.K.: Patient health management system using e-health monitoring architecture. In: IEEE International Advance Computing Conference, pp. 400–405 (2014).
15. Reddy, B.E., Kumar, T.V.S., and Ramu, G.: An efficient cloud framework for health care monitoring system. In: IEEE International Symposium on Cloud and Services Computing, pp. 113–117 (2012).
16. Shah, S.C., and Kusiak, A.: Data mining and genetic algorithm based gene/SNP selection. In: Artificial intelligence in medicine 31(3), 183–196 (2004).
17. Shannon, P., Markiel, A., Ozier, O., Baliga, N.S., and Wang, J.T.: Cytoscape: a software environment for integrated models of biomolecular interaction networks. In: Genome research, 13(11), 2498–2504 (2003).

18. Smoot, M.E., Ono, K., Ruscheinski, J., Wang, P.L., and Ideker, T.: Cytoscape 2.8: new features for data integration and network visualization. In: Bioinformatics 27(3), 431–432 (2011).
19. Venkatraman, S., Bala, H., Venkatesh, V., and Bates, J.: Six strategies for electronic medical records systems. In: Communications of the ACM 51(11), 140–144 (2008).
20. Yadav, V.P., and Kumari, M.: Name Entity Conflict Detection in Biomedical Text Data Based on Probabilistic Topic Models. In: Proceedings of the ACM International Conference on Information and Communication Technology for Competitive Strategies, pp. 61 (2014).

Approach to Reduce Operational Risks in Business Organizations

Heena Handa, Anchal Garg and Madhulika

Abstract Identifying and managing risks in business processes is of major concern for almost all organizations. Risks can be a major threat to the organization and can hamper its reputation and cause them financial loss. Organizations need ways to deal with various types of risks. This paper provides an approach to deal with operational risks. The proposed approach will reduce the time to resolve an incident. Robobank dataset has been used to determine the risks in their incident and problem management systems and then an algorithm has been proposed to minimize risks.

Keywords Risk aware business process management · Risk identification
Risk measurement · Risk optimization · Incident and problem management

1 Introduction

Nowadays, almost all organizations rely on Information Systems (IS) for execution of their business processes. But in today's dynamic environment, organizations are exposed to different type of risks like operational risks, market risks, business risks, etc. [1]. Therefore, it is imperative to find ways to handle such risks. Hence, researchers have started integrating risk management with business process development thus creating way to development of Risk Aware Business Process Management Systems [1]. Risk management includes accessing, identifying and developing strategies to overcome the risks that threaten the organization. The basic

H. Handa (✉) · A. Garg · Madhulika
Department of Computer Science & Engineering, Amity School of Engineering & Technology, Amity University, Uttar Pradesh Noida, India
e-mail: Heenhanda70@gmail.com

A. Garg
e-mail: agarg@amity.edu

Madhulika
e-mail: mbhadhauria@amity.edu

D.K. Mishra et al. (eds.), *Information and Communication Technology for Sustainable Development*, Lecture Notes in Networks and Systems 10,
https://doi.org/10.1007/978-981-10-3920-1_13

steps towards risk management include [2]: (i) identifying risk that can affect the workflow of a business; (ii) measuring the values of risks affecting the organization measured in terms of frequency of occurrence of losses; (iii) formulating actions to minimize the risks. This approach can be integrated with IS to analyze the business processes and provide a way to minimize the risks. For this purpose, we have used the dataset of Robobank, a Dutch bank having its headquarters in Utrecht, Netherlands and is a leader in food and agricultural finance. Robobank may encounter all of the following risks [3]. (i) Credit risks (ii) Transfer risk (iii) Operational risks (iv) Market risks (v) Interest rate risk (vi) Liquidity risks (vii) Insurance risks (viii) Business risks.

Incident and Problem management system of Robobank is used for identifying operational risks and provide a solution to reduce such risks in the system. In this paper, we are dealing operational risks faced by Robobank's IT System. Operational risks are risks due to failure of internal or external processes or events resulting in the loss to an organization in terms of time and quality. We try to reduce the time taken by the service agent or an executive to resolve an issue. Delayed response is an operational risk for any organization as it may hamper its reputation.

This paper is organized as follows. Section 2 covers the literature survey of the previous research done in this area of risk management. In Sect. 3 context description is done. Section 4 describes the steps of risk management. Section 5 gives cybernetics approach and Sect. 6 is conclusion.

2 Literature Review

Conforti et al. [4] used decision tree and linear programming techniques to minimize business risks. Conforti et al. [5] presented the framework for risk management including three major aspects, i.e., risk monitoring, risk prevention, and risk mitigation. Conforti et al. [6], introduced a technique where the participants take decisions regarding risks thus decreasing the frequency of risk occurrence. Conforti et al. [7] gave an approach to verify the risk in all the phases of BPM lifecycle starting from design phase to execution phase. Conforti et al. [8] used the concept of sensors to check risks at runtime; sensor manager is notified by sensor when the risk is being detected in the system and then it is the responsibility of manager to further stop the process. Jakoubi et al. [9] has done many researches in integration of BPM and risk management. The outcome of research Tjoa and Goluch comprises of modeling and simulation threats, detection, and recovery measures along with counter measures. The paper by Goluch et al. [10] provides roadmap to risk aware BPM. His research signifies the steps towards process management and risk assessment. Tamjidyamcholo and Al-Dabbagh [11], described briefly the Risk-Oriented Process Evaluation to IT security and functionality of BPM and Risk Management.

It can be inferred from the literature survey that very little work has been done in the area of risk management in context to business process management. This paper attempts to provide an approach to reduce operational risk in business organizations.

3 Context Description

This paper presents the case study on Rabobank dataset taken from 3TU.datacentrum. The major components of dataset are Interaction Management, Incident Management, and Change Management.

(i) Interaction Management: In this the Service Desk Agent (SDA) manages all calls and emails from the customers and relates them effectively to affected Configuration Item (CI). SDA either directly resolves the issue or creates an incident record for the same. The incident record is then analyzed by group having more technical knowledge to resolve the customer issues. If similar problem is faced by some other customer, then SDA considers it as multiple interaction and one incident record. Figure 1 shows the events that happen when a customer calls SDA [14].

(ii) Incident Management: The incident record created by SDA is resolved based on Impact and Urgency of the incident. Team leader assigns the record to appropriate Operator, who will resolve the issue or reassign the record to some other colleague. Once the issue is resolved, the operator will record the CI due to which it was caused, i.e., CausedBy CI and customer will be informed about resolved issue. Figure 2 represents the flowchart of incident management [14].

(iii) Change Management: If the same issue repeatedly occurs more often than usual then the issue investigation starts. This leads to discovery of main cause of problem reoccurrence, find ways to resolve the problem, implement solution and thus leading to improvement of services. Before improving the plan, a request for change is send on CausedBy CI. Implementation Manager also conducts Risk Impact Analysis to change specific service component [14].

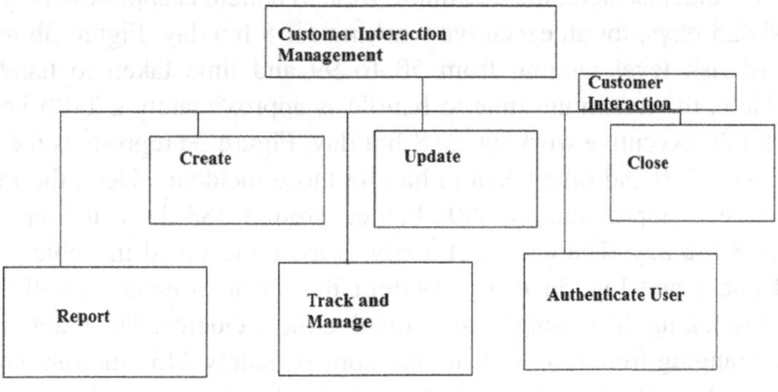

Fig. 1 Interaction management events [12]

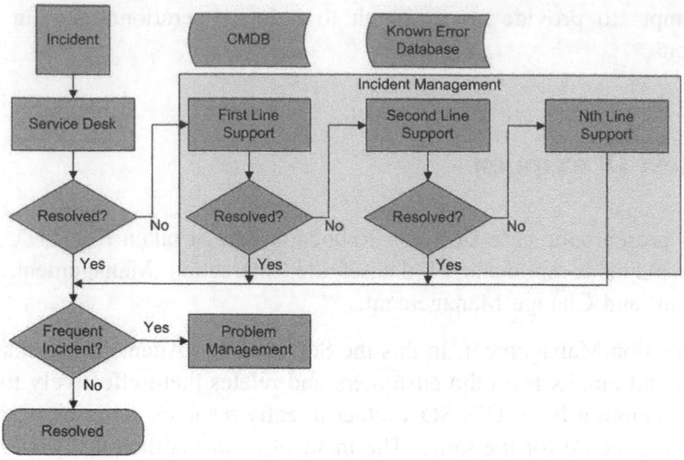

Fig. 2 Incident management flowchart [13]

4 Steps to Risk Management

(a) Risk Identification and Measurement

This paper deals with operational risks confronting in Rabobank data. Risk of each incident is calculated by the formula [15]

$$\text{Risk} = \text{Impact of incident} * \text{Frequency of occurrence of incident} \quad (1)$$

We have divided risks into the following three levels: (i) Low-Level Risk that ranges from 0 to 49, (ii) Medium-Level Risk that ranges from 50 to 99, (iii) High-Level Risk, i.e., above 100.

Figure 3a–c show the comparison between Risk level of incidents and handle time taken to resolve these incidents.

Figure 3a represents the graph of risk level ranging from 0 to 49 and time to handle those incidents. Here, the maximum time to handle is approximately 3400 h, i.e., around 425 days, for an executive working for 8 h a day. Figure 3b represents the graph of risk level ranging from 50 to 99 and time taken to handle those incidents. Here, the maximum time to handle is approximately 2700 h i.e. around 338 days, for an executive working for 8 h a day. Figure 3c represents the graph of risk level above 100 and time taken to handle those incidents. Here, the maximum time to handle is approximately 3900 h, i.e., around 488 days, for an executive working for 8 h a day. The results of analysis are represented in Table 1.

From Table 1 and Fig. 3a–c it is evident that even to handle small risks; the executives are taking fairly long time to resolve the incidents. The main concern is for the risks ranging from 0 to 49 that take approximately 445 working days for an executive to solve. The reasons for this might be that incidents with low risks are

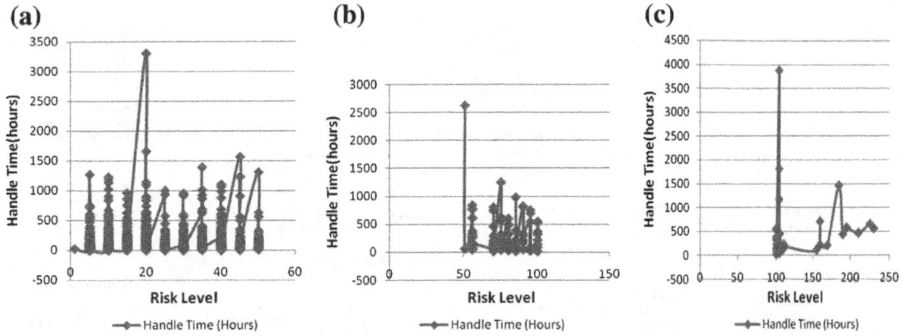

Fig. 3 Comparison between risk level and handle time (h). **a** Comparison between risk level (0–49) and handle time (h). **b** Comparison between risk level (50–99) and handle time (h). **c** Comparison between risk level above 100 and handle time (h)

Table 1 Comparison of risk level and handle time

	Handle time (in h)	Handle time (in day)
Low level risks	3400	425 approx.
Medium level risks	2700	338 approx.
High level risks	3900	488 approx.

either taken for granted and not readily resolved or due to wrong assignment of executive to handle the incidents. Both of these situations may lead to: (i) increase in number of Backlog incidents, (ii) reopening of activities due to inefficient mistake done by the service team, (iii) reassignment of activities that are not correctly assigned by the service desk agent.

Thus, it can be concluded that some way is needed to overcome such operational risks faced by Rabobank. Incidents decrease the efficiency of the employees and the bank as most of their time is spent in resolving the issues further causing customers dissatisfaction. To resolve this issue, an algorithm can be used to calculate the operational risk level and then minimizing the risk level to an adequate limit. This algorithm should also allocate the agent stepwise with least time utilization.

(b) Risk Optimization

An algorithm is proposed to reduce the risks and resolve the issues in the minimum time. All the incidents recorded by SDA are solved with minimum possible operational risk and least time in stepwise manner. Figure 4 represents the flowchart of the algorithm for risk optimization and assigning appropriate executive for each incident. Terminology associated with the algorithm is as follows:

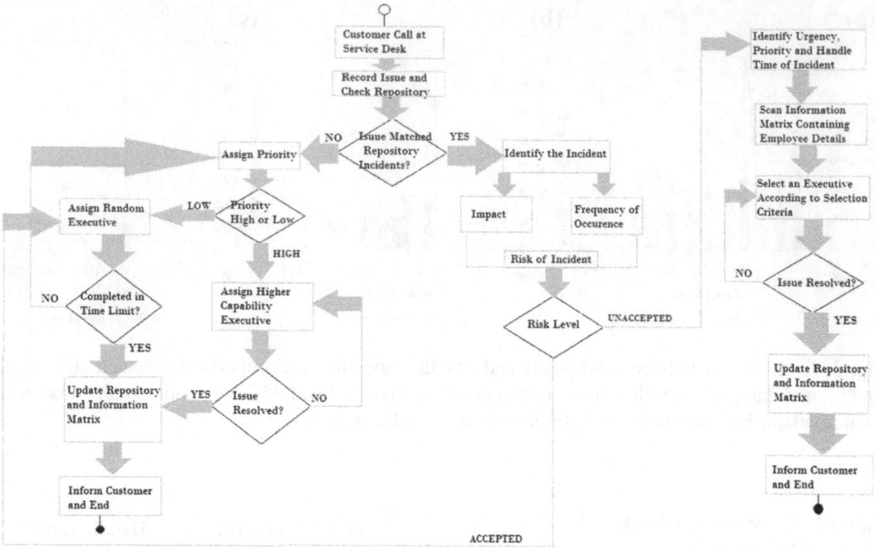

Fig. 4 Algorithm for risk assessment

(i) Repository: The initial dataset of Rabobank containing information regarding incidents their priority, urgency, impact, frequency of occurrence and handle time.

(ii) Information Matrix: It is a matrix containing information of each executive with their capability and area of interest. Each executive is selected to handle the incident based on the capability and area. Process mining technique can be used to mine information regarding capabilities of the executives. Process mining is a technique which allows the analysis of business processes on the basis of event logs stored in the IS of the organizations. This technique enables us to get an idea of executives handling different risk level.

(iii) Selection Criteria: Executive is selected after scanning the information matrix. For each incident there are at least ten executives but the selection out of ten is done on the basis of capability. If the executive with highest capability is busy then the executive with second best capability is chosen and so forth. There is no random assignment. This selection procedure can reduce time to solve the incident.

(iv) Risk Level: Risk level of the incident represents the maximum risk value which will be safe for the system to handle. Risk level can vary for each organization. For this system we have considered risk level of each incident to be 5.

(c) Steps of Algorithm

1. Customer calls at service desk.
2. SDA answers the customer call and records the issue.
3. SDA check the repository if it is an already existing incident or not.

   If it matches the repository record, go to step 4 Else, go to step 13.

4. Identify the Impact and Frequency of the incident from repository data.
5. Calculate the risk of the incident by multiplying the values of impact and frequency of occurrence, i.e., Risk of Incident = Impact * Frequency.
6. Compare the values of Risk of Incident and risk level (i.e., 5).

   If Risk of incident > Risk level then move to step 7 Else move to step 15.

7. Identify the values of urgency, priority, and handle time of incident from repository.
8. Scan information matrix containing employee details regarding capability and area of interest.
9. Select the best employee to handle the incident according to selection criteria.
10. If incident is resolved then move to step 11, Else move to step 9.
11. Update the repository with new incident and also the information matrix.
12. Inform the customer that the issue is resolved and end the process.
13. Now, after step 3 if the incident does not match the repository then SDA assign priority to the incident.
14. The priority more than 3 it is considered as high priority.
15. If priority < 3 move to step 15, Else go to step 18.
16. Assign random executive to solve the incident.
17. If the incident is completed in time limit go to step 17 Else go to step 13.
18. Repeat step 11 and 12.
19. When the priority > 3 then assign an executive with higher capability to solve the incident.

    If incident is solved, go to step 17 Else move to step 18.

5 Cybernetic Approach to Improve Selection Criteria

Cybernetics is an automatic control system in both machines and living things. The organization can fix the maximum time to resolve an issue based on risk. If the issue is not resolved between that stipulated time, information matrix will be improved to get better capability executives to resolve it. This process will continue till the organization is able to resolve all the issues within the specified time [16]. Figure 5 represents the collaboration of cybernetics with risk assessment to improve selection criteria.

Fig. 5 Cybernetics flow

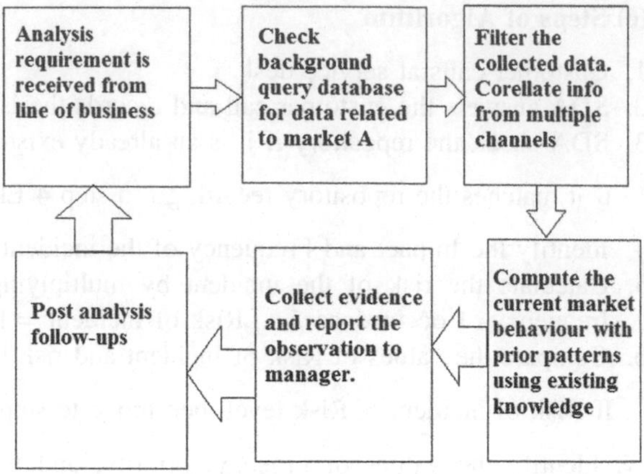

6 Conclusion

Any risk can prove harmful for the organization. Hence, it becomes necessary that preventive measures are taken to overcome these risks. The approach provided in this paper will help organizations to reduce the operational risks by quick allocation of the incident to the most capable executive. The executive capabilities are not based on perceptions but on actual information extracted through mining of event logs using process mining tools. A cybernetic approach may be used to improve the system over time. The algorithm proposed above is a solution to operational risks and can be implemented in future to provide faster service to the priority customers.

References

1. Haggag, M. H., Montasser, H. S., Khedr A. E.: A Risk-Aware Business Process Management Reference Model and Its Application in an Egyptian University. International J. of Computer Science & Engineering Survey. 6(2), 11 (2015).
2. Shrivastva, S., Garg, A.: Data Mining For Credit Card Risk Analysis: A Review, Credit Card Risk Analysis. International J. of Computer Science Engineering and Information Technology Research. 3(2), 193–200 (2013).
3. Risk Management, https://www.rabobank.com/en/investors/risk-management/index.html.
4. Conforti, R., de Leoni, M., La Rosa, M., van der Aalst, W. M., ter Hofstede, A. H.: A recommendation system for predicting risks across multiple business process instances. Decision Support System. International J. of Computer Science & Engineering Survey. 69, 1–19 (2015).
5. Conforti, R., La Rosa, M., Fortino, G., Ter Hofstede, A. H., Recker, J., Adams, M.: Realtime risk monitoring in business processes: A sensor-based approach. J. of Systems and Software. 86(11), 2939–2965 (2013).
6. Conforti, R., De Leoni, M., La Rosa, M., van der Aalst, W. M.: Supporting risk informed decisions during business process execution. Advanced Information Systems Engineering. 116–132 (2013).

7. Conforti, R., La Rosa, M., ter Hofstede, A. H., Fortino, G., de Leoni, M., van der Aalst, W. M., Adams, M. J.: A software framework for risk-aware business process management. In: Proceedings of 25th International Conference on Advanced Information Systems Engineering, pp. 130–137. CEUR Workshop Proceedings, SITE Central Europe (2013).
8. Conforti, R., Fortino, G., La Rosa, M., ter Hofstede, A.H.M.: History-Aware, Real-Time Risk Detection in Business Processes. In: On the Move to Meaningful Internet Systems, pp. 100–118, Springer Berlin Heidelberg (2011).
9. Jakoubi, S., Neubauer, T., Tjoa, S.: A Roadmap to Risk-aware Business Process Management. In: Fourth IEEE Asia-Pacific Services Computing Conference, pp 23–27. IEEE Computer Society, Los Alamitos, CA (2009).
10. Goluch, G., Ekelhart, A., Fenz, S., Jakoubi, S., Tjoa, S. and Mück, T.: Integration of an Ontological Information Security Concept in Risk-Aware Business Process Management. In: 41st International Conference on System Sciences, pp. 7–10. IEEE, Hawaii (2008).
11. Tamjidyamcholo, A., Al-Dabbagh, R. D.: Genetic Algorithm Approach for Risk Reduction of Information Security. International J. of Cyber-Security and Digital Forensics. 1(1), 59–66 (2012).
12. Customer Interaction Management, https://www.tmforum.org/Browsable_HTML_Frameworx_R15.5/main/diagramb9129ce904851 1e3ac5202b09a168801.htm.
13. Thoughts on Cloud Computing and ITIL, https://computingnebula.wordpress.com/category/itil/incident-management/.
14. Thaler, T., Knoch, S., Krivograd, N., Fettke, P., Loos, P.: ITIL Process and Impact Analysis at Rabobank ICT. BPI Challenge (2014).
15. Analyse and Evaluate the Impact of Risks, https://www.business.qld.gov.au/business/running/risk-management/risk-management-plan-business-impact-analysis/analyse-evaluate-impact-risks.
16. Mishra, A. Garg, A., Dhir, S.: A Cybernetic Approach to Control Employee Attitude for Implementation of Green Organization. In: World Congress on Sustainable Technologies, pp. 49–53. IEEE, London (2012).

Web-Based Condition and Fault Monitoring Scheme for Remote PV Power Generation Station

Shamkumar Chavan and Mahesh Chavan

Abstract Photovoltaic power generation stations are located at remote places and are unmanned. Electronic power processing circuitry plays vital role in conversion of electrical energy generated by PV module. The components of such circuitry work under stress and are prone to failure due to numerous reasons. Considering condition monitoring and fault monitoring issues, this communication proposes web-based condition and fault monitoring scheme for PV power generation systems. Scheme proposed here monitors present condition and nature of fault occurring in power generation station. Use of CAN-based controller is suggested to transmit data to master controller which is interfaced with dedicated server. Master station communicates recent information to dedicated server, which sends web page to remote node on request arrival. Nature and type of fault in any equipment or device can be displayed on web page. Further the scheme can be used to study the daily performance of PV power generation station.

Keywords PV station fault monitoring • ICT in renewable energy Web-based fault monitoring • Remote condition monitoring • Remote PV system supervision

1 Introduction

Photovoltaic power generation stations of large capacity are located at remote places and are unmanned. Literature published in recent years on power electronics systems has discussed reliability problems in power processing circuits. Many

S. Chavan (✉)
Department of Technology, Shivaji University, Kolhapur, India
e-mail: sbc_tech@unishivaji.ac.in

M. Chavan
KIT College of Engineering, Kolhapur, India
e-mail: maheshpiyu@gmail.com

D.K. Mishra et al. (eds.), *Information and Communication Technology for Sustainable Development*, Lecture Notes in Networks and Systems 10,
https://doi.org/10.1007/978-981-10-3920-1_14

researchers are working on development of fault diagnosing schemes for such circuits.

Problems and challenges in PV power generation systems are discussed [1] authors say that PV power processing circuits should have lower failure rates and higher efficiency. Environmental conditions, transients, etc., affect the performance of components in PV systems. An industrial survey [2] for reliability assessment of power electronic systems reported that power switches have higher failure rates, capacitors, gate drivers, inductors, resistors; connectors are also prone to fail. H. Wang et al. depicted need of conditioning monitoring circuits [3] because degradation of one component affects performance of others and of system. Physics of Failure (PoF) based approach, opportunities, and challenges in converter reliability improvement are discussed [4]. Reliability oriented assessment of PV system is performed [5] based on Markov model, authors states that topological change, capacitor voltages, frequency, power losses should be considered for reliability improvement. Model is developed for reliability assessment of PV systems [6], authors have classified faults in four categories. Methods for fault detection in PV systems are suggested in [7]. MATLAB/Simulink model for fault detection in PV module is presented [8]. Model-based approach for fault diagnosis in full bridge DC–DC converter of a PV system is discussed [9]. Reliability oriented assessment of interleaved boost converter is presented [10]. Due to failure prone nature of power processing circuits researchers have developed fault diagnostic, tolerant schemes for converters, inverters, etc., Open circuit fault diagnosis and tolerant scheme for three-level boost converters in PV system is presented [11], another fault tolerant scheme for interleaved DC–DC converter in PV application is discussed [12].

Multilevel H-bridge converter topology for fault tolerance application is presented in [13]. Open circuit power switch fault tolerant scheme for phase shift full bridge DC–DC converter is presented in references [14, 15]. For boost converter in PV systems open and short circuit power switch fault tolerant scheme is discussed in [16]. Literature referred above demonstrates that reliability of power electronics systems is important issue and nowadays researchers are working on development of fault diagnosis and tolerant methods.

GSM and voice channel based technique for data communication in PV systems condition monitoring application is presented [17]. Zig Bee based monitoring system for PV application is developed [18] in which condition of PV modules, converters, and batteries are monitored. In this application good or bad status of PV modules and converters can be detected. Central coordinator circuitry collects all information and displays on board. LabVIEW-based system for monitoring condition of PV power plant is developed [19] in which all information can be seen through the GUI developed in LabVIEW. Control and supervising system based on Internet for photovoltaic application is developed in [20]. Use of embedded web server is discussed [21] for monitoring environmental parameters remotely.

Considering varying output nature of PV module, reliability, and component failure issues in power processing circuits, web-based scheme is proposed here for condition and fault monitoring of PV power generation station. Since PV power stations are located remotely and acquire huge land area, the system proposed here will provide information to remote places via web pages. The data like nature and type of fault, malfunction, efficiency degradation; exact fault location, etc., occurring in PV modules, converters, and inverters can be displayed on web pages.

2 Proposed Web-Enabled PV Power Generation Station

Figure 1 shows typical string based distributed architecture of PV power generation system. Multiple PV modules are connected in series or parallel to obtain desired PV array voltage and current. DC–DC, DC–AC converters are its integral components.

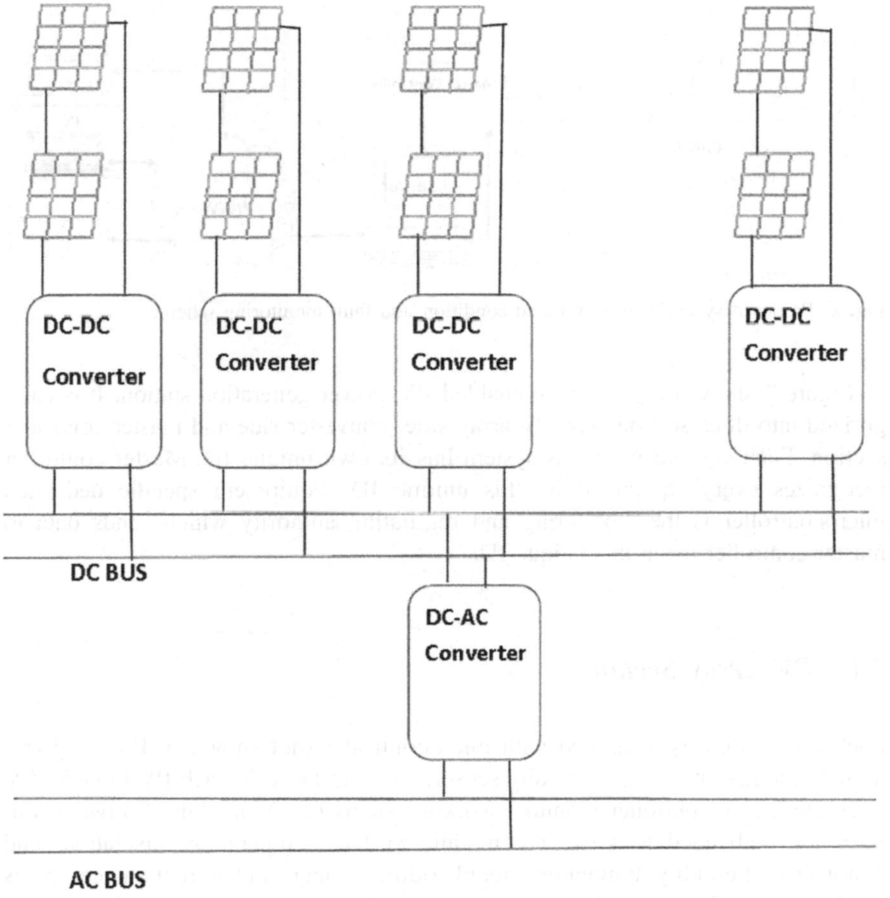

Fig. 1 Typical architecture of PV power generation station

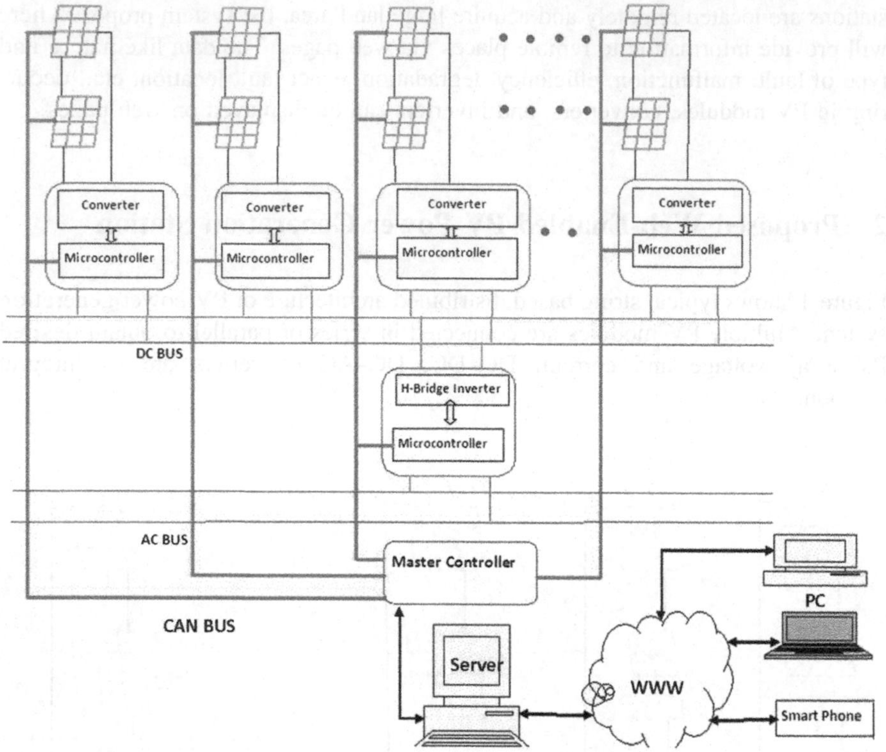

Fig. 2 Proposed system with web based condition and fault monitoring scheme

Figure 2 shows proposed web enabled PV power generation station. It is categorized into three sections viz. PV array side, converter side and master controller section. Each equipment in this system has its own unique ID. Master controller recognizes every equipment by this unique ID. Equipment specific dedicated microcontroller is the monitoring and regulating authority which sends data to master controller using this unique ID.

2.1 PV Array Section

Each PV module is integrated with microcontroller (not shown in Fig. 2). Pyranometer, temperature, and humidity sensors are interfaced to each PV module. PV integrated microcontroller monitors working status of PV module. Software routines are embedded to sense PV module surface temperature, insolation, and atmospheric humidity. It monitors module output voltage and current and compares it with known database. This ensures whether PV modules are providing desired output voltage and current at given environmental conditions.

Microcontroller also monitors fault signatures for faults like shading, module OC/SC, grounding, line to line faults, etc. It have on chip CAN controller facility through which it communicates all details in working as well as in faulty condition to master controller. Flow chart in Fig. 3 shows operation to be carried out by PV module integrated microcontroller.

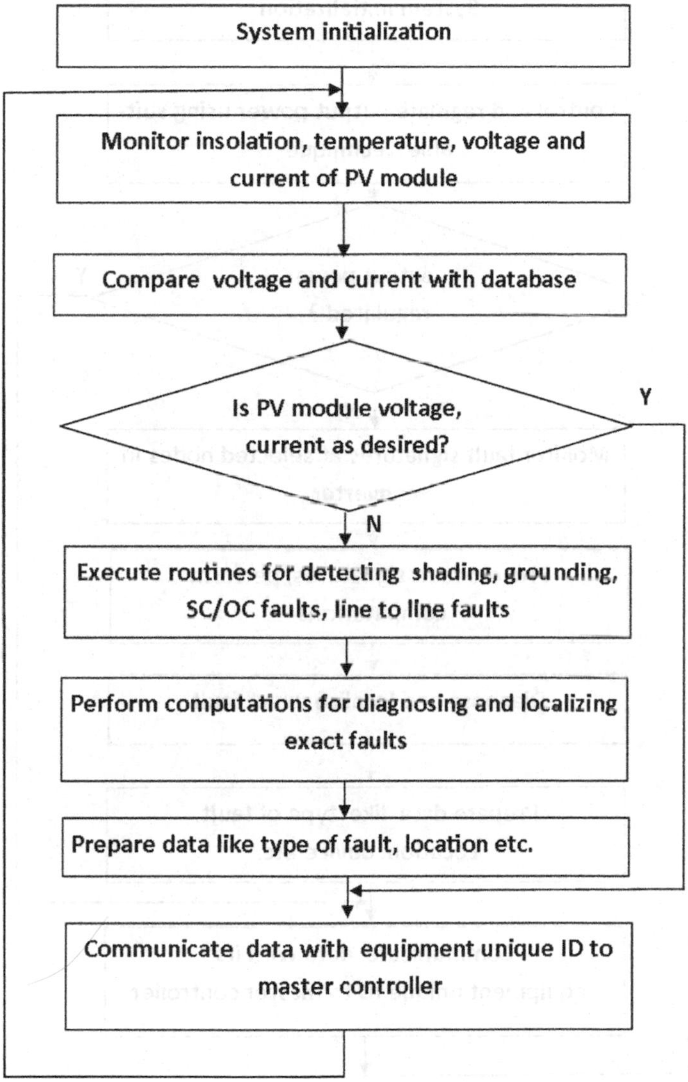

Fig. 3 PV module condition and fault monitoring routine

2.2 Converter Section

There are several categorizes of DC–DC converters. Depending upon power requirements converter types are selected. Microcontroller is regulating and fault diagnosing authority for the converter. It regulates converter output by using suitable scheme like PWM, MPPT, etc. It is equipped with software routines for

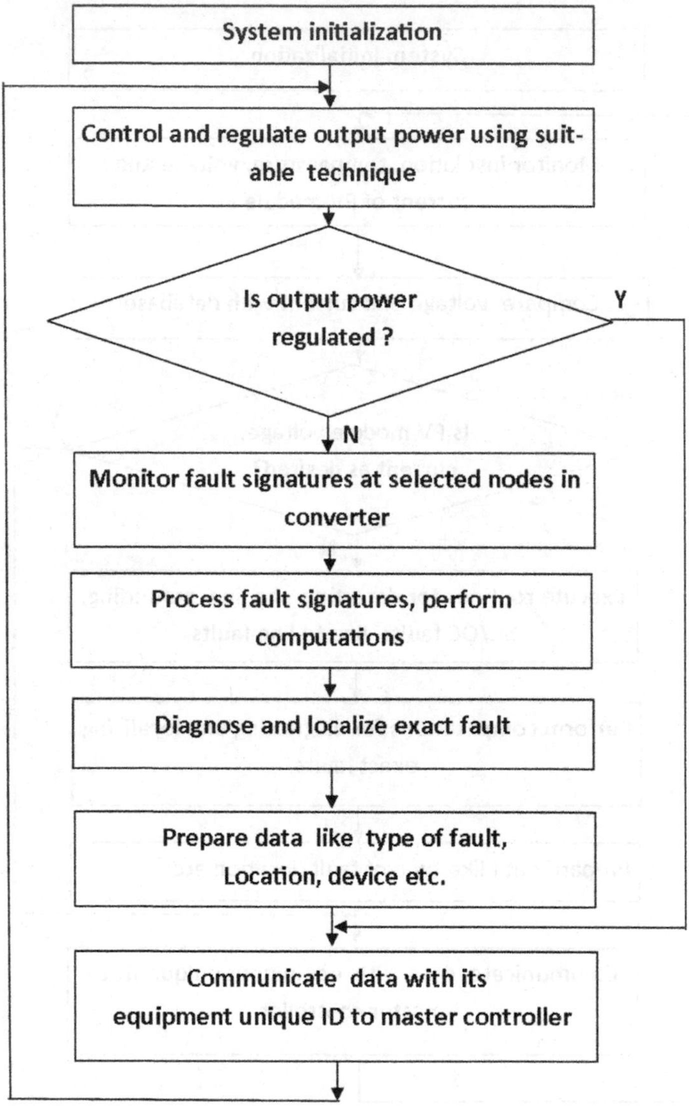

Fig. 4 Converter condition and fault monitoring routine

processing fault signatures and localizing the faults. The controller has on chip CAN controller through which it communicates all details to master controller.

The microcontroller in DC–DC and DC–AC converters perform same type of operation for condition and fault monitoring as shown in flow chart in Fig. 4.

2.3 Master Controller and Server Section

Master controller collects data from all microcontrollers. It is equipped with RTC facility, whenever data is received from microcontrollers; it maintains record of day, date, time, etc. It categorizes information according to their unique ID. Further it organizes the information according to their working status like normal, faulty; other details like type and nature of fault, its location, etc. This information is sent to dedicated server periodically which displays this information in tabular form on web page. Server regularly receives information from master controller and updates the web page. Whenever request arrives from remote node, server is supposed to provide web page after user authentication. User on distant location accesses web page to know the status of PV power generation station. Server supports protocols required for communication with remote nodes.

3 Discussions

Microcontrollers with on chip ADC, PWM modules, and CAN controller facility are good choice for this application. Ethernet controller based embedded web servers have memory limitations, support static web pages and cannot handle multiple requests; therefore are suitable for small applications like domestic standalone PV power generation system. Dedicated server is able to handle large number of requests, large data, dynamic web pages, and have larger bandwidth. Due to these merits dedicated server is preferred for the application proposed in this work.

The merits of proposed system are

(1) Facilitates system performance study remotely

Condition of PV modules, converters can be monitored regularly. Effect of change in environmental parameters on module performance can be studied daily. Performance degradation due to environmental factors, component aging, etc., can be studied. Power generation capacity in case of cloudy/rainfall conditions can be predicted in advance.

(2) Ease in fault localization

Type of fault occurred can be identified, exact fault location can be detected. Repair or replacement decision can be taken in advance. Time required for troubleshooting and fault diagnosing can be minimized.

(3) Preparation of database

Database of system can be maintained for fault oriented research.

System demerits are

(1) Need of additional hardware

Sensing and signal conditioning circuitry is required for sensing fault signatures. Fault diagnosing routines are required to be embedded into controllers, further they need CAN transceiver built on chip for data communication. Server installation and maintenance cost will also be added.

4 Conclusions

Web-enabled condition and fault monitoring system for remote located PV power generation station is proposed in this communication. Using Internet facility such type of power generation stations can be monitored remotely and decision regarding maintenance; component/equipment replacement can be taken well in advance. Further damage or malfunction of equipment due to aging or any other reason can be known earlier. Continuous performance of entire system can be examined regularly.

Acknowledgements Authors are thankful to Department of Technology, Shivaji University, Kolhapur India for providing necessary facilities for completion of this communication.

References

1. Giovanni P., Giovanni S., Remus T., Mummadi V., Massimo V.: Reliability issues is photovoltaic power processing systems, IEEE transactions on Industrial Electronics, 55, 2569–2580 (2008).
2. Shayong Y., Angus B., Philip M., Dawei X., Li R., Peter T.: An industry based survey of reliability in power electronic converters, IEEE transactions on Industry Applications, 47,1441–1451 (2011).
3. Wang H., Ma K., Blaabjerg F., “Design for reliability of power electronic systems”, in Proc. 38th Annual conference of the IEEE Industrial Electronics Society 2012, pp. 33–44.
4. Wang H., Blaabjerg F., Ma K., Wu R.: Design for reliability in power electronics in renewable energy systems – status and future, 4th International Conference on Power Engineering, Energy and Electrical Drives 1846–1851 (2013).
5. Dhopale S.V., Davoudi A., Dominguez A.D., Chapman P.L.: A unified approach to reliability assessment of multiphase DC-DC converters in photovoltaic energy conversion systems, IEEE transactions on power electronics, 27, 739–751 (2012).
6. Firth S.K., Lomas K.J., Rees S.J.,: A simple model of PV system performance and its use in fault detection, Solar Energy, 84,624–635 (2010).
7. Zhao Y., Lehman B., Ball R., Mosesian J., Palma J.: Outlier detection rules for fault detection in solar photovoltaic arrays, 28th IEEE applied Power electronics conference and exposition 2913–2920 (2013).
8. Houssein A., Heraud N., Souleiman I., Pellet G.: Monitoring and fault diagnosis of photovoltaic panels, IEEE International Energy Conference and Exhibition 389–394 (2010).

9. Chavan S.B., Chavan M.S.: A model based approach for fault diagnosis in converter of photovoltaic systems, IEEE global conference on Wireless Computing and Networking, (2014) 112–115.
10. Callega H., Chan F., Uribe I.: Reliability oriented assessment of a DC/DC converter for photovoltaic applications, IEEE power electronics specialists conference, 1522–1527 (2007).
11. Ribeiro E., Cardoso A.J.M., Boccaletti C.: Fault tolerant strategy for a photovoltaic DC-DC converter, IEEE transactions on power electronics, 28, 3008–3018(2013).
12. Ribeiro E., Cardoso A.J.M., Boccaletti C.: Open circuit fault diagnosis in interleaved DC-DC converters, IEEE transactions on power electronics, 29, 3091–3102 (2014).
13. Ambusaidi K., Pickert V., Zahawi B.: New circuit topology for fault tolerant H-bridge DC-DC converter, IEEE transactions on power electronics, 25,1509–1516 (2010).
14. Pei X., Nie S., Chen Y., Kang Y.,: Open circuit fault diagnosis and fault tolerant strategies for full bridge DC-DC converters, IEEE transactions on Power Electronics, 27, 2550–2565 (2012).
15. Pei X., Nie S., Kang Y.: Switch short circuit fault diagnosis and remedial strategy for full bridge DC-DC converters, IEEE transactions on power electronics, 30, 996–1004 (2015).
16. Jamshidpour E., Poure P., Saadate S.: Photovoltaic systems reliability improvement by real time FPGA based switch failure diagnosis and fault tolerant DC-DC converter, IEEE Transactions on Industrial Electronics, 62,7247–7255 (2015).
17. Tejwani R., Kumar G., Solanki C.: Remote monitoring for solar photovoltaic systems in rural application using GSM voice channel, 2013 ISES Solar world congress, Energy Procedia, 57,1526–1535 (2014).
18. V. Katsioulis, E. Karapidakis, M. Hadjinicolaou, A. Tsikalakis: Wireless monitoring and remote control of PV systems based on the ZigBee protocol, DoCEIS (2011) 297–304.
19. Zahran M., Atia Y., Al-Hussain A., El-Sayed I.: LabVIEW based monitoring system applied for PV power station, 12th WSEAS International Conference on automatic control, modelling & simulation (2010) 65–70.
20. Meliones A., Apostolacos S., Nouvaki A.: A web based three tier control and monitoring application for integrated facility management of photovoltaic systems, Applied computing and informatics, 20,14–37(2014).
21. Chavan S.B., Kadam P.A., Sawant S.R.: Embedded web server for monitoring environmental parameters, Instruments and experimental techniques, 52,784–786(2009).

Handling User Cold Start Problem in Recommender Systems Using Fuzzy Clustering

Sugandha Gupta and Shivani Goel

Abstract Recommender engines have become extremely important in recent years because the count of people using Internet for diverse purposes is growing at an overwhelming speed. Different websites work on recommender systems using different techniques like content-based filtering, collaborative filtering, or hybrid filtering. Recommender engines face various challenges like scalability problem, cold start problem and sparsity issues. Cold start problem arises when there is no sufficient information for the user who has recently logon into the system and no proper recommendations can be made. This paper proposes a novel approach which applies fuzzy c-means clustering technique to address user cold start problem. Also, a comparison is made between fuzzy c-means clustering and the traditional k-means clustering method based on different set of users and thus it has been proved that the accuracy of fuzzy c-means approach is better than k-means for larger size of dataset.

Keywords Collaborative filtering · Cold start · Recommender system
Fuzzy clustering

1 Introduction

Recommender engines fall under the sub-category of information filtering systems that aim to forecast preferences or ratings given to the item by the user. With the developing technology, the influence of technology on everyone's life is increasing.

S. Gupta (✉) · S. Goel
Department of Computer Science and Engineering, Thapar University,
Patiala, India
e-mail: sugandhagupta_92@yahoo.in

S. Goel
e-mail: shivani@thapar.edu

D.K. Mishra et al. (eds.), *Information and Communication Technology for Sustainable Development*, Lecture Notes in Networks and Systems 10,
https://doi.org/10.1007/978-981-10-3920-1_15

Therefore, recommender engines are now-a-days an integral portion of E-commerce sites which helps in recommending items or products of interest to people all around the world. The major assistances of having a recommender systems are customer retention, information retrieval, personalization, and many more. Also recommender systems can be used on products such as music, books, restaurant, TV shows, and movies and presently are used in commercial websites successfully such as Movielens, Amazon, MovieFinder, ebay, Linkedln, Jinni, Facebook, and Myspace.

Basically, recommender systems compare the profile of a user to some basic characteristics and try to predict ratings given by a user to an item they had not yet well thought-out.

Recommender systems are categorized into the following three basic categories based on the way recommendations are generated:

- Collaborative Filtering Recommender Systems (also known as social filtering): The information is filtered by using the recommendations from different people. It works on the notion that people who agree with the evaluation of certain items or similar tastes or preferences in the past would agree in the future too (Linkedln[1]).
- Content-based Recommender Systems (also known to as cognitive filtering): It recommends items similar to those items which the user liked previously. Each item's content is symbolized by set of terms, usually the words that appear in a document. These terms represent the profiles of users, which are made after analyzing the contents of items seen by the user (Jinni[2]).
- Hybrid Recommender Systems: These use integration of two techniques, i.e., content-based filtering along with collaborative filtering which could be more effective in some cases.

Cold start problem, also known as new user problem or new item problem, is a special type of sparsity problem when a user or item has no ratings. Due to the lack of purchase history or rating information of a new user or item, it becomes a challenging task for recommender engines to generate appropriate recommendations for the new users and new items. Cold start problem is further divided into two categories

(1) New User cold start problem: It occurs when there are no ratings for the user who has arrived into the system.

[1]https://in.linkedin.com.

[2]http://www.jinni.com.

(2) New item cold start problem: It occurs when an item has just been added to the system and has not been rated as of yet.

The paper is divided into the following sections: Sect. 2 gives the details of previous work done on cold start problem. Section 3 gives the description of experiments applied to the user dataset and Sect. 4 gives the experimental results and their analysis. Section 5 concludes the paper.

2 Related Work

In the section described below, we review few prior studies linked to our suggested approach. Many researchers have done a plenty of work in the field of recommender engines. In this work, Adomavicius and Tuzhilin have given a view of collaborative, content, and hybrid recommender systems along with limitations of current recommendation techniques and their possible extensions [1]. The general concepts of collaborative filtering, their drawbacks and a clustering approach for huge datasets has been proposed by Sarwar et al. [2]. Sanchez et al. analyzed Pearson correlation metric and cosine metric together with the less common mean-squared difference in order to discover their advantages and disadvantages [3]. Also, Schein et al. have given a new evaluation metric known as the CROC curve and explained numerous components of testing strategies empirically in order to obtain better performance of recommender systems [4]. A comparative analysis of three different clustering methods, c-means clustering, k-means clustering, and SOM is proposed by Budayan et al. [5]. The work done by Ling Yanxiang et al. explains how to address cold start problem using character capture and clustering method [6]. Probabilistic neural network method to deal with cold start issues in the traditional collaborative filtering recommender engines is explained by Devi et al. [7]. Shaw et al. has given a technique which uses association rules to get rid of cold start problems in recommender systems [8]. Gupta and Patil proposed an efficient technique for recommender systems based on Chameleon Hierarchical clustering algorithm [9]. Ontology-based method to address cold start issue is proposed by Middleton et al. [10]. Son et al. proposed an application of fuzzy geographically clustering technique for handling the problem of cold start in recommender engines [11]. A novel hybrid approach to handle new item cold start issue in collaborative filtering uses both ratings and content information, as given by Sun et al. [12]. A new hybrid approach using ordered weighted averaging operator of exponential kind is used to get rid of cold start problem of recommender systems given by Basiri et al. [13]. A hybrid system for enhancing correlation using association rules and perceptron learning neural network to handle cold start issue in recommender systems was kept forward by Dang et al. [14].

3 Proposed Work

In this section, a sketch of our planned technique for addressing the user cold start problem is explained.

Our approach is to combine the two techniques in serial order one after the other. First we apply fuzzy clustering technique on the different attributes of user's demographic data which is entered by the user at the time of login into the system. Using the clustered data which we will get after applying fuzzy c-means clustering algorithm, we will apply Mysql using phpmyadmin for generating recommendations, aiming at new user recommendations. These clustering techniques help a new user to find a similar neighborhood. Thus recommendations generated are on the basis of the cluster the he is placed in. Top-N recommendations can be generated by querying in MySQL using aggregate analysis, i.e., by considering highest rating frequency or highest average rating for a movie. This would help to improve the quality of recommendations for a new user who enters the system.

The proposed system architecture is given in Fig. 1.

3.1 Clustering

Collaborative filtering recommender systems give best results when the user-item matrix is extensive and the dataset has high matching information according to the new user. The work done is based upon exploiting user's demographic data for finding similarity between the already existing user and the new user. Demographic information includes different user features like gender, occupation, religion, age, zip-code, race, locality, hobbies, marital-status, and many more. As compared to the existing k-means technique, fuzzy c-means clustering approach works by giving a membership value to every piece of data or data point equivalent to every cluster center based on the distance that lies between the cluster center and the data point. This technique allows one data point to be a part to two clusters or more. Therefore, if the data is nearer to cluster center, its association with specific cluster center is more.

As shown in the algorithm there are two important parameters a_{ij} and b_j. a_{ij} that represent association between *i*th data point and the *j*th cluster center and b_j represents the *j*th cluster center.

Algorithm:

Require: User demographic data D=$\{d_1, d_2, d_3, \ldots\ldots, d_n\}$, Set of cluster center C=$\{c_1, c_2, c_3, \ldots\ldots\ldots, c_m\}$. Assume fuzziness index 'f' $(1 \leq f \leq \infty)$ and Euclidean distance between the ith data point and the jth cluster center $\|d_i - b_j\|^2$

Ensure: New user groups: User Cluster

1. Arbitrarily select 'm' cluster centers.
2. Calculate fuzzy association between the data points and the cluster centers $'a'_{ij}$ using:

$$a_{ij} = 1 / \sum_{k=1}^{m} (e_{ij}/e_{ik})^{(2/f-1)}$$

3. Compute fuzzy centers of the clusters $'b'_j$ using:

$$b_j = (\sum_{i=1}^{n} (a_{ij})^f d_i) / (\sum_{i=1}^{n} (a_{ij})^f)$$

4. Repeat steps [2] and [3] till the objective function value O is minimized :

$$O(A, B) = \sum_{i=1}^{n} \sum_{j=1}^{m} (a_{ij})^f \|d_i - b_j\|^2$$

3.2 Recommendations

In Sect. 3.1, users are clustered into different groups or clusters by applying fuzzy c-means clustering approach based on their features (i.e., gender and age) which they will enter at the time of login into the system. Thus, we obtain the cluster's recommendation to which the user belongs to by using the preprocessed data which will be stored in Mysql database for generating the recommendations. Therefore, the user gets the appropriate recommendations as we assume that the users with identical tastes shares same clusters and therefore tend to rate in the similar fashion.

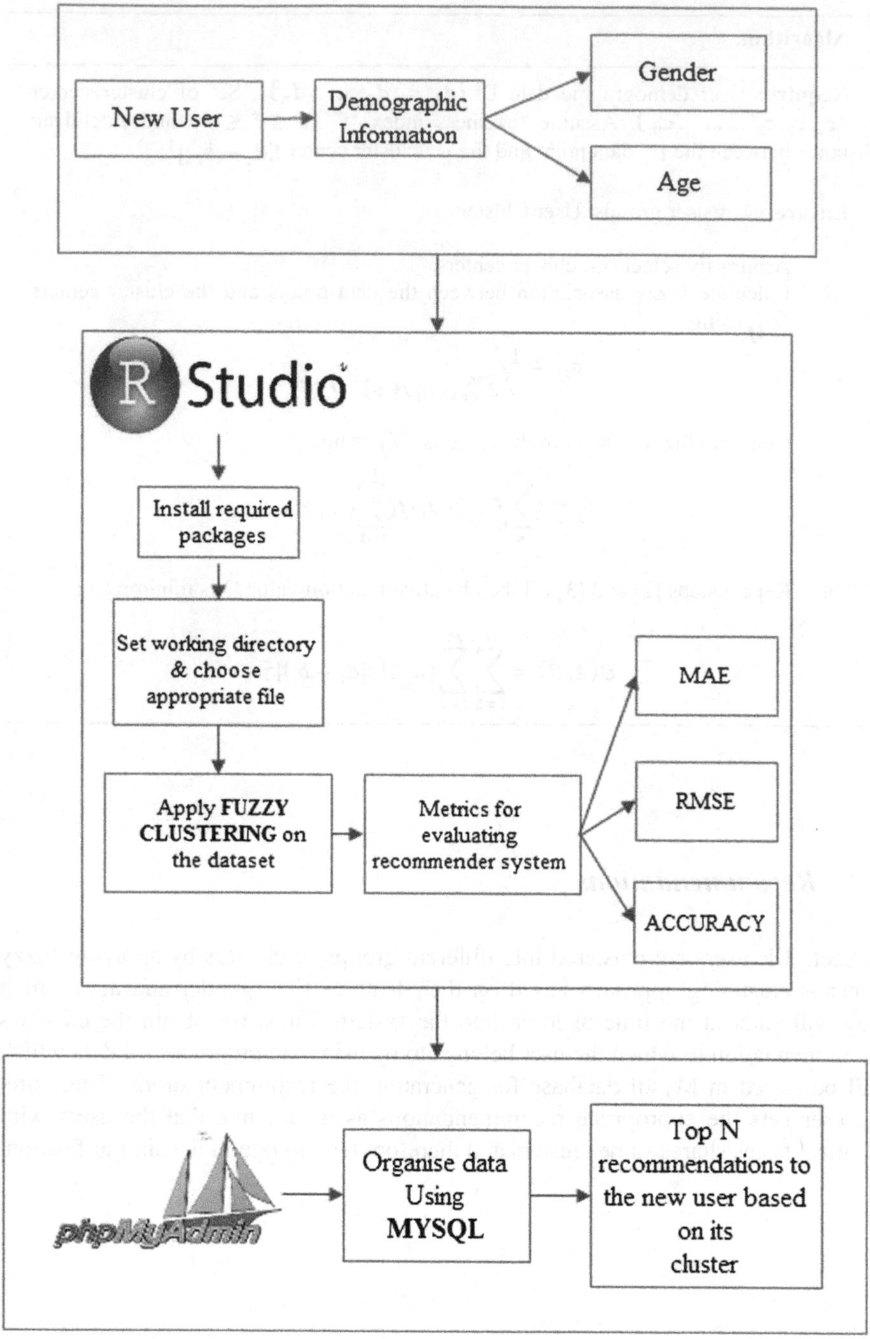

Fig. 1 Architecture of the collaborative filtering recommender system using fuzzy clustering technique

4 Experimental Results

4.1 Experimental Setup

To demonstrate the given approach, we apply it on MovieLens dataset, selecting 6040 users (where user features include userId, gender, occupation, and age) and their respective ratings. We preprocess the dataset using fuzzy c-means clustering technique which uses Euclidean distance to calculate similarity in Rstudio software and cluster the data using three clusters in order to evaluate our technique's performance on user cold start problem in collaborative filtering recommender systems. Therefore, a new user is recommended top-N recommendations according to his/her taste for which the preprocessed data is stored in Mysql database.

4.2 Evaluation Metrics

We have used three evaluation metrics: Mean Absolute Error (MAE), Root Mean Square Error (RMSE) and accuracy for evaluating the predictions. Also, we compare the accuracy fuzzy c-means clustering approach with the traditional k-means clustering approach. The recommendation accuracy is better if we obtain lower values of MAE and RMSE.

$$MAE = mean(abs(x\$cluster - \overline{x})) \quad (1)$$

$$RMSE = sqrt(mean(x\$cluster - \overline{x}) \wedge 2) \quad (2)$$

$$Accuracy = mean(abs(x\$cluster - \overline{x}) \leq 1) \quad (3)$$

where $x\$cluster$ is predicted rating and $\overline{x}$ is value of actual rating.

4.3 Results

Since we have compared fuzzy c-means clustering approach with the traditional k-means clustering approach on the basis of evaluation metrics by varying the number of user present in the dataset, thus finding accuracy. The results demonstrate that as the number of users increase, fuzzy c-means technique comes out to be more accurate as compared to k-means approach. Therefore, it has been proved that fuzzy c-means approach is more accurate than k-means for large datasets as shown in Fig. 2.

The accuracy value for k-means clustering approach and fuzzy c-means clustering approach is shown is Table 1.

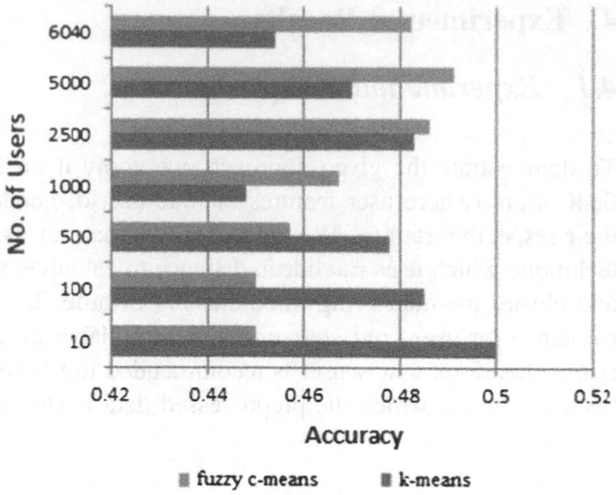

Fig. 2 Accuracy comparison of fuzzy c-means and k-means clustering approach

Table 1 Accuracy values for k-means and fuzzy c-means clustering for different number of users

Number of users	Accuracy	
	k-means	Fuzzy c-means
10	0.5	0.45
100	0.485	0.45
500	0.478	0.457
1000	0.448	0.467
2500	0.4828	0.4858
5000	0.4703	0.4912
6040	0.4542	0.4822

5 Conclusion and Future Scope

In this paper, a method has been proposed to address user cold start problem using fuzzy clustering technique. The basic idea behind our paper is to build clusters using fuzzy c-means clustering technique based on the attributes of the user. Hence, top-N recommendations for new users are successfully generated on the basis of the cluster he/she is placed into. The corresponding recommendation list is offered to the user (no matter old or new user) according to the cluster he/she is categorized into, based on the clustering algorithm used.

Also, a comparison is being made between fuzzy c-means clustering algorithm and the traditional k-means clustering algorithm on the basis of accuracy by varying number of users. It has been proved that as the number of users increase, fuzzy c-means clustering approach gives better accuracy as compared to the k-means clustering approach. The future scope will be to improve the accuracy of different clustering techniques by comparing it on different datasets.

References

1. Adomavicius G. & Tuzhilin A.: Toward The Next Generation of Recommender Systems: A Survey of the State-Of-The-Art and Possible Extensions: IEEE Transactions on Knowledge and Data Engineering, vol. 17, No. 6, pp. 734–749 (2005).
2. Sarwar B., Karypis J., Konstan J. & Riedl J.: Item Based Collaborative Filtering Recommendation Algorithms. In Proceedings of the 10th International Conference on World Wide Web, ser WWW'01. pp. 285—295. ACM, USA (2001).
3. Sanchez J.L., Serradilla F., Martinez E. & Bobadilla J.: Choice of Metrics Used In Collaborative Filtering and Their Impact on Recommender Systems. In: Digital Ecosystems and Technologies: 2nd IEEE International Conference, pp. 432–436 (2008).
4. Schein A.I., Popescul A., Ungar L.H., & Pennock D.M.: Methods and Metrics for Cold-Start Recommendations. In: Proceedings of the 25th Annual International ACM SIGIR Conference on Research and Development in Information Retrieval, ser. SIGIR '02, pp. 253–260. ACM, New York, USA (2002).
5. Budayan C., Dikmen I. & Birgonul M. T.: Comparing the Performance of Traditional Cluster Analysis, Self-organizing Maps and Fuzzy c-means Method for Strategic Grouping. In: Expert Systems with Applications, vol. 36, pp. 11772–11781 (2009).
6. Yanxiang L., Deke G., Fei C., & Honghui C.: User-based Clustering with Top-N Recommendation on Cold-Start Problem. In: Intelligent System Design and Engineering Applications (ISDEA), Third International Conference, pp. 1585–1589 (2013).
7. Devi M.K.K., Samy R.T., Kumar S.V. & Venkatesh P.: Probabilistic Neural Network Approach to Alleviate Sparsity and Cold Start Problems in Collaborative Recommender Systems. In: Computational Intelligence and Computing Research (ICCIC), IEEE International Conference, pp. 1–4 (2010).
8. Shaw G., Xu Y., & Geva S.: Using Association Rules to Solve the Cold-Start Problem in Recommender Systems. In: Lecture Notes in Computer Science, vol. 6118, pp. 340–347 (2010).
9. Gupta U. & Patil N.: Recommender System Based on Hierarchical Clustering Algorithm Chameleon. In: Advance Computing Conference (IACC), IEEE International, pp. 1006–1010 (2015).
10. Middleton S.E., Shadbolt N.R., & De Roure D.C.: Ontological User Profiling in Recommender Systems. In: ACM Trans. Inf. Syst., vol. 22, pp. 54–88 (2004).
11. Son L.H., Cuong K.M., Minh N.T.H., & Canh N.V.: An Application of Fuzzy Geographically Clustering for Solving the Cold-Start Problem in Recommender Systems. In: Soft Computing and Pattern Recognition (SoCPaR), International Conference pp. 44–49 (2013).
12. Sun D., Luo Z., & Zhang F.: A Novel Approach for Collaborative Filtering To Alleviate the New Item Cold-Start Problem. In: Communications and Information Technologies (ISCIT), 11th International Symposium, pp. 402–406 (2011).
13. Basiri J., Shakery A., Moshiri B., & Hayat M.Z.: Alleviating the Cold Start Problem of Recommender Systems Using a New Hybrid Approach. In: Telecommunications (IST), 5th International Symposium, pp. 962–967 (2010).
14. Dang T.T., Duong T.H., & Nguyen H.S.: A Hybrid Framework for Enhancing Correlation to Solve Cold-Start Problem in Recommender Systems. In: Computational Intelligence for Security and Defense Applications (CISDA), Seventh IEEE Symposium pp. 1–5 (2014).

Text Extraction from Images: A Review

Nitin Sharma and Nidhi

Abstract Multimedia, natural scenes, images are sources of textual information. Textual information extracted from these sources can be used for automatic image and video indexing, and image structuring. But, due to variations in text style, size, alignment of text, as well as orientation of text and low contrast of the image and complex background make challenging the extraction of text. From the past recent years, many methods for extraction of text are proposed. This paper provides with analysis, comparison of performance of various methods used for extraction of text information from images. It summarizes various methods for text extraction and various factors affecting the performance of these methods.

Keywords Text extraction · Text localization · Text segmentation
Connected component · Edge-based approach

1 Introduction

Text is the source of information that can be embedded into documents or in images. Computers and humans can easily understand text-based information. In this digital era, images have become a good means for communication. Today people use to send images with text embedded in it which they want to send along with the image. This is very trendy in social medias like Facebook, whatsapp, twitter. Textual images prove good in describing your emotion rather than sending only text messages on social media. Relevant information can be extracted from WWW images on the Internet which can be used in making web search efficient. Text in images is helpful in many content-based image applications such as image searching on web, mobile-based text analysis, video indexing, and human and computer interaction systems.

N. Sharma (✉) · Nidhi
Department of IT, UIET, PU, Chandigarh, India
e-mail: sharma.niti243@gmail.com

D.K. Mishra et al. (eds.), *Information and Communication Technology for Sustainable Development*, Lecture Notes in Networks and Systems 10,
https://doi.org/10.1007/978-981-10-3920-1_16

This paper is arranged as follow: Steps involve in extraction of text is described in Sect. 2. Section 3 analyzed various real-time application of text extraction from images. Section 4 describes different challenges involved in extraction of text in images. Section 5 explains briefly the various methodologies used in extraction of text. Finally summary and scope of text extraction in future is portrayed in Sect. 6.

2 Steps of Text Extraction

The text extraction problem is divided into following steps [1]:

(i) Text detection,
(ii) Text localization,
(iii) Text tracking,
(iv) Text extraction and enhancement, and
(v) Text recognition.

2.1 Text Detection

As there is no prior information that the images contain text or not so, text detection step determines whether text is present in given image or not. It can be done by making use of pixel intensity. It is assumed that text has higher pixel intensity than background pixels so, pixels with value less than predefined threshold value and having significant color difference from neighboring pixel are considered as text pixel. In videos, text is detected by using scene change information between two consecutive video frames.

2.2 Text Localization

It locates the position of text in given image. If the shape of text region is rectangular then text can be efficiently located but text can be aligned in any shape, i.e., rectangular or circular so, it is difficult to locate text. Text is located by using similarity in various text features like color, size, shape, intensity, and distance between two text pixels. Color histogram is used to figure out similarity in color of text pixels (Fig. 1).

Fig. 1 **a** Rectangular text. **b** Circular text

2.3 *Text Tracking*

If the text is not located in text localization step then text tracking is helpful in locating that text. Text tracking maintains the integrity of position across adjacent frame and reduces the processing time for text localization. This step is used to verify text localization results.

Text localization, detection, and extraction are often interchangeably used [1].

2.4 *Text Extraction and Recognition*

It describes how to separate text from images. This can be done by separating text pixels from other non-text pixels. By using various text features like similar font, orientation and stroke width, text can be segmented from image. OCR is the one of common method to extract and recognize text. It can easily recognize text in text document but it is unable to recognize text information in images efficiently because of presence of noise and distortion in images. In order to extract the text, different methods are used like region-based- and texture-based method. Region-based methods involve connected component and edge-based methods.

2.5 *Image Enhancement*

It is done in order to improve the clarity for human viewers of information in images, or to provide better results for the input of other automated image processing techniques. Image is enhanced by removing noise, blurriness. Image smoothening is done to remove the effects of camera noise and missing pixel values. Image sharpening is used to intensify the detail that has been blurred due to noise or other effects, such as unfocused camera, motion of object. Filters are used to remove noise from the image like median filter, average filter, low pass, and high pass filters (Fig. 2).

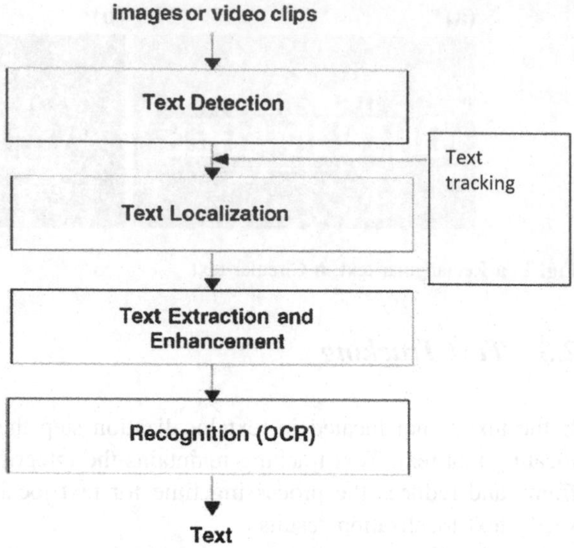

Fig. 2 Text extraction flow chart [1]

3 Applications of Text Extraction

Extraction of text from images can be utilized in many applications. Text extraction proves helpful in indexing of images and videos, structuring of images and in classification of images and videos. Following are some real-time applications of text extraction from images.

3.1 Applications for Disable People

Text in images or in natural scenes provides significant information that is utilized in text to speech devices. These devices assist visually impaired people in understanding grocery product, signs and pharmaceutical labels, and currency, instructions of ATM and in path navigation.

3.2 Navigating Systems

Text on sign board and natural scenes are used in navigation systems embed in robots and automatic geolocators to navigate their path.

3.3 *Texts in Web Images*

Relevant information can be provided by extraction of text from images on Internet. Indexing of images is performed using this information which will lead to efficient web mining.

3.4 *Industrial Automation*

Industrial automation related to large number of applications that used the recognized text on packages, containers, houses, and maps.

4 Challenges in Text Extraction

Text extraction process faces many challenges in segmenting the text from images. Some of these challenges are analyzed as follows:

- Text in images has different font, more than one color, and nonuniform size.
- Natural scene consists of complex background which contains a large number of objects like buildings, signs, bricks, grasses, fences, poles, etc., which resemble text characters causing error and confusion.
- Blurred image can occur with non-focused camera and with moving objects.
- Uneven lightening is very common problem during capturing of image.
- OCR is not used in text recognition in natural scene images as these images contain complex background containing many objects like bricks, fences, signs, etc. This complex background makes OCR to predict wrong word as it considers any object as character if that character resembles any character or alphabet.

5 Various Text Extraction Approaches

To locate the text, three types of approaches are used as follows:

5.1. Connected component based approach
5.2. Edge-based approach
5.3. Texture-based approach

5.1 Connected Component Approach

This is a bottom-up approach in which image is divided into small subparts and text components are extracted from this subparts and then these small text components are grouped up into larger text components until all textual regions are identified in the image. Various connected component approaches are used to extract text from images.

Yin et al. [2] proposed a Maximally Stable Extremal Regions (MSERs) based method to extract text form images. Extremal region can be viewed as a connected component in an image whose pixels can have either lower or higher intensity than its outer boundary pixels. In this method, first character candidates, i.e., pixels containing text are extracted based on their difference in variance. Extremal regions are extracted in the form of a rooted tree for the whole image. Second, character candidates are merged into text candidates by the single-link clustering algorithm. Distance metric learning algorithm can automatically learn about distance weights, i.e., distance between two text pixels and clustering threshold and these are used in clustering character candidates. Third, non-text candidates are eliminated by using character classifier. The width, height, smoothness, and aspect ratio of Text region are used by character classifiers.

Advantage: It proposes a method to detect text in images of low resolution.
Disadvantage: It is unable to detect text in blur images and images with multilingual text

Le et al. [3], present a learning-based approach which involves three steps. First, in preprocessing the image is binarized using Otsu binarization and the connected component are extracted based on various features like elongation, solidity, height, and width of the connected component, Hue moment which describes the shape of the connected component, stroke width of connected components for discriminating text from non-text. These features provide shape and location information of connected components. Then Adaboosting Decision Trees is used to label these connected components into non-text or text component. Then post processing to correct some connected components which are labeled incorrectly by the classifier.

Advantage: Le et al. [3] method easily locates the text in table, charts, and pie charts in documents which are considered as non-text zone.
Disadvantage: this method is not suitable for text extraction in natural scenes (Fig. 3).

Vidyarthi et al. [4] purposed a method in which first colored image is converted into gray-level image and histogram is drawn. Otsu method uses this histogram to find global threshold value. This value is further used to find out connected components. The components which are close to each other are merged to form one component. The height histogram is constructed by using the heights of the bounding boxes. Variance is selected for verification of the text and non-text. Finally, filtered components are verified a text on text.

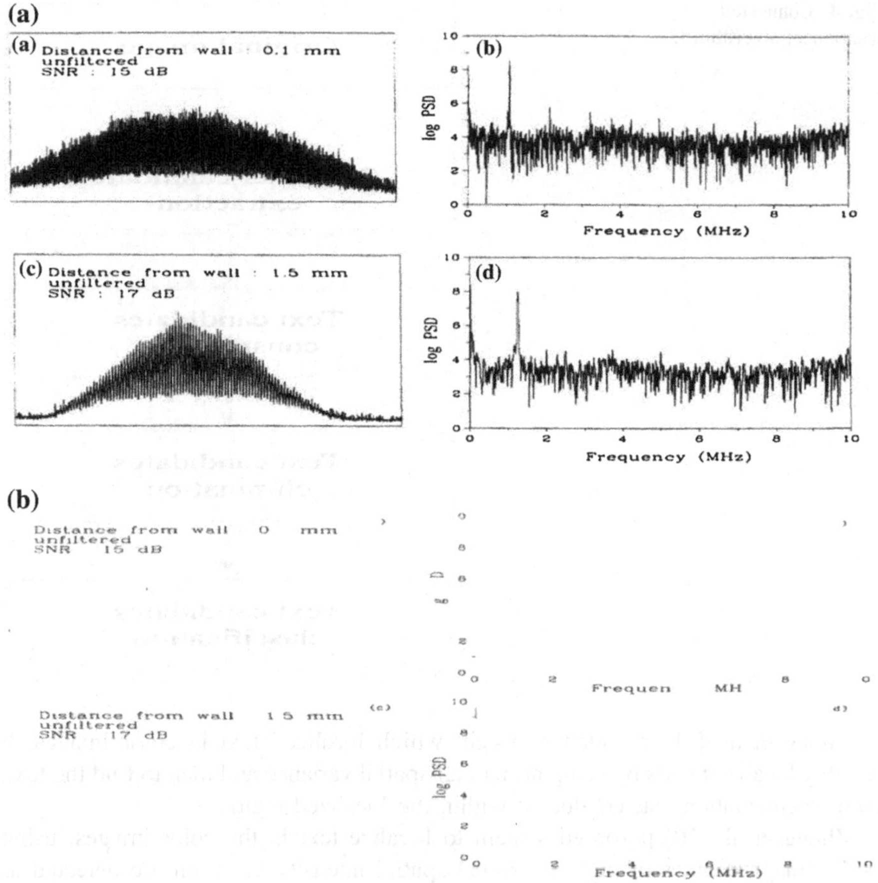

Fig. 3 **a** Shows original image. **b** Text extracted from image [3]

Advantage: This method locates the text in graphical area of documents very easily (Fig. 4).

5.2 *Texture-Based Approach*

Texture-based methodology makes use of the perception that text in images has different textural features that can differentiate text from the image background. These properties are local intensities, variance, filter responses like Gabor filters, and wavelet filters, FFT, etc. These methods usually have high complexity because all locations and scales have to be scanned.

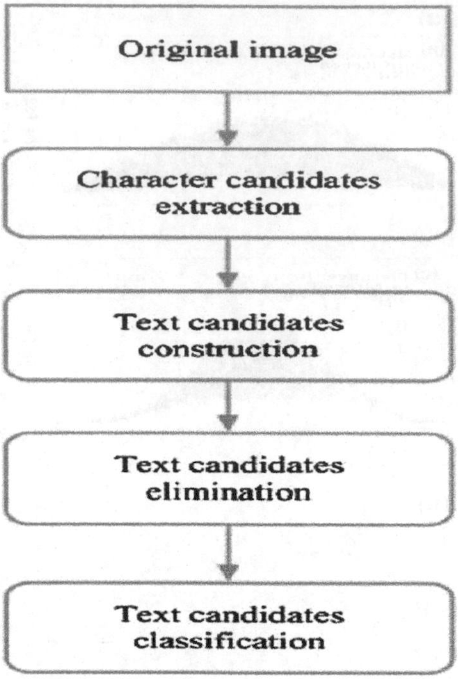

Fig. 4 Connected component algorithm [2]

Zhong et al. [8] presented a system which localized text in color images. It roughly localized texts by using horizontal spatial variance and then to find the text, color segmentation was conducted within the localized regions.

Zhong et al. [10] purposed system to localize text in the color images, using DCT. Image patches of high horizontal spatial intensity variation are detected as text components, morphological operations are applied to those patches and thresholding spectrum energy is used to verify the regions.

Disadvantage: If the size of characters is big and spacing between them is large, then this approach will not work properly.

Li et al. [11] proposed a system for detection and tracking of texts in videos. In this system, the first-order and second-order moments, mean of wavelet coefficients are used as local features to decompose the images. Advantages: The system is able in detection of graphical text and scene text having different font sizes and can also able to track text that undergoes complex motions.

Kim et al. [12] introduce a system to localize the text by using SVMs and texture templates. It classifies the pixels into positives that are connected into text regions by using a mean shift algorithm.

Advantage: This method can extract text from complex and textured background. Disadvantage: It faces problem in extracting small text in low resolution.

5.3 Edge-Based Approach

Edges are the most enhanced part of the image. We can easily distinguish between the objects based on the edges. Edge-based methods generally concern high contrast between the text and background. The edges of the text boundary are recognized and merged, and then various filters are used to verify the text and non-text regions. There are many edge detector filters like canny operator, Roberts operator, sobel operator, prewitt operator, gradient operator.

Yu et al. [13] proposed a system to recognize text through candidate edge recombination. First it separates text edges by using canny operator and then divides this edged image into small edge segments. Then the neighbor edge segment having similar stroke width and color are merged to form one candidate boundary. These candidate boundaries are combined into text chains by using character features and chain based features. Character features include stroke width of characters, aspect ratio, orientation difference of text whereas chain based features include histogram of gradient (HOG) which can be used to describe the boundary of object, structure similarity as it is assumed that text have similar structure (Fig. 5).

Yi et al. [14] proposed a system in which image first operates on canny operator; then the image is partitioned in order to obtain character candidate components by using two methods gradient-based method and color-based method. In gradient-based method, two pixels in opposite direction and having same local gradient magnitude are coupled together by connecting path. Candidate character component are then extracted by computing the distribution of gradient magnitudes at pixels of the connecting path. In color-based method, color histogram is calculated in order to capture the dominant colors around edge pixels. Then K-mean clustering is used to find character candidate components. These are grouped

Fig. 5 Recognition of text [13]

together to detect text strings based on character features like distances between two neighboring characters, character sizes, and character alignment.

Advantages: This system can extract text having different orientation.

Lu et al. [15] introduced a system for extraction of text from the scene images. The image first operates on canny operator to differentiate text edges from the background and then based on three text-specific features like image contrast, horizontal or vertical direction stroke width text is extracted. These text-specific properties are evaluated at images of various scales because it improves various types of image degradation, because some text edges lost at one image scale, due to the complex image background. The text candidate text boundaries are detected by using Global thresholding. For every detected candidate text boundary, one or more candidate characters are determined by using adaptive threshold that can be calculated based on the neighbor image pixels. Support vector regression model which use bag of word (BOW) is used to identify true characters and words.

Advantage: It is used for automatic detection and segmenting the texts from scene images.
Disadvantages: It fails to detect text having small size and when lightening is low.

Khodadadi et al. [17] introduced a method for extraction of text from images based on stroke filters and color histogram. First, stroke filter is applied to the input image and then local and global threshold values are calculated on stroke filter output. The image is then divided into text blocks by making use of horizontal and vertical projections of binary pixels. The text blocks with overlap or close blocks are merged to form a new text block. It is assumed that the text in every block has the same color, so histogram of color channels are used for extraction of the text characters in the candidate blocks for the text and background areas.

Advantages: This proposed method can locate text in natural scene in one iteration.

Kumar et al. [18] suggested an algorithm that consists of three stages: first, images are converted into an edge map by using line edge detector which uses vertical and horizontal masks. Second, for identification of the text regions in images, the vertical and the horizontal projection profiles are analyzed and the edged image is then converted into the localized image. After localization, in order to obtain text area segmentation process is done on the localized image by using median filters.

Algorithm	Precision	Recall	F-measure
Yin et al. [2]	86.29	68.26	76.22
Le et al. [3]	98.37	92.46	–
Yu et al. [13]	78	63	70
Yi and Tian [14]	67.2	58.1	62.3
Lu et al. [15]	89.22	69.58	78.19
Neumann and Matas [19]	85.4	67.5	75.4
Neumann and Matas [21]	73.1	64.7	68.7
Kim et al. [22]	82.98	62.47	71.28
Zhou et al. [25]	69.9	69.2	69.3

6 Summary and Future Scope

Detecting and recognizing texts in natural scenes is becoming important in research areas in computer vision as precise and rich information embedded in text that can help in a large number of real-world applications. This paper provides analysis and comparison of performance of various methods used for extraction of text information from images. Text extraction involves text detection, localization of text, tracking of text, extraction of text, enhancement, and text recognition from a given image. Text detection discovers whether text is present in a given image or not. Normally text detection is applied for a sequence of images. Text localization localizes the text within the image and bounding boxes are generated around the text. If the text is not located in text localization step, then text tracking is helpful in locating that text. Text extraction refers to separate of text from images. In order to extract the text, different methods are used like region-based- and texture-based method. Region-based methods involve connected component and edge-based methods. Connected component based method gives poor performance for merged characters or when the characters are not completely separated from the image background. The texture-based methodology has an inability to recognize the characters that reach below the baseline or above other characters, and this ends up in segmenting a character into two components. Edge-based method also makes false prediction when edge of any object in background of image resembles any character. Text extraction from images can be proved useful information for content-based application. For last few years, many researchers performed their researches to localize and extract the text from images, still there are many areas which are remained untouched. Text extraction can be used in classification of text images which can be used in structuring of images in mobiles and in web search. Text extraction can be used to make a navigation system for unmanned vehicles and can be used by visually impaired persons to find out their way.

References

1. Keechul Jung, Kwang In Kim, Anil K. Jain "Text information extraction in images and video: a survey", Elsevier, Pattern Recognition 37 (2004).
2. Xu-Cheng Yin, *Member*, *IEEE*, Xuwang Yin, Kaizhu Huang, and Hong-Wei Hao "Robust Text Detection in Natural Scene Images" IEEE transactions on pattern analysis and machine intelligence, VOL. 36, NO. 5, (2014).
3. Viet Phuong Le, Nibal Nayef, Muriel Visani, Jean-Marc Ogier and Cao De Trant "Text and Non-text Segmentation based on Connected Component Features" IEEE, 13th International Conference on Document Analysis and Recognition (ICDAR), (2015).
4. Ankit Vidyarthi, Namita Mittal, Ankita Kansal, "Text and Non-Text Region Identification Using Texture and Connected Components", International Conference on Signal Propagation and Computer Technology (ICSPCT), IEEE (2014).
5. Yingying Zhu, Cong Yao, Xiang Bai "Scene text detection and recognition: recent advances and future trends" Front. Comput. Sci., (2016).
6. Qixiang Ye, and David Doermann, "Text Detection and Recognition in Imagery: A Survey" IEEE transactions on pattern analysis and machine intelligence, vol. 37, no. 7, (2015).
7. N. Senthilkumaran and R. Rajesh, "Edge Detection Techniques for Image Segmentation – A Survey of Soft Computing Approaches", International Journal of Recent Trends in Engineering, Vol. 1, No. 2, (2009).
8. Zhong Y, Karu K, Jain A K. "Locating text in complex color images." in Proceedings of the 3rd IEEE Conference on Document Analysis and Recognition, pp-146–149, IEEE (1995).
9. Kim K I, Jung K, Kim J H. "Texture-based approach for text detection in images using support vector machines and continuously adaptive mean shift algorithm." IEEE Transactions on Pattern Analysis and Machine Intelligence, pp-1631–1639, IEEE(2003).
10. Y. Zhong, H. J. Zhang, and A. K. Jain, "Automatic caption localization in compressed video," IEEE Trans. Pattern Anal. Mach. Intell., vol. 22, no. 4, pp. 385–392, IEEE (2000).
11. Li H, Doermann D, Kia O. "Automatic text detection and tracking in digital video.", 9(1): 147–156, IEEE Transactions on Image Processing, (2000).
12. K. I. Kim, K. Jung, and H. Kim, "Texture-based approach for text detection in images using support vector machines and continuously adaptive mean shift algorithm," IEEE Trans. Pattern Anal. Mach. Intell., vol. 25, no. 12, pp. 1631–1639, (2003).
13. Chong Yu, Yonghong Song, Quan Meng, Yuanlin Zhang, Yang Liu, "Text detection and recognition in natural scene with edge analysis", IET Comput. Vis., Vol. 9, Iss. 4, pp. 603–613, (2015).
14. Chucai Yi, Ying Li Tian, "Text String Detection From Natural Scenes by Structure-Based Partition and Grouping", Vol. 20, No. 9, IEEE Transactions on Image Processing (2011).
15. Shijian Lu, Tao Chen, Shangxuan Tian, Joo-Hwee Lim, Chew-Lim Tan, "Scene text extraction based on edges and support vector regression", 18:125–135, IJDAR (2015).
16. K. C. Kim, H. R. Byun, Y. J. Song, Y. W. Choi, S. Y. Chi, K. K. Kim, Y. K. Chung, "Scene Text Extraction in Natural Scene Images using Hierarchical Feature Combining and Verification", 17th International Conference on Pattern Recognition (ICPR'04), IEEE (2004).
17. Mohammad Khodadadi, and Alireza Behrad, "Text Localization, Extraction and Inpainting in Color Images", IEEE, 20th Iranian Conference on Electrical Engineering, (ICEE2012), (2012).
18. Anubhav Kumar "An Efficient Text Extraction Algorithm in Complex Images", IEEE, (2013).
19. Lukáš Neumann, Jiří Matas, "On Combining Multiple Segmentations in Scene Text Recognition", 12th International Conference on Document Analysis and Recognition,, IEEE (2013).
20. L. Neumann and J. Matas, "Text localization in real-world images using efficiently pruned exhaustive search," in Document Analysis and Recognition (ICDAR), 2011 International Conference, pp. 687–691, (2011).

21. L. Neumann and J. Matas, "Real-time scene text localization and recognition," in Computer Vision and Pattern Recognition (CVPR), pp. 3538–3545, IEEE Conference (2012).
22. Hyung Il Ko, Duck Hoon Kim, "Scene Text Detection via Connected Component Clustering and Non-text Filtering", Vol. 22, No. 6, IEEE Transactions on Image Processing (2013).
23. Huizhong Chen, Sam S. Tsai1, Georg Schroth, David M. Chen, Radek Grzeszczuk and Bernd Girod, "Robust Text Detection In Natural Images with Edge-Enhanced Maximally Stable Extremal Regions", 18th IEEE International Conference on Image Processing (2013).
24. Kamrul Hasan Talukder, Tania Mallick, "Connected Component Based Approach for Text Extraction from Color Image", IEEE, 7th International Conference on Computer and Information Technology (ICCIT)(2014).
25. Gang Zhou, Yuehu Liu, Liang Xu1, Zhenhong Jia, "Scene text detection method based on the hierarchical model", Vol. 9, Iss. 4, pp. 500–510 IET Comput. Vis., (2015).

Cross-Domain Sentiment Analysis Employing Different Feature Selection and Classification Techniques

Chakshu Ahuja and E. Sivasankar

Abstract The paramount work of information mustering has been to find out what is the opinion of the people. Sentiment analysis is errand discerning the polarity for the given content which is dichotomized into two categories—positive and negative. Sentiment analysis operates on colossal feature sets of unique terms using bag of words (BOW) slant, in which case discrete attributes do not give factual information. This necessitates the elimination of extraneous and inconsequential terms from the feature set. Another challenging fact is most of the times, the training data might not be of the particular domain for which the perusal of test data is needed. This miscellany of challenges is unfolded by probing feature selection (FS) methods in cross-domain sentiment analysis. The boon of cross-domain and Feature Selection methods lies in significantly less computational power and time for processing. The informative features chosen are employed for training the classifier and investigating their execution for classification in terms of accuracy. Experimentation of FS methods (IG, GR, CHI, SAE) was performed on standard dataset viz. Amazon product review dataset and TripAdvisor dataset with NB, SVM, DT, and KNN classifiers. The paper works on different techniques by which cross-domain analysis vanquishes, despite the lower accuracy due to difference in domains, as better algorithmic efficient method.

Keywords Feature selection · Cross domain · Machine learning Sentiment classification

C. Ahuja (✉) · E. Sivasankar
Department of Computer Science and Engineering, National Institute of Technology, Tiruchirappalli, Tiruchirappalli, India
e-mail: ahuja.chaks@gmail.com

E. Sivasankar
e-mail: sivasankar@nitt.edu

D.K. Mishra et al. (eds.), *Information and Communication Technology for Sustainable Development*, Lecture Notes in Networks and Systems 10,
https://doi.org/10.1007/978-981-10-3920-1_17

1 Introduction

The Internet and Web have made it possible to form a medium for dissemination and exchange of information as well as opinions. The burgeoning of reviews, ratings, recommendations, and opinionated texts, and product assessment websites [4], has caused online opinion to become rather bitcoin for businesses contemplating to trade their commodities, discern recent opportunities and guide their statures [15]. The cause of upswing in sentiment analysis as the trending research domain is the user hanker for and dependency on online recommendations. This calls for processing of unformatted content [5]. The basis of Sentiment Classification (SC) is arbitrating the viewpoint of the user audit as positive or negative. There are basically two ways to superintend sentiment analysis Single-Domain and Cross-Domain Sentiment Classification (CDSC). For Single-Domain, chunks of the dataset train the classifier and remaining chunks are expended for testing. For cross-domain, two distinct unrelated domains are taken into consideration and classifier model is made to train using one and tested on another. The two main points of concern are the depuration of cumbersome data and also model to operate across heterogeneous commercial commodities information [6]. Every sentiment lexicon in the extract is not revealed unequivocally. In-depth analysis is required to be done by SC to unmask the didactic features in the extract. We tackle the problem of high dimensional data by focusing on the importance of FS applied to machine learning (ML) for SC and explore the comparison of disparate FS methods. In order to overcome problem of unavailability of particular domain information and also for the model to not be domain specific, we investigate this combined with CDSC.

2 Related Works

Considering spectrum of decision-making philosophies, one end of spectrum involves open approach involving collective wisdom and thinking of large group of people [15]. The online websites have invigorated the user to participate in discussions and share their reviews with other users. This user data is regarded extremely vital for both the consumers and service providers to increase their productivity as well as to arrive at an appropriate decision from the given feedback [2, 21]. The SC technique called ML approach uses a collection of labeled reviews to train the classifier model that will be employed for to classify the testing reviews as negative or positive [2].
In carrying out the SC for user reviews, since the processing of huge amounts of data is concerned and real time response is also desired, exacting requirements are being faced by technology to obtain better computational efficiency [6]. With availability of wide variety of commodities and services, also the need arises for consummate model for heterogeneous realms. The SC model needs to be trained recurrently to dive into different domain applications, customarily with bulky corpora of sampling

[6, 17]. Moreover, expressed sentiment is different depending on the chosen domain. So annotating corpora for every particular domain involves exorbitant cost. It is here when CDSC takes over. CDSC is a itinerant area of research in ML. Federica Bisio and Erik Cambria researched on the review mining across heterogeneous domains and obtained propitious results [6]. The contemplated approach espoused pragmatic study to implement SC and used distance-based prognostic exemplar to couple computational efficiency and modularity. Bollegala et al. [7] proposed an approach by creating sentiment sensitive distributional thesaurus. The incongruity between features of different domains was handled by using labeled information from multifarious training domains and raw text from training and testing domains to reckon relatedness of attributes and frame sentiment sensitive thesaurus which was used at training and testing times to expand the feature vector.
Further to make SC approach efficient and scalable, another hurdle was to deal with large number of words in domains and select only the most informative ones. For this, the effective FS Techniques are required [10, 21]. These comprise Information Gain (IG), Relief-F, Gain Ratio (GR), CHI-squared, Document Frequency (DF), Significance Attribute Evaluation (SAE) [20], Mutual Information (MI) [10, 11] and Fisher discriminant ratio [22]. Another efficient method to excerpt the consequential attributes using condition probability was proposed to augment the knowledge and also supplement with computational speed [20].
Having researched on these existing techniques gave me a glimpse of CDSC and FS Techniques and fostered to work on improving in this. The experimentation has been performed on Amazon product reviews [7] and TripAdvisor hotel reviews [9]. The basis for choosing these is to gauge cross-domain abilities pertaining to areas disparate from each other and their large volume (8000 reviews from Amazon and 8000 reviews from TripAdvisor).

3 Methodology

This section presents the proposed techniques for analyzing the sentiments associated with reviews in two distinct datasets. Analysis is done in various stages as follows. The Fig. 1 shows the proposed method involved:

3.1 Data Collection

Amazon reviews of Books, Electronics, DVDs and Kitchen Appliances (4000 positive and 4000 negative) and Trip Advisor reviews of Hotels (4000 positive and 4000 negative) have been collected. Using NLTK tokenization method, the review is converted into the BOW. These BOW are considered to be features for a review dataset. Not every feature has meaningful information. The informative features are ranked by efficient FS methods.

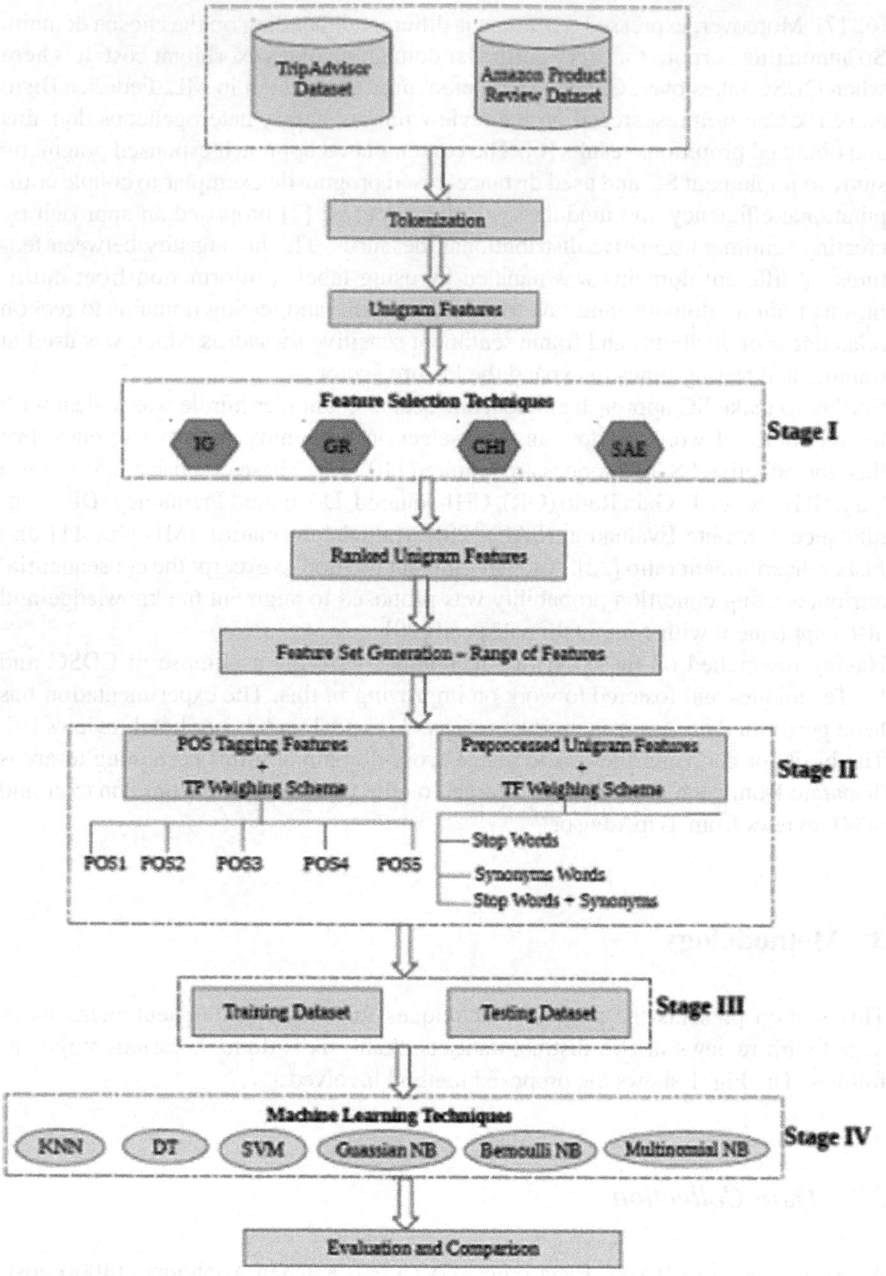

Fig. 1 Proposed architecture for cross-domain sentiment analysis

3.2 Feature Selection

Datasets for analysis holds abounding number of features, but all the features may not be significant for framing classifier model. To improve classifier accuracy, the redundant and irrelevant features have to be taken out [10]. FS mechanisms used in this paper include IG, GR, CHI, and SAE.

Information Gain IG makes use of entropy [18, 19] based attribute evaluation approach to actuate the measure of information on the features required for classification of document review. IG is implanted on reduction in entropy after the text review dataset is split into attributes. It is substance of probability of every label times log probability for that particular label. For classifying the review using training dataset D with m labeled classes, expected amount of information needed is defined as,

$$E(D) = -\sum_{i=1}^{m}(p_l)log_2(p_l), \tag{1}$$

where m denotes the count of classes. The value of m is taken as two for binary classification and the probability of the review belonging to the class C_l, $1 \leq l \leq m$ for the training dataset D is denoted by p_l. If review in D on any feature f_i can take values say, $a_1, a_2, ..., a_v$ then the dataset will split into $D_1, D_2, ..., D_v$ partitions. The expected measure of information required for classifying reviews in the training dataset D, provided reviews are classified as per one peculiar feature f_i is,

$$E(D,f_i) = \sum_{j=1}^{v}\frac{|D_j|}{|D|} \times E(D_j) \tag{2}$$

where $\frac{|D_j|}{|D|}$ represents weightage for jth partition, with $1 \leq j \leq v$ and $E(D_j)$ representing entropy of partition $E(D)$.

$$Information\text{-}gain(f_i) = E(D) - E(D,f_i) \tag{3}$$

represents information gain on feature f_i. The features are graded in accordance with the maximum value for information gain [10].

Gain Ratio GR works on iterative procedure of selecting subspace of attributes [18, 19]. It was first employed in the decision tree and is the extension of IG. The IG measure is swayed by features that have large count of outcomes. GR vanquishes the bias which applies normalization to IG using split information measure. For an attribute f_i, split information is calculated by dividing dataset D into v partitions, where v is effect of a test on feature f_i. Each partition has a number of review instances which represents one of the future events. The split information for specific feature f_i is given as,

$$Split\text{-}information_{f_i}(D) = -\sum_{j=1}^{v} \frac{|D_j|}{|D|} \times log_2 \frac{|D_j|}{|D|} \tag{4}$$

All the partitions are homogenous in the number of review instances in case if the Split Information is high, else review instances contained in smaller number of partitions are many. The gain ratio for attribute f_i in dataset D is given as,

$$Gain\text{-}Ratio_{f_i} = \frac{Information\text{-}gain(f_i)}{Split\text{-}information(f_i)} \tag{5}$$

CHI-Squared CHI selects the most informative features by calculating the association of feature f_i with class C_l [11]. The CHI measure for feature f_i with the class C_l is directly proportional to the competence of the feature f_i to classify the review [10]. For binary classification, CHI value with feature f_i is,

$$\chi^2(f_i, C_l) = \frac{N[N(C_1,f_i) \times N(C_2,\overline{f_i}) - N(C_2,f_i) * N(C_1,\overline{f_i})]^2}{[N(C_1,f_i) + N(C_2,f_i)] * [N(C_1,f_i) + N(C_2,\overline{f_i})] * [N(C_2,f_i) + N(C_1,\overline{f_i})] * [N(C_1,\overline{f_i}) + N(C_2,\overline{f_i})]}, \tag{6}$$

where N represents total text reviews, $N(C_l,f_i)$ represents the count of reviews belonging to class C_l having feature f_i and $N(C_l,\overline{f_i})$ representing the count of reviews which belong to class C_l and not containing the feature [11].

Significance Attribute Evaluation The weight of the attribute is quantified by calculating the probability importance of instances of Class Feature and Feature Class Association [20]. Class Attribute association is calculated by determining the alteration in value of feature along with alteration in class label. Likewise, Attribute Class association cumulates values for all features along with their relationship to the class label. The feature is notable when Class Feature as well as Feature Class association is high and accordingly these important attributes are graded. Feature Class Association is represented by, $P_i^r(w) + P_i^{\sim r}(\sim w)$, where, w represents the binary class labeled data set fragment, $P_i^r(w)$ is the probability calculating frequency enumeration for samples of w having rth feature value class as that of ith feature value of sample in entire document and $P_i^{\sim r}(\sim w)$ is probability for frequency enumeration of class instances with rth feature values, not same as the ith value of feature of sample in the entire dataset. Class Feature association is represented as, $P_i^j(V) + P_i^{\sim j}(\sim V)$, where, V is the subspace of values of features of F_i, $P_i^j(V)$ is probability of frequency number of instances possessing the class j and $P_i^{\sim j}(\sim V)$ represents probability of frequency enumeration of samples not categorized into class j having the values of feature of F_i different from the values of features in V.

3.3 Classification

Having generated the matrix, based on selected features and data as Term Presence or Term Frequency, the formed matrix on training domain is given as input to classifiers and sentiment of the testing domains reviews is predicted. The analysis is performed on K-nearest neighbor (N = 15) (KNN), Support Vector Machine (SVM) having Radial Basis Function (RBF) and Linear kernel, Decision Tree (DT), Gaussian Naive Bayes (GNB), Multinomial Naive Bayes (MNB), Bernoulli Naive Bayes (BNB).

Naive Bayes NB is conditional probabilistic model based on Bayes theorem for classification [3, 13]. For review r with class as c that can be classified into positive or negative based on probability value $p(c \mid r)$ as denoted in Bayes Theorem,

$$p(c \mid r) = \frac{p(c) \times \prod_{i=1}^{m} p(f_i \mid c)}{p(r)} \tag{7}$$

where f_i denotes the attributes in the given review r. In GNB, incessant values linked to every class are scattered in accordance with Gaussian distribution.

$$p(x = v \mid c) = \frac{1}{\sqrt{2\pi\sigma_c^2}} e^{\frac{(v-\mu_c)^2}{2\mu_c^2}}, \tag{8}$$

where c is a class, x represents steady attribute, μ is mean, σ_c^2 is the variance of values in x linked with class c. In case of model of multinomial event, feature vectors denote frequencies of generation of particular events by multinomial. The likelihood in a histogram X is denoted as,

$$p(x \mid C_k) = \frac{(\sum_i x_i)!}{\prod_i x_i!} \prod_i p_{k_i}^{x_i}, \tag{9}$$

where p_i is the probability of occurrence of event i. The feature set $X = (x_1, \ldots, x_n)$ represents a histogram, where X_i represents the count of times particular review shows occurrence of event i.

In BNB, features are individualistic Booleans outlining the inputs. If X_i is Boolean that represents absence or presence of ith feature, then odds of a review denoted a class C_k is represented by,

$$p(x \mid C_k) = \prod_{i=1}^{n} p_{k_i}^{x_i} (1 - p_{ki})^{(1-x_i)} \tag{10}$$

where p_{ki} represents probability of class C_k that generates the term w_i.

K-Nearest Neighbor KNN is a nonparametric classifier relying on training reviews class labels that are identical to test review [10]. The reviews are classified in accordance with k nearest neighbors. The weightage of class of neighboring review is defined by the similarity of every nearest neighbor review to test data r. The weighted sum in KNN is,

$$Score(r, c_l) = \sum_{r_i \in KNN(r)} sim(r, r_i) \times \delta(r_i, c_l), \quad (11)$$

where $KNN(r)$ denotes the review vector of k neighbors nearest to review. If r_i has class c_l, then $\delta(r_i, c_l)$ is 1, else 0. The test review is assigned the class of review having highest resultant weighted sum.

Decision Tree DT predictive model [14] uses decision tree learning and maps experimentations of an object to outcome of the object's target. The tree is constructed by taking into consideration the best splitting feature at every stage. Branches of tree structures denote conjunctions of features leading to class labels which in turn are represented by Leaves [18].

Support Vector Machine SVM is non-probabilistic classification model [3, 12, 13]. It has been built by plotting points in space thereby estimating hyper plane separating the sample reviews into two distinct classes that have frame of maximum distance. The review defining hyper plane is known as support vector denoted by $\vec{w}$. The solution to the constraint optimization problem is shown as:

$$\vec{w} = \sum_j \alpha_j C_j \vec{r_j} \quad (12)$$

where α_j represents support vectors estimated by finding solution to dual optimization problem more than zero, and C_j is a precise class for $\vec{r_j}$ review vector. The kernel function is incorporated in SVM that maps the low-dimensional nonlinear values into high-dimensional linear value points. There are many kernel core functions: linear kernel, Gaussian kernel, RBF, and so on. For this paper, we have optimized the SVM parameters using linear kernel function.

3.4 Performance Analysis

The measures for determining the performance in the SC used are Accuracy, Recall, Precision, and F-measure. Assuming 2×2 contingency matrix as in Table 1

$$Accuracy = \frac{TP + TN}{TP + TN + FP + FN}$$

$$Precision = \frac{TP}{TP + FP}$$

Table 1 Contingency table for sentiment classification

		Predicted sentiment	
		Positive	Negative
Actual sentiment	Positive	True positive (TP)	False negative (FN)
	Negative	False positive (FP)	True negative (TN)

$$Recall = \frac{TP}{TP + FN}$$

$$F_{measure} = 2 * \frac{Precision \cdot Recall}{Precision + Recall}$$

4 Results and Analysis

4.1 Dataset Collection

The data set experimented on is the binary labeled class Amazon review and TripAdvisor review dataset. All classifier models and FS techniques have been executed using Python.

4.2 Preprocessing on Collected Unigrams

The significant features are graded and classifiers are trained and tenfold cross validated. The reduction in count of informative attributes derived at end of every processing stage has been summarized as shown in Table 2.

4.3 Comparison of Feature Selection Methods

Different ranges of top informative features are extracted and experimented using KNN classifier. The accuracies obtained with different FS method applied on Amazon and TripAdvisor reviews using KNN classifier are as follows in Tables 3 and 4.

From the analysis as shown in Tables 3 and 4, GR outperforms other methods giving 67.83% for the top 2000 features where the training model was using Amazon dataset and testing using TripAdvisor dataset. Consistency in the results was found when training was done using TripAdvisor and testing using Amazon dataset, where GR

Table 2 Number of features using different processing techniques

Method experimented	Amazon reviews	TripAdvisor reviews
Positive instances	4000	4000
Negative instances	4000	4000
Unique attributes	53,949	40,149
Above threshold 0 (IG, GR, CHI, SAE)	2785	5275
After removal of stop words	2707	5067
After removal of synonyms	2387	4145
After removal of stop words and synonyms	2328	4024
Features using POS1 (adjectives)	266	361
Features using POS2 (adverbs)	128	209
Features using POS3 (nouns)	1793	3388
Features using POS4 (verbs)	488	999
Features using POS5 (POS1+2)	394	570

Table 3 Training-Amazon Testing-TripAdvisor

No. of features	IG	GR	CHI	SAE
500	57.94	57.23	58.05	57.48
1000	58.47	58.60	58.17	58.20
1500	58.01	58.37	57.70	58.51
2000	58.17	67.83	58.23	58.39
2785	58.56	58.40	58.56	58.56

Table 4 Training-TripAdvisor Testing-Amazon

No. of features	IG	GR	CHI	SAE
1000	53.13	57.05	56.03	53.63
2000	53.55	57.64	56.23	53.95
3000	53.43	53.65	55.30	53.03
4000	53.31	58.99	56.45	53.50
5275	53.46	53.44	53.25	53.46

gives the best (58.99%) accuracy for top 4000 features. This has been confirmed by further experimenting even on classifiers other than KNN.

Preprocessed Feature Selection The three kinds of preprocessed feature sets are obtained by removing stopwords, synonyms and both together and performance for each model is found. The number of features obtained in the preprocessed feature set is less compared with the raw feature set which makes implementation of algorithm little faster.

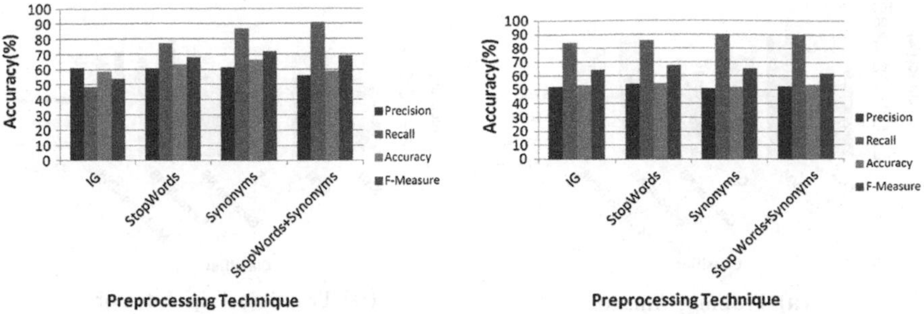

(a) Training-Amazon Testing-Trip Advisor

(b) Training-Trip Advisor Testing-Amazon

Fig. 2 Preprocessed feature selection

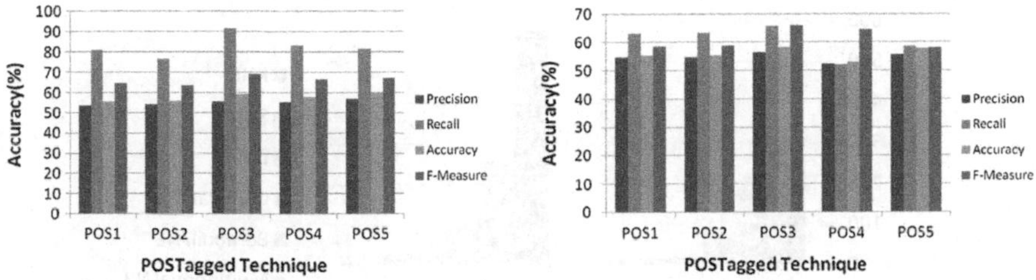

(a) Training-Amazon Testing-Trip Advisor

(b) Training-Trip Advisor Testing-Amazon

Fig. 3 POS tagged feature selection

As shown in Fig. 2, we observe that there is an improvement in the performance after each preprocessing stage and even when there is no substantial modification in the result compared to initial feature set results but the number features considered is reduced.

POS Tagged Feature Selection POS tagging is applied and five different kinds of feature sets are generated. These POS feature sets are used to construct the classifier model on the two different datasets.
Experimental results given in Fig. 3 show POS3 gives better results in both the cases. Finally, the comparison is extended to different classifiers other than KNN classifier used till now, namely GNB, BNB, MNB, SVM, and DT.
Considering performance analysis as in Fig. 4, SVM and BNB outperforms other classifiers with getting accuracy as high as 78.02% and 81.26% respectively. With respect to the comparison between classifiers in time analysis as shown in Fig. 5, KNN takes the maximum time (312.99 s) for training as well as testing. There was drastic decrease in time required in other classifiers to as low as 15.33 s in NB.

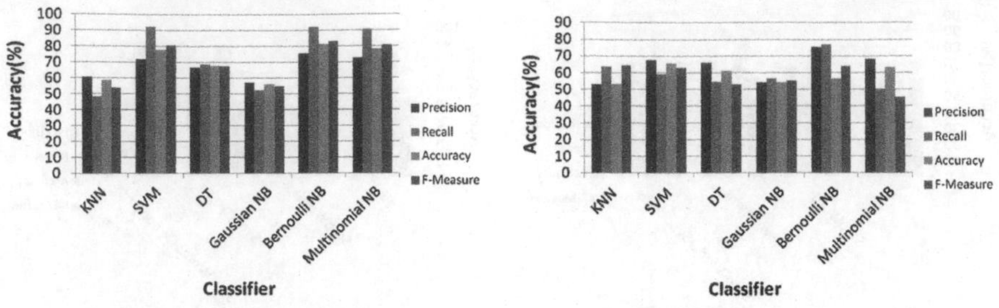

(a) Training-Amazon Testing-Trip Advisor

(b) Training-Trip Advisor Testing-Amazon

Fig. 4 Performance analysis between classifiers

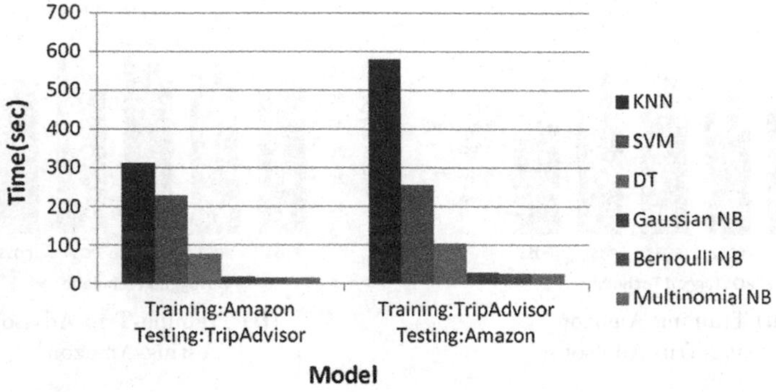

Fig. 5 Time analysis between classifiers

5 Conclusion

The paramount motivation behind this analysis was to build efficient model when it comes to cross domains. The need was to attain better result with respect to computation as well as space. Having experimented on different techniques, we have tried to put forward a better method to accomplish our requirement. Amalgamating Gain Ration FS set with POS3 tagged features and built on SVM and BNB classifier produced the best result. The forthcoming work is to perform SC on unstructured dataset and moving to other ensemble classifiers rather than traditional classifiers. Further ahead, we would like to move to Big Data so as to deal with petabytes of data.

References

1. R. Feldman, Techniques and applications for sentiment analysis, Communications of the ACM 56 (4) (2013) 82–89.
2. Y. Dang, Y. Zhang, H. Chen, A lexicon-enhanced method for sentiment classification: An experiment on online product reviews, Intelligent Systems, IEEE 25 (4) (2010) 46–53.
3. R. Xia, C. Zong, S. Li, Ensemble of feature sets and classification algorithms for sentiment classification, Information Sciences 181 (6) (2011) 1138–1152.
4. B. Liu, Web Data Mining: Exploring Hyperlinks, Contents, and Usage Data, Springer, 2007
5. E. Boiy et al., Automatic Sentiment Analysis in On-line Text, ELPUB2007
6. F. Bisio, P. Gastaldo, C. Peretti, R. Zunino, E. Cambria, Data intensive review mining for sentiment classification across heterogeneous domains, in: Advances in Social Networks Analysis and Mining (ASONAM), 2013 IEEE/ACM International Conference on, IEEE, 2013, pp. 1061–1067.
7. D. Bollegala, D. Weir, J. Carroll, Cross-domain sentiment classification using a sentiment sensitive thesaurus, Knowledge and Data Engineering, IEEE Transactions on 25 (8) (2013) 1719–1731.
8. P. Jambhulkar, S. Nirkhi, A survey paper on cross-domain sentiment analysis, International Journal of Advanced Research in Computer and Communication Engineering 3 (1) (2014) 5241–5245.
9. H. Wang, Y. Lu, C. Zhai, Latent aspect rating analysis on review text data: a rating regression approach, in: Proceedings of the 16th ACM SIGKDD international conference on Knowledge discovery and data mining, ACM, 2010, pp. 783–792.
10. A. Sharma, S. Dey, A comparative study of feature selection and machine learning techniques for sentiment analysis, Proceedings of the 2012 ACM Research in Applied Computation Symposium, 2012, pp. 1–7
11. S. Tan, J. Zhang, An empirical study of sentiment analysis for Chinese documents, Expert Systems with Applications 34 (4) (2008) 2622–2629.
12. M. Annett, G. Kondrak, A comparison of sentiment analysis techniques: Polarizing movie blogs, in: Advances in artificial intelligence, Springer, 2008, pp. 25–35.
13. B. Pang, L. Lee, S. Vaithyanathan, Thumbs up?: sentiment classification using machine learning techniques, in: Proceedings of the ACL-02 conference on Empirical methods in natural language processing-Volume 10, Association for Computational Linguistics, 2002, pp. 79–86.
14. E. Cambria, A. Hussain, Sentic computing: Techniques, tools, and applications, Vol. 2, Springer Science & Business Media, 2012.
15. B. Pang, L. Lee, Opinion mining and sentiment analysis, Foundations and trends in information retrieval 2 (1-2) (2008) 1–135.
16. G. A. Miller, Wordnet: a lexical database for english, Communications of the ACM 38 (11) (1995) 39–41.
17. P. Turney, Thumbs Up or Thumbs Down? Semantic Orientation Applied to Unsupervised Classification of Reviews, Proceedings of the 40th annual meeting of the association for computational linguistics (ACL), 2002, pp. 417–424
18. J. Han, M. Kamber, J. Pei, Data mining, southeast Asia edition: Concepts and techniques, Morgan kaufmann, 2006.
19. D. I. Morariu, R. G. Creulescu, M. Breazu, Feature selection in document classification.
20. A. Ahmad, L. Dey, A feature selection technique for classificatory analysis, Pattern Recognition Letters 26 (1) (2005) 43–56.
21. B. Liu, Sentiment analysis: A multi-faceted problem, IEEE Intelligent Systems 25 (3) (2010) 76–80.
22. Wang S., Li D., Song X., Wei Y., Li H.: A feature selection method based on improved fishers discriminant ratio for text sentiment classification. Expert Systems with Applications 38.7, 8696-8702 (2011)

Efficient Facial Expression Recognition System Based on Geometric Features Using Neural Network

Ankita Tripathi, Shivam Pandey and Hitesh Jangir

Abstract In this paper, facial expression recognition (FER) system is presented using eigenvector to recognize expressions from facial images. One of the distance metric approaches called Euclidean distance is used to discover the distance of the facial features which was associated with each of the face images. A comprehensive, efficient model using a multilayer perceptron has been advanced whose input is a 2D facial spatial feature vector incorporating left eye, right eye, lips, nose, and lips and nose together. The expression recognition definiteness of the proposed methodology using multilayer perceptron model has been compared with J48 decision tree and support vector machine. The final result shows that the designed model is very efficacious in recognizing six facial emotions. The proposed methodology shows that the recognition rate is far better than J48 and support vector machine.

Keywords Facial expression recognition · Facial expressions
Eigenvectors · Eigenvalues

1 Introduction

With the boom in advancement of human and computer interaction, an emotion recognition system is an interesting area to work in. Out of all, facial emotions are one of the nonverbal communication methodologies. Psychology mainly considers facial feature expressions to study [1]. Generally, in images, two categories of

A. Tripathi (✉) · S. Pandey · H. Jangir
Mody University of Science and Technology, CET, Lakshmangarh, Rajasthan, India
e-mail: er.ankita.tripathi02@gmail.com

S. Pandey
e-mail: shivampandey.cet@modyuniversity.ac.in

H. Jangir
e-mail: hiteshjangir.cet@modyuniversity.ac.in

D.K. Mishra et al. (eds.), *Information and Communication Technology for Sustainable Development*, Lecture Notes in Networks and Systems 10,
https://doi.org/10.1007/978-981-10-3920-1_18

features are used for facial expression recognition (FER): Geometric features contain information about the shape and position of the features, such as appearance features of information about the wrinkles, moles, scars, etc. [2]. But one disadvantage of appearance-based approach is that it is difficult to induce appearance among different individuals.

In this paper, the designed method has been proposed for expression recognition of facial images based on eigenvector and distance metric approach "Euclidean distance". For training 20 facial images and for testing 10 facial images are taken into account. All of these images were cropped into five parts: left eye, right eye, nose, lips, lips and nose together [3]. Using MATLAB, for training phase 20 images and for testing 10 images of six basic expressions (anger, happy, disgust, sad, surprise) are processed and specific eigenvectors in form of a matrix are stored. Following this, Euclidean distance between eigenvectors of the training and testing image is calculated. The testing image is categorized as one of the facial expression out of six, if the Euclidean distance between training and testing eigenvectors is obtained as minimum [2]. To make the system more accurate, we need to come up with a very efficient FER result. Multilayer perceptron (MLP) is to classify the facial features data into six expressions [1]. MLP follows supervised learning technique called as back propagation for training the network [3]. To evaluate the performance of the designed methodology, we analyze the designed method through WEKA with three commonly used classifiers and algorithm: J48 Decision tree (J48), Support Vector Machine (SVM), Multilayer Perceptron (MLP). The remaining part of the paper shows the result in percentage for proving the methodology.

2 Related Work

In this section, the focus is on bringing up a review that has been followed for this research.

Khashman et al. [2] proposed model presented a real-life application experimented over ORL database which performs global and local averaging to obtain information within the encoded facial patterns. Following that neural network is trained. Experimental results show that pattern averaging performs fast and efficient face recognition.

Zhang et al. [3] The proposed paper presented a expression recognition system using facial features along with muscle movement. The paper proposed an approach that uses "salient" distance features, which are computed using patch-based 3D Gabor features, and then patch matching operations are performed. The proposed system output displays high recognition rate with significantly improved performance due to consideration of facial features along with muscle movements and fast processing time.

Perveen et al. [4] The proposed methodology presented facial expression recognition system (FER). Basically, a facial image is fed as input to the feature extraction for face detection. Facial features are computed in order to identify one out of six facial expressions. The designed method for expressions like, surprise, fear, and happy gives efficient results, however, for expressions like, neutral, sad, and angry the recognition rate is not so efficient.

S. Mohseni et al. [5] Main aim of the proposed system is to develop a method for facial movement recognition. The approach is based on facial feature movement after any facial expression. The define algorithm constructs a face model graph based on facial expression movement in each frame and extracts facial features by computing graph's edges, size, and various angle. The system classifies seven facial features including neutral face using support vector machine (SVM) and other MMI classifier. The proposed system results show that computing on the basis of facial movements gives accurate and efficient output to identify different facial expressions.

3 Proposed System

For years, researchers are working on facial expression recognition. Significant efforts were made for efficient detection during this period [6]. This designed methodology presents an FER system that uses facial features (left eye, right eye, nose, lips, lips and nose together) [5]. Here, many facial images of an individual with different facial expressions are used where left eye, right eye, nose, lips, lips and nose together facial features are considered. These necessary features from different facial expressions were cropped as five different feature images used for computation of eigenvector. Further, Euclidean distance is calculated and then this information is used to train a multilayer perceptron network [7].

The facial images of persons used for training and testing of the neural network, are of different age group. A total of 30 face images from Karolinska Directed Emotional Faces (KDEF) database with different face emotions are used, where 20 images are for training network and 10 images are for testing [15]. Figure 1 shows a sample of few images that are used in various expressions.

3.1 Image Processing

At first, Image processing is required for training and testing images of the neural network. This is basically done to reduce the computational cost and to obtain a fast

Fig. 1 Sample of facial images with various expressions

recognizing system while representing the neural network with adequate data and information of each face to achieve efficient results. The multilayer perceptron network is trained using six basic facial expressions of each individual, which is attained by computing eigenvectors of each cropped facial feature image, and following that the Euclidean distance between cropped facial feature image is computed (like left eye to left eye) [9]. Once training is finished, the network is tested using the six different expressions.

The algorithm to compute the Euclidean distance of various features for expressions:

To compute the mean

$$\varphi = \frac{1}{N}\sum_{m-1}^{N} \tau_m \tag{1}$$

φ *mean of the training image*
τ_m *vector of training image*
N *number of training images*

Subtracting off the mean for each dimension from vector of training image

$$\varnothing_m = \tau_m - \varphi \tag{2}$$

$\varnothing_m$ *subtracted vector of training image*
τ_m *vector of training image*
φ *mean of the training set*

Calculating the covariance matrix

$$C = \frac{1}{N}\sum_{m-1}^{N} \varnothing_m \varnothing_m^{\tau} \tag{3}$$

C *covariance matrix*

Calculating the eigenvectors and eigenvalues of the covariance matrix

$$D = E.C.E^{-1} \tag{4}$$

C *covariance matrix*
E *matrix of all Eigen vectors of C*
D *diagonal matrix of all Eigen values*

Compute Euclidean Distance: Classification will be performed by correlating the feature vectors of the training images with feature vector of the input testing image.

$$d_e(x,y) = \sqrt{\sum_{i=1}^{n} (x_i - y_i)^2} \tag{5}$$

d *Euclidean distance between two Eigen vectors*

3.2 *Training and Testing in WEKA*

The multilayer perceptron algorithm is used for the execution of the designed FER system, due to its accuracy and improved efficiency. The neural network holds an input layer with five neurons that takes the Euclidean distance values of facial features, a hidden layer with five neurons, and an output layer with five neurons, which is the number of people with the same facial expression [1, 3].

For simulation and comparison of data in Weka obtained from MATLAB, two data sets are prepared. Training dataset will have data of 20 images and the testing dataset will have data of 10 images. These datasets are then used to find output using multilayer perceptron neural network (MLP) [12, 13]. After this, J48 decision tree (J48) and support vector machine (SVM) classifier are applied over the same dataset and the results are compared.

4 Simulation and Evaluation

4.1 *Multilayer Perceptron Neural Network*

The ARFF file of training and testing set is prepared from MATLAB simulation and is loaded to WEKA and various facial features (left eye, right eye, nose, lips, lips and nose together) are displayed in the form of histograms. WEKA calculates a statistical range and displays it as min and max values for all the facial features [1, 9, 13]. On simulating training and testing datasets on a multilayer perceptron neural network with a learning rate of 0.3 rate and training time of 0.12 s and a result is obtained, i.e., correctly classifying recognition rate of 80% while incorrectly classified as 20% [14]. Figure 2, shows MLP neural network and Fig. 3, shows classifier output showing all the details of neural network, respectively.

4.2 *J48 Decision Tree*

The ARFF file of training and testing set simulated on a multilayer perceptron neural network is loaded to WEKA for comparison with J48 decision tree classifier.

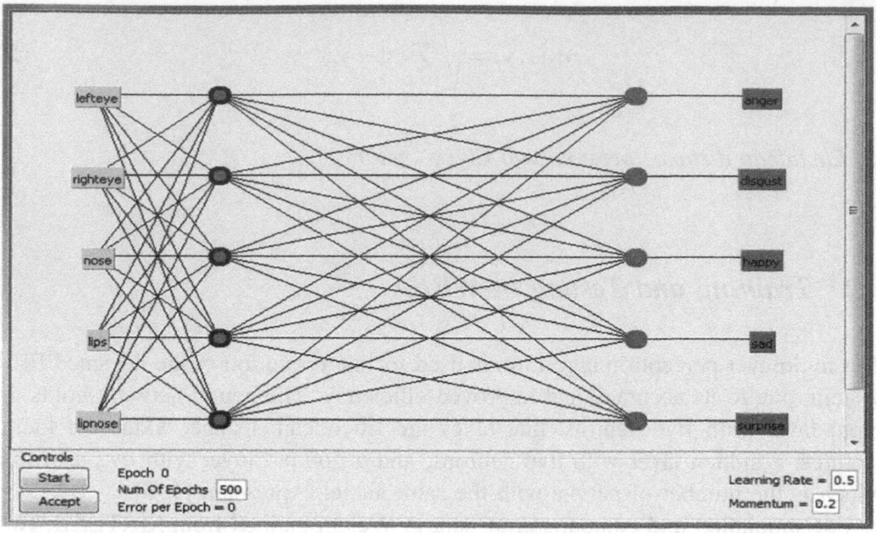

Fig. 2 Shows MLP neural network

```
Classifier output

Time taken to build model: 0.12 seconds

=== Evaluation on test set ===

Time taken to test model on supplied test set: 0 seconds

=== Summary ===

Correctly Classified Instances           8               80      %
Incorrectly Classified Instances         2               20      %
Kappa statistic                          0.75
Mean absolute error                      0.1669
Root mean squared error                  0.2558
Relative absolute error                 52.1414 %
Root relative squared error             63.7736 %
Coverage of cases (0.95 level)         100      %
Mean rel. region size (0.95 level)      66      %
Total Number of Instances               10

=== Detailed Accuracy By Class ===

                 TP Rate  FP Rate  Precision  Recall   F-Measure  MCC      ROC Area  PRC Area  Class
                 1.000    0.000    1.000      1.000    1.000      1.000    1.000     1.000     anger
                 1.000    0.250    0.500      1.000    0.667      0.612    1.000     1.000     disgust
                 1.000    0.000    1.000      1.000    1.000      1.000    1.000     1.000     happy
                 0.000    0.000    0.000      0.000    0.000      0.000    0.750     0.667     sad
                 1.000    0.000    1.000      1.000    1.000      1.000    1.000     1.000     surprise
Weighted Avg.    0.800    0.050    0.700      0.800    0.733      0.722    0.950     0.933
```

Fig. 3 Shows classifier output showing details of MLP neural network

The obtained recognition rate was compared with an MLP neural network for efficient result.

On simulating training and testing datasets on J48 decision tree to compare with MLP recognition rate, the result obtained is, correctly classifying recognition rate of 70% while incorrectly classified as 30% [14]. Figure 4, shows classifier output showing all the details of J48 decision tree classifier and Fig. 5 shows J48 decision tree, respectively.

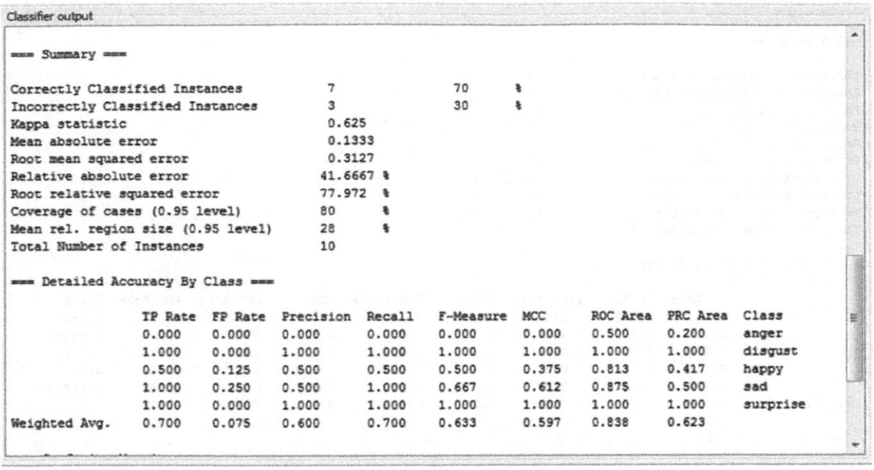
Classifier output

=== Summary ===

Correctly Classified Instances	7	70 %
Incorrectly Classified Instances	3	30 %
Kappa statistic	0.625	
Mean absolute error	0.1333	
Root mean squared error	0.3127	
Relative absolute error	41.6667 %	
Root relative squared error	77.972 %	
Coverage of cases (0.95 level)	80 %	
Mean rel. region size (0.95 level)	28 %	
Total Number of Instances	10	

=== Detailed Accuracy By Class ===

	TP Rate	FP Rate	Precision	Recall	F-Measure	MCC	ROC Area	PRC Area	Class
	0.000	0.000	0.000	0.000	0.000	0.000	0.500	0.200	anger
	1.000	0.000	1.000	1.000	1.000	1.000	1.000	1.000	disgust
	0.500	0.125	0.500	0.500	0.500	0.375	0.813	0.417	happy
	1.000	0.250	0.500	1.000	0.667	0.612	0.875	0.500	sad
	1.000	0.000	1.000	1.000	1.000	1.000	1.000	1.000	surprise
Weighted Avg.	0.700	0.075	0.600	0.700	0.633	0.597	0.838	0.623	

Fig. 4 Shows classifier output showing details of J48 decision tree classifier

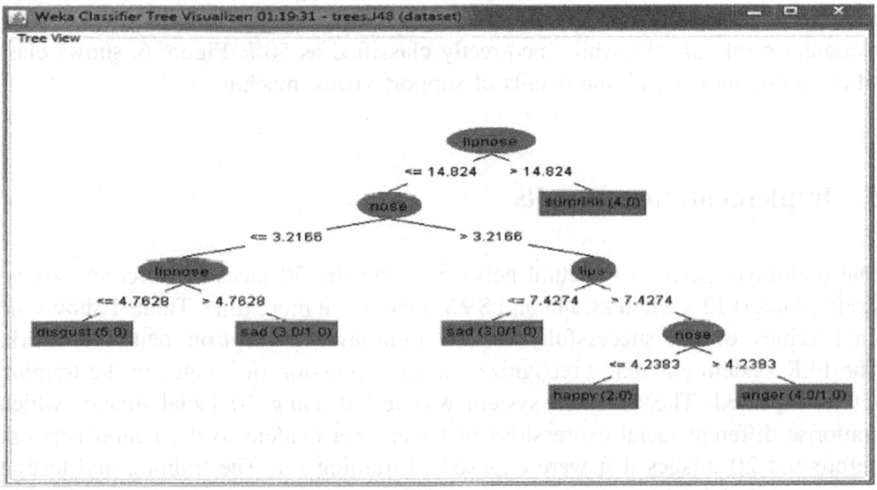

Fig. 5 Shows J48 decision tree classifier

4.3 Support Vector Machine

The ARFF file of training and testing set simulated on multilayer perceptron neural network and J48 decision tree is loaded to WEKA for comparison with support vector machine (SVM). The obtained recognition rate was compared with support vector machine (SVM) for efficient results [13, 14].

```
Classifier output

=== Summary ===

Correctly Classified Instances          5               50      %
Incorrectly Classified Instances        5               50      %
Kappa statistic                         0.375
Mean absolute error                     0.296
Root mean squared error                 0.395
Relative absolute error                92.5     %
Root relative squared error            98.4875 %
Coverage of cases (0.95 level)         90       %
Mean rel. region size (0.95 level)     80       %
Total Number of Instances              10

=== Detailed Accuracy By Class ===

                 TP Rate  FP Rate  Precision  Recall   F-Measure  MCC      ROC Area  PRC Area  Class
                 0.000    0.000    0.000      0.000    0.000      0.000    0.563     0.222     anger
                 1.000    0.375    0.400      1.000    0.571      0.500    0.813     0.400     disgust
                 1.000    0.250    0.500      1.000    0.667      0.612    0.875     0.500     happy
                 0.000    0.000    0.000      0.000    0.000      0.000    0.563     0.222     sad
                 0.500    0.000    1.000      0.500    0.667      0.667    0.750     0.600     surprise
Weighted Avg.    0.500    0.125    0.380      0.500    0.381      0.356    0.713     0.389
```

Fig. 6 Shows classifier output showing details of support vector machine

On simulating training and testing datasets on support vector machine to compare with MLP recognition rate, the result obtained is, correctly classifying recognition rate of 50% while incorrectly classified as 50%. Figure 6, shows classifier output showing all the details of support vector machine.

5 Implementation Results

The multilayer perceptron neural network learnt the 30 faces of different expressions within 0.12 s whereas, J48 and SVM take much more time. Table 1 shows the final values of the successfully trained multilayer perceptron neural network. The FER system efficiently recognizes facial expression of images in the training set as expected. The designed system was tested using 10 facial images which comprise different facial expressions that were not evident to the neural network before and 20 images that were exposed as training set. The training and testing dataset of facial images was fed to multilayer perceptron neural network and obtained correctly identified yielding 80% recognition rate. The same dataset was

Table 1 Final parameters of the successfully trained multilayer perceptron neural network

Input layer nodes	5
Hidden layer nodes	5, 10
Output layer nodes	5
Learning rate	0.5
Training time (s)	0.12
Mean absolute error (%)	0.1669
Root mean square error (%)	0.2558

Table 2 Shows correctly identified recognition rate

Classifier	Correctly identified recognition rate (%)
MLP	80
J48	70
SVM	50

used to compare results with J48 decision tree classifier and support vector machine. And obtained correctly identified yielding 70% and 50% recognition rate, respectively. The system designed to finally yield an efficient and improved result of 10% in terms of accuracy. Table 2 shows correctly identified recognition rate.

6 Future Work

The accuracy is not sufficient as we are considering images that are not much varying in terms of noise. Other than this, designed expression recognition system has been tested with limited file types such as jpg, png, and jpeg extension only. It could be tested with gif, tiff, and bmp types images also for more accurate result. The system is tested by database of simple images in which persons do not have glasses, facial hairs, and any disability in facial features. In the near future, we can emphasis on images with all these changes.

7 Conclusion

This proposed methodology, which is based on eigenvector, described another system for efficient facial expression recognition. The method approximates five essential facial features (left eye, right eye, nose, lips, lips and nose together) from five different facial expressions (anger, disgust, happy, sad, surprise), and trains a multilayer perceptron neural network using the Euclidean distance metric of features to learn the face. Once training is finished, the neural network can identify the faces with various facial expressions or emotions. The result of this method was achieved through a real-life scenario using 30 face images showing six different expressions for each. An overall recognition rate of 80% with accuracy was achieved. The overall time for training neural network is 0.12 s for face recognition. On comparing the results for efficiency check with J48 decision tree and support vector machine (SVM), classifier 70 and 50% recognition was obtained which shows that the method described in our work gives the best result.

References

1. K. Youssef and P. Woo, "A Novel Approach to Using Color Information in Improving Face Recognition Systems Based on Multi-Layer Neural Networks" In: INTECH Open Access Publisher (2008).
2. A. Khashman, "Intelligent face recognition: local versus global pattern averaging." In: Advances in Artificial Intelligence, pp. 956–961, Springer Berlin Heidelberg, (2006).
3. L. Zhang and D. Tjondronegoro, "Facial expression recognition using facial movement features," In: Affective Computing, pp. 219–229, IEEE Transactions (2011).
4. N. Perveen, S. Gupta, and K. Verma, "Facial expression recognition using facial characteristic points and Gini index," In: Engineering and Systems (SCES), Students Conference, pp. 1–6, IEEE (2012).
5. J. Kalita and K. Das, "Recognition of facial expression using eigenvector based distributed features and Euclidean distance based decision making technique" In: International Journal of Advanced Computer Science and Applications, arXiv preprint arXiv:1303.0635 (2013).
6. S. Mohseni, Z. Niloofar and S. Ramazani, "Facial expression recognition using anatomy based facial graph," In: Systems, Man and Cybernetics (SMC), IEEE International Conference, pp. 3715–3719, IEEE (2014).
7. S. Thuseethan and S. Kuhanesan, "Eigenface Based Recognition of Emotion Variant Faces" In: Computer Engineering and Intelligent Systems (2016).
8. P. Carcagnì, D. Marco, L. Marco, and D. Cosimo, "Facial expression recognition and histograms of oriented gradients: a comprehensive study," In: SpringerPlus, pp. 1–25 (2015).
9. P. Belhumeur, N. Peter, J. Hespanha and D. Kriegman, "Eigenfaces vs. fisherfaces: Recognition using class specific linear projection," In: Pattern Analysis and Machine Intelligence, pp. 711–720. IEEE Transactions, (1997).
10. P. Campadelli, R. Lanzarotti, and G. Lipori, "Automatic Facial Feature Extraction for Face Recognition" In: Itech Education and Publishing (2007).
11. Levin, Anat, and Amnon Shashua. "Principal component analysis over continuous subspaces and intersection of half-spaces" In Computer Vision—ECCV, pp. 635–650, Springer Berlin Heidelberg (2002).
12. D. Anifantis, E. Dermatas and G. Kokkinakis, "A Neural Network Method For Accurate Face Detection on Arbitrary Images" In: International Conference on Electronics, Circuits, and Systems, pp. 109–112, IEEE"(1999).
13. K. Curran, X. Li and N. Caughley, "The Use of Neural Networks In Real-time Face detection" In: Journal of Computer Sciences, pp. 47–62, Science Publications (2005).
14. W. Zheng, X. Zho, C. Zou and L. Zhao, "Facial Expression Recognition Using Kernel Canonical Correlation Analysis (KCCA)" In: IEEE Transactions On Neural Networks, pp. 233–238, IEEE Transactions (2006).
15. D. Lundqvist, A. Flykt and A. Ohman, "The Karolinska Directed Emotional Faces-KDEF," CD-ROM from Department of Clinical Neuroscience, Psychology section, Karolinska Institutet, Stockholm, Sweden. ISBN 91-630-7164-9, 1998.

Extended Four-Step Travel Demand Forecasting Model for Urban Planning

Akash Agrawal, Sandeep S. Udmale and Vijay K. Sambhe

Abstract For years, the traditional four-step travel demand forecasting model has helped the policy makers for making decisions regarding transportation programs and projects for metropolitan regions. This traditional model predicts the number of vehicles of each mode of transportation between the traffic analysis zones (TAZs) over a period of time. Although this model does not suggest a suitable region where transportation project can be deployed. Therefore, this paper extends one more step to traditional four-step model for suggesting the zones which are in higher need of a highway transportation program. The severity of traffic load is the basis for results of the added step.

Keywords Urban planning · Travel demand forecasting · Mathematical model · Project region estimation

1 Introduction

The four-step travel demand forecasting model remains the backbone of the transportation and urban planning processes. It helps in predicting the use of existing transport facility by the residents and the vehicles in the system after certain years. The planners take the decisions regarding implementation of transportation services in the zone of the metropolitan region. These decisions are based on the evaluation of the outcomes of implementing the substitute courses of action including such transportation services such as new highways, etc. in those zones. Furthermore, these decisions are often influenced by various factors which are local to that particular region and not based on any mathematical calculations.

The number of vehicles in each available link in the region under consideration is obtained as the output of the four-step travel demand model. This data can be

A. Agrawal (✉) · S.S. Udmale · V.K. Sambhe
Department of Computer Engineering and Information Technology,
Veermata Jijabai Technological Institute, Mumbai, India
e-mail: akashagrawal108@gmail.com

D.K. Mishra et al. (eds.), *Information and Communication Technology for Sustainable Development*, Lecture Notes in Networks and Systems 10,
https://doi.org/10.1007/978-981-10-3920-1_19

utilized to estimate the total number of vehicles using that link which is a part of the shortest path from a source to some destination. This estimation will be useful for the identification of some heavily loaded links in the present roadway system.

Based on the requirements of analysis, a technique is generally chosen from several ones available for travel demand forecasting which vary in accuracy, complexity, cost, and level of effort. For example, the sketch tools deal with more number of alternatives but the specified level of detail is low. On the contrary, microanalysis tools work with great level of detail and consider less number of alternatives as compared to sketch tools. Traditional tools form a balance between these techniques [1]. The four-step model is a major way in traditional tools for forecasting of travel demand analysis [2].

The basic model consists of four steps namely trip generation, trip distribution, mode choice, and trip assignment. Trip generation translates the socioeconomic data of all zones to number of trips generated and attracted by those zones. The relationship between travel and socioeconomic characteristics is quantified in this step. Yao et al. [3] present a modified trip generation model which calculates destination attractiveness as a function of parameters such as distance of a zone from center of city, traffic advantage index, and land acreage. A trip generation model combining OD matrix estimation and land classification is proposed by Liu and Hu [4].

After trip generation, the planner knows how many trips are produced by and how many are attracted to every TAZ in the region of consideration. But, where the produced trips are going and the attracted trips are coming from is evaluated in trip distribution step. This step gives the split of total productions and attractions into trips from every zone to every other zone in the region in the form of origin-destination (OD) matrix. Potential energy theory is taken into account for the process of trip distribution by Huiying and Wenbiao [5]. Goel and Sinha [6] revealed an Adaptive Neural Fuzzy Inference System (ANFIS)-based trip distribution model for Delhi, India.

The next step, i.e., mode split, now splits the zonal trips into the trips undertaken through the use of every particular available mode. Mode choice is influenced by many factors, such as traveler characteristics, trip characteristics, qualitative factors, etc. This step gives different tables for every available mode in terms of trips per person per day. Wan et al. [7] state that traffic resource supply and land use greatly impact the decision of a traveler to choose a mode of transportation. Li et al. [8] explore the relationship between the residential, workplace locations, and the mode choice of a commuter and conclude that they are intricately linked. Again Hu et al. [9] try to analyze the multidimensional-coupling aspects of the spatial-temporal relationship of travel demand and land use.

The fourth step, trip assignment, first calculates the shortest path in terms of generalized travel cost in terms of time from every node to every other node. Then, the number of trips for all hops in every shortest path is assigned to each available link in the region. The final output is the number of vehicles of each type of mode of transportation that will be present in each of the links after a predefined period of time in years. Chen and Yan [10] present the competitive nature of space-time of the traffic zones and use F–W algorithm for trip assignment considering the generalized costs and impedances. Hajbabaie and Benekohal [11] propose a technique for simultaneous traffic signal optimization and optimal traffic assignment with a new objective function as weighted trip maximization considering many factors, such as gridlocks, signal timing, traffic loading, etc.

The aim in this paper is to design a mathematical model for a new step to be added to the traditional travel demand forecasting model. The purpose of this to be added step is to identify the heavily loaded roadways between the TAZs and suggest the zones containing such loaded links as the best zones to implement a highway transportation program from the overall region. The next section, implementation focuses on the details of the mathematical procedures used to implement the added step to the model. Next, results throw light on the outcomes of the implementation in the form of heavily loaded links in the zones. Finally, the conclusion summarizes the findings in this paper and suggest the future scope.

2 Implementation

The data used in this implementation are primarily derived from the study [12], namely, the growth rates of all socioeconomic factors as well as the values of all of the factors for all zones, the cost in terms of time to travel from all zones to all other zones, etc. Also, the current productions and attractions, trip characteristics constants and coefficients, available links between the zones, and other some constants, were also extracted from the same reference paper. Though, some methods apart from the referenced paper were used which resulted in varying values of the constants as compared to those given in the referenced one, some of the constants were purely assumed. As the main objective is not to mirror or enhance the traditional four-step model, but to propose to add a new step in the existing model for project region estimation, the variations in results of the implementation are acceptable for the current study. Also, a hypothetical metropolitan region as shown in Fig. 1 is considered with ten TAZs but with the study data [12]. The result of the trip assignment step to be supplied as the input to the proposed fifth step is as given in the Table 1 in the next section of this text.

At first, from among the shortest paths available to us from the Trip Assignment step, highest number of hops, or intermediate zones, which need to be crossed by any vehicle while traveling from any zone to any other zone is identified and is called the Maximum Path Length denoted by h. Later on, the shortest path, which has the Maximum Path Length while traveling from one zone to another zone, is searched for and is identified as the needful subregion for a transportation project.

If there is more than one shortest path from any zone to any other zone having Maximum Path Length, then such paths are searched for. Say, there are n shortest paths with path length equal to h. Then p_i^h be the ith path from among n paths with path length h. The total traffic load L is calculated at each p_i^h summing up traffic from every mode from K modes of transportation.

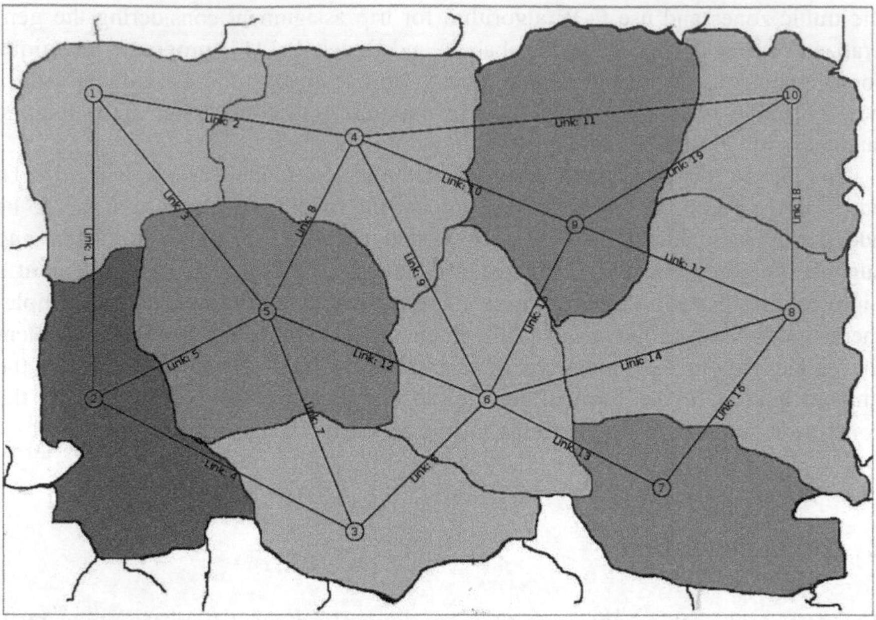

Fig. 1 Map of a hypothetical urban region with ten TAZs

Table 1 Occupancy-based assignment for all modes in each link

Link	Cars	Buses	Rickshaws
1–2	7275	2073	27593
1–4	3405	2499	23035
1–5	101	270	1760
2–3	3909	2354	37840
2–5	397	417	3777
3–6	6321	3569	36557
3–5	3643	1194	15625
4–5	3874	1545	18422
4–6	2162	1599	15480
4–9	3603	1571	17217
4–10	650	1018	7537
5–6	9234	2672	36873
6–7	12481	3505	46500
6–8	4720	2417	25567
6–9	9368	2602	35225
7–8	8351	1838	28072
8–9	15291	1748	27217
8–10	2793	1747	18238
9–10	677	492	5078

$$L_i = \sum_{j=1}^{k} L_i^j$$

This gives the total traffic load at each p_i^h in the region.

Let L_i^{max} be the maximum traffic load at some link p_i and let z represent the zones p_i. Then,

$$\forall z \in p_i$$

are the most loaded zones in terms of vehicular traffic among all the zones. The number of zones in such p_i^h is obviously h, as every p_i for which the sum of loads of every mode was calculated, have h as their path length.

These h zones, as the output of this purposefully added step, are identified as the most loaded zones with the maximum traffic flowing through them in peak hours in the region of interest. And thus, a transportation project is suggested, as the output of this step as well as of the whole modeling process, to be undertaken covering these h zones so that the time and cost required by the travelers to commute between these zones diminishes significantly.

3 Results

The last step, i.e., trip distribution gave us the number of vehicles of each type present in each available link in the region under consideration. The results from the trip assignment step are summarized in the Table 1. The results are obtained after considering the occupancy of the vehicles and total number of trips per person per

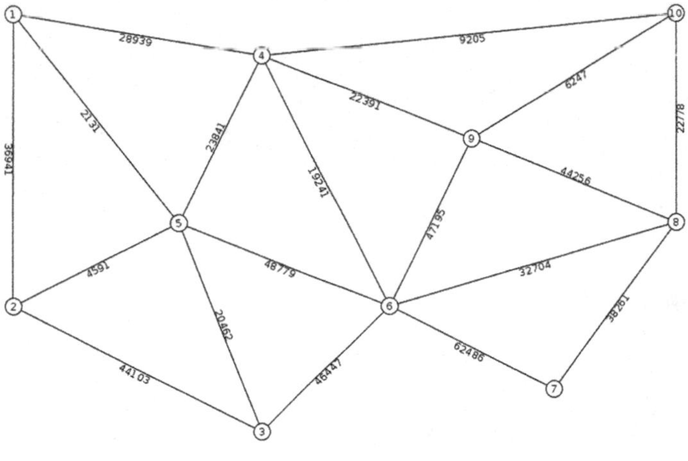

Fig. 2 Total vehicles in each link

day of the peak hours for all nineteen links present between the TAZs. Since we are concerned about the roadways transportation projects, the modes of transportation chosen here are car, bus, and rickshaw.

From this, we can easily calculate and present the total vehicles present in each link in the form of a graph as shown in Fig. 2.

Here, length of the longest shortest path in terms of hops from one zone to another is found out to be four, i.e., no path includes more than four nodes including source and destination and there are 14 such shortest paths with path length equal to four.

Summing up traffic from every mode from K modes of transportation, the path *[2-3-6-7]*, *[2, 3, 6, 9]*, *[2, 3, 6, 8]*, and *[1, 4, 6, 7]* are found out to be very heavily

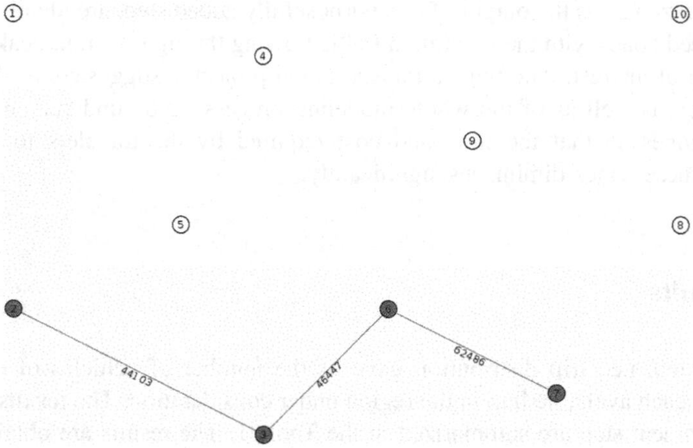

Fig. 3 Heavily loaded link 1

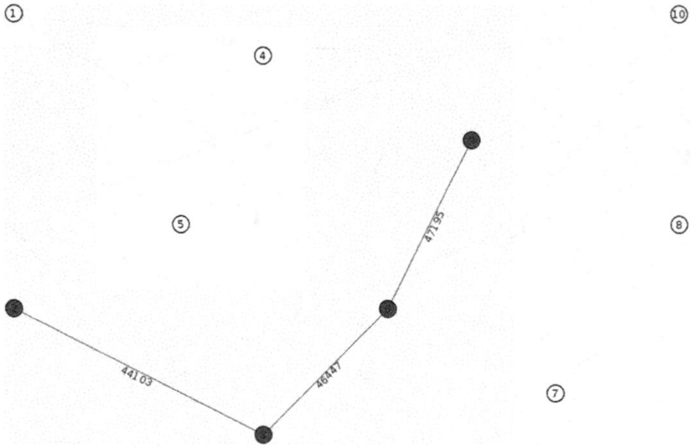

Fig. 4 Heavily loaded link 2

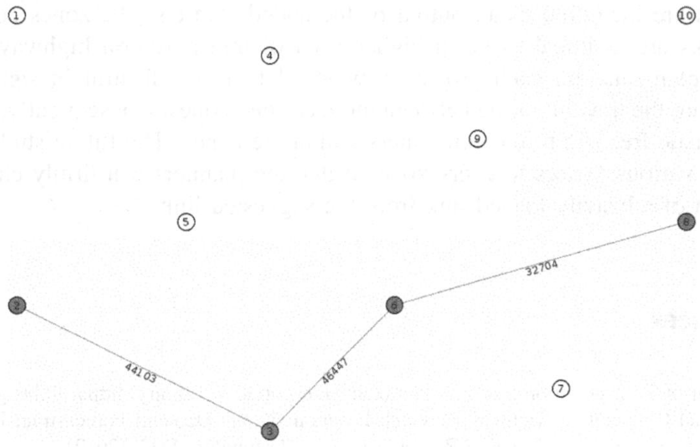

Fig. 5 Heavily loaded link 3

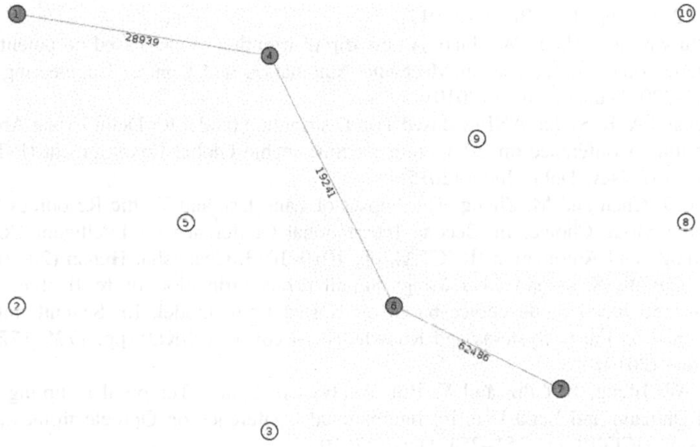

Fig. 6 Heavily loaded link 4

loaded in decreasing order of the total traffic as presented by Figs. 3, 4, 5, and 6, respectively.

4 Conclusion

This paper gives a technique for estimating the zones which are in a need of a highway transportation project. The proposed estimation approach is based on the total number of vehicles passing by the zones. The four most heavily loaded links in the zones are identified as an output of the added step and the zones containing those links are assumed to be in higher need of transportation highway project. Upon implementation, such project is believed to prove fruitful in significantly diminishing the traveling time between the identified zones. Consequently, this will ensure hustle-free trip to the commuters within the zones. The future study should associate various factors to every zone so that the planners can firmly choose the subregion of a heavily loaded link from the suggested links.

References

1. Bureau of Transportation Statistics National Transportation Library, http://nlt.bts.gov.
2. National Cooperative Highway Research Program: Travel Demand Forecasting: Parameters and Techniques. Transportation Research Board. Washington, D.C. (2012).
3. L. Yao, H. Guan and H. Yan: Trip generation model based on destination attractiveness. Tsinghua Science and Technology 13, 632–635 (2008).
4. C. Liu and D. Hu: A forecasting method of trip generation based on land classification combined with OD matrix estimation. In: 7th Advanced Forum on Transportation of China (AFTC 2011), pp. 1–4. Beijing (2011).
5. Wen Huiying and Zhang Wenbiao: A new trip distribution model based on potential theory. In: International Conference on Mechanic Automation and Control Engineering (MACE), pp. 717–720. Wuhan, China (2010).
6. S. Goel and A. K. Sinha: ANFIS Based Trip Distribution model for Delhi Urban Area. In: 2nd International Conference on Computing for Sustainable Global Development (INDIACom), pp. 453–457. New Delhi, India (2015).
7. X. Wan, J. Chen and M. Zheng: The Impact of Land Use and Traffic Resources Supply on Commute Mode Choice. In: Second International Conference on Intelligent Computation Technology and Automation (ICICTA), pp. 1010–1013. Changsha, Hunan (2009).
8. X. Li, Chunfu Shao and Liya Yang: Simultaneous estimation of residential, workplace location and travel mode choice based on Nested Logit model. In: Seventh International Conference on Fuzzy Systems and Knowledge Discovery (FSKD), pp. 1725–1729. Yantai, Shandong (2010).
9. P. Hu, W. Jifeng, J. Yulin and Y. Rui: Analysis of Spatial-Temporal Coupling for Urban Travel Demand and Land Use. In: International Conference on Optoelectronics and Image Processing (ICOIP), pp. 251–254. Haiko (2010).
10. J. Chen and Q. P. Yan: Trip assignment model for urban traffic system based on network equilibrium. In: International Conference on Mechanic Automation and Control Engineering (MACE), pp. 4680–4683. Wuhan (2010).
11. A. Hajbabaie and R. F. Benekohal: A Program for Simultaneous Network Signal Timing Optimization and Traffic Assignment. In: IEEE Transactions on Intelligent Transportation Systems, 16, 2573–2586 (2015).
12. Ahmed, B.: The Traditional Four Steps Transportation Modeling Using Simplified Transport Network: A Case Study of Dhaka City, Bangladesh. IJASETR 1(1) (2012).

Authentication Framework for Cloud Machine Deletion

Pooja Dubey, Vineeta Tiwari, Shweta Chawla and Vijay Chauhan

Abstract Today Digital Investigation on the cloud platform is a challenging task. As cloud follows notion of pay as you demand people are attracted to adopting this technology. But an unclear understanding of control, trust and transparency are the challenges behind the less adoption by most companies. Investigation in the cloud platform is hard to collect the strong evidences because resource allocation, dynamic network policy are facilitated on demand request fulfillment. This is why, it is difficult to perform forensic analysis in such a virtualized environment because the state of the system changes frequently. Even to prevent the deletion of system on cloud is a tough task. This paper will cover all the theoretical concepts of cloud along with the challenges presented in NIST guidelines. Through this paper, we explore the existing frameworks, loopholes, and suggest some possible solutions that will be a roadmap for forensic analysis in future.

Keywords Digital investigation · Cloud · Forensic · Virtualized · NIST

1 Introduction

The importance of Cloud in today's world can be understood by the wide usage in most of the companies. But as everything thing exist with some drawbacks, Cloud also have some such loopholes which are yet to overcome in near future. We have listed out the drawbacks and challenges in our earlier paper [1]. The existing

P. Dubey (✉) · V. Chauhan
Gtu Pg School, Ahmedabad, Gujarat, India
e-mail: pd.0509@gmail.com

V. Tiwari
CDAC-Acts, Pune, India
e-mail: vineetat@cdac.in

S. Chawla
SC Cyber Solutions, Pune, India
e-mail: sccybersolutions@gmail.com

D.K. Mishra et al. (eds.), *Information and Communication Technology for Sustainable Development*, Lecture Notes in Networks and Systems 10,
https://doi.org/10.1007/978-981-10-3920-1_20

frameworks have been discussed in previous paper [1]. Even in the earlier paper, major problems faced in Cloud were discussed and out of that problems we are going to address the problem of deletion on the Cloud [1].

Major problem faced by the investigators in the domain of Cloud is unauthorized deletion of Cloud machines. There is a live example which is the case occurred few months back is shown in Fig. 1.

There was an incident where one person's VM was deleted by his own friend. Assume both persons as Jack and John, respectively. John purchased some amount of cloud space and became CSP himself. Now Jack was in need to get cloud platform to host his mail exchange server. On the request of jack, his friend John hosted server on cloud. Jack started growing good with his business and John due to greedy intentions thought to take over his business. So John somehow was successful in hacking Jack's system and sent a mail requesting to delete his VM and deleted the VM too. Now when investigation was held and investigator asked John to give access to logs saying that he was requested through a mail by Jack to do so. Thus, the situation arises for forensic investigation and without the data evidences it is of no use. And if a case is filed against the criminal it would be a very long process that we cannot even imagine sometime. Therefore, a mechanism is proposed that will prevent the unauthorized deletion of VM on cloud.

The major concerns listed out by NIST that exists in Cloud today which are needed to overcome includes: recovering the overridden data, evidence correlation, log format synchronization, log capture, and most importantly the deletion on Cloud. The deletion in present Cloud Scenario takes place without any Authorization as shown in Fig. 2 and this in turn is believed to be the major drawback in Cloud. So, this issue is considered and feasible solution for it has been given as a proposed solution.

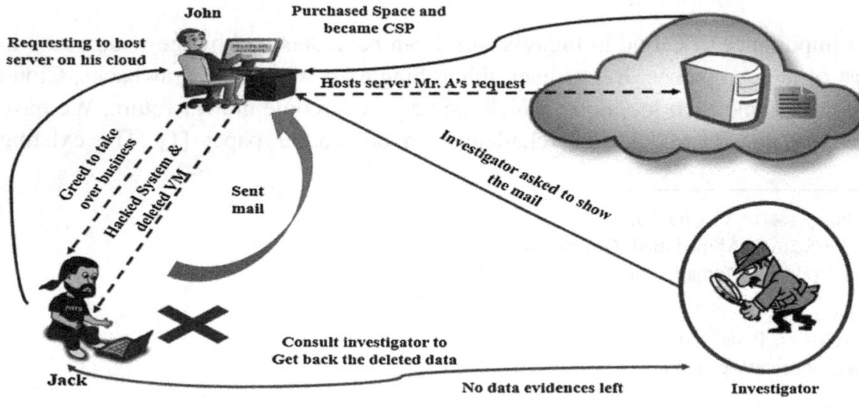

Fig. 1 Live example of current issue in cloud

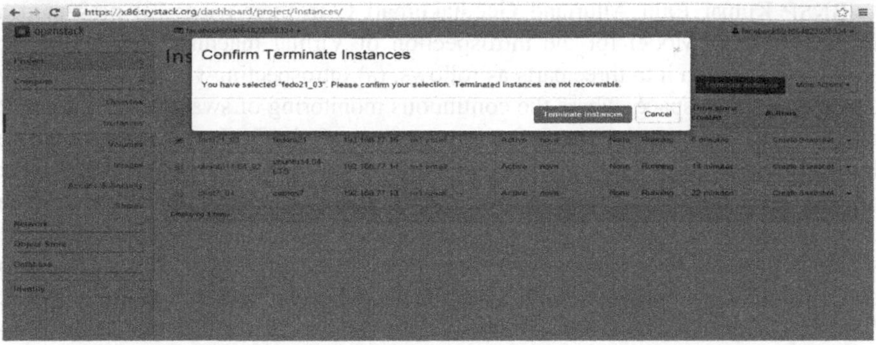

Fig. 2 Challenge in current scenario

The remainder of this paper is organized as follows: In section 2 Literature Review of existing mechanisms has been provided, in Section 3 proposed framework has been discussed and in the next section experimental results have been presented followed by the conclusion and future work.

2 Existing Frameworks

A critical assessment of the work has been done so far on Cloud Forensics to show how the current study related to what has already been done. Numerous companies are now a days migrating to cloud due to greater economic issues. But for small-and medium-sized companies the security of information is the primary concern. For these companies, the best alternative is to use managed service which is also known as outsourced service in which they are provided with the full package of service including antivirus software to security consulting. And the alternative model that provides such outsourced security is known as Security as a service (SECaaS). Scientists and researchers together presented their latest ideas and findings on what the real world scenario is and what all efforts are made but it was found that despite of being so much research work in the field of cloud forensic there is only a fraction part of the total work that has contributed for the wealth of the society. However, cloud came into existence in the mid of 90s yet it is not taken up by everyone fully. There have been lots of works before in this field and variety of methods for the forensic analysis of cloud yet there is a huge room for improvement that needs to be carried forward into the research.

The technique of Forensic investigation of VM using snapshots as evidence was introduced by Deevi Radha Rani and Geethakumari [2]. In that mechanism, software stored and maintained snapshots of running VM selected by the user which acted as a good evidence.

BKSP Kumar Raju Alluriand Geethakumari G in their paper [3] presented a Digital Forensic Model for the introspection of Virtual machine. They split the entire Introspection into three parts as follows. (a) Introspecting virtual machines by considering swap space, where the continuous monitoring of swap space is done. It provides the information about current process of the VM. (b) A conjunctive Introspection approach for virtual instances in cloud computing. In this, three models were used namely out of band, derivation-based and in-band delivery in order to collect as much accurate data evidence can be collected and reduce the semantic gap. But later, out of these three methods in-band method was proved to be less useful for live forensic **as** it modified the data at the time of collection phase. (c) A Terminated Process-based Introspection for virtual machines in cloud computing. This captured every process that was terminated and later was improvised to capture only the processes that were found doubtful.

In Switzerland, Hubert Ritzdorf, Nikolaos Karapanos, and Srdjan Capkun presented a system called IRCUS (Identification of Related Content from User Space) which assisted the users to delete the related content in a very secure way [4].

A mechanism referred as Digital Evidence Detection in Virtual Environment for Cloud Computing was presented by Mr. Digambar Powar and Dr. G. Geethakumariat Hyderabad [5]. In this paper, the traditional digital forensic analysis techniques were used to find and analyze the digital evidence in virtual environment for cloud computing. Also, the ways in which forensic analysis can be possible in virtual environment that is omnipresent in the cloud has been discussed.

In India, Saibharath S and Geethakumari G proposed the solution of remote evidence collection and preprocessing framework using Struts and Hadoop distributed file system [6]. In [7], the author tried to decrease the workload at client side and provide integrity and security to the stored data. In USA, Scenario-based Design for a Cloud Forensics Portal was proposed that was used for data collection from the virtual environment to improve the ability of hypervisor that could provide the effective monitoring to portals [8].

In summary, the work presented in this paper is built on previous research to explore how security of data stored on cloud relates to people's trust. While earlier work focused on the data storage impacts people, we focus on its impact on the world wide acceptance of cloud. Further, we are able to study the cloud security with the keen point of view and make efforts toward securing the unauthorized deletion of data on cloud by asking to give strong and dual authentication before deleting the material.

3 Proposed Framework

In this work, the main focus is on to prevent the unauthorized deletion on Cloud. Preservation of evidences is an ultimate goal behind performing cloud forensics. In the virtual scenario, virtual machines contain evidences. If once VMDK (Virtual Machine Disk file) is destroyed, it is impossible to recover your VM. At present there does not exist a single mechanism that can recover a destroyed (deleted) VM again which is the flaw in VM itself. All the activities on the VM is logged in VM, whereas activities of CSP (Cloud Service Provider) is logged on the server. So even if someone deleted the VM, all the evidences will be lost. This creates a disaster for the user and acts as a barrier for a forensic investigator to dig out the private crucial data of user that was stored in the VM sometime.

Thus, through this work, keeping in mind the limitations of existing systems and the burning challenge in the current Cloud Scenario, a framework is being proposed as shown in Fig. 3 that will enable the secure deletion on cloud. The two main factors that would be taken into consideration while implementing the proposed system to provide the two-factor authentication in order to prevent the unauthorized deletion would be:

1. Email Verification
2. One-Time Password (OTP).

This will ensure that whenever a deletion of VM on cloud is requested before the fulfillment of request two-factor authentication will be performed. If and only if the authentication is successful the deletion would be proceeded else it would be discarded.

Thus, the proposed work not only looks at security but also forensic perspective and would aid both.

This work needs to first establish a mail server so that Cloud user can immediately be acknowledged about the uncertain mishap that can not only be dangerous for them but also can ruin their lives to a great extent. For such circumstances, the work proposed uses two mediums as mentioned above to ensure the correct and proper authorization of the person who requested for VM deletion.

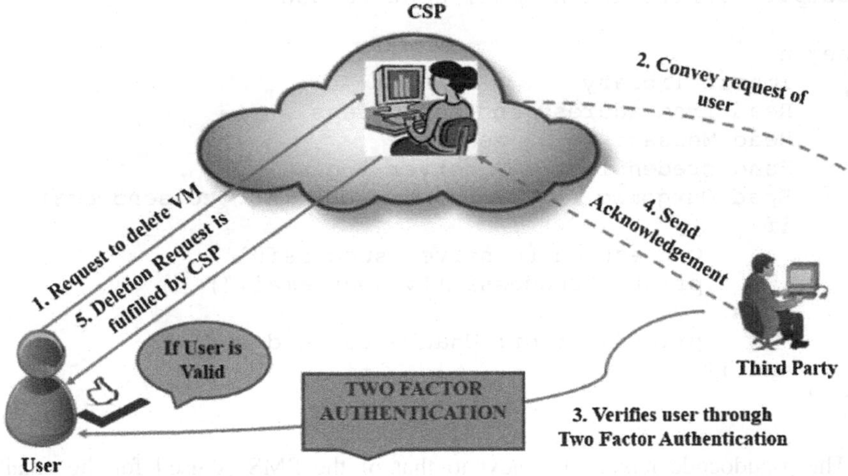

Fig. 3 Proposed model

Pseudocode for OTP verification:

```
Input: username and password of SMS gateway
Output: SMS sent to registered user
begin
import library
Read username and password
prompts for the number to send SMS
Message = "Hi"
generate otp
            Logs into the SMS gateway using provided
credentials

try:
Open Url
print ("Url connected -SMS sent ")
except:
Handle all error
            print ("Url not connected - SMS not sent ")
end try
end
```

The proposed work uses the pseudocode represented above for the SMS verification of the registered user. An OTP is appended in the SMS and is sent to the user to ensure the integrity of the user.

Pseudocode for Email verification:

```
 Input: Login credentials of mail server
 Output: Verification email sent to user

 begin
      import library
      Read from_address and to_address
      Read Message to be sent
      Read credentials for server login
      Read Outgoing server Url and Port No. to send email
      if
            Connection to server successful
            print ("Successfully sent email")
      else
            print ("Error: Unable to send email")
      end if
 end
```

The pseudocode represented next to that of the SMS is used for the email verification of the registered user.

4 Experimental Results

As we had hoped, our experiments with sending and authenticating verification messages proved successful and met our expectations. The proposed model was intended to prevent the unauthorized machine deletion on Cloud. But as for now the two-factor authentication mechanism for both SMS via payment gateway and email verification via mail server are favorably working individually. Registered user is able to receive the verification alert via email and SMS. Now, the further concern is to integrate the entire work with Cloud.

5 Conclusion and Future Work

Undoubtedly, Cloud Forensic is a burning topic and much work is being done in that area. This is also found to be an interesting field for which even the researchers are also making efforts to emerge with feasible and optimized solutions. As per the expectations the task of modules for verification in proposed work has been successfully accomplished. So in future, based on this review, the experiments carried out till now will be integrated further to overcome one of the major challenges that is to prevent the deletion on cloud without proper authorization.

Acknowledgements The Trademarks, product names, company names, screenshots referred in this paper are acknowledged to their respective owners.

References

1. Pooja Dubey, Vineeta Tiwari, Shweta Chawla.: Proposing an authentication mechanism for Prevention of Cloud machine Deletion. In proceedings of Internation Journal of Advance Research and Innovative Ideas in Education, Vol. 2, Issue. 3, India (2016).
2. Deevi Radha Rani, G. Geethakumari.: An Efficient Approach to Forensic Investigation in Cloud using VM Snapshots. In proceedings of International Conference on Pervasive Computing, India (2015).
3. BKSP Kumar Raju Alluri, Geethakumari G.: A Digital Forensic Model for Introspection of Virtual Machines in Cloud Computing. IEEE, India 2015.
4. Hubert Ritzdorf, Nikolaos Karapanos, Srdjan Capkun.: Assisted Deletion of Related Content. ACM, Switzerland (2014).
5. Mr. Digambar Powar, Dr. G. Geethakumari.: Digital Evidence Detection in Virtual Environment for Cloud Computing. ACM, India (2012).
6. Saibharath S, Geethakumari G.: Cloud Forensics: Evidence Collection and Preliminary Analysis. IEEE, (2015).

7. Mr. Chandrashekhar S. Pawar, Mr. Pankaj R. Patil, Mr. Sujitkumar V. Chaudhari.: Providing Security and Integrity for Data Stored In Cloud Storage. ICICES, India (2014).
8. Curtis Jackson, Rajeev Agrawal, Jessie Walker, William Grosky.: Scenario-based Design for a Cloud Forensics Portal. IEEE, (2015).
9. NIST, NIST Cloud Computing Forensic Science Challenges. National Institute of Standards and Technology Interagency or Internal Report 8006, (2014).

Predicting the Size of a Family Based upon the Demographic Status of that Family

Kanika Bahl and Rohitt Sharma

Abstract Size of a family is very much affected by the demographic factors of that family like education and occupation of family members. Socio-Economic status and type of a family also play a major role in determining the family size. Data of more than half a thousand families was studied to find an association between family size and the above listed factors. A neural network was trained using TRAINGD and LEARNGDM as the training and the learning adaptive functions respectively. Later, a GUI was developed for the easy prediction of family size.

Keywords Data mining · Demographic status · Family size
Literacy rate · Prediction · Preprocessing · Neural network nntool

1 Introduction

Processing large amount of data to come up with new results is called Data Mining [1]. It enables us to analyze a given dataset and draw relationships between the parameters that are present. A sensible shopkeeper organizes his shop items according to the results that he obtains after mining his shop's data. Data Mining is applied when a doctor asks medical history to patients in order to diagnose what kind disease he may be suffering from.

Data Mining was applied in our research work to analyze the data of more than half a thousand families from a village named Badyal Brahmna in Jammu and Kashmir (J&K), India which was collected by personally visiting the village with the help of Dr. Ritesh Khullar of Health Services, Jammu (J&K) and his team. These families will be thoroughly studied on the factors like education and occu-

K. Bahl (✉) · R. Sharma
Department of Computer Science and Engineering, Lovely Professional University, Phagwara, India
e-mail: kb.kanikabahl@gmail.com

R. Sharma
e-mail: rohit.17458@lpu.co.in

D.K. Mishra et al. (eds.), *Information and Communication Technology for Sustainable Development*, Lecture Notes in Networks and Systems 10,
https://doi.org/10.1007/978-981-10-3920-1_21

pation of family members, their type of family, Socio-economic status and age of female at the time of marriage. Association was found out between these variables and size of the family. Bringing out this association was necessary to predict and hence control size of the family which will further control the population of a country. Predicting size of a family is very important to study various aspects of a family like if the family is prone to mal-nutrition and poverty.

2 Current Work

Y. Dharani Kumari did analysis on the variables: Education, Occupation, Annual Income of Family, Property and Son Preference, which she mentioned in her paper "Women's Position and Their Behaviors Towards Family Planning In Two Districts of Andhra Pradesh [2]. She deduced that women who are well educated are more tend to adopt methods of family planning as compared to relatively less educated women and it was also concluded that working women prefer adopting family methods more than the housewives. Higher annual income was also found to be a factor to promote family methods. The families which had strong son preference were found to have a very low rate of adopting family methods of 19% [4]. Similarly, Aniceto C. Orbeta, Jr. gave facts to show that a family, if larger in size proves to be a threat to that family. He discussed them in detail in his paper "Poverty, Vulnerability and Family Size: Evidence from the Philippines [3]". It was deduced that a family economy growth is restricted if the family is larger. Additional children hinder a mother to take employment was also an inference [4].

Furthermore, there are various factors to determine the size of a family based upon Socio-Economic status and/or type of the family, education level and occupation of family members, age of wife at the time of marriage etc. This determination was done using Chi-square test which will find association between the variables and size of family. Later nntool was used to train the network.

3 Work Plan

Hypothesis formed was:

- High educated females go for less number of children
- People in white collared jobs tend to go for smaller families
- Families with higher Socio-economic Status go for less number of children.

The research work was carried out in following manner.

- Collection of data followed by its preprocessing.

- Chi-square Test was applied and then its analysis and interpretations is done.
- Application of ANN by nntool. Network with the minimum error is chosen.
- Comparison of the trends obtained by Chi-square and initially formed hypothesis is done.
- Developing the Graphical User Interface (GUI).
- Efficiency Analysis to measure the accuracy of the AI developed.

4 Results and Discussion

Preprocessing of data was done and the data was then ready for the analysis. The data contained no mismatch between the measuring units and categorization which was necessary to bring out more generalized results. In addition to this the columns which were not required for the analysis purpose were dropped. Chi-square test was applied on the preprocessed dataset and hence formed association was used to extrapolate new results.

The results achieved are discussed below. The total number of families taken into consideration for the study is 587 out of which one family was taken as outlier and hence not taken for the study.

4.1 Socio-Economic Status and Family Size

Socio-economic status appears to be one of the significant factors affecting the family size. Better socio-economic status shows positive impact on the family size. As shown in Table 1 families with lower Socio-economic status tend to have 3 (30.4%) and 2 (34.57%) number of children whereas in case of Lower Middle and medium middle families, people tend to have 2 or 1 child only. Maximum number of families in Lower middle (46.9%) and Medium middle (78.9%) prefer to have 2 children only. One child is opted by families with Lower Middle and Higher Middle status with percentages 24.1 and 15.7 respectively.

4.2 Religion and Family Size

Results obtained from Chi-square test are insignificant and thus we say that family size is independent of the religion of family (Table 2).

Table 1 Socio-economic status and number of children

Socio-economic status and number of children									
Number of children	Lower		Lower Middle		Medium Middle		Higher Middle		
	Count	Percentage	Count	Percentage	Count	Percentage	Count	Percentage	
Nil	12	4.460	16	5.369	1	5.263		0	
1	45	16.728	72	24.161	3	15.789	1	100	
2	93	34.572	140	46.979	15	78.947		0	
3	82	30.483	54	18.120		0		0	
4 and above	37	13.754	16	5.369		0		0	
Total	**269**		**298**		**19**		**1**		**587**

X^2 results 46.16 and is highly significant at 0.0000065 or 0.00065% (degree of freedom, df = 12)
Note Number of children = 5 and above were taken in 4 and above

Table 2 Religion of family and number of children

Religion and number of children							
Number of children	Hindu		Sikh		Christian		
	Count	Percentage	Count	Percentage	Count	Percentage	
Nil	22	5.365	7	3.977	0	0	
1	86	20.975	35	19.886	0	0	
2	160	39.024	88	50	0	0	
3	96	23.414	39	22.159	1	100	
4	42	10.243	7	3.977	0	0	
5	3	0.731	0	0	0	0	
More than 5	1	0.243	0	0	0	0	
Total	**410**		**176**		**1**		**587**

χ^2 results 14.98 and is insignificant at 0.24 (df = 12)

4.3 *Literacy of Wife and Family Size*

Women in India are hardly given any opportunity to study. This has a very bad impact on population of India. We found out that literate women in India tend to have less number of children as compared to the illiterate women. We found in Table 3, that 37.7% and 23.3% of illiterate women have 3 and 2 number of children respectively. These figures are very high as compared to those in educated women. Most of the middle educated women (43.0%), higher educated women (49.3%) and graduate women (51.3%) tend to have 2 number of children. We see that only one child is also preferable in highly educated women. 24.3% of higher educated women and 32.4% of graduate women tend to have only one child.

4.4 *Occupation of Husband and Family Size*

Occupation of a husband in a family is a very crucial factor that decides the fate of a family. In our society a man still holds a better position and has a better say in a family than his lady. Occupation of husband was categorized into two categories: white and non-white collared jobs. Daily wagers and business men were categorized into non-white collared jobs. It was found that husbands into white collared jobs tend to have less number of children than those in non-white jobs. Maximum number of white collared job husbands prefer 2 (46.0%) or 1 (26.2%) child only whereas husband in non-white collared jobs may prefer 2 (39.0%) or 3 (28.1%) number of children as shown in Table 4.

Table 3 Literacy of wife and number of children

Number of children	Illiterate		Middle		Higher		Graduate		
	Count	Percentage	Count	Percentage	Count	Percentage	Count	Percentage	
Nil	1	1.111	17	5.592	9	5.769	2	5.405	
1	11	12.222	60	19.736	38	24.358	12	32.432	
2	21	23.333	131	43.092	77	49.358	19	51.351	
3	34	37.777	78	25.657	22	14.102	2	5.405	
4 and above	23	25.555	18	5.921	10	6.410	2	5.405	
Total	**90**		**304**		**156**		**37**		**587**

χ^2 results 72.07 and is highly significant at 0.0000001 or 0.00001% (df = 12)
Note Primary level literacy count and percentage were taken as 0

Table 4 Occupation of husband and number of children

Occupation of husband					
Number of children	White collared		Non-white collared		
	Count	Percentage	Count	Percentage	
Nil	15	5.617	14	4.375	
1	70	26.217	51	15.937	
2	123	46.067	125	39.062	
3	46	17.228	90	28.125	
4	12	4.494	37	11.562	
5	0	0	3	0.937	
More than 5	1	0.374	0	0	
Total	**267**		**320**		**587**

χ^2 results 29.48 and is highly significant at 0.000049 or 0.0049% (df = 6)

4.5 *Age of Wife at Marriage and Family Size*

Age of wife at the time of marriage is also a very important while predicting the size of her family. It was found out (as shown in Table 5) that women if get married before the age of 20 years general have either 2 (38.49%) or 3 (30.0%) number of children. Whereas it was found that if women marry after getting mature enough like at the age of 20–30 years, they prefer 2 or 1 child only. Most of the women who marry at the age of 20–30 years (approx. 45%) prefer 2 children only. Also, most of the women who marry at a later age of above 30 years (50.0%), they prefer only one child.

4.6 *Nntool*

Later, nntool was used to train a neural network. Many combinations of Training and Learning Adaptive functions with the Transfer function were tried and tested as shown in the Table 6. Out of all the networks tried the Network12 gave the best efficiency and so this was chosen as the optimum network. The training, learning adaptive and the transfer functions chosen were TRAINGD, LEARNGDM and TANSIG. The number of neurons in the hidden layer was 16. Performance was shown with the help of Mean Squared Error (mse).

The performance graph of the Network12 is shown in Fig. 1. As can be seen from the figure, the error decreases as the number of epochs increase. The best validation performance was shown at epoch 793. The Training state graph showing the Gradient and the Validation checks is shown in Fig. 2. The figure shows that with the increase in the number of epochs, the gradient decreases and the minimum gradient is observed at the epoch number 799. Also the figure depicts that the 6 validations failed from Epoch 793–799.

Table 5 Age of wife at marriage and number of children

Age of wife at marriage (years)	15–20		20–25		25–30		30 and above	
Number of children	Count	Percentage	Count	Percentage	Count	Percentage	Count	Percentage
Nil	6	2.654	12	4.444	9	12.676	2	10
1	30	13.274	65	24.074	16	22.535	10	50
2	87	38.495	122	45.185	33	46.478	6	30
3	68	30.088	55	20.370	12	16.901	1	5
4 and above	35	15.486	16	5.925	1	1.408	1	5
Total	**226**		**270**		**71**		**20**	**587**

χ^2 results at 58.34 and is significant at 0.0000001 (df = 12)

Table 6 Tried and tested combinations

Network name	Training function	Learning adaptive function	Transfer function	Performance function	Number of neurons
Network1	TRAINBR	LOGSIG	TANSIG	MSE	16
Network2	TRAINBR	LEARNGDM	TANSIG	MSE	16
Network3	TRAINBR	LEARNGDM	TANSIG/PURELIN	MSE	16
Network4	TRAINBR	LEARNGDM	TANSIG	MSE	14
Network5	TRAINLM	LOGSIG	TANSIG	MSE	16
Network6	TRAINLM	LEARNGDM	TANSIG	MSE	16
Network7	TRAINLM	LEARNGDM	TANSIG	MSE	14
Network8	TRAINLM	LEARNGDM	LOGSIG	ME	16
Network9	TRAINLM	LEARNGDM	LOGSIG	SSE	16
Network10	TRAINLM	LEARNGD	TANSIG	MSE	16
Network11	TRAINGD	LEARNGD	TANSIG	MSE	16
Network12	TRAINGD	LEARNGDM	TANSIG	MSE	16

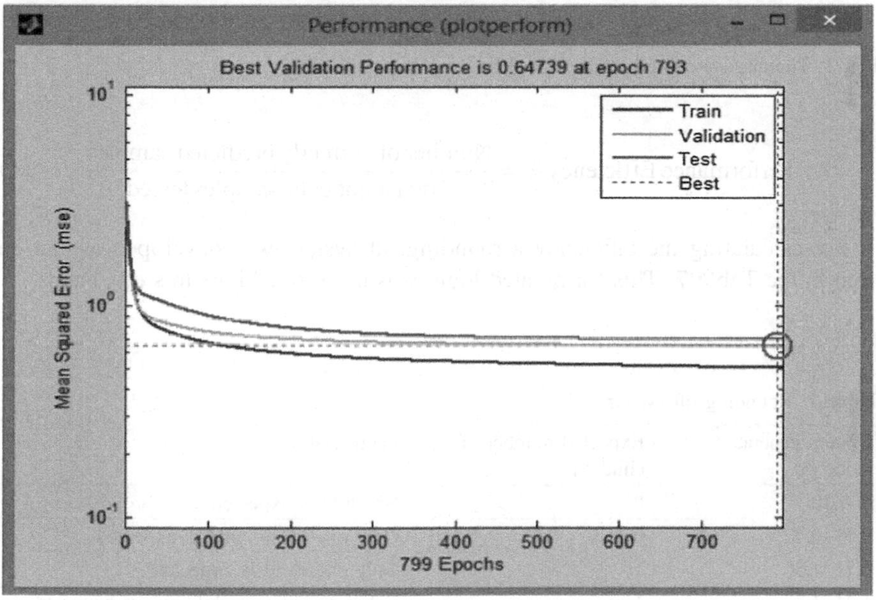

Fig. 1 Performance graph of Network12

Network12 was tested on a randomly selected set of inputs. 51 samples out of 60 were correctly predicted which gives us the efficiency of 85%. The efficiency was calculated as:

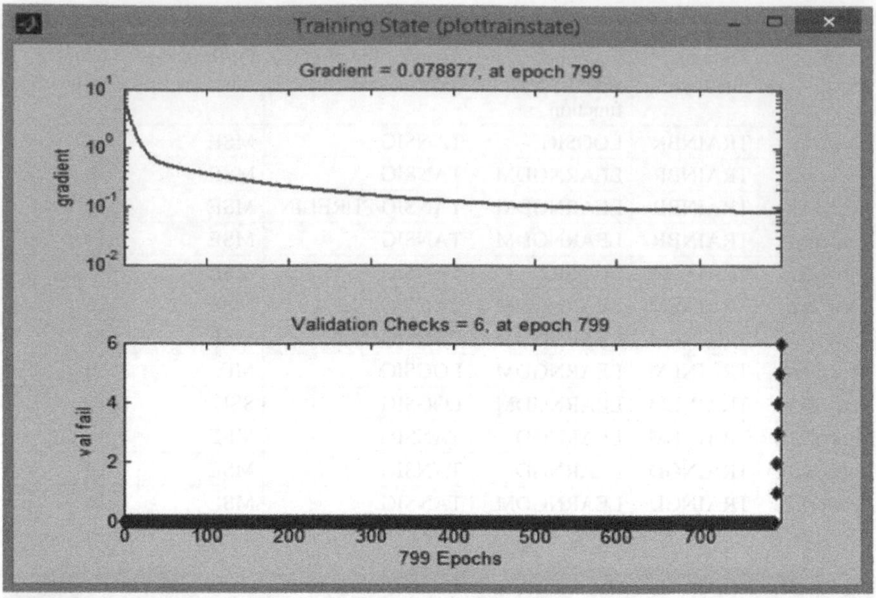

Fig. 2 Training state graph of Network12

$$\text{Performance Efficiency \%} = \frac{\text{Number of correctly predicted samples}}{\text{Total number of samples tested}} \quad (1)$$

For calculating the efficiency a rounding off system was developed as can be seen in the Table 7. This formulated logic was used to achieve this efficiency.

Table 7 Rounding off system

System generated value (a)	Expected number of children	Interpretations
0–0.30	0	No child is expected
0.31–0.70	0 or 1	Only one or no child is expected
0.71–1.30	1	Only one child is expected
1.31–1.70	1 or 2	Either one or two children are expected
1.71–2.30	2	Two children are expected
2.31–2.70	2 or 3	Two or three children are expected
2.71–3.30	3	Expected number of children is three
3.31–3.70	3 or 4	Three or four children are expected
3.71–4.30	4	Four children are expected (not recommended)
4.31–4.70	4 or 5	Four or five children are expected (not recommended)

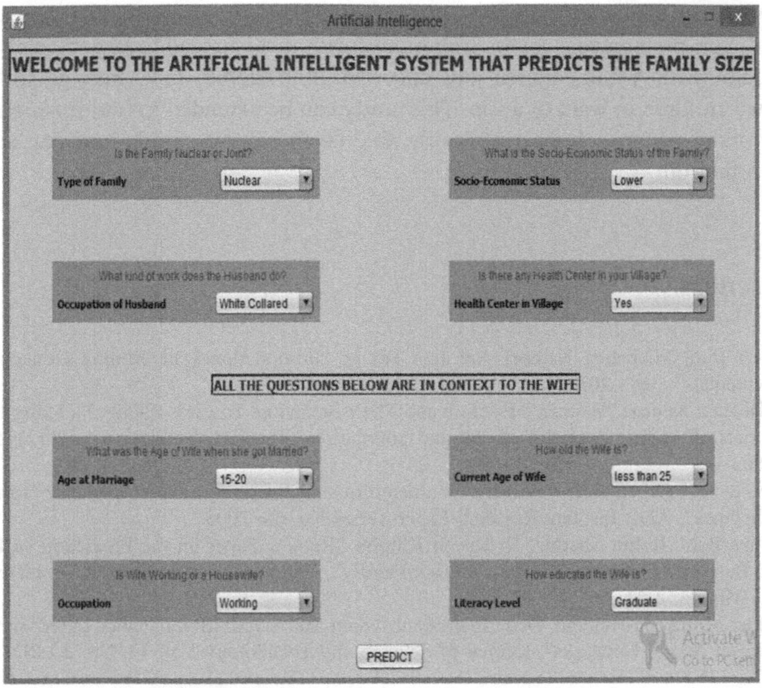

Fig. 3 A Screenshot of the GUI developed

The Network12 was used to predict the size of a family using a GUI that was developed in Java. A screenshot of the GUI is shown in Fig. 3. This GUI was developed in Java and it predicts the size of a family by calling the Network12 at the backend.

5 Conclusion and Future Scope

According to the results that have been discussed above we infer that family size is very much dependent on the demographic factors of a family. We found that a literate wife and husband in white collared jobs have a positive impact on the family size. It was also found that women if get married at a very young age (less than 20 years) they tend to have a larger family size. Better Socio-economic status was found to have a positive impact on the family size whereas Religion of a family was found to have no impact on its size. The network best suited for our data was the one with TRAINGD and LEARNGDM as its training and learning adaptive functions. TANSIG was the transfer function that was employed with 16 neurons in the hidden layer. The efficiency of network was calculated as 85%. This efficiency can be increased in future with enhancements in the functions used in nntool. Also to increase the efficiency many other factors like psychology of a family needs to be studied. Son-preference is a very important factor that can lead to larger family size. Even families in white-collared jobs and with high literacy level may go for more than two children in want of a son. This study can be extended to future generations to find more patterns. Impact of family size on interpersonal relations can also be tried to find out.

References

1. Jiawei Han, Micheline Kamber and Jian Pei in "Introduction Data Mining Concepts and Terchniques", 3ed., 2012.
2. Y. Dharani Kumari "Women's Position and Their Behaviors Towards Family Planning In Two Districts of Andhra Pradesh", Health and population—perspectives and issues 28(2): 58–70, 2005.
3. Aniceto C. Orbeta, Jr. "Poverty, Vulnerability and Family Size: Evidence from the Philippines", ADB Institute Research Paper Series No. 68, 2005.
4. Kanika Bahl, Rohitt Sharma, Dr Ritesh Khullar "Review Paper on the Prediction of Family Size Based on Demographic Status of that Family", National Conference on Recent Trends in Data Mining, 2015.
5. Eltaher Mohamed Hussein, Dalia Mahmoud Adam Mahmoud "Brain Tumor Detection Using Artificial Neural Networks", Journal of Science and Technology Vol. 13, No. 2 ISSN 1605–427X.

A New Approach Toward Sorting Technique: Dual-Sort Extraction Technique (DSET)

Darpan Shah and Kuntesh Jani

Abstract There is a huge demand for efficient and high scalable sorting techniques to analyze data. Earlier many researchers have proposed various sorting algorithms such as heap sort, merge sort, and bubble sort,. Some of them are very popular in achieving efficient data sorting at a great extent like heap sort and merge sort. Increase in the amount of data has also lead to increase in complexity of sorting algorithms leading to decreased speed and efficiency. For this a new sorting technique named "Dual-Sort Extraction Technique (DSET)", with two levels of data sorting extraction is proposed which enhances the performance and efficiency of the algorithm.

Keywords Sorting · Extraction · Swapping · Iteration

1 Introduction

Sorting algorithms are an intelligent piece of code which rearranges dataset in ascending or descending order. Today, continuous researches are going to handle large scale of data in the sorted form. Analytic results describing the performance of the programs are summarized. A variety of special situations are considered from a practical standpoint.

In this research paper, a improved sorting technique "Dual-Sort Extraction Technique (DSET)" is proposed in which two-level sorting is performed with considerably large amount of data. In the first level of this two-level technique, we move the largest number at the end of the dataset [1]. While in the second level, we move the smallest number at the start of the dataset [2]. This procedure is run on the remaining unsorted dataset until the series come into sorted form. Various test cases have been taken care of.

D. Shah (✉) · K. Jani
Government Engineering College, Gandhinagar, Sector-28, Gujarat 382026, India
e-mail: darpanshah5@gmail.com

K. Jani
e-mail: kunteshjani@gecg28.ac.in

D.K. Mishra et al. (eds.), *Information and Communication Technology for Sustainable Development*, Lecture Notes in Networks and Systems 10,
https://doi.org/10.1007/978-981-10-3920-1_22

2 Related Work

Sorting means arranging the various data either in ascending or descending order as per the requirement [3]. To accomplish this task, a certain amount of data in the form of number has been taken and then sorting operation will be performed on all elements using various sorting algorithms. This process continues until all the numbers are sorted in a format [4]. For this purpose, a new algorithm DSET is presented.

3 Proposed Dual-Sort Extraction Technique (DSET)

3.1 The DSET Has the Following Features

- It is a two-level based technique,
- Performs sorting on large amount of data, i.e., in lakhs,
- Discards extra iteration or passes,
- Two elements are sorted in single pass,
- Except the sorted elements in each pass it extracts the remaining elements.

3.2 Initialization

Step 1: Initialize variable swap=true, k1=0, k2=0, i=0, element, n=10000, array[n];

Step 2: Perform iteration until swap is false
While (swap==true) then
Follow step 3
Otherwise
Follow step 4

Step 3: Make swap false.
Follow Level 1 and Level 2

Step 4: Stop iteration.
And Display Result

3.3 Level 1 Sorting

Step 1: Perform iteration until element is greater than or equal to n-1-k2

For element=i to n-2-k2 step by 1
Follow step 2

If above condition false then
Follow step 4

Step 2: Check is there any swapping of element

If element is swapping? Then
Follow step 3
Else
Follow step 1

Step 3: Make swap true.
Swap=true
Follow step 1

Step 4: Check swap is true or false
If (swap==true) then
Follow Step 5
Else
Follow Step 6

Step 5: Increment k1 by 1
k1++
Perform Level 2 sorting

Step 6: Stop iteration using break and avoid Level 2 sorting.
Display Result

(Fig. 1).

3.4 Level 2 Sorting

Step 1: Initialize min=i, element=i+1;

Step 2: Perform iteration until element is greater than or equal to n-k1

For element to n-1-k1 step by 1
Follow step 3

If above condition false then
Follow step 5

Step 3: Check if value at index of element is less than value at index of min

If (array [element] < array [min]) Then
Follow step 4
Else
Follow step 2

Step 4: Assign j to min
Min = element
Follow step 2

Step 5: perform swapping of element at index i and min
Swap element at i to min and vice versa.

Increment i by 1 and K2 by 1.
I++
K++
Follow Step 2 in section 3.2 until swap becomes false

(Fig. 2).

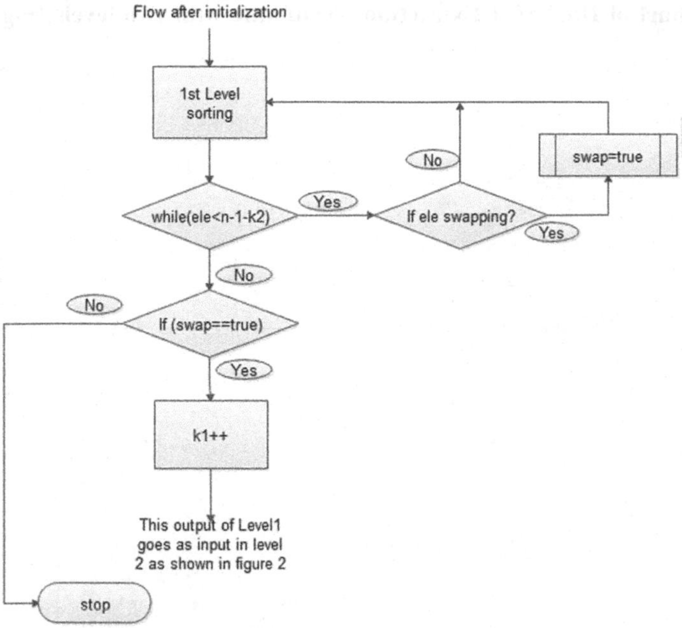

Fig. 1 Level 1 sorting of Dual-Sort Extraction Technique (DSET)

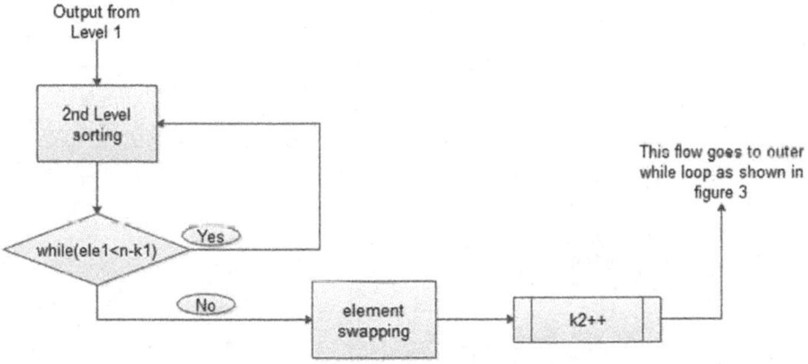

Fig. 2 Level 2 sorting of Dual-Sort Extraction Technique (DSET)

Flowchart of Dual-Sort Extraction Technique with two levels (Fig. 3).

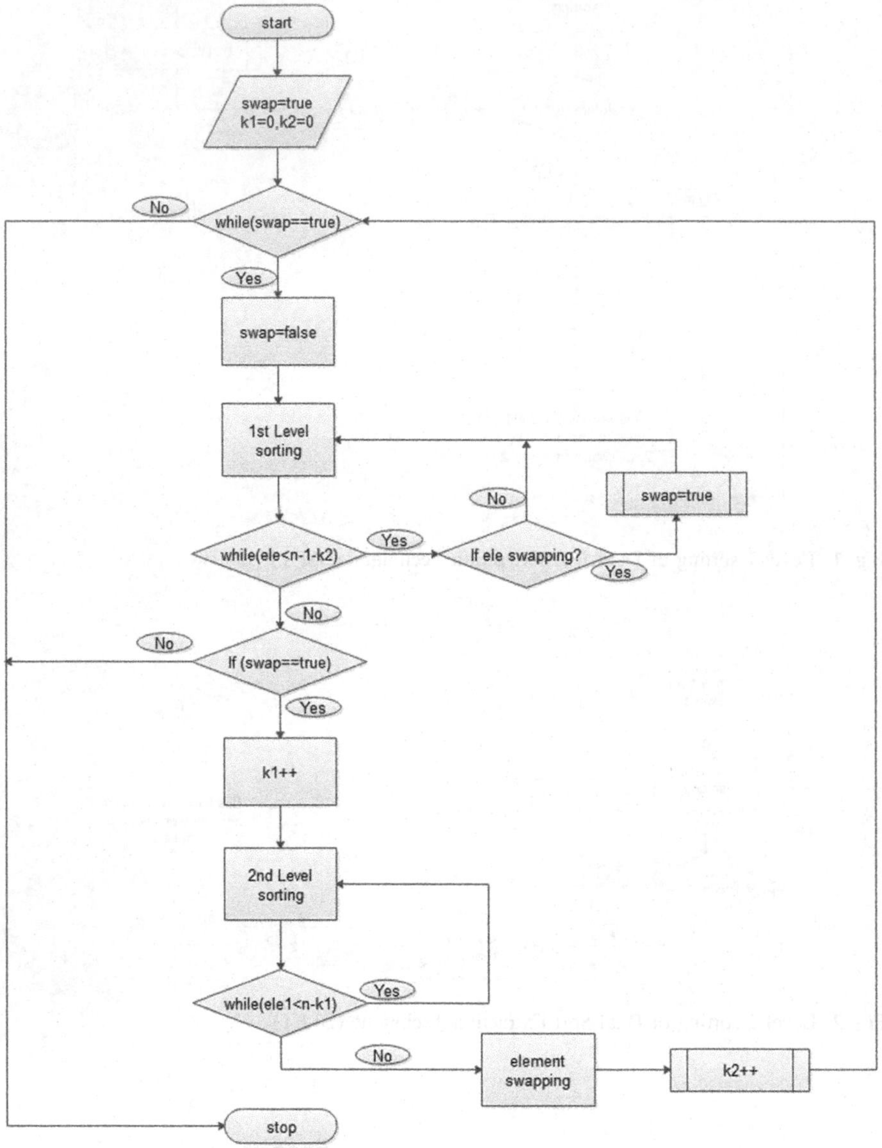

Fig. 3 Flowchart of DSET with two-level sorting

4 Average Time Comparison Between Various Techniques

Comparison of different sorting techniques with different sizes of input

Processor: i3 2.50 GHz, RAM: 6 GB, HDD: 500 GB, OS: Windows 10 64 bit unit of below average results are in nanoseconds

Size of input	DSET	Bubble	Selection	Insertion	Merge	Heap
13,50,000	6996482.8	7827558.2	8835160.6	9333296.8	105313272.8	238605909.2
15,50,000	6878743.6	8090296.8	28985979.2	10471447.2	121925824.8	268886124
20,00,000	6944181.8	8060902.6	8615691.6	10920976.8	138097140.4	335991454.4
22,50,000	7274493.2	8679405.8	10184652	12658746.4	153337211.6	389137318.2
25,00,000	7146572.2	9344791.4	10000159.8	12147966	169382256.2	428395968.4
50,00,000	9600222.4	15242534.2	16989664	22059944.6	369637269	907026246.2
1,00,00,000	14559906.8	24942269.4	27705212.2	36705323.4	783687544.6	1917055472
10,00,00,000	109611430.2	304586800.4	332062608.6	429948671.6	10025880493	22243257942

5 Pseudocode of DSET

Time complexity of above DSET is [5, 6]: (Figs. 4 and 5)

$$T(n) = O\left(n^2\right) \tag{1}$$

```
int k1=0,k2=0,i11=0,min;   Boolean swap=true;
while (swap) {
      swap = false;
      for (int j1 = i11; j1 < n - 1 - k2; j1++) {
        if (a_ele [j1] > a_ele [j1 + 1]) {
          temp_store = a_ele [j1];

          a_ele[j1] = a_ele[j1 + 1];
          a_ele[j1 + 1] = temp_store;
          swap = true;
        }
      }
      if (swap == false) {
        break;
      }
      k1++;

      min = i11;
      if (swap) {
      for (int j1 = i11 + 1; j1 < n - k1; j1++)      {
         if (a_ele[j1] < a_ele[min]) {
              min = j1;
              swap = true;
         }
      }
      temp_store = a_ele[i11];

      a_ele[i11] = a_ele[min];
      a_ele[min] = temp_store;
      i11++;
      k2++;
    }
}
```

Fig. 4 Pseudocode of DSET [7]

6 Future Work

This algorithm can be easily applied on various data for sorting without modifying it. We would test DSET on various operating systems with different hardware configurations. We can use this technique to sort text numeric data. We can implement this technique on the web to give the result in the sorted format.

7 Conclusion

Every algorithm has its own advantages and disadvantages, our technique contains high LOC, but proposes a good strategy to perform sorting on large scale of data with improved efficiency and performance than the other well-known techniques.

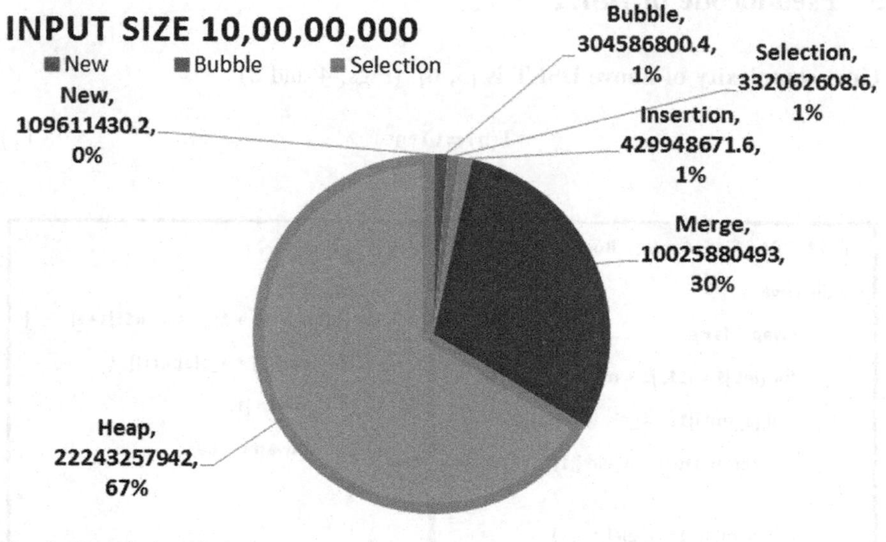

Fig. 5 Graphical analysis

Acknowledgements First of all, I would like to thanks my parents who encourage me for my every effort on this precious work. I am thankful to Mr. Kuntesh K Jani who is always standing beside me and appreciate my work. He always motivates me to do this type of job. I am also grateful to many people for their help in writing this research paper.

References

1. Bubble sort technique: http://www.personal.kent.edu/~rmuhamma/Algorithms/MyAlgorithms/Sorting/bubbleSort.htm.
2. OXFORD University Press "Data Structure Using C" (2nd Edition) by Reema Thareja, Assistant Professor in Department of computer science at Shyama Prasad Mukherjee College for Women, University of Delhi.
3. Technical Publication "Design & Analysis of Algorithm" by Mrs. Anuradha A. Puntambekar. Formerly Assistant Professor in P.E.S. Modern College of Engineering, Pune.
4. SAMS, "Data Structure & Algorithm in Java" by Robert Lafore (1998).
5. Design and Analysis of Algorithms, Dave and Dave, Pearson.
6. Algorithm Design: Foundation, Analysis and Internet Examples, GoodRich, Tamassia, Wiley India.
7. Balaguruswamy for Programming in C (4th Edition).
8. Addison Publication Company "Algorithms" by Robert Sedgewick at Brown University.
9. Introduction to Algorithms, Thomas H. Cormen, Charles E. Leiserson, Ronald L. Rivest and Clifford Stein, PHI.
10. Fundamental of Algorithms by Gills Brassard, Paul Bratley, PHI.
11. Introduction to Design and Analysis of Algorithms, Anany Levitin, Pearson.

Student's Performance Evaluation of an Institute Using Various Classification Algorithms

Shiwani Rana and Roopali Garg

Abstract Machine learning is the field of computer science that learns from data by studying algorithms and their constructions. The student's performance based on slow learner method plays a significant role in nourishing the skills of a student with slow learning ability. The performance of the students of Digital Electronics of University Institute of Engineering and Technology (UIET), Panjab University (PU), Chandigarh is calculated by applying two important classification algorithms (Supervised Learning): Multilayer Perceptron and Naïve Bayes. Further, a comparison between these classification algorithms is done using WEKA Tool. The accuracy of grades prediction is calculated with these classification algorithms and a graphical explanation is presented for the BE (Information Technology) third semester students.

Keywords Classification · WEKA · Naïve Bayes · Multilayer perceptron

1 Introduction

The slow learner prediction is the branch of the automatic predictive method for the students learning abilities. For minimizing the adverse future effects of the slow learning problem, slow learner methods are important [1]. The early stage detection can help the institutions to identify and evaluate the individual performance of the students to incorporate the special care on the slow learners.

This prediction can be done using some supervised learning algorithms. This type of learning algorithm is used for generating a function which is used for mapping the inputs to the desired outputs [2]. These functions are based on the training data set. A supervised technique uses a dataset with known classification.

S. Rana (✉) · R. Garg
Department of IT, UIET, Panjab University, Chandigarh, India
e-mail: shiwanirana40@gmail.com

R. Garg
e-mail: roopali.garg@gmail.com

D.K. Mishra et al. (eds.), *Information and Communication Technology for Sustainable Development*, Lecture Notes in Networks and Systems 10,
https://doi.org/10.1007/978-981-10-3920-1_23

Various algorithms like Naïve Bayes, logistic regression, neural networks, linear regression, and decision trees are highly dependent on the information given by the predetermined classifications [3].

The paper is organized in the following section. Section 2 describes the classification algorithms (supervised machine learning algorithms) which further discusses the two algorithms: Naïve Bayes and Multilayer Perceptron. Section 3 deals with the implementation and comparison in which the performance of 58 students is analyzed with both the classification algorithms by using WEKA Tool, Sect. 4 deals with the results, comparing both the algorithms, Sect. 5 describes the output and Sect. 6 describes the conclusion and future work.

2 Classification Algorithms

(i) Naïve Bayes algorithm (NB)

Naive Bayes algorithm is a simple method which uses Bayesian theorem for classification. Bayes theorem can be written as: [4]

$$\mathbf{P(A|B)} = \frac{P(B|A)P(A)}{P(B)}, \quad (1)$$

where

P (A)	is the probability of A
P (B)	is the probability of B
P (A|B)	is the probability of A given B
P (B|A)	is the probability of B given A

It is called naïve because it simplifies problems relying on two important assumptions:

- It assumes that the predictive attributes are conditionally independent with known classification, and
- It supposes that there are no hidden attributes that could affect the process of prediction.

Naive Bayes algorithm is for classification. It is a type of supervised learning where the class is known for a set of a training data points (already known data sets) and needs to propose the class for any other given data points. The complexity for Naïve Bayes algorithm is O (log n).

(iii) Multilayer Perceptron (MLP):

MLP is a classification algorithm which contains basic three layers: the input layer, the hidden layer, and the output layer. Hidden layer contains nodes and each node is a function of the nodes in the input layer and output node is a function of the node in the hidden layer [5] (Fig. 1).

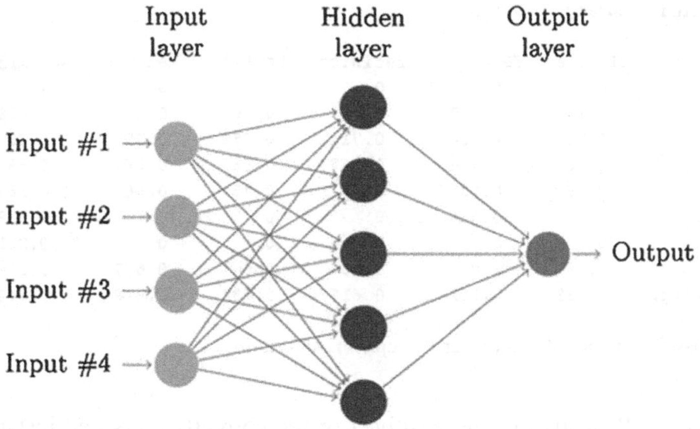

Fig. 1 Multilayer perceptron

The advantages of multilayer perceptron are:

- It has the ability to learn nonlinear models.
- It has the capability to learn models in real time (online learning).

3 Implementation and Comparison

Four classification algorithms are used to analyze the performance of BE (Information Technology) third semester students of UIET, PU, Chandigarh. The academic data of 58 students from the data set (given in the appendix) are taken which include eight attributes namely total marks, grade, attendance, major, minor1, minor2, institution, area, and then both the approaches are applied on this data set using WEKA Tool [6]. Classification in WEKA includes some major terminologies which can be seen in Figs. 2 and 3:

=== Detailed Accuracy By Class ===

	TP Rate	FP Rate	Precision	Recall	F-Measure	ROC Area	Class
	0	0.017	0	0	0	?	A+
	1	0.019	0.857	1	0.923	1	A
	0.895	0.051	0.895	0.895	0.895	0.991	B+
	0.8	0.023	0.923	0.8	0.857	0.952	B
	1	0.087	0.75	1	0.857	0.935	C+
	0	0	0	0	0	0.402	C
	0	0	0	0	0	0.649	D
	0.667	0	1	0.667	0.8	0.721	F
Weighted Avg.	0.845	0.043	0.827	0.845	0.829	0.93	

Fig. 2 Classifier output of Naïve Bayes Algorithm

=== Detailed Accuracy By Class ===

	TP Rate	FP Rate	Precision	Recall	F-Measure	ROC Area	Class
	0	0	0	0	0	?	A+
	0.167	0.019	0.5	0.167	0.25	0.888	A
	0.789	0.154	0.714	0.789	0.75	0.835	B+
	0.667	0.116	0.667	0.667	0.667	0.854	B
	0.75	0.13	0.6	0.75	0.667	0.884	C+
	0	0.036	0	0	0	0.589	C
	0	0	0	0	0	0.211	D
	0.667	0.018	0.667	0.667	0.667	0.976	F
Weighted Avg.	0.638	0.112	0.617	0.638	0.616	0.844	

Fig. 3 Classifier Output of Multilayer Perceptron Algorithm

- **TP Rate**: TP stands for true positives or the correctly classified instances.
- **FP Rate**: FP stands for false positives or the incorrectly classified instances.
- **Precision**: It is defined as the ratio of the instances of a class to the total instances classified as that class [7].
- **Recall**: Ratio of the proportion of instances classified as a given class to the actual total in that class (equivalent to TP rate)

$$\text{F} - \text{Measure: F} - \text{Measure} = \frac{2 * precision * recall}{(precision + recall)} \quad (2)$$

(i) Naïve Bayes:

In Fig. 2, Naïve Bayes Algorithm is applied which is used for classifying the BE third semester students and then predicting their grades.

(ii) Multilayer Perceptron:

Figure 3 shows the implementation of multilayer perceptron algorithm which is used for predicting the grades [8].

4 Result

Comparison between Naïve Bayes and multilayer perceptron algorithms is shown in Table 1. After applying both the algorithms over the data set of 58 students belonging to BE third semester of UIET, a striking outcome is obtained which shows the accuracy of Naïve Bayes is much higher than that in the multilayer perceptron for the given data set [9].

The accuracy results can be explained in more detail by having a look on both the Figs. 4 and 5, which depict the confusion matrix. Correctly and incorrectly

Table 1 Comparison of the two approaches

Features/name of approaches	Naïve Bayes	Multilayer perceptron
Instances	58	58
Attributes	8	8
Test mode	10-fold cross-validation	10-fold cross-validation
Time taken to build model (seconds)	0.02	0.68
Accuracy using all the attributes (8) (%)	84.48	63.79

Fig. 4 Confusion matrix of Naïve Bayes Algorithm

```
=== Confusion Matrix ===

 a  b  c  d  e  f  g  h   <-- classified as
 0  0  0  0  0  0  0  0 |  a = A+
 0  6  0  0  0  0  0  0 |  b = A
 0  1 17  1  0  0  0  0 |  c = B+
 0  0  2 12  1  0  0  0 |  d = B
 0  0  0  0 12  0  0  0 |  e = C+
 0  0  0  0  2  0  0  0 |  f = C
 0  0  0  0  1  0  0  0 |  g = D
 1  0  0  0  0  0  0  2 |  h = F
```

Fig. 5 Confusion matrix of multilayer perceptron

```
=== Confusion Matrix ===

 a  b  c  d  e  f  g  h   <-- classified as
 0  0  0  0  0  0  0  0 |  a = A+
 0  1  5  0  0  0  0  0 |  b = A
 0  1 15  3  0  0  0  0 |  c = B+
 0  0  1 10  4  0  0  0 |  d = B
 0  0  0  2  9  0  0  1 |  e = C+
 0  0  0  0  2  0  0  0 |  f = C
 0  0  0  0  0  1  0  0 |  g = D
 0  0  0  0  0  1  0  2 |  h = F
```

classified instances shown in the matrix are the actual result for accuracy/prediction of grades of the BE third semester students of UIET [10].

In case of Naïve Bayes algorithm,

$$\begin{aligned}\text{Correctly classified instances, (CCI)} &= aa + bb + cc + dd + ee + ff + gg + hh\\ &= 0 + 6 + 17 + 12 + 12 + 0 + 0 + 2\\ &= 49\end{aligned} \tag{3}$$

$$\text{Correctly classified instances (accuracy in \%)} = \frac{\text{CCI}}{\textit{Total no.ofinstances}} \times 100$$
$$= \frac{49}{58} \times 100 = 84.48\% \quad (4)$$

The sum of all the remaining values in the confusion matrix gives the incorrectly classified instances [11].

In case of multilayer perceptron,

$$\begin{aligned}\text{Correctlyclassified instances, (CCI)} &= aa + bb + cc + dd + ee + ff + gg + hh \\ &= 0 + 1 + 15 + 10 + 9 + 0 + 0 + 2 \\ &= 37\end{aligned}$$

$$\text{Correctly classified instances (accuracy in \%)} = \frac{\text{CCI}}{\textit{Total no.of instances}} \times 100$$
$$= \frac{37}{58} \times 100 = 63.79\%$$

5 Output

X-axis represents the predicted grade and Y-axis represents the major (the marks scored in the final exam) [12]. The actual grades are shown with different colors depicting the output of the graph.

The different colors given to the actual grades are:

A+: Blue	A: Red
B+: Green	B: Cyan
C+: Pink	C: Magenta
D: Yellow	F: Orange

In the outputs, the cross sign indicates that the predicted and the actual values are equal whereas the box sign indicates that the predicted value is different from that of the actual value.

Figure 6 shows the output of Naïve Bayes algorithm which is used for applying the classification for predicting the grades of BE third semester students [13]. For example, according to the prediction of grades, most of the students fall under A, B +, B, and C+ and F.

In Fig. 7, multilayer perceptron algorithm is used. This approach gives a more clear prediction of grades, as according to the marks in major, the predicted grades

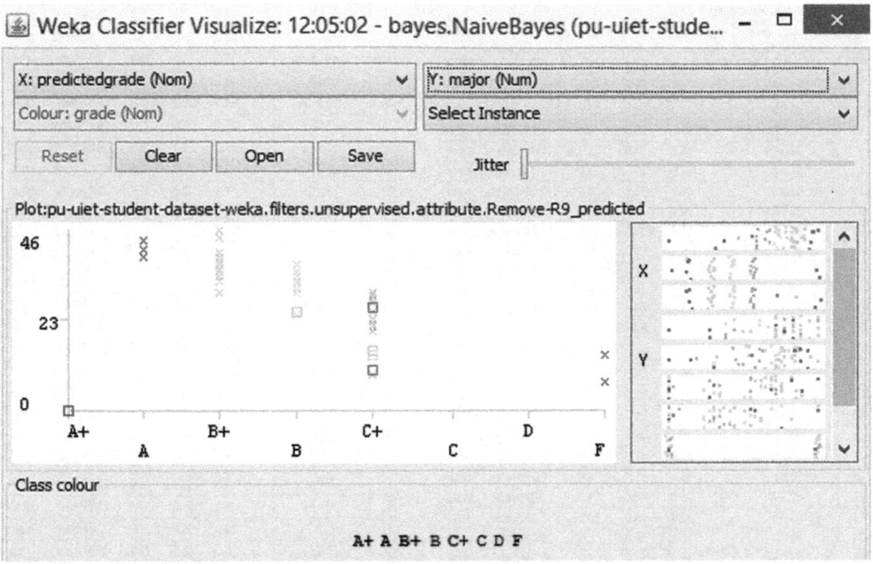

Fig. 6 Visualize classifier errors—Naïve Bayes Algorithm

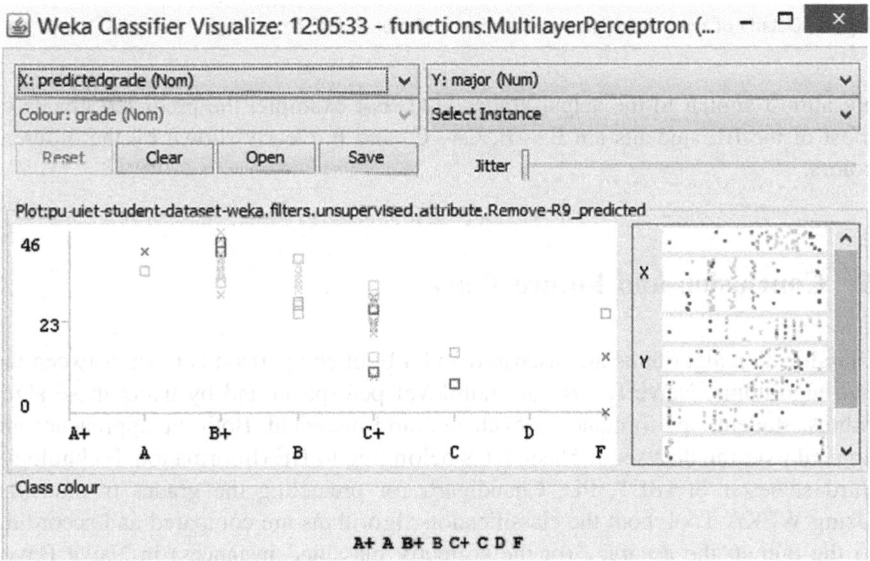

Fig. 7 Visualize classifier errors—multilayer perceptron

Relation: pu-uiet-student-dataset

No.	totalMarks Numeric	grade Nominal	attendance Numeric	major Numeric	minor2 Numeric	minor1 Numeric	Institution Nominal	Area Nomin
1	71.0	B+	13.0	34.0	7.0	24.0	Private	urban
2	50.0	C+	7.0	27.0	16.0	10.0	Governm...	rural
3	63.0	B	7.0	34.0	22.0	0.0	Governm...	urban
4	50.0	C+	15.0	25.0	10.0	0.0	Governm...	urban
5	72.0	B+	15.0	35.0	22.0	9.0	Private	rural
6	61.0	B	16.0	24.0	21.0	0.0	Governm...	rural
7	40.0	D	15.0	15.0	1.0	10.0	Private	rural
8	55.0	C+	14.0	9.0	22.0	9.0	Governm...	urban
9	60.0	B	14.0	14.0	10.0	12.0	Governm...	urban
10	32.0	F	5.0	7.0	20.0	0.0	Governm...	rural
11	85.0	A	15.0	41.0	29.0	16.0	Private	rural
12	27.0	F	5.0	14.0	8.0	0.0	Governm...	urban
13	61.0	B	7.0	37.0	17.0	11.0	Private	urban
14	53.0	C+	7.0	30.0	16.0	0.0	Governm...	rural
15	81.0	A	14.0	41.0	26.0	0.0	Private	urban
16	70.0	B+	16.0	33.0	21.0	7.0	Governm...	rural
17	50.0	C+	8.0	21.0	21.0	0.0	Private	rural
18	77.0	B+	16.0	39.0	22.0	10.0	Private	rural
19	78.0	B+	15.0	39.0	24.0	11.0	Governm...	urban
20	64.0	B	11.0	33.0	20.0	16.0	Private	urban
21	70.0	B+	15.0	36.0	19.0	9.0	Private	urban
22	82.0	A	15.0	43.0	24.0	17.0	Private	urban
23	68.0	B	12.0	38.0	18.0	7.0	Private	rural
24	83.0	A	15.0	39.0	0.0	29.0	Private	rural
25	73.0	B+	15.0	37.0	21.0	10.0	Private	rural
26	76.0	B+	15.0	44.0	0.0	17.0	Governm...	urban
27	73.0	B+	15.0	40.0	18.0	15.0	Governm...	urban
28	56.0	C+	13.0	27.0	16.0	0.0	Private	rural
29	72.0	B+	13.0	40.0	19.0	13.0	Private	rural
30	73.0	B+	9.0	46.0	0.0	18.0	Private	urban
31	48.0	C	13.0	26.0	9.0	1.0	Governm...	rural
32	64.0	B	13.0	31.0	20.0	2.0	Private	urban
33	61.0	B	11.0	30.0	20.0	9.0	Private	rural
34	58.0	C+	14.0	28.0	16.0	8.0	Private	urban
35	73.0	B+	18.0	35.0	20.0	9.0	Governm...	rural
36	56.0	C+	18.0	29.0	8.0	9.0	Private	urban
37	60.0	B	13.0	32.0	9.0	15.0	Private	urban
38	58.0	C+	14.0	28.0	16.0	2.0	Private	urban
39	51.0	C+	15.0	20.0	16.0	0.0	Governm...	rural
40	72.0	B+	13.0	33.0	26.0	13.0	Private	rural
41	65.0	B	7.0	33.0	0.0	25.0	Governm...	urban
42	67.0	B	15.0	35.0	17.0	8.0	Private	rural
43	50.0	C+	16.0	23.0	10.0	11.0	Governm...	rural
44	65.0	B	16.0	33.0	16.0	7.0	Private	rural
45	70.0	B+	18.0	36.0	10.0	16.0	Private	urban
46	80.0	A	15.0	43.0	0.0	22.0	Governm...	rural
47	62.0	B	13.0	34.0	15.0	0.0	Governm...	urban
48	0.0	F	0.0	0.0	0.0	0.0	Private	rural
49	70.0	B+	17.0	36.0	17.0	12.0	Private	rural
50	50.0	C+	7.0	22.0	21.0	12.0	Private	urban
51	72.0	B+	17.0	30.0	25.0	13.0	Governm...	urban
52	70.0	B+	16.0	25.0	0.0	29.0	Private	rural
53	47.0	C	14.0	10.0	23.0	0.0	Private	urban
54	78.0	B+	18.0	38.0	22.0	18.0	Private	urban
55	74.0	B+	10.0	39.0	25.0	0.0	Private	rural
56	80.0	A	14.0	39.0	27.0	0.0	Governm...	urban
57	63.0	B	8.0	35.0	0.0	20.0	Governm...	urban
58	67.0	B	13.0	32.0	0.0	22.0	Private	rural

Fig. 8 Details of marks of digital electronics students of UIET

are almost similar to the actual grades [14]. For example, the predicted grades of most of the BE students are B+, B, C+, C, and F, clearly shown by the different colors.

6 Conclusion and Future Work

Classification algorithms are discussed and a brief comparison is made between the two algorithms, Naïve Bayes and multilevel perceptron and by using these algorithms, student's performance is evaluated and predicted. Both the approaches are applied over the data set of 58 students belonging to BE (Information Technology) third semester of UIET, PU, Chandigarh for predicting the grades of students. Using WEKA Tool, both the classification algorithms are compared and according to the output, the accuracy (or the correctly classified instances) in Naïve Bayes algorithm is more than that in the multilayer perceptron and the time taken to build a model for the given data set in Naïve Bayes algorithm is less (0.02 s) than that in multilayer perceptron (0.68 s). Naïve Bayes predicts the grades of students more accurately than the other approaches for the given data set. Further in future, with the use of these algorithms one can compare the marks of different subjects of a student with a large data set. The research could be extended over various subjects, the student studies in his/her 4-year under graduation. These algorithms can be coded in Python to analyze and discuss the data.

Appendix

(See Fig. 8).

References

1. Khan, Irfan Ajmal, and Jin Tak Choi. "An Application of Educational Data Mining (EDM) Technique for Scholarship Prediction." *International Journal of Software Engineering and Its Applications* 8, no. 12 (2014): 31–42.
2. Goyal, Monika, and Rajan Vohra. "Applications of data mining in higher education." *International journal of computer science* 9, no. 2 (2012): 113.
3. Ayodele, Taiwo Oladipupo. *Types of machine learning algorithms*. INTECH Open Access Publisher, 2010.
4. Kaur, Gurneet, and Er Neelam Oberai. "NAIVE BAYES CLASSIFIER WITH MODIFIED SMOOTHING TECHNIQUES FOR BETTER SPAM CLASSIFICATION." (2014).
5. Fukumizu, Kenji. "Active learning in multilayer perceptrons." *Advances in Neural Information Processing Systems* (1996): 295–301.
6. Ahmed, Abeer Badr El Din, and Ibrahim Sayed Elaraby. "Data Mining: A prediction for Student's Performance Using Classification Method." *World Journal of Computer Application and Technology* 2, no. 2 (2014): 43–47.
7. Kumar, Vaneet, and Vinod Sharma. "Student"s Examination Result Mining: A Predictive Approach." *International Journal of Advanced Computer Science and Applications (IJSER)* 3, no. 11 (2012).
8. Mohammad, Thakaa Z., and Abeer M. Mahmoud. "Clustering of Slow Learners Behavior for Discovery of Optimal Patterns of Learning."*LITERATURES* 5, no. 11 (2014).
9. Anuradha, C., and T. Velmurugan. "A data mining based survey on student performance evaluation system." In *Computational Intelligence and Computing Research (ICCIC), 2014 IEEE International Conference on*, pp. 1–4. IEEE, 2014.
10. Bydovska, Hana, and Lubomir Popelinsky. "Predicting Student Performance in Higher Education." In *Database and Expert Systems Applications (DEXA), 2013 24th International Workshop on*, pp. 141–145. IEEE, 2013.
11. de Morais, Alana M., Joseana MFR Araujo, and Evandro B. Costa. "Monitoring student performance using data clustering and predictive modelling." In *Frontiers in Education Conference (FIE), 2014 IEEE*, pp. 1–8. IEEE, 2014.
12. Singh, Sushil, and Sunil Pranit Lal. "Educational courseware evaluation using Machine Learning techniques." In *e-Learning, e-Management and e-Services (IC3e), 2013 IEEE Conference on*, pp. 73–78. IEEE, 2013.
13. Lopez, Manuel Ignacio, J. M. Luna, C. Romero, and S. Ventura. "Classification via Clustering for Predicting Final Marks Based on Student Participation in Forums." *International Educational Data Mining Society* (2012).

14. Borkar, Suchita, and K. Rajeswari. "Attributes Selection for Predicting Students' Academic Performance using Education Data Mining and Artificial Neural Network." *International Journal of Computer Applications* 86, no. 10 (2014).
15. Pradeep, Anjana, Smija Das, and Jubilant J. Kizhekkethottam. "Students dropout factor prediction using EDM techniques." In *Soft-Computing and Networks Security (ICSNS), 2015 International Conference on*, pp. 1–7. IEEE, 2015.

Speaker Identification in a Multi-speaker Environment

Manthan Thakker, Shivangi Vyas, Prachi Ved and S. Shanthi Therese

Abstract Human beings are capable of performing unfathomable tasks. A human being is able to focus on a single person's voice in an environment of simultaneous conversations. We have tried to emulate this particular skill through an artificial intelligence system. Our system identifies an audio file as a single or multi-speaker file as the first step and then recognizes the speaker(s). Our approach towards the desired solution was to first conduct pre-processing of the audio (input) file where it is subjected to reduction and silence removal, framing, windowing and DCT calculation, all of which is used to extract its features. Mel Frequency Cepstral Coefficients (MFCC) technique was used for feature extraction. The extracted features are then used to train the system via neural networks using the Error Back Propagation Training Algorithm (EBPTA). One of the many applications of our model is in biometric systems such as telephone banking, authentication and surveillance.

Keywords Speaker identification • Neural network • Multi-speaker
Mel frequency cepstral coefficients (MFCC)

M. Thakker (✉) · S. Vyas · P. Ved · S. Shanthi Therese
Information Technology, Thadomal Shahani Engineering College, Mumbai, India
e-mail: manthanthakker40@gmail.com

S. Vyas
e-mail: shivangivyas.1812@gmail.com

P. Ved
e-mail: prachiv2612@gmail.com

S. Shanthi Therese
e-mail: shanthitherese123@gmail.com

D.K. Mishra et al. (eds.), *Information and Communication Technology for Sustainable Development*, Lecture Notes in Networks and Systems 10,
https://doi.org/10.1007/978-981-10-3920-1_24

1 Introduction

Speaker recognition is defined as identifying a person based on his/her voice characteristics. This is useful in applications for authentication to identify authorized users i.e., enable access control using voice of an individual. Most of the times there are scenarios where multiple speakers speak simultaneously [1, 2]. Single speaker identification systems fail to handle such audio signals. Therefore, there it is essential to make the speaker recognition systems to handle multi-speaker audio files and classify them [1, 2].

2 Review of Literature

The paper that we have chosen as the foundation of our project is a technical paper [3] written by Wei-Ho Tsai and Shih-Jie Liao from the National Taipei University of Technology. The paper highlights the issue of identifying separate speakers in a multi-speaker environment.

The paper introduces 'Single Speaker Identification' which has seen a lot of development and success and goes on to explain the problem of multiple speakers and their identification in a conversation.

The important applications of multi-speaker identification are also listed, which include the likes of suspect identification in police work and automated minuting of meetings. The paper further explains two approaches to solving this problem.

1. A two stage process where the signal is first tested to identify whether it contains speech from a single speaker or from multiple speakers.
2. The second approach is a single stage process that carries out the single speaker and multi-speaker identification in parallel.

3 System Design

3.1 Mel Frequency Cepstrum Coefficients

This method was implemented as a feature extraction technique [4, 5]. To pre-emphasize speech signal, a high pass filter is implemented in this process. As speech is a non-stationary signal, which means the statistical properties of such speech is not constant all the time, we assume that the signal is made stationary by using a window of frame size 25 ms and frame shift of 10 ms [5]. We then apply the MFCC algorithm to determine 20 coefficients for the data set (Fig. 1).

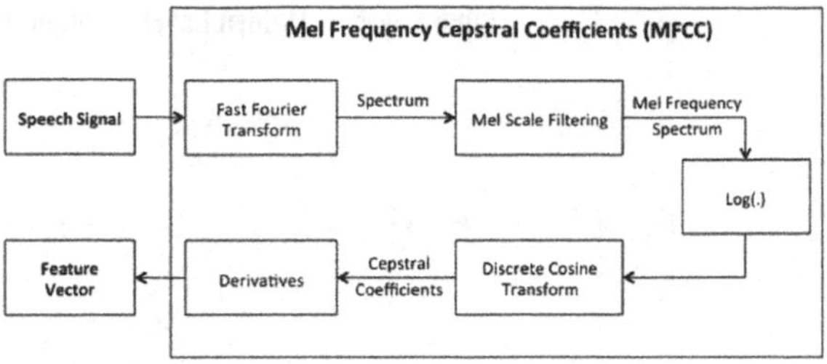

Fig. 1 MFCC process

3.2 *Normalization*

In order to make the neural network training more efficient, we have normalized the input features [6]. In this system, the values are normalized in the range of 0–1.

Therefore, the normalization technique used is:

1. Select the maximum value from the input data set.
2. Divide all data set values by the maximum value to get the normalized matrix.

Suppose X is the input matrix and x is the normalized matrix then

$$n = \max(X);$$
$$x = X/n;$$

3.3 *Neural Networks*

We have used neural networks for training and testing the data set.

1. The neural network consists of one input layer, one hidden layer and one output layer of neurons [7, 6]. The input neurons correspond to the features extracted (MFCC) per frame. An input matrix consisting of all features is given as input to the neural network. There are nine neurons in our output layer for nine speakers to be recognized, i.e. one neuron for one speaker.
2. The basic neural network having four outputs is as shown in Fig. 2. If the identified speaker is speaker 1, the first output neuron gives an output of 1 and the rest output neurons give an output of −1. Similarly, for second, third and fourth speaker, output neurons 2, 3 and 4 are fired and they give an output of 1 respectively.
3. The number of hidden neurons depends upon the number of hyperplanes required to correctly classify the input set into individual speakers in n-dimensional space [6]. (In our case 20 dimensional space).

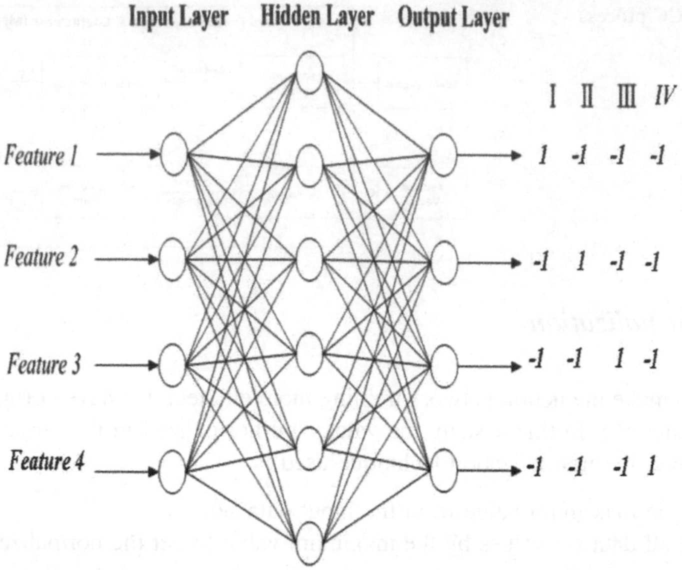

Fig. 2 Back propagation algorithm methodology

3.4 Tools Used

1. Audacity:
   This tool was used to digitally mix the audio files for multi-speaker recognition. It was also used to pre-process the audio files before using the audio files for training.
2. Text2speech.org
   This site was used to generate audio files which were used for testing and training. The data set consisted of 10 audio files per speaker i.e., each speaker speaking the digits 0–9 and some words [8]. There are 25 recordings which serves as multi-speaker files [9].
3. Matlab
   This software was used to acquire MFCC and to train and test the system using Error Back Propagation Training Algorithm (EBPTA).

4 Result Analysis

See Table 1 and Fig. 3.

Table 1 Result obtained

Sr. no.	Training set (no. of recordings)	Test set (no. of recordings)	Accuracy (%)
1. Single speaker	9	110	95.45% (105 recordings identified correctly)
2. Multi-speaker	–(Same training set as single speaker)	25	88% (22 recordings identified correctly)

Fig. 3 Result bar chart

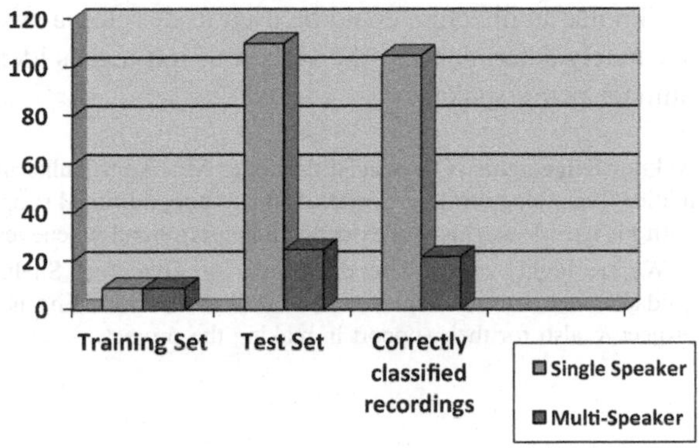

5 Conclusion

We were successful in identifying all the speakers using the Mel Frequency Cepstral Coefficients technique for feature extraction and Error Back Propagation Training Algorithm (EBPTA) for feature matching. To make our system more robust and adaptable to real life application, our system also identifies speakers in a multi-speaker environment. As the algorithm used is neural network that basically tries to mimic the working of a human brain, it is always adaptable to learning with new datasets. The multispeaker environment detection and learning capability of our system are the novel and user friendly features of our propesd system.

6 Future Work

1. Speaker identification using large data sets:

To make an application for identifying speakers in real-time, it is necessary to use cluster of computers for training to utilize parallel computing in neural networks. Various technologies such as 'MapReduce' [10] could be used for large datasets for training.

2. Speech diarization:

As the system proposed by us gives us frame by frame classification, one can easily perform speech diarization i.e. identifying who speaks when.

3. Speech isolation:

Once the frames are identified, one can also isolate the speech in multi-speaker environment so as to understand what each individual said.

4. Speech recognition (speech to text conversion):

Similar architecture could be used to develop a system wherein speech could be accurately determined by the system which means identifying the letters, words and number being spoken.

Acknowledgements Our special thanks to Mr. Arun Kulkarni, our Head of Department (Information Technology) for his cooperation and unconditional support. As our teacher he provided us with his useful insights and extended a helping hand whenever it was required.

We are highly indebted to the faculty of Thadomal Shahani Engineering College for their guidance and constant supervision as well as for providing necessary information regarding the project & also for their support in making the project.

References

1. Barry Arons, "A Review of The Cocktail Party Effect", MIT Media Lab.
2. M.K. Alisdairi,"Speaker Isolation in a "Cocktail-Party" Setting", Term Project Report, Columbia University, 2002.
3. Wei-Ho Tsai and Shih-Jie Liao, "Speaker Identification in Overlapping Speech", Journal Of Information Science and Engineering paper published in 2010.
4. Amit Sahoo and Ashish Panda, "Study of Speaker Recognition Systems", National Institute of Technology, Rourkela, 2011.
5. Lindasalwa Muda, Mumtaj Begam and I. Elamvazuthi,"Voice Recognition Algorithms using Mel Frequency Cepstral Coefficient (MFCC) and Dynamic Time Warping (DTW) Techniques", Volume 2, Issue 3, Journal of computing, March2010.
6. Noor Khaled, Saad Najiam Al Saad, "Neural Network Based Speaker Identification System Using Features Selection", Department of Computer Science, Al-Mustansiriyah University, Baghdad, Iraqi.
7. PPS Subhashini, Dr. M. Satya Sairam, Dr. D Srinivasarao, "Speaker Identification with Back Propagation Neural Network Algorithm", International Journal of Engineering Trends and Technology paper published in 2014.
8. Douglas A. Reynolds, "Automatic Speaker Recognition Using Gaussian Mixture Speaker Models", Volume B, Number 2, The Lincoln Laboratory Journal, 1995.
9. Ms. Asharani V R, Mrs. Anitha G, Dr. Mohamed Rafi, "Speakers Determination and Isolation from Multispeaker Speech Signal", Volume 4 Issue 4, International Journal of Computer Science and Mobile Computing, April 2015.
10. Changlong Li, Xuehai Zhou, "Implementation of Artificial Neural Networks in MapReduce Optimization "University of Science and Technology of China, 2 Texas Tech University, US, 2014.

Fault Detection and Classification Technique for HVDC Transmission Lines Using KNN

Jenifer Mariam Johnson and Anamika Yadav

Abstract In this paper, we have introduced a novel fault detection and classification technique for high-voltage DC transmission lines using K-nearest neighbours. The algorithm makes use of rectifier end AC RMS voltage, DC line voltage and current measured at both poles. These signals are generated using PSCAD/EMTDC and are further analysed and processed using MATLAB. For fault detection, the signals are continuously monitored to identify the instant of occurrence of fault, whereas for fault classification, the standard deviations of the data over a half cycle (with respect to AC signal) before and after the fault inception instant are evaluated. The algorithm hence makes use of a single-end data only, sampled at 1 kHz. The technique has proven to be 100% accurate and is hence reliable.

Keywords HVDC transmission system · Rectifier end signals
PSCAD/EMTDC · Fault detection · Faulty pole identification
KNN

1 Introduction

HVDC transmission systems, although not yet capable of replacing the conventional AC transmission systems, are extensively being used these days due to its superiority over HVAC transmission systems based on various aspects. The most important of these are controllability of power flow, economy while transmitting bulk power over very long distances, lower transmission losses and absence of stability issues [1–3].

J.M. Johnson (✉) · A. Yadav
Department of Electrical Engineering, National Institute of Technology, Raipur 492010, CG, India
e-mail: jeniferjohns91@yahoo.com

A. Yadav
e-mail: ayadav.ele@nitrr.ac.in

D.K. Mishra et al. (eds.), *Information and Communication Technology for Sustainable Development*, Lecture Notes in Networks and Systems 10,
https://doi.org/10.1007/978-981-10-3920-1_25

The transmission line is the entity of a power system that enables the interconnection of the generating station to the end user, in case of HVDC transmission system, this being the receiving end AC system. The protection of this entity is of utmost importance as any maloperation of the concerned protective relay would lead to prolonged disruption in the power flow and associated economic losses.

Most of the works that have been reported for detection and classification of faults in HVDC transmission lines are based on travelling wave algorithms [4–6]. These algorithms tend to be accurate, yet cannot be considered completely reliable since their accuracy depends on the accuracy with which the wave head is detected. The fault analysis using KNN approach has been proposed in [10] for three-phase transmission system. To the best of our knowledge, KNN has not yet been applied for detection or classification of faults in HVDC transmission lines.

2 HVDC Transmission System

The monopolar HVDC CIGRE benchmark model provided in PSCAD/EMTDC was modified to make it bipolar. Figure 1. shows the schematic diagram of the bipolar HVDC model that was used to carry out the fault analysis. Table 1 shows the details of the system used.

2.1 *Simulation of Fault on DC Line*

Pole 1 to ground, pole 2 to ground and pole to pole faults were simulated at intervals of 10 km from the rectifier end to the inverter end of the transmission line. The sampling frequency used is 1 kHz. The faults are simulated at 0.3 s and are permanent faults. The signals that need to be recorded are the AC RMS voltage, DC line voltage and current on both poles, at the rectifier end only. Consideration of the inverter end readings too would make the technique more reliable, but would

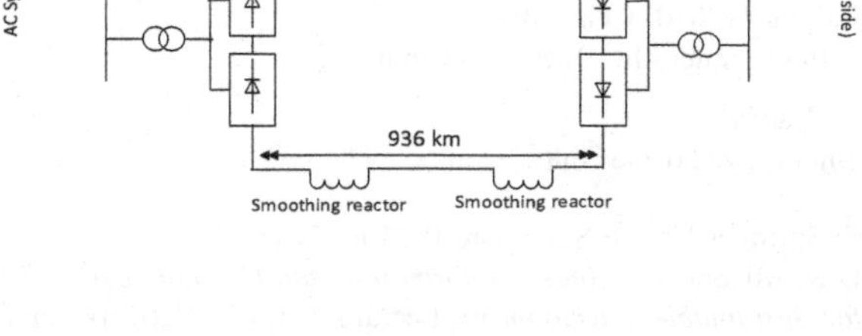

Fig. 1 Schematic diagram of HVDC system

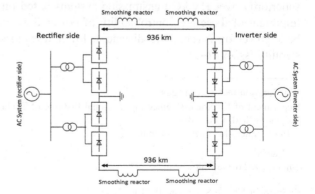

Table 1 Details of HVDC system modelled

Module	Details
Rectifier end AC system	345.0 kV, SCR = 2.5 @ 84.0 deg, 50 Hz
Inverter end AC system	230.0 kV, SCR = 2.5 @ 75.0 deg, 50 Hz
Transmission line	±500 kV, 936 km
Smoothing reactor	5 mH on both ends of DC line

require the two-end data to be synchronised at one end where the rest of the manipulations are done. This would increase the complexity and cost. Single-end data gives sufficiently accurate results as shown in this paper.

2.2 *Waveforms*

Figure 2a–e shows the waveforms of the signals that have been used as inputs for the KNN-based detection and classification algorithm. The classifier needs to correctly identify as to which class each waveform belongs.

3 Proposed Methodology

As seen from the waveforms, the constant current controllers at the converters will bring down the current from the actual high faulty values to near to normal values within a very less time, depending on the speed of response of the controllers. Hence, once a DC fault has occurred, it needs to be detected as soon as possible, for the fault clearance to be carried out effectively. The speed of the detection algorithm is hence of utmost importance. Moreover, the faulty pole too needs to be identified so that the maintenance can be carried out as early as possible to ensure continuity of power flow.

3.1 *K-Nearest Neighbours Classifier*

KNN is among the simplest of all machine learning algorithms. In this algorithm, the input vector consists of the k closest training samples in the feature space, where k is an integer. Classification of data is done by identifying the most common class among the k-nearest neighbours. These neighbours belong to a data set with which the algorithm is first trained and are determined by the distance from the test sample. In other words, while testing, the class which occurs most frequently among the neighbouring classes of the test sample under consideration becomes the class to which this individual test sample belongs.

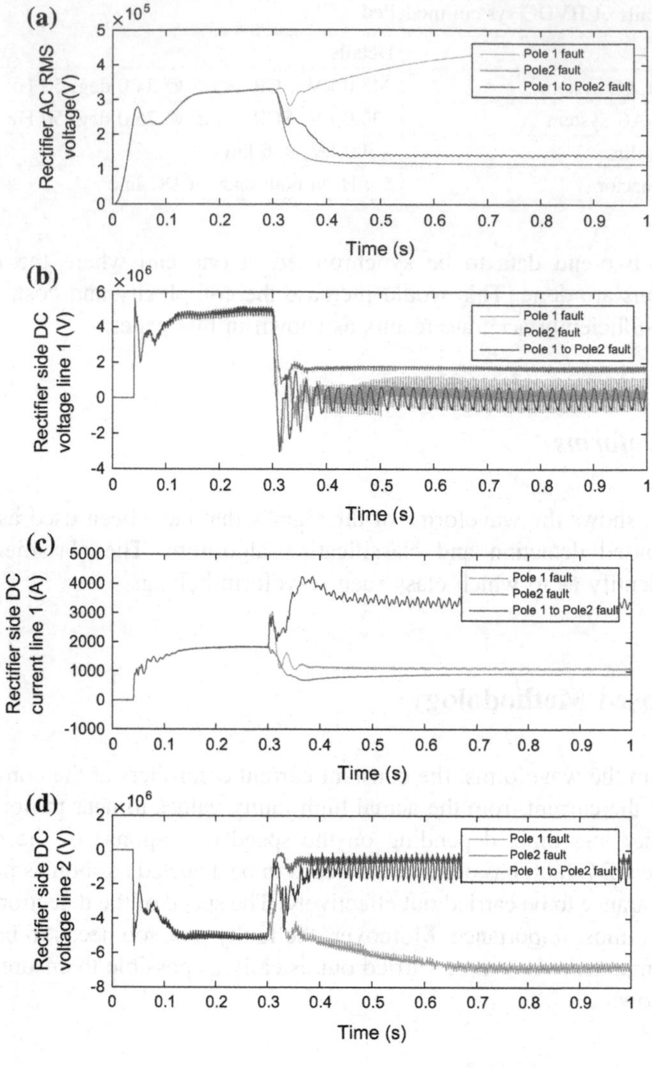

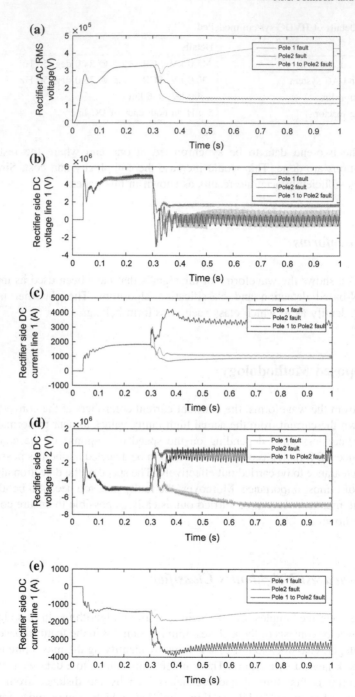

Fig. 2 **a–e** Waveforms used for relay algorithm

There are six different KNN Classifiers available in MATLAB that can be used to classify our data [11]. The Fine, Medium and Coarse KNN algorithms make use of Euclidian distance to determine the nearest neighbours. Their details, as given in MATLAB are:

(i) Fine KNN: A nearest neighbour classifier that makes finely detailed distinctions between classes with the number of neighbours set to 1.
(ii) Medium KNN: A nearest neighbour classifier that makes fewer distinctions than a Fine KNN with the number of neighbours set to 10.
(iii) Coarse KNN: A nearest neighbour classifier that makes coarse distinction between classes, with the number of neighbours set to 100.
(iv) Cosine KNN: A nearest neighbour classifier that uses the cosine distance metric.
(v) Cubic KNN: A nearest neighbour classifier that uses the cubic distance metric.
(vi) Weighted KNN: A nearest neighbour classifier that uses distance weighting.

For implementing the fault detector, the entire time domain signals are used whereas for fault classification, the standard deviation of the signals over a time extending from half cycle before the instant of fault to half cycle after the instant of fault is used to train the KNN-based algorithm. The Euclidian distances are

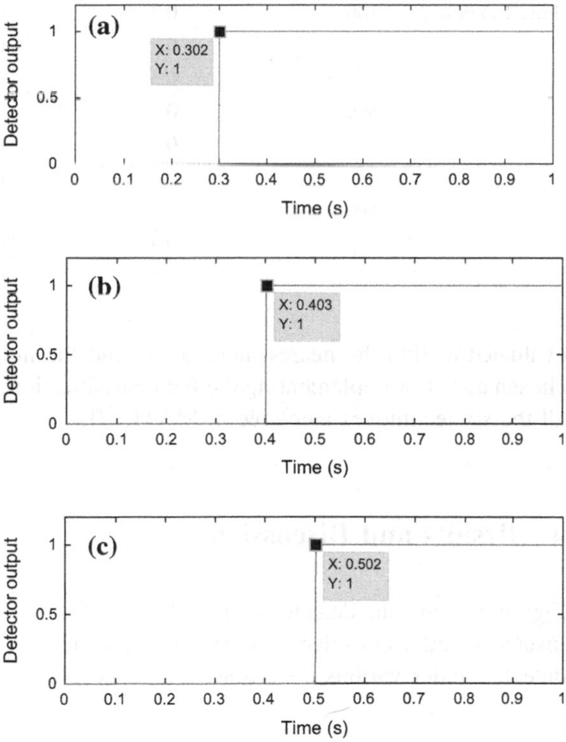

Fig. 3 Detector output obtained using KNN for fault at **a** 0.3 s, at 500 km on pole 1, **b** 0.4 s, at 500 km on pole 1, **c** 0.5 s, at 500 km on pole 1

Table 2 Delay in detecting fault at various conditions

Fault type	Fault location (km)	Fault instant (s)	Detection instant (s)	Delay (ms)
Pole 1 to ground	100	0.3	0.301	1
		0.4	0.401	1
		0.5	0.502	2
	500	0.3	0.302	2
		0.4	0.403	3
		0.5	0.502	2
	800	0.3	0.303	3
		0.4	0.403	3
		0.5	0.504	4
Pole 2 to ground	100	0.3	0.302	2
		0.4	0.401	1
		0.5	0.503	3
	500	0.3	0.301	1
		0.4	0.403	3
		0.5	0.502	2
	800	0.3	0.303	3
		0.4	0.403	3
		0.5	0.503	3
Pole 1 to pole 2	100	0.3	0.303	3
		0.4	0.404	4
		0.5	0.504	4
	500	0.3	0.302	2
		0.4	0.403	3
		0.5	0.503	3
	800	0.3	0.301	1
		0.4	0.403	3
		0.5	0.502	2

evaluated to find the nearest neighbours and the number of nearest neighbours is chosen as 1. For implementing the fault classification, the KNN model is trained for all the six techniques available in MATLAB.

4 Results and Discussion

Figure 3 shows the detector output obtained for fault on pole 1 under various fault instances under consideration. Table 2 tabulates the delay with which the fault is detected under various conditions. The maximum delay is 4 ms. This is very less

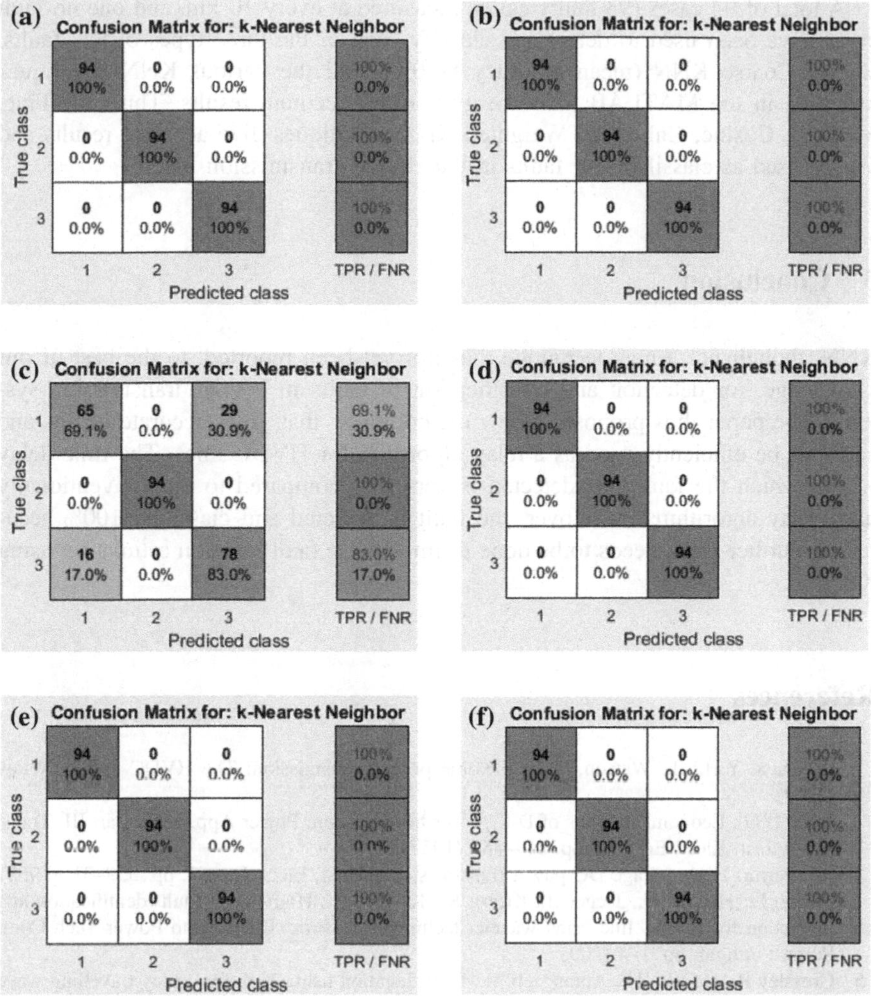

Fig. 4 Confusion matrix for **a** Fine KNN **b** Medium KNN **c** Coarse KNN **d** Cosine KNN **e** Cubic KNN **f** Weighted KNN

compared to the time delay in conventional relays which lies in the range 30–40 ms.

Figure 4a–f show the confusion matrices obtained using Fine KNN, Medium KNN, Coarse KNN, Cosine KNN, Cubic KNN and Weighted KNN, respectively. The diagonal elements represent the test samples that have been correctly classified by the classifier, whereas the off-diagonal elements represent the number of test samples that have been wrongly classified. TPR stands for True Positive Rate and FNR stands for False Negative Rate.

A total of 94 cases (93 faulty cases, measured at every 10 km, and one no-fault case) have been used to detect and classify each of the three types of DC faults. Except Coarse KNN (mean accuracy 84.03%), all the various KNN techniques available in the MATLAB software give 100% accurate results. Thus, the Fine, Medium, Cosine, Cubic and Weighted KNN techniques give accurate results and can be used as classifiers for faults in the HVDC transmission line.

5 Conclusion

KNN, though not a new technique, has not yet been reported, to the best of our knowledge, for detection and classification of faults in HVDC transmission systems. The paper thus proposes a new methodology that gives accurate results and can thus be efficiently used as a relay algorithm for HVDC lines. The time delay within which the fault gets detected is very low compared to the conventionally used relay algorithms. Moreover, the fault is detected and classified 100% accurately. Further work needs to be done to implement fault location estimation using KNN.

References

1. Arrillaga, Y.H.L.J., Watson, N.R.: Flexible power transmission: The HVDC option, Wiley (2007)
2. John, H.G.: Economic aspects of D-C power transmission. Power App. Syst., Part III. Trans. Amer. Inst. Elect. Eng. 76, pp. 849–855 (1957)
3. J.F. Perrin: High-voltage DC power transmission, J. Inst. Elect. Eng. 7, pp. 559–31, (1961)
4. Shang, L., Herold, G., Jaeger, J., Krebs, R., Kumar, A.: High-speed fault identification and protection for HVDC line using wavelet technique. In Proc. IEEE Porto Power Tech Conf. Porto, Portugal, pp. 1–4 (2001)
5. Crossley P.A., Gale P.F, Aurangzeb M.: Fault location using high frequency travelling wave measured at a single location on a transmission line. Developments in Power System Protection, Amsterdam, Holland, pp. 403–406 (2001)
6. Chen, P., Xu, B., Li, J.: A traveling wave based fault locating system for HVDC transmission lines, in Proc. Int. Conf. PowerCon, Chongqing, China, pp. 1–4 (2006)
7. Burek, A., Rezmer, J.: Evolutionary algorithm for detection and localization of faults in HVDC systems. Int. Conf. On Environment and Electrical Engineering, pp. 1317–1322 (2015)
8. Cui, H., Tu, N.: HVDC transmission line fault localization base on RBF neural network with wavelet packet decomposition. Int. Conf. on Service Systems and Service Management, pp. 1–4 (2015)
9. Narendra, K. G., Sood, V. K., Khorasani, K., Patel, R.: Application of a Radial Basis Function (RBF) for fault diagnosis in an HVDC system. Proc. of the 1996 Int. Conf. on Power Electronics, Drives and Energy Systems for Industrial Growth, vol. 1, pp. 140–146 (1996)

10. Yadav, A., Swetapadma, A.: Fault Analysis in Three Phase Transmission Lines Using k-Nearest Neighbours. Int. Conf. on Advances in Electronics, Computers and Communications, pp. 1–5 (2014)
11. Matlab R2015a software

Retinal Disease Identification by Segmentation Techniques in Diabetic Retinopathy

Priyanka Powar and C.R. Jadhav

Abstract Detection of microaneurysms before diabetes increases is an essential stage in diabetic retinopathy (DR) which damages the eye, so it is big medical problem. A need arises to detect it at an early stage. It is not showing any symptoms, so it can only be a diagnosed by oculist. This paper presents the study and review of various techniques used in detection of microaneurysms as well as new approach to increasing sensitivity and reducing computational time for detection and classification of microaneurysms from the diabetic retinopathy images. This new strategy to detect MAs is based on (1) Elimination of nonuniform part of an image and standardize grayscale content of original image. (2) Performed Morphological operations for detection of Region of Interest (ROI) and elimination of blood vessels. (3) To identify real MAs two features extracted where one feature of shape which discriminates normal eye image and abnormal eye image and second feature of texture. So, this increases sensitivity and also availability. So for this whole process of new technique used publically available database called DiaretDB1 database. (4) To discriminate normal and abnormal images different clustering algorithm used.

Keywords Image processing · Diabetic retinopathy · Fundus images Microaneurysms detection · Retinal image

P. Powar (✉) · C.R. Jadhav
Department of Computer Engineering, D.Y. Patil Institute of Engineering & Technology, Pimpri, Pune, India
e-mail: priya.powar27@gmail.com

C.R. Jadhav
e-mail: chaya123jadhav@gmail.com

D.K. Mishra et al. (eds.), *Information and Communication Technology for Sustainable Development*, Lecture Notes in Networks and Systems 10,
https://doi.org/10.1007/978-981-10-3920-1_26

1 Introduction

This diabetic retinopathy disease comes only when effect of diabetes on small blood vessels increases as well as damaging the part of eye called the retina. The risky part is that patients are not able to get that whether he has diabetes because usually it does not cause any observable symptoms in early stage. If we are able to detect diabetic retinopathy in this stage, then proper treatment can stop worse condition of patients. Otherwise, it is very difficult to overcome the diabetes after getting some observable symptoms which become much more untreatable. Diabetes turns out to be one of the quickly increasing diseases worldwide. Diabetic retinopathy is one of the main reasons for blindness. In the previous stage of the diabetic retinopathy (DR), there are no symptoms present in the eye and when disease starts, the presence of microaneurysms (MAs), soft exudates and hard exudates and new damaged blood vessels increases. Any disease is treatable if it detected and monitored properly by ophthalmologist through fundus camera as well as more efficient detection and monitoring saves costs. Taking pictures or photographs of retina by Fundus camera has an important role to control diabetes and the presence of retinal abnormalities is common but consequences of it are serious.

In this whole process of detecting diabetes, screening plays a central role in which capturing high resolution or accurate pictures of retina and also reliable or more beneficial use of algorithms and techniques can help to get success in detecting abnormalities. In this paper, a new technique, method and algorithm, is surveyed for efficient detection of microaneurysms from fundus images. The first mark of DR is microaneurysms (MAs) that are not able to damage vision. The number of MAs has been taken into account by the specific tool which gives progress of diabetic retinopathy in particular candidate. A number of different methods or techniques are used for detecting MA automatically, these methods were tested on different databases which are publically available like DirectDB1 database but still there is need to increase sensitivity and availability.

Figure 1 depicts both the normal retina structure at left most side for those who does not have diabetes and abnormal retina structure at right most side for those who has diabetic retinopathy. Most of existing techniques includes number of systems which use hardware and software. Some of those software systems are partially automated systems which are time consuming. Currently, most of the hospitals using hardware machines for detection provide accurate results but time required for it are in hours. Also, automated system of detection uses clustering methods such as, k-NN classifier for classifying normal retina and abnormal retina. Whereas, proposed system uses SVM classification technique to differentiate normal and abnormal retina.

Section 2 describes related work of microaneurysms detection in diabetic retinopathy disease and related information. Section 3 depicts proposed system and methodologies which are used. Section 4 describes the expected results. Section 5 describes the extracted conclusions.

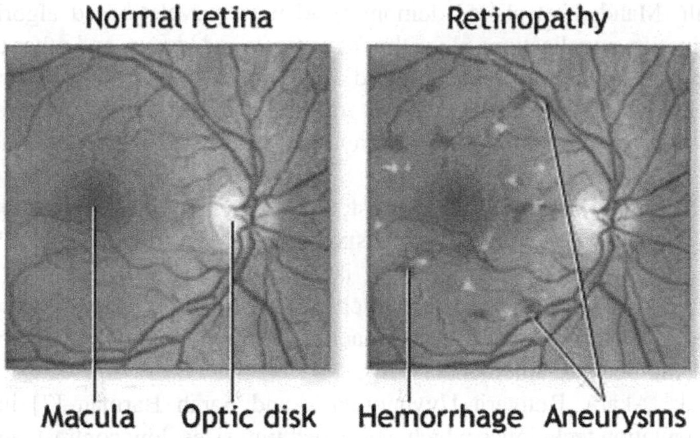

Fig. 1 Normal retina and abnormal retina [10]

2 Related Work

Spencer Timothy and John Olson [1] describe different segmentation techniques which detect microaneurysms present in fluorescein angiograms. Microaneurysms segmentation is done using bilinear top-hat transformation and matched filtering technique. A novel region-growing algorithm fully depicts each marked object and analysis of the size, shape, and energy characteristics as well as fully grown objects accumulated in separate binary image. This approach is valuable for increasing accuracy of monitoring the progress of diabetic retinopathy.

Niemeijer, Meindert, et al. [2] demonstrated hybrid approach used to mark the red lesions which are present in digital color fundus photographs. Proposed pixel classification method separates red lesions from background of an image. Also K-nearest neighbor classifier is used to classify detected objects.

Walter, Thomas, et al. [3] proposed a new algorithm which is divided into four stages for detection of MAs automatically in fundus images. In first step, image enhancement, shade correction and image normalization are carried out. Second stage designed for detection of candidate that all patterns originated which relates to MA which were achieved by diameter closing and automatic threshold scheme. In last stage, feature is extracted to do the classification of candidate into real MA and other objects in automatic manner.

Fleming, Alan D, et al. Fleming, Alan D., et al. [4] used watershed transform method as well as image contrast normalization to derive no vessels or other lesions of region and differentiate MAs and other injuries which are present in retina, respectively. The dots within the blood vessels are handled successfully by using local vessel detection technique. A genetic algorithm helps to optimize a process and analyze some images.

Esmaeili, Mahdad, et al. [5] demonstrated new curvelet-based algorithm and new illumination equalizations algorithm to separate red lesions and other unwanted part and improve nonuniform background of an image. In next stage, to produce an enhanced image applied digital curvelet transform and to modified curvelet coefficient in order to do red objects to zero and also gives sparse representation of the object.

Antal, Blint, and Andrs Hajdu [6] first, select set of preprocessing methods for candidate extractor and its measure in six different MA categories and for each different category best preprocessing is method selected. Second, adaptive weighting approach is presented for which actual works on spatial locations were categorized into: near to vessel, in the macula, on the periphery and also works on contrast of the detected microaneurysms.

Sopharak, Akara, Bunyarit Uyyanonvara, and Sarah Barman [7] introduced some segmentation techniques which can detect tiny sizes, low contrast, and similar blood vessels of microaneurysms. There used two segmentation techniques, one coarse segmentation using mathematical morphology which identifies MA candidate in retinal images and fine segmentation using naive Bayes classifier.

Tavakoli, Meysam, et al. [8] proposed a novel method based on random transform (RT) and multi-overlapping for early detection of microaneurysms. Initially, optic nerve head (ONH) was detected or masked and to remove the background top-hat transformation and averaging filter were applied in preprocessing stages. After detecting and masking retinal vessels and ONH, microaneurysms were detected and numbered by using RT and thresholding.

Adal, Kedir M., et al. [9] designed detection of microaneurysm is same problem as finding interest region or blobs from image. Characterized these blob regions by scale adapted region descriptor. Semi-supervised-based learning approach is proposed to detect true MAs, there is need to train a classifier. A Gaussian mixture model based clustering combined and classification of microaneurysms at pixel level logistic regression classification method have been used.

3 Proposed Scheme

Detection of microaneurysms or any injuries which are observable in fundus retinal images has to be improved. And it will achieve by analyzing the images which are captured by fundus camera. This detection process starts with preprocessing of the images after that extraction of different feature which will use for more effective treatment. In an early stage need to be detect the diabetic retinopathy automatically to diagnose it completely without any delay. At the time of first screening of any candidate, the ophthalmologists have to examine a large number of retinal images in order to achieve successful or accurate treatment. To do large number of screening required more cost. Solution to this is to develop an automated screening method

Table 1 List of symbols and description

Symbol	Description
MA	Microaneurysms
FP	False Positive
DR	Diabetic Retinopathy
ROI	Region of Interest
SVM	Support Vector Machine
FA	Fluoresce in Angiography
RT	Random Transformation
ONH	Optic Nerve Head
RGB	Red Green Blue

for retinal images in diabetic retinopathy in early stage. This kind of system should be able to distinguish between an image which has true microaneurysms (MA) and normal retinal images so that workload of ophthalmologists will get reduced and more number of candidates will get diagnosed. This new system gives early detection of microaneurysms as well as it increases sensitivity and efficiency than before. The proposed detection process consists of three main modules such as preprocessing module, feature extraction module and clustering and classification module (Table 1).

3.1 Preprocessing Module

The proposed approach helps to solve the problem of detecting candidates on retinal fundus images, where candidates are regions possibly corresponding to microaneurysms (MAs), which is separated into two stages. As shown in Fig. 2, this module takes high definition (HD) retina images from publically available database as an input to do the preprocessing. The first stage is reduction of nonuniform illumination and to analyze gray scale content images. And second stage is to perform morphological operations to eliminate blood vessels and detect Region of Interest (ROI).

3.2 Grayscale Conversion

Original images are captured by fundus camera in which the value of each pixel is a single sample which holds only intensity information. These images are converted into grayscale image like in color of black and white also it looks more in gray color so that it is called grayscale image. To convert an image based on an RGB color model to a grayscale form. Weighted sums must be calculated in a linear RGB

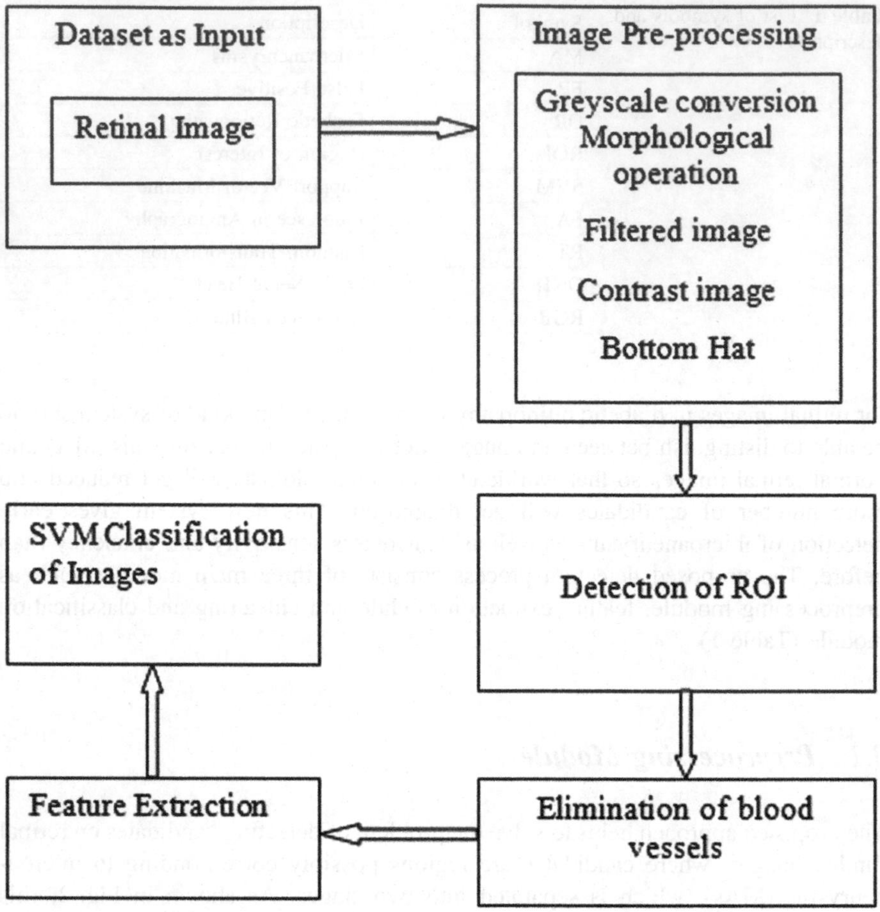

Fig. 2 System architecture

space, that is, after the gamma compression function. For RGB color space. Mathematically this can be represented as: [10]

$$Y = 0.2126R + 0.7152\,G + 0.0722\,B$$

Figure 3 shows response of grayscale conversion after applying it on original retinal image as well green, red, blue planes are created because they give more information. In Fig. 3, left most images have taken as original retinal image and right most image is responsible for grayscale conversion after calculating RGB values.

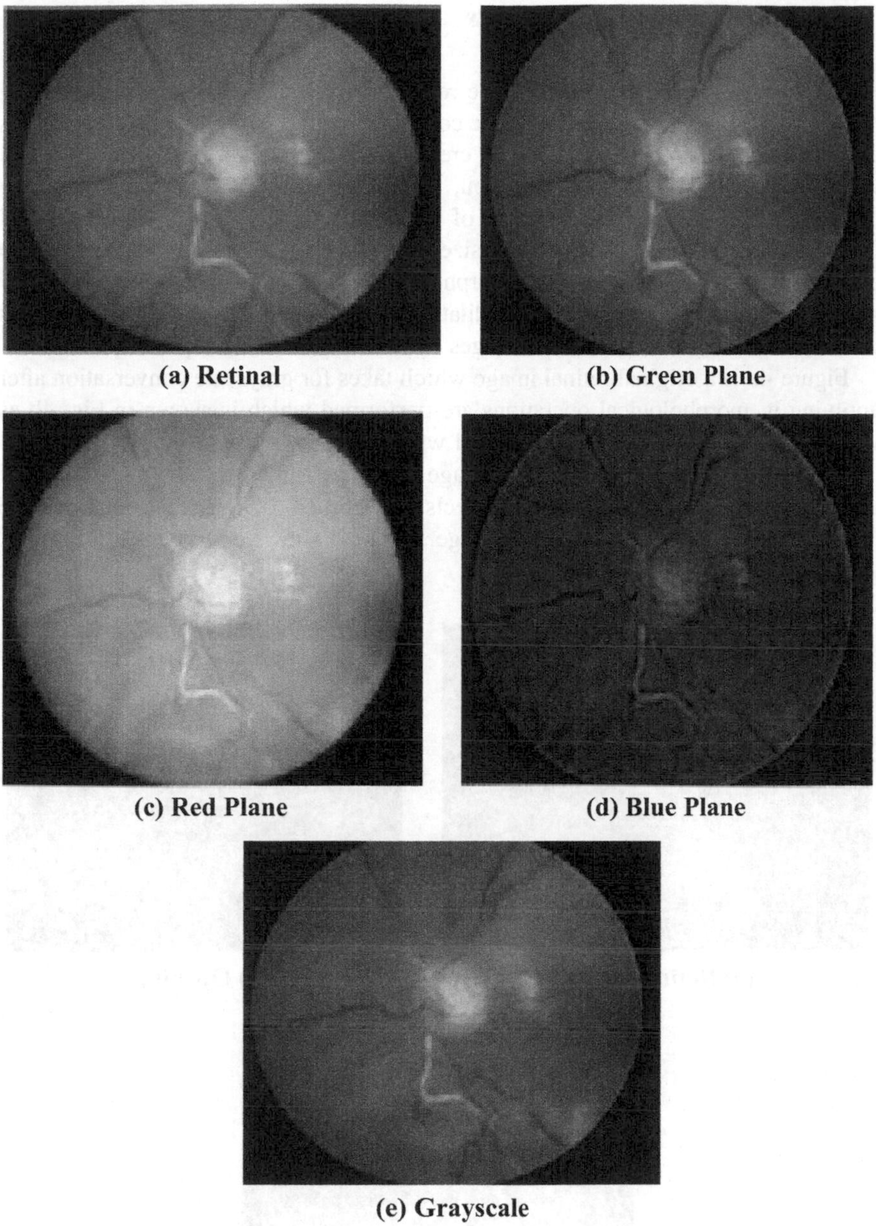

Fig. 3 Grayscale conversion

3.3 Morphological Operations

Figure 4 shows how blood vessels are removed from original image by applying morphological operations on gray scale converted image. Morphological operations apply to original images of retina to create an image which has removed blood vessels. In a morphological operation, the value of each pixel after applying operations is based on a comparison of the corresponding pixels. Morphological operations are done by choosing the size and shape of the neighborhood pixel to detect microaneurysms properly. Morphological operation can perform on gray scale images. Some operations like dilate, erosion, opening, closing, hit or miss transformation used to on retinal images to remove blood vessels.

Figure 4a is a original retinal image which takes for grayscale conversation after applying it, morphological operations are performed which is shown in Fig. 4b as well as bottom hat transform performed which is shown in Fig. 4c.

Enhancement process enhances image by increasing intensity of objects of image. This method highlights the objects and borders of image. It equalizes the intensities of ROI present in retinal image. Figure 5 shows enhancement of image.

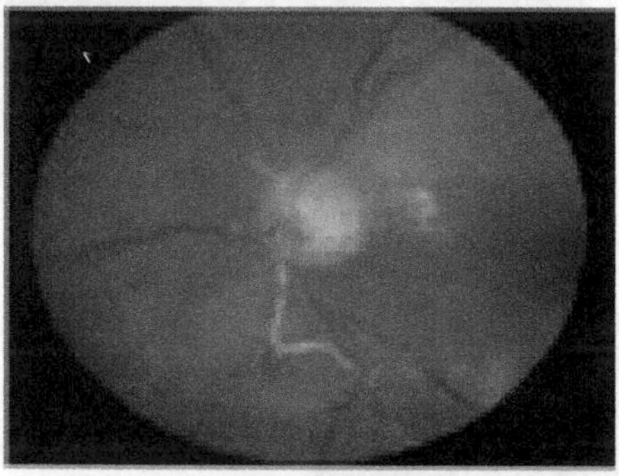

(a) Retinal Image

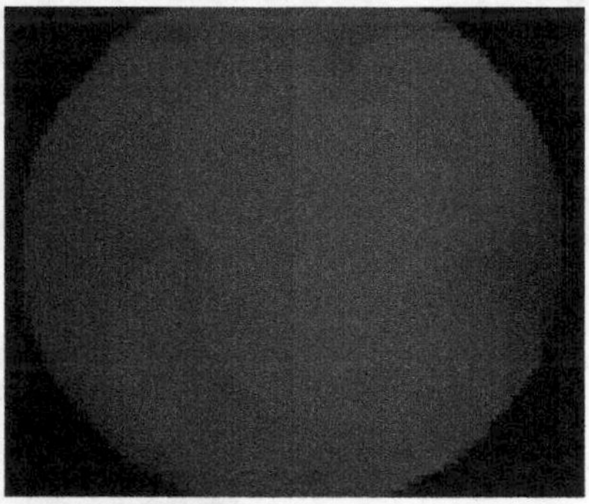

(b) Opening

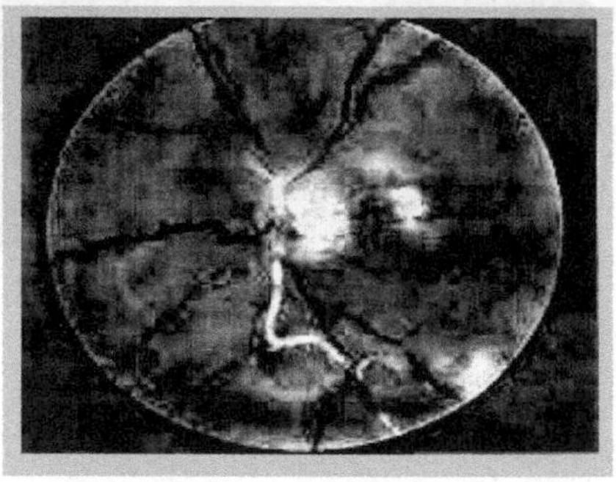

(c) BHT

Fig. 4 Morphological operations and Bottom Hat Transform

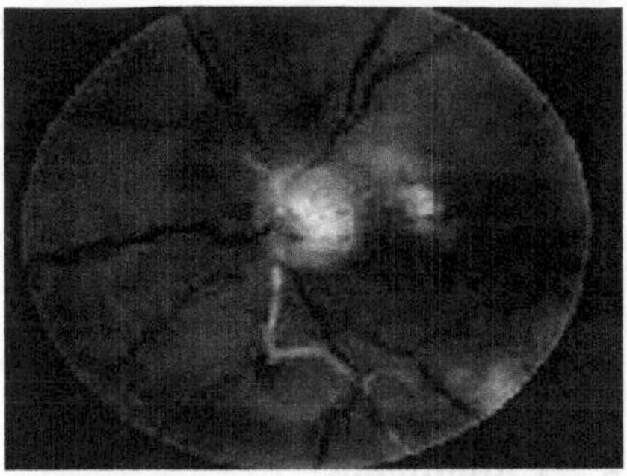

Fig. 5 Image enhancement

3.4 *Feature Extraction*

This module performs feature extraction by extracting features from the original retinal image which may be in the form of .jpeg or .img or digital image. In order to understand whether a particular person has diabetes or not, there is a need to extract some feature like shape color, size, and texture. Whether round-shaped candidates or elongated shaped ones which distinguish real MAs from FPs [11].

3.5 *Classification*

In this classification module preprocessed images are used instead of original images. The images obtained after grayscale images and morphological operations are only useful for correct classification which gives fast availability of particular candidate record as well as increase in sensitivity of diabetic retinopathy system application. For better classification, SVM (Support Vector Machine) [10] is the suitable classifier

4 Expected Result

Nowadays, a number of diabetes patients are more than number of ophthalmologist because of lack of automated detection system which detects the microaneurysms in early stage. Accuracy in detection of lesser age candidates plays a vital role. Need to remove high computation for each retinal image with high sensitivity. This new approach of automated detection system removed above problems which use computerized segmentation and classification techniques that are automated schemes. This proposed new system takes input of original retinal image for pre-processing which eliminates nonuniform illumination or unwanted things. Grayscale conversion helps to transform an image in the form black and white color to easily detect the MAs from FPs. In order to remove blood vessels morphological operations performed on grayscale images. Features like shape, color, texture are extracted for better accuracy and to increase sensitivity of diabetic retinopathy detection system. Further, it reduces high computation time for each high definition images by using SVM classifier. Also distinguishes normal and abnormal retina images by using k-means clustering algorithm. The proposed MA detection method achieves sensitivity almost 92.50% for DiretDB1 database images.

5 Conclusion

The objective or main task of this approach is to detect microaneurysms in retinal fundus image through segmentation followed by feature extraction in diabetic retinopathy disease. Shape features of MAs are considered for better accuracy of detection. MAs detection with extracted features classified using k-means clustering and comparative study of algorithm and classification using support vector machine (SVM). Automatic segmentation and classification scheme produce accurate results by detecting MAs using retinal image and classifying MAs in round and elongated shapes. Feature extraction process extracts shape, color and texture features for identification. Furthermore, the system should be fitted accurately in any environment and increases sensitivity. Moreover, it is important to study and analyze the use of new features that may be decisive for this type of analysis.

References

1. Pencer, Timothy, et al. "An image-processing strategy for the segmentation and quantification of microaneurysms in fluorescein angiograms of the ocular fundus." Computers and biomedical research 29.4 (1996): 284–302.
2. Niemeijer, Meindert, et al. "Automatic detection of red lesions in digital color fundus photographs." Medical Imaging, IEEE Transactions on 24.5 (2005): 584–592.
3. Alter, Thomas, et al. "Automatic detection of microaneurysms in color fundus images." Medical image analysis 11.6 (2007): 555–566.
4. Uellec, Gwnol, et al. "Optimal wavelet transform for the detection of microaneurysms in retina photographs." Medical Imaging, IEEE Transactions on 27.9 (2008): 1230–1241.
5. Smaeili, Mahdad, et al. "A new curvelet transform based method for extraction of red lesions in digital color retinal images." Image Processing (ICIP), 2010 17th IEEE International Conference on. IEEE, 2010.
6. Antal, Blint, and Andrs Hajdu. "Improving microaneurysm detection in color fundus images by using context-aware approaches." Computerized Medical Imaging and Graphics 37.5 (2013): 403–408.
7. Opharak, Akara, Bunyarit Uyyanonvara, and Sarah Barman. "Simple hybrid method for fine microaneurysm detection from non-dilated diabetic retinopathy retinal images." Computerized Medical Imaging and Graphics 37.5 (2013): 394–402.

8. Avakoli, Meysam, et al. "A complementary method for automated detection of microaneurysms in fluorescein angiography fundus images to assess diabetic retinopathy." Pattern Recognition 46.10 (2013): 2740–2753.
9. Dal, Kedir M., et al. Automated detection of microaneurysmsusing scale-adapted blob analysis and semi-supervised learnin.
10. http://www.google.com.
11. Leming, Alan D., et al. "Automated microaneurysm detection using local contrast normalization and local vessel detection." Medical Imaging, IEEE Transactions on 25.9 (2006): 1223–1232.
12. Adzil, MH Ahmad, Lila Iznita Izhar, and Hanung Adi Nugroho. "Determination of foveal avascular zone in diabetic retinopathy digital fundus images." Computers in biology and medicine 40.7 (2010): 657–664.
13. Iemeijer, Meindert, et al. "Retinopathy online challenge: automatic detection of microaneurysms in digital color fundus photographs." Medical Imaging, IEEE Transactions on 29.1 (2010): 185–195.
14. Ae, Jang Pyo, et al. "A study on hemorrhage detection using hybrid method in fundus images." Journal of digital imaging 24.3 (2011): 394–404.
15. Kram, M. Usman, Shehzad Khalid, and Shoab A. Khan. "Identification and classification of microaneurysms for early detection of diabetic retinopathy." Pattern Recognition 46.1 (2013): 107–116.
16. Aade Mahesh k, "A Survey of Automated Techniques for Retinal Disease Identification in Diabetic Retinopathy". International Journal of Advancements in Research & Technology, May-2013 ISSN 2278–7763.

High Dimensionality Characteristics and New Fuzzy Versatile Particle Swarm Optimization

Shikha Agarwal and Prabhat Ranjan

Abstract Technological developments have reshaped the scientific thinking, since observation from experiments and real world are massive. Each experiment is able to produce information about the huge number of variables (High dimensional). Unique characteristics of high dimensionality impose various challenges to the traditional learning methods. This paper presents problem produced by high dimensionality and proposes new fuzzy versatile binary PSO (FVBPSO) method. Experimental results show the curse of dimensionality and merits of proposed method on bench marking datasets.

Keywords High dimensionality · Particle swarm optimization
Fuzzy logic · Feature selection · Classification

1 Introduction

Ultra-high-dimensional data is characterized by very high dimensionality as compared to the number of samples. For the point of view of medical care, stock market and many other users, high dimensionality is a blessing rather than curse because twenty-first century is a data-driven society and we are consumers of data. This high dimensionality brings three unique features [1] namely; noise accumulation [2], spurious correlations [3], and incidental endogeneity [4, 5]. Analyzing high-dimensional data requires us to simultaneous study of many features. Therefore, if a number of variables are more than number of samples then it will impose statistical,

S. Agarwal (✉) · P. Ranjan
Department of Computer Science, Central University of South Bihar,
Patna 800014, India
e-mail: shikhaagarwal@cub.ac.in

P. Ranjan
e-mail: prabhatranjan@cub.ac.in

D.K. Mishra et al. (eds.), *Information and Communication Technology for Sustainable Development*, Lecture Notes in Networks and Systems 10,
https://doi.org/10.1007/978-981-10-3920-1_27

mathematical, and computational challenge. Dimensionality reduction using feature selection is one of the ways to handle high dimensionality. This paper experimentally shows the failure of classical computational methods when applied on high dimensional data and proposes the new fuzzy versatile binary particle swarm optimization (FVBPSO) for feature selection. In FVBPSO, particle is made up of 0, 1 and # string, where, # position is resolved using fuzzy logic.

This paper is organized as follows. Section 2, presents overview of high dimensionality and importance of dimensionality reduction. Section 3, presents the overview of PSO and small survey on different variants of PSO. Section 4, proposes fuzzy versatile binary PSO. Section 5, presents the experimental setup. Section 6, presents the results on benchmarking datasets and discusses the findings. Section 7, concludes the paper.

2 High Dimensionality

High dimensionality means that data is placed in a space in which numbers of axes are very high. In other words, each sample has many attributes or variables. For example, in NMR spectroscopy data, microarray experiment data, text document data ("bag of words"). In case of high dimensionality, simultaneously estimating many features accumulates errors which are called noise accumulation due to which traditional analysis methods perform no better than random results [6]. Increase in dimension leads to an exponential increase in demand of sample data points to develop any learning model [7]. Silverman also illustrated some other consequences of dimensionality on the number of kernels problem [8]. According to Donoho, "concentration of measure" is a phenomenon in which the norm of any vector remains constant in high-dimensional spaces while average of any vector increases with the dimension [9]. According to Scott et al. [10], the volume of sphere decreases to zero when dimension increases. Scott et al. [10] has justified it by plotting Gaussian function in high dimension. Shikha and Rajesh have also pointed out some difficulties of high dimensionality [11]. Dimensionality reduction is important to handle high-dimensional datasets to neutralize the effects of the curse of dimensionality, identifying the behavior and performance of the complex system.

3 Particle Swarm Optimization

Particle swarm optimization (PSO) is a population-based stochastic bio-inspired search method which is inspired by swarm behavior of animals and birds. In PSO, swarm is made up of particles. Each particle is defined using position vector (X) and velocity vector (V). The optimal particle of the swarm represents the global solution (*gbest*) and best position of any particle based on its fitness is represented as local best (*pbest*) position of each particle. Based on the fitness, *pbest* and *gbest* are calculated which are used to update velocity and position using Eqs. (1) and (2) given by Kennedy and Eberhart [12, 13].

$$V(t+1) = V(t) + c_1 r_1 (pbest(t) - X(t)) + c_2 r_2 (gbest(t) - X(t)) \tag{1}$$

$$X(t+1) = X(t) + V(t+1) \tag{2}$$

Kennedy and Eberhart have published two books on PSO [14, 15] and applied PSO for the dimensionality reduction in 1997 using Binary PSO [16]. Chuang et al. improved the binary PSO for the feature selection problem [17]. Agarwal and Rajesh have proposed some modifications in BPSO for feature selection [18, 19]. Some variants of PSO which are hybrid with fuzzy logic, are, fuzzy discrete PSO by Emami and Derakhshan hybridized the fuzzy k mean clustering with PSO [20]. Abdelbar et al. proposed Charisma-based PSO [21]. Chuang et al. developed the fuzzy adaptive Catfish PSO [22]. Fuzzy particle swarm optimization with cross mutation [23], type-2 fuzzy adaptive [24], parameter adaptation through fuzzy logic [25].

4 Proposed Method

In this section, we propose a novel fuzzy versatile BPSO (FVBPSO). In FVBPSO, particles are string of (0, 1, #). # is defined as a state in which attribute is considered neither accepted nor rejected (uncertain state). In FVBPSO, each particle has been handled in dual mode, acceptance mode denoted by X_A, where 70% of # are replaced with 1 and rest 30% with 0. In rejection mode particle is represented as X_R, in which 70% of # are replaced with 0 and rest 30% with 1. Fitness of each mode is calculated using K-nearest neighbor classifier with leave one out cross-validation method. Fitness of X_A (acceptance mode) is denoted by F_A and fitness of X_R (rejection mode) is denoted by F_R. Jaccord distance between global best and particle's current position is calculated. J_A and J_R are Jaccord distance for acceptance and rejection mode, respectively. In order to construct the fuzzy rules, F_A, F_R, J_A, and J_R are converted into fuzzy variables. The fuzzy linguistic terms of the inputs, namely, F_A, F_R, J_A, and J_R are shown in Fig. 1. Six fuzzy rules are formed to predict the actual mode of particle as shown in the Fig. 2, where the output variable take singleton values (H takes 0.8 and L takes 0.2). If FIS output is greater than 0.5 then particle is further treated in acceptance mode and if FIS output is less than 0.5 then rejection mode of particle is carried forward. The detail algorithm is given in Algorithm 1.

Algorithm 1: Algorithm of Fuzzy Versatile BPSO

```
Input: High Dimensional Data S ={S1, S2, ..., SN},(Sn ∈ R^d) and associated class C={C1, C2, ..., CN}
Output: Selected features {D∈N^l: 1<l≤d}
Initialize particles X={X1, X2, ..., XN},(Xn∈{1,0,#}^d)
/* {1, 0,#} elements of a particle */
Initialize Velocity, pbest and gbest randomly
Repeat until termination is not met
   for p = 1:no of particles do
/* XA= randomly replace 70 % of # with 1 and rest 30% with 0, XR= randomly replace 70% of # with 0 and rest 30% with 1*/
   [XA, XR] = ACCREJ(Xp)
   JA = Jcoff (XA, gbest)
   JR = Jcoff (XR, gbest)
   FA=fitness of XA
   FR=fitness of XR
   FIS OUTPUT=FIS(FA, FR, JA, JR)
         if FIS OUTPUT ≥ 0.5 then
               Fp=FA and Xp= XA
         else
               Fp=FR and Xp= XR
         end if
end for
/*Update gbest particle*/
   if Fitness of any particle ≥ F^gbest then
         gbest = position of particle
   end if
   for p=1 to no of paricles do
         if Fp ≥ Fp^pbest then
               pbestp=Xp
         end if
               for d=1 to number of features do
               /*Update the Particle*/
```

$$v_p^d(t+1) = v_p^d(t) + c_1 r_1 (pbest_p^d - X_p^d(t)) + c_2 r_2 (gbest^d - X_p^d(t))$$

$$Vsig_p^d = 1/(1 + e^{-v_p^d(t+1)})$$

$$X_p^d = \begin{cases} 1 & : VSig_p^d \geq 0.6 \\ \# & : 0.4 < VSig_p^d < 0.6 \\ 0 & : Vsig_p^d < 0.4 \end{cases}$$

```
               end for
   end for
end
```

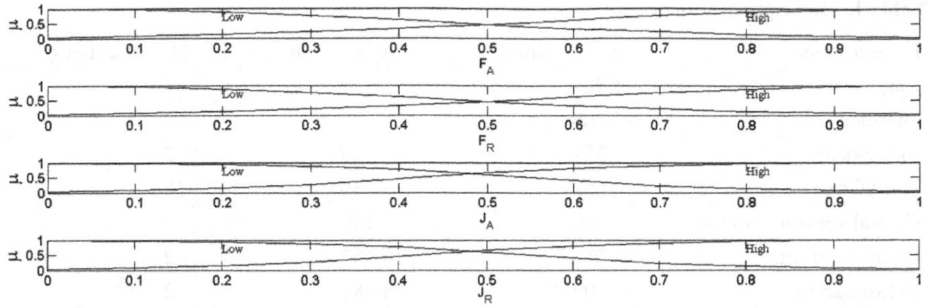

Fig. 1 Membership plot of four inputs F_A, F_R, J_A, and J_R

1. If (Fitness_A is High) and (Fitness_R is Low) and (J_A is Low) and (J_R is high) then (output is 0.8) (1)
2. If (Fitness_A is High) and (Fitness_R is High) and (J_A is Low) and (J_R is high) then (output is 0.8) (1)
3. If (Fitness_A is High) and (J_A is High) and (J_R is low) then (output is 0.2) (1)
4. If (Fitness_A is Low) and (Fitness_R is Low) and (J_A is Low) and (J_R is high) then (output is 0.8) (1)
5. If (Fitness_A is Low) and (Fitness_R is High) and (J_A is Low) and (J_R is high) then (output is 0.2) (1)
6. If (Fitness_A is Low) and (J_A is High) and (J_R is low) then (output is 0.2) (1)

Fig. 2 Fuzzy rules for FVBPSO

5 Experimental Setup

5.1 First Experiment

To show the curse of dimensionality, eight datasets from UCI repository [26] having different dimension have been classified using Naive Bayes Classifier [27], Adaboost Decision [28], K-Nearest Neighbor [29], Instance-based learning [30], Locally weighted Classifier [31], C4.5 Classifier [32], and Random forest [33].

5.2 Second Experiment

Three microarray gene expression profile data sets (SRBCT, DLBCL, Prostate Tumor) are taken from the Gene Expression Model Selector (GEMS) [34]. Fitness of each particle is obtained using K-nearest neighbor classifier with leave one out cross validation. Value of k for k-NN classifier is put to 1. The performance of algorithms is justified according to classification accuracy and selected features metrics. All the shown results are average of multiple runs. In this experiment, our proposed FVBPSO are compared with the k-NN (classifier with all features) and IBPSO. All parameters have been kept the same for both the PSO variants; $\omega = 0.5$, $c_1 = 2$ and $c_2 = 2$, $[V_{min}\ V_{max}] = [-6\ 6]$, number of particles = 40, and stopping

Table 1 Description of datasets

Dataset name	No. of samples	No. of features	No. of classes
Car	1,728	7	4
Diabetes	768	9	2
Ionosphere	351	35	2
Arrhythmia	452	280	16
Central nervous system	60	7129	2
Ovarian cancer	253	15,154	2
Breast cancer	97	24,481	2
SRBCT	83	2308	4
DLBCL	77	5469	2
Prostate tumor	102	10,509	2

Table 2 Classification accuracy of different classifiers

Classifier	Naïve Bayes	Adaboost decision stump	KNN	Instance based	K*	LWL	C4.5	Random forest
Car	85.53	70.02	77.25	93.51	87.55	70.02	92.36	94.50
Diabetes	76.3	74.34	70.18	70.18	69.18	71.22	73.82	74.86
Ionosphere	82.62	90.88	86.32	86.32	84.61	82.33	91.45	93.16
Arrhythmia	62.38	55.53	52.87	52.87	55.97	57.30	64.38	69.02
Central nervous system	**61.66**	**63.33**	**56.66**	**56.66**	**35**	**76.66**	**58.33**	**58.33**
Ovarian cancer	**58.89**	**55.33**	**56.12**	**55.73**	**58.89**	**61.66**	**55.33**	**52.56**
Breast cancer	**54.63**	**64.94**	**53.60**	**60.82**	**62.88**	**56.70**	**62.88**	**64.94**

criteria have been set to max iteration = 150 or max accuracy = 100. The details of data sets of both the experiments are given in Table 1.

6 Results

Table 2 and Fig. 3 show the classification accuracy of classifiers. From results in Table 2, it is revealed that as the dimension increases, performance of the classifiers decreases. In some cases (Ovarian and breast cancer) where number of features is much higher than the sample size, there is a noticeable decrease in performance of learning algorithms. Figure 3 shows that the performance of classifier is decreasing as the number of dimension is increasing beyond the normal range of expectation. Table 3 shows the average classification accuracy, selected features and standard

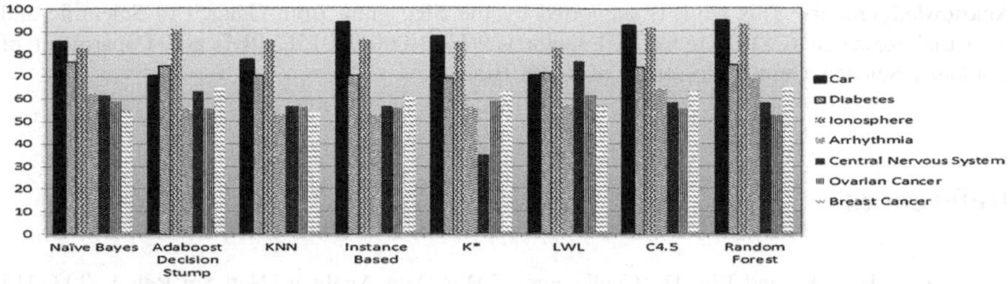

Fig. 3 Graphical representation of results of second experiment

Table 3 Average of classification accuracy, selected features, and standard deviation using FVBPSO

Data set	Method	Accuracy	Selected features	Standard deviation
SRBCT	KNN	91.57	**ALL**	
	IBPSO	97.59	**1124**	0.5040
	FVPSO	**98.31**	512	**0.2451**
DLBCL	KNN	87.01	**ALL**	
	IBPSO	94.81	**2697**	$2.3e^{-16}$
	FVPSO	**91.45**	112	**0.0113**
Prostate tumor	KNN	76.47	**ALL**	
	IBPSO	91.14	**1029**	0.646
	FVPSO	**97.02**	689	**0.0356**

deviation of the accuracy obtained from KNN, IBPSO, and FVBPSO. From Table 3 it is clearly visible that in almost all cases, FVBPSO is outperforming the other methods in terms of feature selection. Table 3, reveals the blessings of dimensionality reduction with increased classification accuracy after feature selection as compared with the classification with all features.

7 Conclusion

Among many surprising facts about high dimensionality, concentration of measure and empty space phenomenon are most important. These characteristics of high dimensionality cause the deviation of classical learning methods from normal performance. Results of experiment 1 show the decrease in the performance of classical learning algorithms in case of high-dimensional data which clearly shows the need of some dimensionality reduction method to prevent the curses of dimensionality. Therefore, in this paper a new fuzzy versatile BPSO for feature selection has been proposed and experiment has shown the merits of FVBPSO on high-dimensional data sets.

Acknowledgements This work is supported by the SRF grant from Council of Scientific and Industrial Research (CSIR), India, SRF grant (09/1144(0001)2015EMR-I) and Department of Computer Science, Central University of South Bihar.

References

1. Fan, J., Han, F., and Liu, H.: Challenges of Big Data Analysis. Natl Sci Rev 1, 293–314 (2014) doi:10.1093/nsr/nwt032.
2. Fan, J., and Fan, Y.: High dimensional classification using feature annealed independence rules. Ann. Stat. 36(6), 2605–2637 (2008).
3. Fan, J., Guo, S. and Hao, N.: Variance estimation using refitted cross validation in ultrahigh dimensional regression. J R Stat Soc Series B Stat Methodology, vol. 74, no. 1, pp. 37–65 (2012).
4. Liao Y. and Jiang, W.: Posterior consistency of nonparametric conditional moment restricted modeld. Ann. Stat. 39(6), 3003–3031 (2011).
5. Fan, J. and Liao, Y.: Endogeneity in ultrahigh dimension. Ann. Stat. 42(3), 872–917 (2014).
6. Hall P., Pittelkow, Y. and Ghosh, M.: Theoretical Measure of relative performance of classifiers for high dimensional data with small sample size. J R Stat Soc Series B Stat Methodology, vol. 70, no. 1, pp. 159–173 (2008).
7. Bellmann R.: Adaptive control processes: A Guided Tour. Triceton University Press (1961).
8. Silverman B.: Density estimation for statistics and data analysis. Chapman and Hall, (1986).
9. Donoho, D.: High-Dimensional Data Analysis: The Curse and Blessings of dimensionality. AMS Math Challenges Lecture, pp. 1–32 (2000).
10. Scott, D. W. and Thompson, J. R.: Probability density estimation in higher dimensions. In: Computer Science and Statistics: Proceedings of the Fifteenth Symposium on the Interface, J.E. Gentle Eds. Amsterdam, New York (1983).
11. Agarwal, S. and Rajesh, R: Difficulty in Handling High Dimensional Data-Please stand UP. JRET: International Journal of Research in Engineering and Technology. 3(14), 101–104 (2014).
12. Kennedy, J. and Eberhart, R. C.: Particle swarm optimization. In: IEEE International Conference on Neural Networks, Perth, Australia, pp. 942–1948 (1995).
13. Shi, Y. and Eberhart, R. C.: A modified particle swam optimizer, IEEE Word Congress on Computational Intelligence, pp. 69–73 (1998).
14. Kennedy, J., Eberhart, R. C. and Shi, Y.: Swarm Intelligence. Morgan Kaufman, San Mateo, CA (2001).
15. J. Kennedy, Eberhart R. C. and Y. Shi, Computational Intelligence, Morgan Kaufman, San Mateo, CA (2007).
16. J. Kennedy and Eberhart R. C.: A discrete binary version of the particle swarm algorithm. In: IEEE conference on systems, man, and cyber, vol. 5, pp. 4104–4108 (1997).
17. Chuang, L. Y., Chang, H. C., Tu, J. and Yang, C. H.: Improved binary PSO for feature selection using gene expression data. Comp. Bio. Chem. 32, 29–38 (2008).
18. Rajesh, R and Agarwal, S: Some Modification in Particle Swarm Optimization. The 18th Online World Conference on Soft Computing in Industrial Applications (2014).
19. Agarwal, S. and Rajesh. R.: Enhanced Velocity BPSO and Convergence Analysis on Dimensionality Reduction. In proceedings of Recent Advances In Mathematics, Statistics and Computer Science (2015) in press.
20. Emami, H. and Derakhshan, F.: Integrating Fuzzy K-Means, Particle Swarm Optimization, and Imperialist Competitive Algorithm for Data Clustering. Arab Journal Sci. Eng. 40, 3545–3554 (2015).

21. Abdelbar, A. M., Abdelbar, S., Wunsch, D. C.: Fuzzy PSO: A Generalization of Particle Swarm Optimization. In: Proc. Intl. Joint Conf. on Neural Networks, Montrael, Canada, pp. 1086–1091 (2005).
22. Chuang, L. Y., Tsai, S. W., Yang, C. H.: Fuzzy adaptive catfish particle swarm optimization. J. Artif. Intell. Res. 1, 149–170 (2012).
23. Chai, R., Ling, S. H., Hunter, G. P., Tran, Y., Nguyen, H. T.: Brain Computer Interface Classifier for Wheelchair Commands Using Neural Network With Fuzzy Particle Swarm Optimization. IEEE J. Biomed. Health Inform. 18, 1614–1624 (2014).
24. Soeprijanto, A., Abdillah, M.: Type 2 fuzzy adaptive binary particle swarm optimization for optimal placement and sizing of distributed generation. Proc. of the 2nd Intl Conf. on Instrumentation, Communications, Information Technology, and Biomedical Engineering (ICICIBME), IEEE, pp. 233–238 (2011).
25. Olivas, F., Valdez, F. and Castillo, O.: Fuzzy Classification System Design Using PSO with Dynamic Parameter Adaptation Through Fuzzy Logic. Fuzzy Logic Augmentation of Nature - Inspired Optimization Met heuristics, Studies in Computational Intelligence, Springer. 574, 29–47 (2015).
26. Lichman, M.: UCI Machine Learning Repository [http://archive.ics.uci.edu/ml]. Irvine, CA: University of California, School of Information and Computer Science.
27. John, G. H.: Pat Langley, Estimating Continuos Distribution in Bayesian. In: Proc. of the eleventh conf. on uncertainty in artificial intelligence, Morgan Kaufman Publisher, San Mateo (1995).
28. Freund, Y., Schapire, R. E.: Experiment with a New Boosting Algorithm Machine Learning. In: Proc. of the Thirteenth Int. Conf. (1996).
29. Aha, D. W., Kibler, D., Albert, M. K.: Instance based Learning. Machine Learning. 6, 37–66 (1991).
30. Cleary, J. G., Trigg, L. E.: K*: An Instance-based Learner Using an Entropic. In: Proc. of the 12th Int. Conf. on Machine Learning Distance Measure (1995).
31. Atkesonm, C. G., Moore, A. W. and Schaal, S.: Locally Weighted Learning. College of Computing, Georgia Institute of Technology (1996).
32. Quinlan, J. R.: C4.5: programs for machine learning. Morgan Kaufmann Publishers Inc., San Francisco, CA (1993).
33. Breiman, L.: Random Forest. Machine Learning. 45, 5–32 (2001).
34. Statnikov, A.: Gene expression model selector. www.gems-system.org (2005).

E-commerce Recommendation System Using Improved Probabilistic Model

Rahul S. Gaikwad, Sandeep S. Udmale and Vijay K. Sambhe

Abstract Recommender system is the backbone of e-commerce marketing strategies. Popular e-commerce websites use techniques like memory-based collaborative filtering approach based on user similarity with only rank as an attribute. This paper proposes a model-based collaborative filtering recommender system based on probabilistic model using improved Naive Bayes algorithm. Proposed system uses Naive Bayes algorithm with bigram language model to improve search query analysis. Therefore, search query, click time and query time are used as features for Naive Bayes algorithm model. This model is trained on 1.2 million customer data over a 3-month period for 1.8 million products. Proposed system predicts the probability of products and products will be recommended to the user to make top-N recommendations. Results of the proposed system show the model recommends products with 14% more accuracy as compared to simple Naive Bayes model.

Keywords Recommendation · Naive Bayes · Bigram language model
Collaborative filtering · Item-to-item based approach

1 Introduction

E-commerce websites provide millions of products from where users can browse and purchase. These huge array of choices can often prove overwhelming to users. Recommender systems guide users toward products that they might find interesting. Many recommender systems work by suggesting products that similar people have purchased in the past.

R.S. Gaikwad (✉) · S.S. Udmale · V.K. Sambhe
Department of Computer Enginering and Information Technology, VJTI, Mumbai, India
e-mail: gaikwadrahul20@gmail.com

S.S. Udmale
e-mail: ssudmale@vjti.org.in

V.K. Sambhe
e-mail: vksambhe@vjti.org.in

D.K. Mishra et al. (eds.), *Information and Communication Technology for Sustainable Development*, Lecture Notes in Networks and Systems 10,
https://doi.org/10.1007/978-981-10-3920-1_28

E-commerce websites use recommendation systems for improving its sales. These recommendations depend on customer behavior, customers historical data, and stock keeping unit/item. Based on these, there are mainly two types of recommendation system algorithms, Content-Based Recommendation and Collaborative Filtering Recommendation. Collaborative Filtering works on the concept of similarities between users whereas Content-Based algorithms depend on the features of items or products. Collaborative filtering method is mainly classified as model-based collaborative filtering and memory-based collaborative filtering algorithms.

Collaborative filtering (CF) [1–4] work by building a list of product preferences for consumers. Nowadays, websites use item-to-item based collaborative filtering algorithm which recommends products based on how similar the products of other user are to the target user. Products that other users who are similar to the user have purchased, are generated as recommendations to the user. This algorithm uses rank of the page as feature. Rank of the page can be calculated from number of hits of that particular page, this algorithm is similar to K-Nearest Neighbor algorithm. Recommendations are obtained by considering only one attribute as a feature. Sometimes, this may result in repetitive recommendations.

To overcome this repetitive recommendation, this paper proposed a model-based collaborative filtering approach. This approach uses three features, unique identity of the user, query that user searched and the time difference between the time at which user fired the query and the time at which user clicked the item from the search results. Naive Bayes with bigram language model to improve the accuracy of the product recommendation is applied on these features. A model is trained using these features. Probability that an item is purchased given a specific user, query, and query time is calculated. Products are ranked based on the probability of being clicked by the user. Products with higher probabilities are then recommended to the user.

The rest of the paper is organized as follows. Section 2 gives a brief overview of related work and collaborative filtering based recommender systems. Section 3 discusses the architecture of the proposed system, with detailed explanation of all the modules. Section 4 describes the experimental setup, results, and discussion. Section 5 concludes the work.

2 Related Work

Many approaches have been applied to the problem of making accurate and efficient recommender and data mining systems. Automatic recommender systems use a wide range of techniques, ranging from nearest neighbor algorithms to Bayesian analysis. Naive Bayes algorithm uses Bayes theorem as an underlying approach which was introduced in 1950. Linden et al. [5] suggest that amazon.com uses item-to-item collaborative filtering approach which is also referred as customer who viewed this also viewed? algorithm. This algorithm calculates the similarity of the products between the users. Huang et al. [6] describes the comparison of six collaborative filtering approaches of which the proposed model used the user-based collaborative filtering algorithm. This algorithm calculates the similarity of the users based on the attributes. This paper discusses a detail comparison of various recommendation algorithms. Yang et al. [7] also improved the Naives Bayes model for movie recommender system by considering the fact that conditional probability assumption is not obeyed strictly. In 2013, Wang et al. [8] proposed a new Bayesian-based recommender system, which computes the conditional probability of a rating similarity of two friends as measured by Bayesian network. In paper [9], author He et al. designed recommender system using Hidden Markov Model and business rules. Shi and Wang [10] give the effectiveness of three types of online recommender system such as best sellers, the personalized recommendation, and reviews of customer. Hybrid recommender system is also designed for web service recommendation by author Lina Yao et al. [11]. It is the combination of collaborative and content-based recommendation tested on 3,693 web services.

This work proposes a model-based collaborative filtering recommender system based on probabilistic model. The approach accounts for various features of user behavior on e-commerce sites and builds a training model by extending the Naive Bayes approach, and improving it using the bigram language model.

2.1 Collaborative Filtering Based Recommender System

Collaborative filtering (CF) [2–4] is the one of the earliest recommender system technologies, and is used in many of the most successful recommender systems on the Web. The fundamental idea of collaborative filtering is that: users who have similar preference in the past are similar and are likely to have similar preferences in the future. The idea is to predict products to users that they may find interesting based on what similar users have liked in the past. These similar users are called as neighbors. Statistical methods are used to generate the neighborhood of users for all users.

Collaborative filtering methods widely fall into the two categories: memory-based and model-based methods. The primary difference between the two approaches is that memory-based algorithms generates recommendations by performing statistical computations on the entire data every time, whereas model-based algorithms build a training model using the data, which can later be used to generate recommendations. This means that the memory-based algorithms require that all data be in memory, whereas model-based can make fast predictions using generated model, after model building.

3 Implementation

The system architecture of proposed approach is shown in Fig. 1. It mainly consists of three module, preprocessing of data, training the model using improved probabilistic technique and predicting the products.

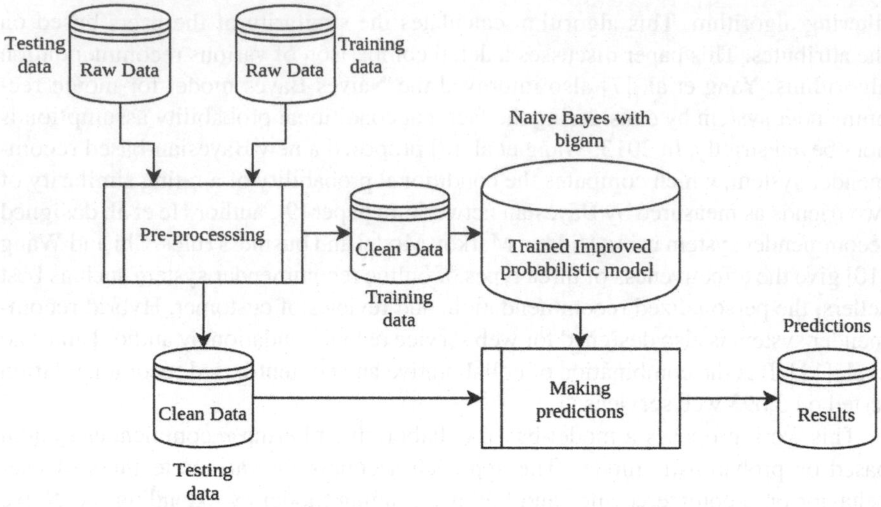

Fig. 1 Proposed system architecture: improved probabilistic model

3.1 Data Preprocessing Module

Data preprocessing is the process by which raw data is cleaned to bring data which can easily be processed by the core algorithm. Often, data that is acquired is not in a form that can be readily analyzed. This data has to be put through a variety of pre-processing techniques to bring it into a form suitable for analysis. Data preprocessing module includes the following:

- Query Preprocessing: Data preprocessing module applied on "query" attribute includes conversion of query to lowercase, removal of punctuations and special characters, splitting of words and numbers and lemmatization. Lemmatization is a linguistic transformation that changes a word to its base form, also known as lemma. Lemmatization brings all the variants of a word to the same underlying lemma.
- Processing "time" Variable: The difference between the "query time" and "click time" is an indicator of how relevant the clicked product was to the search query.

Table 1 Raw data

Query	Click time	Query time
"Televisions Panasonic 50 pulgadas"	"2011-09-01 23:44:52.533"	"2011-09-01 23:43:59.752"
"Children44"	"2011-09-07 15:54:47.956"	"2011-09-07 15:53:24.353"
"Watch The Throne"	"2011-09-04 10:55:20.427"	"2011-09-04 10:55:10.874"

Table 2 Preprocessed data

Query	Time difference
Television panasonic 50 pulgadas	21
Child 44	27
Watch throne	24

In this module, the difference between query time and click time is calculated and normalized. Table 1 shows the query and corresponding query time and click time. Table 2 shows the preprocessed query and normalized time difference.

3.2 Training Model—Naives Bayes with Bigram

Naive Bayes algorithm is a simple probabilistic classifier. Probabilities are calculated by using Bayes theorem with the assumption that the independence between features is strong, i.e., naive. It uses supervised learning to train the system. By using Bayes theorem, the formula for conditional probability as given in Eq. 1 is

$$p(S_i|F) = \frac{p(S_i)p(F|S_i)}{p(F)}, \tag{1}$$

where

$F = \{Q, T\}$ is the set of features that has been considered,
Q is the search query,
T is the normalized time difference.

The feature set F does not depend on item S_i, due to this we can neglect the denominator, and the numerator term is equal to joint probability $p(S_i, Q_1, \ldots, Q_n, T)$ After applying chain rule, Eq. 1 becomes

$$p(S_i, Q_1, \ldots, Q_n) = p(Q_1|Q_2, \ldots, Q_n, S_i), p(Q_2|Q_3, \ldots, Q_n), \ldots, p(Q_n|S_i)p(T|S_i)p(S_i) \tag{2}$$

Since the features are independent of each other, joint expression can be expressed as

$$p(S_i|Q_1, \ldots, Q_n) \; \alpha \; p(S_i)p(T|S_i)\prod_{j=1}^{n} p(Q_j|S_i) \tag{3}$$

Equation 3 shows the equation of Naive Bayes, where S_i is the "sku", $p(S_i)$ is the likelihood, $p(Q_j|S_i)$ is the probability of appearance of word Q_j when there is an item S_i, $p(T|S_i)$ is the probability of that item S_i was clicked during time difference T. Bigram language model is used to improve the accuracy of the Naive Bayes algorithm. Probability of item being recommended based on the query; using both unigram and bigram is calculated separately. A linear combination of unigram prediction and bigram prediction is computed to get the final probability as in Eq. 4.

$$p(total) = c_1 \times p(unigram) + c_2 \times p(bigram) \tag{4}$$

where, s_1 and s_2 are smoothing constants.

4 Results and Discussion

4.1 Datasets

Datasets for the experiments include user behavior on an e-commerce site ranging over 3 months, 1.8 million bought products and 1.2 million customers. The training dataset contains 1.8 million records each having fields: "user", "sku", "category", "query", "click time", and "query time". The model is tested for 1.2 million records on the accuracy of predicting Top-5 items for each user.

4.2 Evaluation Results

The model has been tested for the following metrics:

Precision: It is the ratio of recommended items that are relevant to the total number of recommended items.

Recall: It is the ratio of recommended items that are relevant to the total number of relevant items.

F_1 Measure: F_1 Measure is the weighted average of the Precision and Recall.

$$F_1\,Measure = 2 \times \frac{precision \times recall}{precision + recall} \tag{5}$$

Figure 2 shows the improvement in F_1 Measure of the proposed Improved Probabilistic Model against Simple Naive Bayes approach explained in [12] for varying number of recommendation; i.e., K = 1, 2, 3, 4, 5. For increasing number of recommendations, the precision and recall of the model increases, thus increasing the F_1 Measure.

Average Precision: Average Precision (AP) is a measure of the number of relevant products recommended in the first "K" results. AP is calculated using precision and recall at every position in the ranked sequence of recommendations.

$$AP = \sum_{1}^{x} (precision\ at\ i) \times (change\ in\ recall\ at\ i) \tag{6}$$

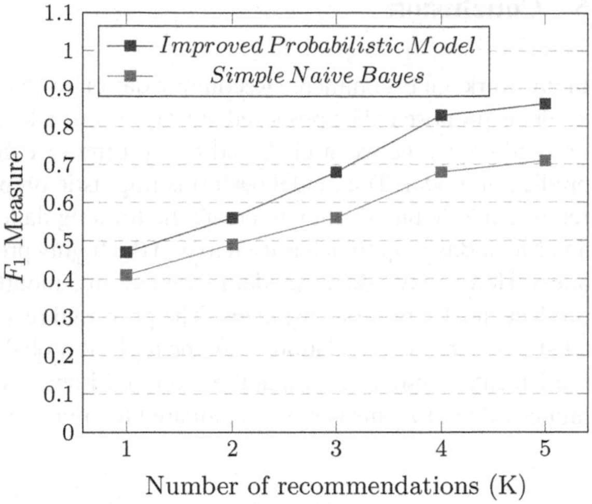

Fig. 2 F_1 measure of improved probabilistic model versus simple Naive Bayes model

Precision at "i" is a percentage of correct items among first i recommendations. Change in recall is "1/x" if item at "i" is correct (for every correct item), otherwise zero.

Mean Average Precision (MAP) is the average of Average Precisions for all users and is given as

$$\frac{\sum_{u=1}^{N} AP_u}{M} \tag{7}$$

Figure 3 shows the MAP for top-K recommendations; K ranging from 1 to 5. It can be seen that for increasing number of recommendations the MAP of the system decreases.

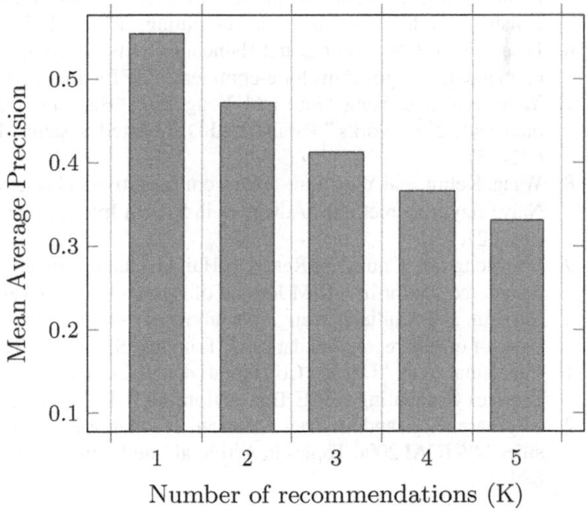

Fig. 3 Mean Average Precision (MAP) versus number of recommendations

5 Conclusion

In this work, an E-commerce Recommendation System using improved probabilistic model is proposed. The proposed approach uses additional features like user query, time difference between click and query time of e-commerce browsing to train a prediction model. The model performs linguistic preprocessing tasks like stopword removal and lemmatization to create the training dataset. A combination of Bigram and Naive Bayes approach is used to find the highly probable product based on user's query. Here, Naive Bayes model is trained on 3 months of data, 1.8 million bought product and 1.2 million customers. The proposed recommendation model generates a list of five recommendations with the highest probability for every user. Results of implemented system show that the proposed improved probabilistic model has 14% higher value of F_1 measures as compared to simple Naive Bayes model [12].

References

1. Herlocker, J. L., Konstan, J. A., Borchers, A., & Riedl, J. (1999, August). An algorithmic framework for performing collaborative filtering. In Proceedings of the 22nd annual international ACM SIGIR conference on Research and development in information retrieval (pp. 230–237). ACM.
2. Konstan, J. A., Miller, B. N., Maltz, D., Herlocker, J. L., Gordon, L. R., & Riedl, J. (1997). GroupLens: applying collaborative filtering to Usenet news. Communications of the ACM, 40(3), 77–87.
3. Resnick, P., Iacovou, N., Suchak, M., Bergstrom, P., and Riedl, J. (1994). GroupLens: An Open Architecture for Collaborative Filtering of Netnews. In Proceedings of CSCW 94, Chapel Hill, NC.
4. Shardanand, U., & Maes, P. (1995, May). Social information filtering: algorithms for automating word of mouth. In Proceedings of the SIGCHI conference on Human factors in computing systems (pp. 210–217). ACM Press/Addison-Wesley Publishing Co.
5. Linden, Greg, Brent Smith, and Jeremy York. "Amazon. com recommendations: Item-to-item collaborative filtering." Internet Computing, IEEE 7.1 (2003): 76–80.
6. Huang, Zan, Daniel Zeng, and Hsinchun Chen. "A comparison of collaborative-filtering recommendation algorithms for e-commerce." IEEE Intelligent Systems 5 (2007): 68–78.
7. Yang, Xiwang, Yang Guo, and Yong Liu. "Bayesian-inference-based recommendation in online social networks." Parallel and Distributed Systems, IEEE Transactions on 24.4 (2013): 642–651.
8. Wang, Kebin, and Ying Tan. "A new collaborative filtering recommendation approach based on Naive Bayesian method." Advances in Swarm Intelligence. Springer Berlin Heidelberg, 2011. 218–227.
9. He, Mengxun, Chunying Ren, and Haijun Zhang. "Intent-based recommendation for B2C e-commerce platforms." IBM Journal of Research and Development 58.5/6 (2014): 5-1.
10. Shi, Lin, and Kanliang Wang. "An laboratory experiment for comparing effectiveness of three types of online recommendations." Tsinghua Science & Technology 13.3 (2008): 293–299.
11. Yao, Lina, et al. "Unified Collaborative and Content-Based Web Service Recommendation." Services Computing, IEEE Transactions on 8.3 (2015): 453–466
12. Miyahara, Koji, and Michael J. Pazzani. "Collaborative filtering with the simple Bayesian classifier." PRICAI 2000 Topics in Artificial Intelligence. Springer Berlin Heidelberg, 2000. 679–689.

Extraction of Top-k List by Using Web Mining Technique

Priyanka Deshmane, Pramod Patil and Abha Pathak

Abstract In present days, finding relevant and desired information in less time is very crucial, however, problem is that very small proportion data on internet is interpretable and meaningful and need lot of time to extract. The paper provides solution to problem by extracting information from top-k websites, which consist top-k instances of a subject. For Example "top 5 football teams in the world". In comparison with other structured information like web tables top-k lists contains high quality information. It can be used to enhance open-domain knowledge base (which can support search or fact answering applications). Proposed system in paper extract the top-k list by using title classifier, parser, candidate picker, ranker, content processor.

Keywords Top-k list · Information extraction · Top-k web pages
Structured information

1 Introduction

WWW is very important source for getting information and huge amount of information is available. Top-k pages are rich source of valuable information available on internet and these pages exist in small percentage. It is important to extract top-k list from such pages for getting correct information but it is difficult to extract knowledge from information explained in natural language and unstructured format. Some information over internet is present in organized or semi-organized form for

P. Deshmane (✉) · P. Patil · A. Pathak
Department of Computer Engineering, Dr. D.Y. Patil Institute of Engineering and Technology (DYPIET), Pimpri, Pune, India
e-mail: priyanka.deshmane8218@gmail.com

P. Patil
e-mail: pdpatiljune@gmail.com

A. Pathak
e-mail: abhanayak@gmail.com

D.K. Mishra et al. (eds.), *Information and Communication Technology for Sustainable Development*, Lecture Notes in Networks and Systems 10,
https://doi.org/10.1007/978-981-10-3920-1_29

example, as records coded with specific names, e.g. html5 pages. As per a large measure of new technique has to be dedicated for understanding structured information on the web, (like web tables) especially from internet platforms [1]. Quantity of web tables is large but slight proportion of them includes helpful information. A very small amount of table includes data interpretable without context. Many tables are not relational, since it is easy to interpret relational table with rows referring to entities and columns referring to characteristics of these entities [2].

Consider a table which has four rows and three columns, where the three columns are marked as "bikes", "model" and "prize", respectively. We do not understand why these four bikes are gathered together (e.g. are these most expensive, or fastest). We do not know the definite situations for which information is useful. The context is very important for extracting information, but in many of the cases, context is represented in such a manner that the machine could not understand it. This paper concentrates on context that is easily understandable. Then, context is used to interpret free text and less structured information [3].

Top-k page title should consist minimum three section of information: (i) k , e.g. 12, ten. Means number of items does page contain. (ii) A topic or idea items is associated with, e.g. artists, players. (iii) Ranking criterion, e.g. fastest, tallest, best seller. Sometime title contain two optional elements time and location

2 Literature Survey

2.1 Automatic Extraction of Top-k List from Web

Zhixian Zang, Kenny Q. Zhu, Haixun Wang, Hong Song Li [1].

Author proposed a method for information extraction from web pages that contain top-k sample. This method gives improved performance by providing domain specific lists and focusing more on the content. It does not focus only on the visual area of the lists. If list is divided into more than one page it may not get included completely. Author demonstrated algorithm which not only extract list but also discovers its structure [1].

2.2 System for Extraction of Top-k List from Web

Z. Zhang, K. Q. Zhu, H. Wang [4].

Author defined list extraction problem which concentrated on finding and extracting top-k lists. The problem was different from other as top-k lists are easily understood and contain high quality of information. Probase can be enhanced with the help of knowledge stored in lists and use for developing an efficient search engine.The four stage framework demonstrated by author has high precision to extract top-k list [4].

2.3 *Extracting General List from Web Document*

F. Fumarola, T. Weninger, R. Barber, D. Malerba and J. Han [5].

Author proposed a new hybrid technique for general lists extraction from internet. Method uses basic assumption on visible assimilation of list and architectural arrangement of element. The aim of system was to overcome the limitations of existing work which deals with the generality of extracted lists. Several visual and structural characteristics were combined for achieving goal. To find and extract the general list on web both information on visual list item structure and invisible data like DOM tree skeleton of structure of visible line up element were used.

2.4 *Short Text Conceptualization Using Probabilistic Knowledge Base*

Y. Song, H. Wang, Z. Wang, H. Li, W. Chen [6].

Author proposed a technique to improve text understanding by making use of a probabilistic knowledge. Conceptualization of short words is done by Bayesian interference mechanism. Comprehensive experiments were performed on conceptualizing textual terms and clustering short segments of text such as Twitter messages. Compared with purely statistical technique like latent semantic topic modelling or methods that use existing knowledge base (e.g. WordNet, Freebase and Wikipedia) approach brings notable improvements in short text conceptualization as shown by the clustering accuracy [6].

2.5 *Extracting Data Records from Web Using Tag Path Clustering*

G. Miao, J. Tatemura, W.P. Hsiung, A. Sawires [7].

Author proposed a technique for extraction of record that recognizes the list in powerful fashion on basis of detail study of online page. The focus is on frequent appearance of noticeable tag path in DOM tree. It correlates tag path pattern pair (visual signal) to calculate similarity between two tag paths. Data record clustering of tag path is done on basis of similarity measure. Algorithm efficiently extracts atomic level and nested level data records in comparison with state-of-art algorithm. For practical data sets algorithm has linear execution time [7].

2.6 *Towards Domain Independent Information Extraction from Web Table*

W. Gattterbaur, P. Bohunsky, Herzog, B. Krupal, B. Pollak [8].

Author mentioned the difficult task of extraction by moving focus from representation in tree format of web page to variety of visual box model which are multi-dimensional and used by web browsers to show the information on screen. The gap formed by missing domain specific knowledge about content and table templates can be fill by topological information obtained [8].

3 Problem Statement

To extract a top-k list from structured as well as unstructured information on web by using web mining algorithms

4 Proposed System

See Fig. 1.

4.1 *Title Classifier*

Online page title helps us to identify top-k page. First reason to use title is, for many situation page titles gives introduction about the subject. Second body of page could have composite pattern but titles have identical pattern.

Analysis of title is light weight and economical. The pages which are not top-k in the result are skipped. Example of top-k title is shown in Fig. 2. Title may contain additional segments like time and location which are optional in addition to k, concept and ranking criterion. Segments may be separated with "-" or "—". Main segment contains the topic and other segments contain additional information. Title is split and the part which contains number is obtained. Number k is important for representing topic concept [1].

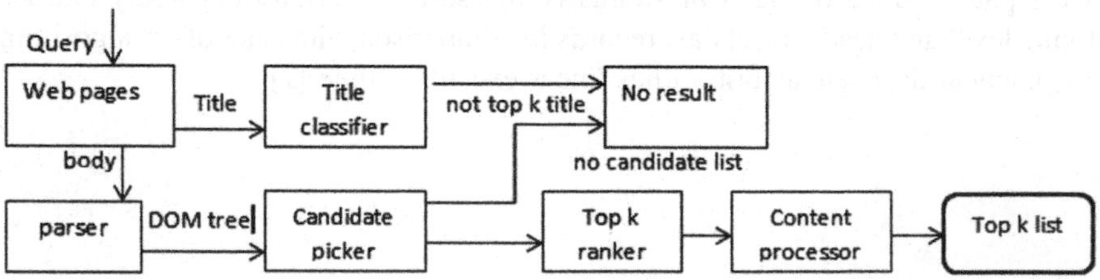

Fig. 1 System architecture

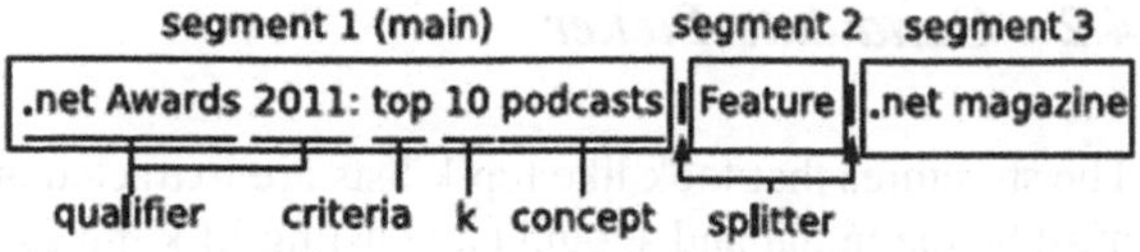

Fig. 2 Example of top-k title [1]

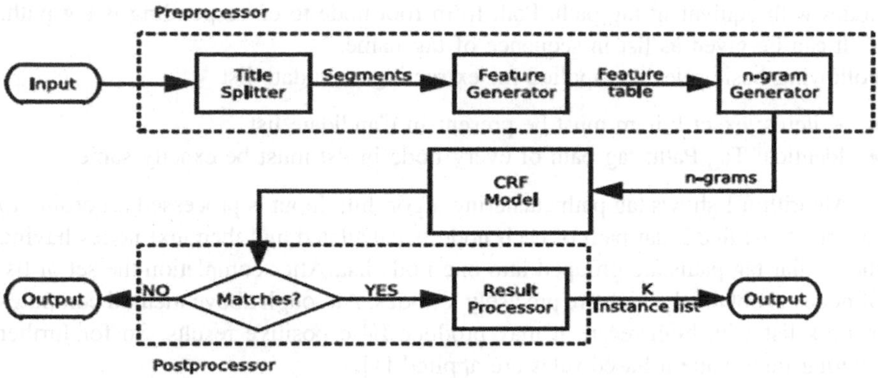

Fig. 3 Working of title classifier [1]

CRF Model

We define X as a word sequence and label sequence is defined as Y, Y_i{TRUE, FALSE}.

Conditional probability for linear chain CRF is calculated as

$$P(Y|X) = \frac{1}{Z(x)}\exp\left(\sum_{i=1}^{n}\sum_{j=1}^{m}\lambda_j f_j(y_{i-1}, y_i, x, i)\right)$$

Normalization factor $Z(x)$ one of the m function is f_j feature weight to be trained is λ_j.

Creating Training Data Set

"*top CD*": top word with number, e.g. top 10 singers

"*top CD*": *without word* "*top*"

"*CD JJS*": "*JJS*" means superlative adjective, e.g. tallest building

"*CD RBS JJ*": "RBS" and "JJ" stands for superlative adverbs and adjective, respectively, e.g. most expensive

Feature Extraction

Feature extraction of title is done in fixed size window which is centred around number k, four features are selected *Word*, *Lemma*, *POS tag*, *Concept*

Working of Title Classifier

Figure 3 shows how title classifier works. (1) Feature generated by pre-processor. (2) n gram pattern are labelled as TRUE or FALSE by classifier. (3) if value is true, then post processor extracts k, concepts, ranking criterion from title.

4.2 Candidate Picker

The structures that look like top-k lists are extracted at this stage. A top-k candidate must be the initial and should have listing of k things. Visually it must recognize as k horizontal and k vertical aligned in regular pattern. Structurally it is check list of nodes with equivalent tag path. Path from root node to corresponding is tag path.

It can be given as list in sequence of tag name.

Following basic rules are applied for extracting candidate list

- K items: exact k item must be present in Candidate list
- Identical Tag Path: tag path of every node in list must be exactly same.

Algorithm 1 shows tag path clustering algorithm. Input is processed according to two rules. At line 2 tag path of each node is calculated and then text nodes having the similar tag paths are grouped into one node list. After completion the set of list of nodes is obtained which are precisely k nodes. Although above method computes a top-k list with high recall, it may produce false positive results. So for further filtering three pattern-based rules are applied [1].

(1) index: number like 1, 2, 3 must be in order and in the range of (1, k)
(2) Highlighting tag: for highlighting purpose candidate list tag path should surrounded by <strong> <b> <h1–h6>
(3) Table: list is expressed in table format.

List satisfying at least one rule remain in candidate list.

Algorithm 1 Tag Path Clustering Method

```
1: procedure TAGPATHCLUSTERING(n,table)
2: n.TagPath ←n. Parent TagPath + Spliter + n.TagName;
3:  if n is text node then
4:       if table contains the key n.TagPath then
5:          list ← table[n.TagPath];
6:       else
7:           list ← new empty lists;
8:          table[n.TagPath] ← list
9:     end if
10:     insert n into list
11:     return
12: end if
13:    for each node i ∈ n.children do
14:        TagPathClustering(i,table);
15:    end for
16:    return
17:end procedure
```

4.3 Top-k Ranker

It select top ranked list by ranking the candidate set and using scoring function which is weighted sum of two feature scores.

P-Score

P-score is used to measure the interaction of the title and list. Basic concept is extracted from title because it is essential that elements in the list must contain instance of concept. For example if concept is "museum" then items in the main list should be instances of "museum" concept.

Each candidate's p-score in list L is calculated by:

$$P-score = \frac{1}{k}\sum_{n\in L}\frac{LM1(n)}{len(n)}$$

where LM1(n) is longest match instance word count in text node n while len(n) is entire text world count. To normalize p-score [0, 1] the LM1 is divided to len [1].

V-Score

Visual area inhabited by list is calculated V-score. The main and important list of the page is preferably bigger and outstanding than smaller lists. The sum of visual area of each node is v-score and is calculated by following formula.

$$Area(L) = \sum_{n\in L}(Textlength(n)\times Fontsize(n)^2)$$

Above method is known as rule-based ranker limitation is that it lack of flexibility and depends upon two features. An improved approach is learning-based ranker in which Bayesian model is trained from bigger training candidate list set. Then discretization method handle numeric feature.

4.4 Content Processor

Attribute\value pairs for every element are extracted from its explanation. Objective is to get skeletal info for every element in list. By analyzing the title valuable information can be obtained. Three important steps in the content processor.

Infer the Structure of Text Node

In most of the situations the text node describing every element might contain structure (e.g. XXX(YYY)).

Inner structure of text is inferred by computing frequency distribution for each possible separated tokens "by", ":", ":" from every element of top-k list [1].

Conceptualize the List Attribute

It is beneficial to infer the schema for the attribute after distributing the list item into attribute value. In system three methods are utilized to conceptualize list attributes:

Table head: The table head is included in the <th> tags.

Attribute/value pair: The list might contain explicit attribute/value pairs I.

Column is considered as the attribute column and the column to the right as the value column. The attribute name is used to conceptualize the corresponding values [1].

Column Content: If both table head and attribute/value pairs are not present, the basic technique is to conceptualize the extracted column content using probase and Bayesian model.

Detect When and Where

Important semantic information for extracted top-k lists is time and location. The page title is use to investigate information. A state-of-art NER tools are applied to solve (NER) problem.

5 Comparison of Similar Systems

	Extracting general list hybrid approach	Extracting data records using tag path clustering	Proposed approach
Working/Algorithm	HyLiEn (**Hy**brid approach for automatic **L**ist discovery and **E**xtraction on the Web)	Tag path clustering spectral clustering algorithm	Tag path clustering
Advantage	Employees both general assumptions on the visual rendering of lists and the structural representation of items	Can also detect nested data records Template tags and decorative tags are distinguished naturally	Extract top-k list with high performance 92.0% precision and 72.3% recall
Limitation	The computation time for HyLiEn is 4.2 s on average	Extracts data records from single web pages	Can not extract web pages that are interlink
Complexity	Bounded on the structural complexity	$O(M \times L) + O(M^3)$ computation time 0.3 s	Computation time is much less

6 Implementation Details

Result of Top-k Search depends on Searching techniques. Mainly Search to Number of K Elements. Number K means number like 0, 1, 2, 3, 4, …, n, User Search the content by inserting query in search box, if user insert the “Top 5 mobile BRANDS” in search box searching result as Google API link is displayed and from displayed link. Titles of web pages are extracted and classified by title classifier. After title classifier list on web page is parsed and extracted. User get more correct result if query is inserted in more specific and in detail way.

7 Conclusion

The problem of top-k list extraction from web is very important because top-k list are easy to understand and contain high quality information. The paper gives the survey of various previous approaches and proposed a system which can efficiently extract top-k list from web using tag path clustering a threshold algorithm. The system is interesting search system in which user enter the top query and get the top-k list as output. More work can be done in future as data on internet is increasing and more use of internet gives rise to new demands.

References

1. Zhixian Zhang, Kenny Q. Zhu, Haixun Wang Hong song Li, "Automatic top k list extraction from web" IEEE, ICDE Conference, 2013, 978-1-4673-4910-9.
2. J. Wang, H. Wang, Z. Wang, and K. Q. Zhu, "Understanding tables on the web," in ER, 2012, pp. 141–155.
3. M. J. Cafarella, E. Wu, A. Halevy, Y. Zhang, and D. Z. Wang, "Web tables: Exploring the power of tables on the web," in VLDB, 2008.
4. Z. Zhang, K. Q. Zhu, and H. Wang, "A System for extracting top k list from web" in KDD, 2012.
5. F. Fumarola, T. Weninger, R. Barber, D. Malerba, and J. Han, "Extracting general lists from web document: A hybrid approach," in IEA/AIE (1), 2011, pp. 285–294.
6. Y. Song, H. Wang, Z. Wang, H. Li, and W. Chen, "Short text conceptualization using a probabilistic knowledgebase," in IJCAI, 2011.
7. G. Miao, J. Tatemura, W.-P. Hsiung, A. Sawires, and L. E. Moser, "Extracting data records from the web using tag path clustering," in WWW, 2009, pp. 981–990.
8. W. Gatterbauer, P. Bohunsky, M. Herzog, B. Krupl, and B. Pollak, "Towards domain-independent information extraction from web tables," in WWW. ACM Press, 2007, pp. 71–80.

Improving the Accuracy of Recommender Systems Through Annealing

Shefali Arora and Shivani Goel

Abstract Matrix factorization is a scalable approach used in recommender systems. It deals with the problem of sparse matrix ratings in datasets. The learning rate parameter in matrix factorization is obtained by using numerical methods like stochastic gradient descent. Learning rate affects the accuracy of the system. In this paper, we make use of annealing schedules which will impact the value of learning rate. Five annealing schedules namely exponential annealing, inverse scaling, logarithmic cooling, linear multiplicative cooling and quadratic multiplicative cooling have been used to affect the learning rate and thus the accuracy of our recommender system. The experimental results on Movielens (http://www.grouplens.org/datasets/movielens) dataset with different sizes show that minimum mean absolute error for the system is obtained by exponential annealing at a lower value of learning rate and by linear multiplicative cooling at higher learning rate values. Apache Mahout (http://www.mahout/apache.org) 0.9 is chosen as the platform for conducting the experiments.

Keywords Recommender system · Matrix factorization · Annealing · Mahout

1 Introduction

There are a vast variety of items available on internet in all categories like books, music CDs, movies etc. Thus customers find it tough to choose from a vast set of overwhelming choices. Personalized recommender systems help the users to choose the necessary items as per their taste.

Collaborative filtering (CF) is widely used in recommender engines. The recommendations are made on the basis of preferences and ratings of users who have

S. Arora (✉) · S. Goel
Department of Computer Science and Engineering, Thapar University, Patiala, India
e-mail: arorashef@gmail.com

S. Goel
e-mail: shivani@thapar.edu

D.K. Mishra et al. (eds.), *Information and Communication Technology for Sustainable Development*, Lecture Notes in Networks and Systems 10,
https://doi.org/10.1007/978-981-10-3920-1_30

already submitted the ratings. This is called user based CF. If recommendations are based on item ratings, it is called item based CF. A number of algorithms have been used to evaluate recommender systems based on collaborative filtering [1].

Collaborative filtering fails to deal with certain issues. If no user has rated an item beforehand, then it becomes a challenging task to make recommendations for a new user. This causes cold start problem [2]. If the users introduce biasing in recommendations, then there is problem of shilling attack. In this situation only specific items will get higher ratings and will be recommended by the filtering algorithm [3].

Another issue faced in CF algorithm is sparsity of rating matrix. The user-item rating matrix is generally sparse in terms of entries due to less number of customers with respect to a huge number of items. Matrix factorization has become an efficient approach being used these days to get rid of sparse matrix problems [4]. Matrix factorization uses a latent space to map users as well as items. This is done on the basis of latent features. These latent features are obtained by characterizing items and users. For example, latent features in case of movies would be genres like comedy/drama/action etc. The latent feature with respect to item would describe the extent to which an item has this feature. With respect to the user, these features define the amount of interest a user has in this item. Stochastic gradient descent has become a popular minimization technique in matrix factorization method these days as it is highly effective. The entire training data is looped through in this technique.

1.1 Matrix Factorization

Matrix factorization introduces a set of latent features in a latent space d in an incomplete rating matrix R. For each item i, a vector $q_i \in Rd$ is associated, which gives the value to which an item possesses a particular feature. Similarly, a vector $p_u \in Rd$, is associated with every user, which shows the amount of interest a user takes in that particular feature. The dot product $p_u q_i^T$ gives us the user-item interaction according to a particular feature.

The overall objective of recommender systems is to learn by minimizing the errors in predicted values. The function is defined as:

$$\min p, q\ \Sigma(u,i \in K)(r_{ui} - p_u\, q_i^T) + ʎ(||q||^2 + ||p||^2) \quad (1)$$

where K is the index of element in the existing matrix R, r_{ui} is the user rating for an item i, ʎ is regularization parameter which can be obtained by cross validation and is used to avoid overfitting. || || is a Euclidean normalization standard.

This algorithm iterates through all the ratings and generates a prediction rating $\hat{r}_{ui}$ and prediction error is found.

$$e_{ui} = r_{ui} - \hat{r}_{ui} \quad (2)$$

where e_{ui} is the prediction error. Here the learning objective for the recommender systems is to reduce e_{ui}.

For a given training case r_{ui},

$$q_i \leftarrow q_i - \eta(\lambda q_i - e_{ui} p_u) \quad (3)$$

$$p_u \leftarrow p_u - \eta(\lambda p_u - e_{ui} q_i) \quad (4)$$

1.2 Related Work

It is essential to choose an appropriate value of learning rate in matrix factorization so that mean absolute error obtained is minimum. Learning rate has a great impact on the convergence rate of each iteration and also on the computational cost and complexity of the system.

Luo et al. worked on the adaptation of learning rate by integrating three strategies-Deterministic Step Size Adaptation (DSSA), Incremental Delta Bar Delta and Stochastic Meta Descent. They chose regularized matrix factorization to determine learning rate's effect in matrix factorization [5]. Chin et al. proposed a technique to introduce a new schedule in matrix factorization known as per coordinate schedule. By using the concept of twin learners, the time spent in selecting learning rate is considerably reduced [6].

Singular value decomposition is an approach used to implement matrix factorization. It reduces the sparse matrix to a lower dimensionality and latent features are represented as concepts Thus matrix, considered as R is split to give two matrices U and V. With p features and q items we obtain U: q × r (item × concepts), a matrix r with strength of concepts given by λ and another matrix V: m × r (concept × features). Figure 1 shows the concept of SVD.

Sarwar et al. compared the accuracy of SVD recommender systems against recommenders using collaborative filtering [7]. The comparison was made on the basis of quality of the best recommendations obtained based on explicit user

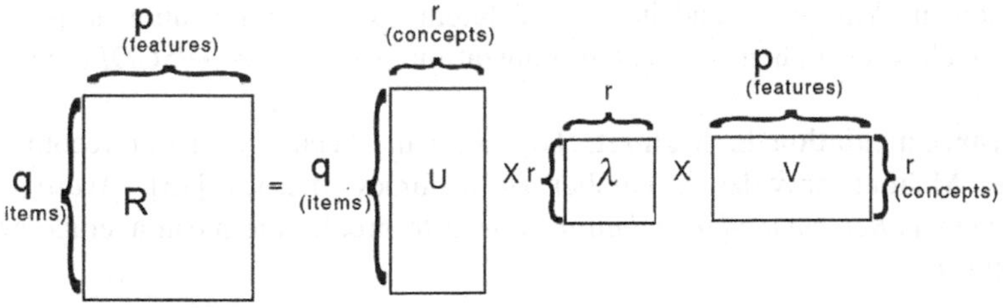

Fig. 1 Concept of SVD algorithm

ratings. Bounded SVD factorization was done by Le et al. in which comparison of SVD was done against bounded matrix factorization, by binding the ratings in constraints [8]. Paterek et al. worked on the improvements focused on regularized SVD by applying kernel ridge regression and using fewer parameters [9]. Incremental models of SVD are being widely accepted these days.

Improvements in matrix factorization have been critical in improving the accuracy of recommender engines. Yu et al. gave coordinate descent approach to parallelize and optimize matrix factorization [10]. Weimer et al. gave an improvement in maximum margin matrix factorization (MMMF) which aims at introducing offset terms and regularization factors for improving the existing MMMF approach [11].

In this paper, Apache Mahout has been used to implement matrix factorization in recommender systems.

Bamnote and Agrawal implemented collaborative filtering in recommender systems using this platform [12].

In this paper, we have used stochastic gradient descent to predict accuracy of our system. The quality of Movielens dataset is predicted using five annealing schedules and mean absolute error is used as the evaluation metric of the recommender system.

2 Proposed Work

2.1 Apache Mahout

Apache Mahout is an open source platform which provides the appropriate characteristics to develop recommender systems Mahout is used to implement recommender systems in two categories—distributed and non-distributed. MapReduce is used in the distributed version, which makes the system highly scalable and capable of handling large sized datasets. To implement non-distributed recommender systems, Mahout provides a number of classes and key interfaces. UserSimilarity, UserNeighborhood, SVDRecommender are some classes that have been used in this paper as the building blocks to construct recommender systems.

SVDRecommender: SVDRecommender provides an interface in Mahout which helps to build matrix factorization recommender systems. Using dimensionality reduction, it maps users and items to a latent space. Optimization is performed using stochastic gradient descent minimization (using *ParallelSGDFactorizer in Mahout*).

Error Calculation in Mahout: To gauge the accuracy of our recommender engine, Mahout provides a number of accuracy metrics [13]. *AverageAbsoluteDifferenceRecommenderEvaluator* is used to predict the mean average error in the ratings.

$$\text{MAE} = \sum_{i=1}^{n} \frac{|Actual\,Rating - Predicted\,Rating|}{n} \tag{5}$$

where n gives the total number of predicted ratings.

2.2 Annealing Schedules

At the start of the process, simulated annealing chooses a random combination of values associated with an initial energy. The combination is better if the value of energy is low. A variable chosen for temperature decreases over time and tests the changes that occur in the initial combination. For the change in each iteration, a cost factor can be used to determine the difference in energy between the current iteration and the previous one. This approach is adapted in our recommender system model, where an initial learning rate is chosen and is decremented by a value for each iteration. Thus in order to find minimum absolute error for our recommender system, annealing schedules have been applied on an initial learning rate.

For every epoch or iteration, the following five schedules have been implemented:

(i) **Exponential Annealing (EA)**—This annealing schedule sets the learning rate value by multiplying it with a factor that decreases exponentially with regard to iteration i [14].

$$\eta = \eta_0 \ * \text{pow}(\text{base}, \text{i}) \tag{6}$$

where η_0 is the initial learning rate and i is the corresponding iteration.

(ii) **Inverse Scaling (IS)**—This annealing schedule lowers the rate much quickly for the initial iterations and then slows down for the later ones. The annealing rate α is fixed by the user. It sets the learning rate based on the formula:

$$\eta = \frac{\eta_0}{1 + \alpha} \tag{7}$$

(iii) **Logarithmic Cooling (LC)**—It determines the learning rate by multiplying the initial learning rate by a factor that decreases inversely with respect to natural logarithm of iteration i.

$$\eta = \frac{\eta_0}{1 + \beta * \log(\text{i} + \text{d})} \tag{8}$$

where d is constantly set to 1. This schedule generally leads to a random search in the state space. The factor β is set to be greater than 1.

(iv) **Linear Multiplicative Cooling (LMC)**—The initial learning rate is decreased by a factor which is inversely proportional to iteration i. The bias factor β is set between 0 to 1.

$$\eta = \frac{\eta_0}{1 + \beta * i} \tag{9}$$

(v) **Quadratic Multiplicative Cooling (QMC)**—The initial learning rate is decreased by a factor inversely proportional to square of iteration i. The bias factor β is set between 0 to 1.

$$\eta = \frac{\eta_0}{1 + \beta * i * i} \tag{10}$$

2.3 *Proposed Framework*

The figure shows the proposed framework of implementing our recommender system in Mahout, by using the described annealing schedules (Fig. 2).

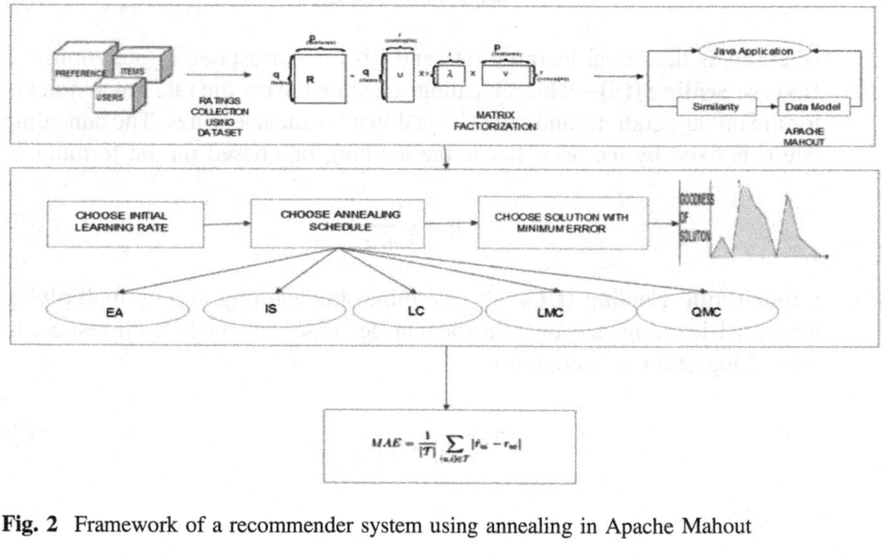

Fig. 2 Framework of a recommender system using annealing in Apache Mahout

3 Experiments

Dataset description—In this paper, experiments have been conducted on Movielens dataset of size 1 million with user ratings from 1 to 5.

Evaluation Metrics. The annealing strategies have been implemented and the mean absolute error (MAE) has been chosen for finding the accuracy of the system.

Experimental Setup. The recommender systems are evaluated on the basis of five annealing strategies to observe the magnitude of MAE in each case. Each dataset is divided into 3 different samples and the value of η is ranged from 0.008 to 0.08. The regularization parameter and number of features are set to 0.05 and 3 respectively.

Results. Tables 1, 2 and 3 give the trend of mean absolute error in case of our recommender systems with η ranging from 0.008 to 0.08 for Movielens dataset.

From the experimental results, it is evident that exponential annealing is the best annealing approach at a lower value of learning rate. Linear multiplicative cooling beats exponential annealing as the learning rate values get higher. The mean absolute error is minimum for our recommender system in these cases. Figures 3, 4 and 5 show the results.

Table 1 Mean Absolute Error (MAE) statistics for n = 10,000

η	EA	IS	LC	LMC	QMC
0.008	0.77356	0.91123	0.80910	0.81496	0.85293
0.01	0.77797	0.90732	0.80323	0.80912	0.84533
0.015	0.80152	0.89841	0.79294	0.79853	0.83276
0.018	0.81310	0.89395	0.78863	0.79395	0.82766
0.02	0.81945	0.89752	0.78630	0.79138	0.82480
0.03	0.84318	0.87738	0.77894	0.78268	0.81394
0.05	0.86419	0.85956	0.77363	0.77546	0.80042
0.08	0.87485	0.84639	0.78724	0.77560	0.78770

Table 2 Mean Absolute Error (MAE) statistics for n = 100,000

η	EA	IS	LC	LMC	QMC
0.008	0.77356	0.91123	0.80910	0.81496	0.85293
0.01	0.77797	0.90732	0.80323	0.80912	0.84533
0.015	0.80152	0.89841	0.79294	0.79853	0.83276
0.018	0.81310	0.89395	0.78863	0.79395	0.82766
0.02	0.81945	0.89057	0.78630	0.79138	0.82480
0.03	0.84318	0.87738	0.77849	0.78268	0.81394
0.05	0.86419	0.85956	0.77363	0.77546	0.80042
0.08	0.87485	0.84369	0.78724	0.77560	0.78770

Table 3 Mean Absolute Error (MAE) statistics for n = 1,000,000

η	EA	IS	LC	LMC	QMC
0.008	0.70356	0.88184	0.73761	0.74300	0.78875
0.01	0.70490	0.87482	0.73284	0.73765	0.77890
0.015	0.70925	0.85889	0.72572	0.72944	0.76302
0.018	0.71131	0.85035	0.72305	0.72638	0.75674
0.02	0.71251	0.84509	0.72165	0.72476	0.75335
0.03	0.71698	0.82426	0.71541	0.71931	0.74198
0.05	0.72719	0.79817	0.70349	0.70683	0.73081
0.08	0.72598	0.77684	0.70686	0.70424	0.72322

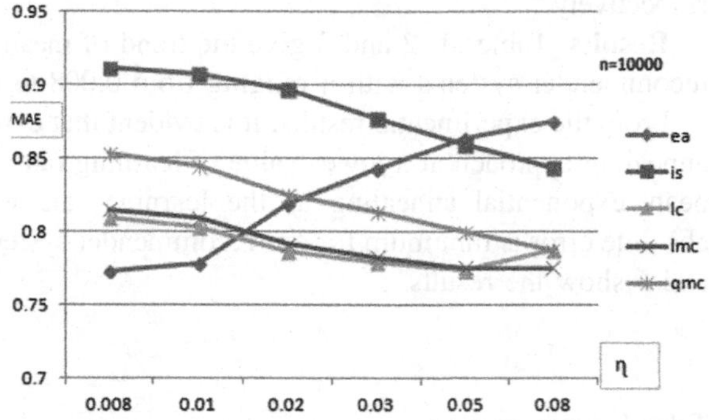

Fig. 3 Annealing schedules for n = 10,000

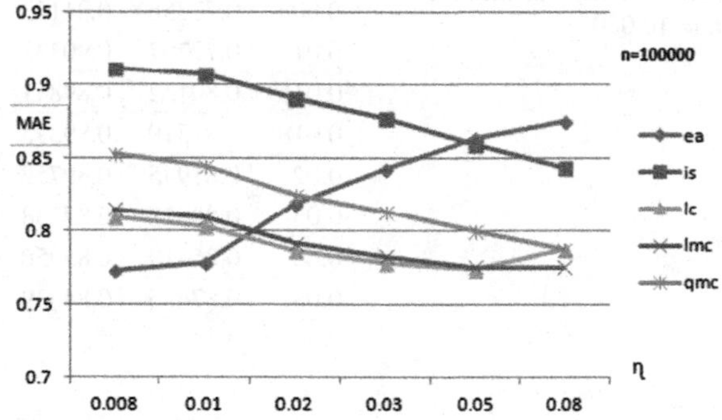

Fig. 4 Annealing schedules for n = 100,000

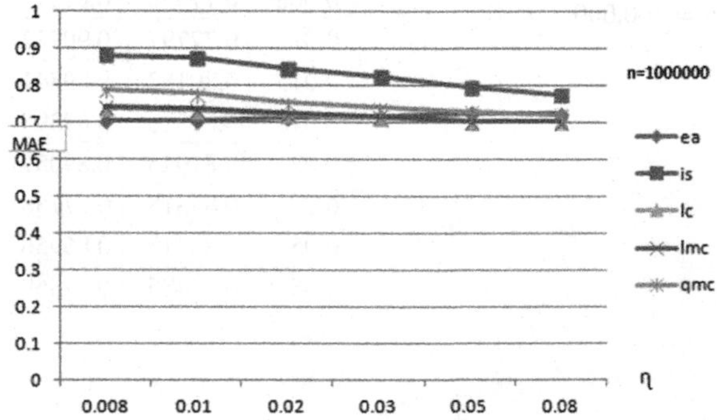

Fig. 5 Annealing schedules for n = 1,000,000

4 Conclusion

Learning rate selection is important for minimizing the mean absolute error in recommendations made by any recommender system. In this paper, we have used Apache Mahout framework and made recommendations successfully for Movielens dataset sizes n = 10000, 100000 and 1000000. At a learning rate value = 0.008, exponential annealing gives the best accuracy, with MAE 0.70356 (for n = 1,000,000). For a higher value = 0.08, LMC gives an error of 0.70424 (for n = 1,000,000) which is the lowest against all the five annealing schedules. Further, a new annealing schedule could be designed to suggest the best learning rate to give an improved value of the accuracy of our system.

References

1. Ekstrand, M.D., Riedl, J.T., Konstan J.A.: Collaborative Filtering in Recommender Systems.: Foundations And Trends In Human Computer Interaction, vol. 4, No. 2, pp. 81–173 (2011).
2. Zhang, M., Tang, J., Zhang, X. & Xue X.: Addressing Cold Start in Recommender Systems: A Semi Supervised Co-Training Algorithm. In: Proceedings of the 37th International ACM SIGIR Conference On Research And Development in Information Retrieval, SIGIR'14, pp. 73–82. ACM, New York, NY, USA (2014).
3. Bhebhe, W., Kogeda, O.P.: Shilling Attack Detection in Collaborative Recommender Systems Using a Meta Learning Strategy. In: Emerging Trends In Networks and Computer Communications (ETNCC), International Conference on IEEE, pp. 56–61 (2015).
4. Koren, Y., Bell, R., Volinsky, C.: Matrix Factorization Techniques for Recommender Systems.: Computer, vol. 42, No 8, pp. 30–37 (2009).
5. Luo, X., Xia, Y., Zhu Q.: Applying the Learning rate Adaptation to the Matrix Factorization Based Collaborative Filtering.: Knowledge Based Systems, vol. 37, pp. 154–164 (2013).
6. Chin, W-S., Zhuang, Y., Juan, Y-C., Lin C-J.: A Learning Rate Schedule for Stochastic Gradient Descent Methods to matrix Factorization. In: Advances In Knowledge Discovery And Data Mining, 19th Pacific Asia Conference PAKDD, Ho Chi Minh City, Vietnam, Proceedings, Part I. pp. 442–455 Cham: Springer International Publishing. (2015).
7. Sarwar, B., Karypis, G., Riedl, J.: Application Of Dimensionality Reduction In Recommender System-A Case Study.: ACM WebKDD 2000 Workshop, pp. 1–12. ACM SIGKDD (2000).
8. Le, B.H., Nguyen, K.Q., Thawonmas, R.: Bounded SVD: A Matrix Factorization Method With Bound Constraints For Recommender Systems. In: Emerging Information Technology and Engineering Solutions, pp. 23–26 (2015).
9. Paterek, A.: Improving Regularized Singular Value Decomposition for Collaborative Filtering. In: Proceedings KDD Cup Workshop at SIGKDD'07, 13th ACM International Conference on Knowledge Discovery And Data Mining pp. 39–42. ACM (2007).
10. Yu, H.-F., Hsieh, C.-J., Dhillon I.S.: Scalable Coordinate Descent Approaches to Parallel Matrix Factorization For Recommender Systems. In: IEEE International Conference Of Data mining, pp. 766–774. IEEE (2012).
11. Weimer, M., Karatzoglou, A., Smola, A. Improving Maximum Margin Matrix Factorization. In: Machine Learning, vol. 72, No. 3, pp. 263–276 (2008).
12. Bamnote, G.R., Agrawal S.S. Evaluating and Implementing Collaborative Filtering Systems Using Apache Mahout. In: Computing Communication Control And Automation (ICCUBEA), pp. 858–862 (2015).

13. Andresen, B., Nourani, Y.A.: A Comparison of Simulated Annealing Cooling Strategies: Journal of Physics: Mathematics and General, vol. 31, No. 41, pp. 8373–8380 (2011).
14. Gemulla, R., Nijkamp, E., Hass, P.J, Sismanis, Y.: Large Scale Matrix Factorization With Distributed Stochastic Gradient Descent. In: Proceedings of the 17th ACM SIGKDD International Conference on Knowledge Discovery and Data Mining, ser. KDD'11, pp. 69–77. New York, NY, USA (2011).
15. Semainario, C.E., Wilson, D.C.: Case Study Evaluation of Mahout As a Recommender Platform. In: 6th ACM conference on recommender engines, pp. 45–50. ACM (2012).

Classifying Gait Data Using Different Machine Learning Techniques and Finding the Optimum Technique of Classification

Anubha Parashar, Apoorva Parashar and Somya Goyal

Abstract The Classification of Humanoid locomotion is a troublesome exercise because of nonlinearity associate with gait. The high dimension feature vector requires a high computational cost. The classification using the different machine learning technique leads for over fitting and under fitting. To select the correct feature is also the difficult task. The hand craft feature selection machine learning techniques performed poor. We have used the deep learning technique to get the trained feature and then classification we have used deep belief network-based deep learning. Classification is utilized to see Gait pattern of different person and any upcoming disease can be detected earlier. So in this paper we first selected the feature and identify the principle feature then we classify gait data and use different machine learning technique (ANN, SVM, KNN, and Classifier fusion) and performance comparison is shown. Experimental result on real time datasets propose method is better than previous method as far as humanoid locomotion classification is concerned.

Keywords GAIT · Classification · Biometric identification · Feature selection

A. Parashar (✉)
Vaish College of Engineering, Rohtak, India
e-mail: anubhaparashar1025@gmail.com

A. Parashar
Maharshi Dayanand University, Rohtak, India
e-mail: apoorvaparashar0000@gmail.com

S. Goyal
PDM College of Engineering, Bahadurgarh, India
e-mail: somyagoyal1988@gmail.com

D.K. Mishra et al. (eds.), *Information and Communication Technology for Sustainable Development*, Lecture Notes in Networks and Systems 10,
https://doi.org/10.1007/978-981-10-3920-1_31

1 Introduction

Human gait is supposed as unique biometric identification like thumb print. So in GAIT analysis to examine the human walking moment we collect and analyze the data [1]. Human gait can be utilized to identify people for diverse security reason and exercised before to detect many abnormalities. For disable person the classified data is used to implement the prosthetic leg.

The GAIT data can be useful to help in walking of old people and is used in pattern analysis and classification of different walk. By seeing the abnormality in GAIT cycle, it is worth to do research for prognostication for many upcoming diseases. Differently abled and elderly people can be helped by any research regarding GAIT. During human movement, top and bottom limb of human body yield the periodic motion. Due to this particular motion human beings get a particular motion and a unique walking pattern. The biometric identity like gender, age and race based on human locomotion can be obtained by the use of GAIT analysis shown in Fig. 1.

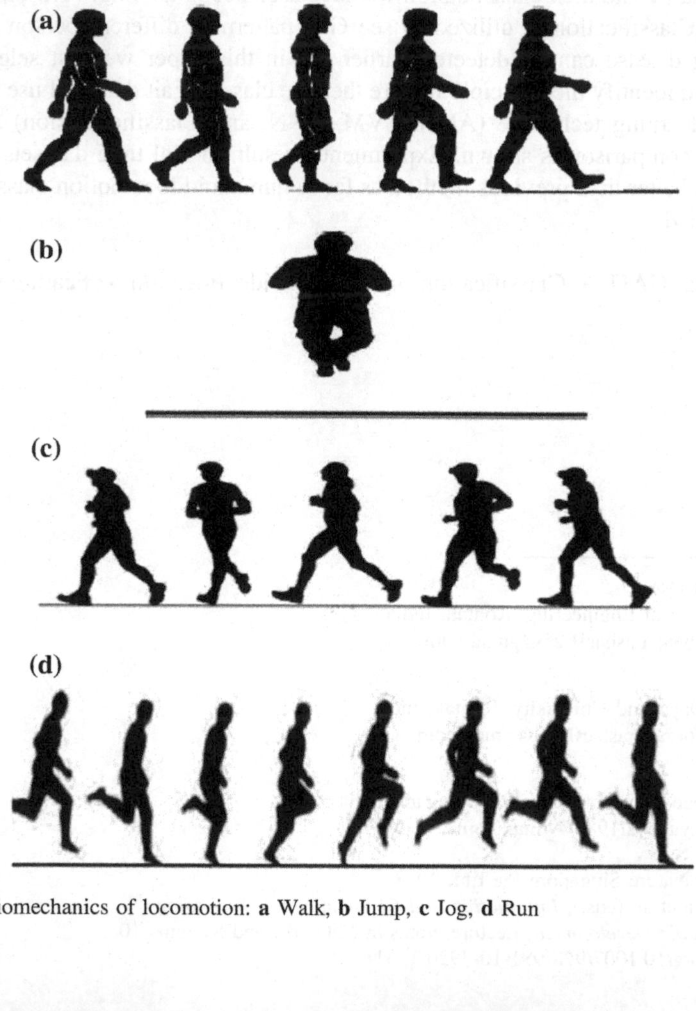

Fig. 1 Biomechanics of locomotion: **a** Walk, **b** Jump, **c** Jog, **d** Run

A lot of precursory research work about various machine learning techniques is gathered from the paper given [2–6] below. The classification of GAIT pattern is very accurate and unique due to rhythmic pattern. The new arena of biometric research is different category of GAIT classification. The GAIT can be helpful for security reason as it can identify from distance. In comparison with other biometric identification technique such as face reorganization, i.e., iris, finger print which is considered as first generation biometric, the GAIT is divergent as it does not require any subject contact other than walking. This can be considered as behavioral authentication technique. Therefore identification using gait is becoming very popular in security. The development of highly sophisticated and more accurate humanoid can be made easy with the study of human GAIT data. So the foremost requirement is the feature reduction.

2 Proposed System

Proposed system consists of seven phases. They are collection of gait data, detection of gait data, trajectory smoothening, feature extraction, feature selection and classification. Figure 2 tells systematic proposal. In the following subsections there are details of each phase (Fig. 3).

$$w_i = \frac{1}{\sum_{k \neq i} \|x_k - x_i\|_2}$$

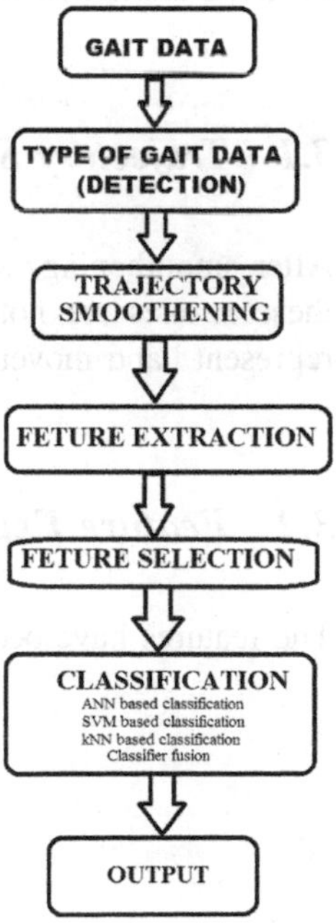

Fig. 2 Architecture of the proposed system

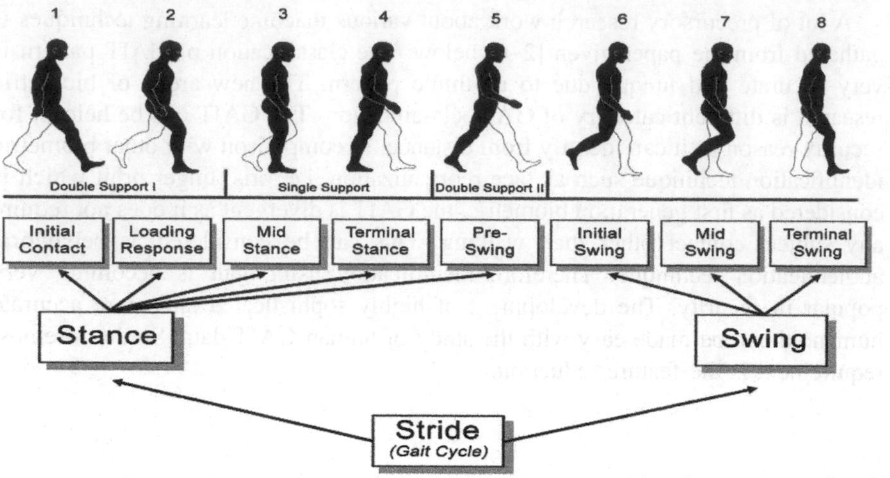

Fig. 3 Eight phases of gait cycle

3 Methodology

Firstly, we have collected data of different gait using accelerometer [7]. Secondly, features were selected on the basis of ANOVA and IFA (Incremental Feature Analysis) to get optimal features from the existing features. Then different combinations of features are made by IFA using ANN, SVM, and KNN classifier. Lastly, classifier fusion technique is used.

3.1 *Trajectory Smoothening*

After smoothening, we detect and remove the self-co-articulated strokes. And as these strokes are not part of gesture therefore are discarded here as these just represent hand movements.

3.2 *Feature Extraction*

The features have been collected from [8] (Table 1).

Table 1 Crouch dataset description

Feature category	Feature name	Feature category	Feature name
F1	Pelvis_tx	F10	knee_angle_r
F2	Pelvis_ty	F11	ankle_angle_r
F3	Pelvis_tz	F12	hip_flexion_l
F4	Pelvis_tilt	F13	hip_adduction_l
F5	Pelvis_list	F14	hip_rotation_l
F6	Pelvis_rotation	F15	knee_angle_l
F7	hip_flexion_r	F16	ankle_angle_l
F8	hip_adduction_r	F17	lumbar_extension
F9	hip_rotation_r		

Table 2 ANN results

Exp No.	Hidden units	Network structure	Iteration	Training accuracy (%)	Testing accuracy (%)
1	50	33-50-40	100	84	80
			500	86	80
2	52	33-52-40	100	86	80
			500	90	82
3	54	33-54-40	100	86	82

3.3 *Feature Selection*

For selecting the final optimal features here we are using 2-stage process. In first stage we perform the ANOVA test to obtain feature [9–11]. In second stage after the significant features were obtained from Stage 1, the IFS technique is used to determine the optimal features.

4 Classification

Classification is done using following machine learning techniques.

4.1 *Classification Based on ANN*

The network is trained many times till root mean squared error reaches an optimum level. To adjust weights of the network, we train system through back-propagation algorithm. Results of ANN classification is shown in Table 2.

Table 3 SVM results

Exp No.	Function kernel	Training accuracy (%)	Test accuracy (%)
1	Linear function	68	71
2	Quadratic function	70	76
3	Polynomial function	84	84
4	Radial basis function	88	84

Table 4 KNN results

Exp No.	K	Training accuracy (%)	Testing accuracy (%)
1	3	74	70
2	5	79	80
3	7	78	74
4	9	84	81

4.2 Classification Based on SVM

SVM is supervised learning method. We optimize class separation hyper plane so we get the maximum distance between hyper plane and pattern separating the classes. Now dataset is trained with different kernel functions like polynomial, quadratic, linear and radial basis function. Results of SVM classification is shown in Table 3.

4.3 Classification Based on KNN

KNN classification depends on associating the anonymous instances according to distance function. To train *KNN* we use different *k* values as it is an essential parameter. In training phase we used different number of k values such as 9, 7, 5, and 3. Results of KNN classification is shown in Table 4.

4.4 Classifier Fusion

Classifier fusion combines many classifiers and provides the result by combining the results of the individual classifiers. In this paper, we have combined the results of KNN, SVM, and ANN models so as to achieve result of classifier fusion. The experiment of classifier fusion was done by combining the other classifiers. Thus, the performance of classifier fusion is likely superior than the individual classifiers used in isolation.

Once the training is done, testing is done by fivefold cross validation which is shown in table number 5 (Table 5).

Table 5 Result of fivefold cross validation

Exp No.	ANN classifier	SVM classifier	kNN with k = 5	Classifier fusion
	Accuracy (%)	Accuracy (%)	Accuracy (%)	Accuracy (%)
1	89.67	91.62	80.76	89.67
2	92.91	91.92	78.89	90.75
3	87.82	88.52	91.82	87.82
4	93.60	85.60	90.40	89.60
5	91.01	89.21	92.21	86.75

Table 6 Comparisons of different classifiers using fivefold cross validation and showing their success rate

Classifiers	Rate of success (%)
SVM	80.11
KNN	87.72
ANN	82.59
Classifier fusion	93.56

5 Experimental Results and Discussions

5.1 Data Set Used

In order to evaluate our proposed approach of human activity recognition, we have used crouch datasets. The data set have four classes. Each has three different data base so total 12 dataset for all four classes. It has total 17 features (Times series data). The dimension of data is 17 * 300 for a particular data set, in which we choose one gait cycle for our experiment. So our dataset has 17 * 100 elements each. Hence total 17 * 4 * 3 * 100 element.

5.2 Performance of the Classifiers

The training and testing accuracies are evaluated for different parameters of ANN such as iterations and hidden units. The highest train and test accuracy was observed for network structure 40L54N40L (where L is the linear layer and N is the nonlinear layer), iterations = 500. These network parameters were used for performing fivefold cross validation to determine the performance and validity of the model. Results of cross validation process are shown in Table 6.

5.3 Comparative Analysis

In our proposed system we perform comparative analysis by selecting the optimal features. Now the comparison is done with various set of gestures as seen in Fig. 4.

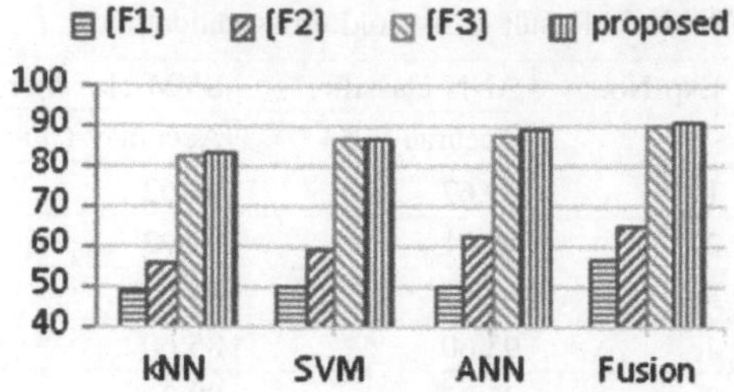

Fig. 4 Classification of different GAIT pattern: comparison of proposed optimal features using ANN, SVM, KNN, and classifier fusion: **a** Crouch_2, **b** Crouch_3, **c** Crouch_4, **d** Normal [10]

It was noted that from all the other listed individual classifier for various feature set performed, the performance of the classifier fusion is best. Therefore we get the highest overall accuracy of 93.56% by using the combination of optimal features and classifier fusion as compared to other feature and classifier combinations.

5.4 Conclusion

With the help of proposed methodology we can develop gesture controlled hexadecimal keyboard through which human–computer interaction becomes easier. From the existing literatures, a total of 10 features were selected. To check the statistical significance of the 10 features ANOVA test was performed. We arrange the 10 significant features in descending order of F-static value, later it is given as input to the IFS so that optimal features can be selected. Total six features of ANN were detected, seven features of SVM were detected and eight features of KNN were detected. We combined the results of three individual classifier to get the classifier fusion result. Then to obtain overall accuracy of the system we used fivefold cross validation technique. Overall accuracy of system has come out 93.56% for classifier fusion, 87.72% using KNN, 80.11% using SVM, and 82.59% for ANN.

References

1. Rattiya Thanasoontornerk, Chatkaew Pongmala, Areerat Suputtiada: Tree Induction for Diagnosis on Movement Disorders Using Gait Data. In: 2013 5th International Conference on Knowledge and smart technology (KST).
2. Kiichi_Katoh, *Makoto Katoh*.: Using FS in Tuning a Biped Locomotion Surrogate Robot for Walking Training. In: 2011 4th International Conference on Biomedical Engineering and Informatics (BMEI).
3. Mitsuru Yoneyama, Yosuke Kurihara, Kajiro Watanabe: Accelerometry-Based Gait Analysis and Its Application to Parkinson's Disease Assessment—Part 2: IEEE Transactions on Neural Systems and Rehabilitation Engineering, vol. 21, no. 6, November 2013.

4. M. Ogata, T. Karato, T. Tsuji, H. Kobayashi and K. Irie.: Development of Active Walker by using Hart Walker. In: SICE-ICASE International Joint Conference 2006 Oct. 18–21, 2006 in Bexco, Busan, Korea.
5. Hiroshi Kobayashi, Takeo Karato, and Toshiaki Tsuji.: Development of an Active Walker as a New Orthosis. Proceedings of the 2007 IEEE International Conference on Mechatronics and Automation August 5–8, 2007, Harbin, China.
6. Daniel T.H. Lai, Ahsan Khandoker, Rezaul K. Begg and M. Palaniswami.: A hybrid Support Vector Machine and autoregressive model for detecting gait disorders in the elderly. In Proceedings of International Joint Conference on Neural Networks, Orlando, Florida, USA, August 12–17, 2007.
7. Anubha Parashar, Somya Goyal, Apoorva Parashar, Bharat Sahjalan: Push Recovery for Humanoid Robot in Dynamic Environment and Classifying the Data Using K-Mean. ICTCS 2016 - CSI-Udaipur Chapter.
8. Daniel T.H. Lai, Ahsan Khandoker, Rezaul K. Begg and M. Palaniswami: A hybrid Support Vector Machine and autoregressive model for detecting gait disorders in the elderly. Proceedings of International Joint Conference on Neural Networks, Orlando, Florida, USA, August 12–17, 2007.
9. Semwal, Vijay Bhaskar, et al: Biped model based on human Gait pattern parameters for sagittal plane movement. Control, Automation, Robotics and Embedded Systems (CARE), 2013 International Conference on. IEEE, 2013.
10. Semwal, Vijay Bhaskar, et al.: Biologically-inspired push recovery capable bipedal locomotion modeling through hybrid automata. Robotics and Autonomous Systems 70 (2015): 181–190.
11. Semwal, Vijay Bhaskar, Manish Raj, and Gora Chand Nandi.: Biometric gait identification based on a multilayer perceptron. Robotics and Autonomous Systems 65 (2015): 65–75.

Adaptive Kalman Filter Approach and Butterworth Filter Technique for ECG Signal Enhancement

Bharati Sharma, R. Jenkin Suji and Amlan Basu

Abstract About 15 million people alive today have been influenced by coronary illness. This is a major and critical issue in recent days. There are so many people have been lost their lives due to heart attack and other heart related issues. So, early on analysis and proper cure of heart disease is required to minimize the death rate due to heart disease. For better diagnosis we need exact and consistent tools for determine the fitness of human hearts to analysis the disease ahead of time before it makes around an undesirable changes in human body. For heart diagnosis one of the tools is Electrocardiogram (ECG) and the obtained signal is labeled ECG signal. This ECG signal contaminated by an amount of motion artifacts and noisy elements and deduction of these noisy elements from ECG signal must important before the ECG signal could be utilized for illness diagnosis purpose. There are various filter methods available for denoising ECG signal and select the best one on the dependence of performance parameter like signal to noise ratio (SNR) and power spectrum density (PSD).

Keywords Electrocardiogram (ECG) · Kalman filter · Butterworth filter Denoising · Signal to noise ratio (SNR) · Power spectrum density (PSD)

B. Sharma (✉) · R. Jenkin Suji · A. Basu
Department of Electronics and Communication Engineering, ITM University, Gwalior 474001, Madhya Pradesh, India
e-mail: bharatisharma30@gmail.com

R. Jenkin Suji
e-mail: sujijenkin@gmail.com

A. Basu
e-mail: er.basu.amlan@gmail.com

D.K. Mishra et al. (eds.), *Information and Communication Technology for Sustainable Development*, Lecture Notes in Networks and Systems 10,
https://doi.org/10.1007/978-981-10-3920-1_32

1 Introduction

Observation of the ECG (Electrocardiogram) has quite some time been utilized as a part of clinical practice. As of late, the relevance area of ECG observation is extending to regions outer the laboratory [1]. Home observing of patients with rest apnea is one of an example of such an area. There are various ECG monitoring technology available, a move in ECG observing applications is occurring. With advancement in sensor innovation such as material and capacitive terminals, sensors are fused in pieces of the incubator have ended up accessible [2].

Another sensor technology brings the solace of the patient is enhancing continuously. While a few years back the patient needed to accommodate itself according to discomforts of the only available technology, but now a day's patient used to new technology for ECG monitoring and goes with a comfortable treatment of heart diseases [3].

Sometimes these ECG monitoring technologies contaminated due to breathing, and mismatching measurement and explanation of the signal's components therefore, noise artifacts generated into acquired ECG signal and corrupt it. Due to certain stress test, the noise artifacts vary unpredictably. Elimination of Interference in the ECG signal by using various filtering method such as, Adaptive filter and Weiner filter are utilized for removal artifacts from ECG signal [4]. An adaptive Wavelet Weiner filtering of ECG signals has been proposed with stationary Wavelet Transform (SWT) and Wavelet Filtering method (WF) compared by different thresholding strategies [5].

This work presents an Adaptive Kalman filter and butter worth filter approach for the estimation and denoising ECG signal. These IIR methods will be utilized for approving the proposed Kalman filter approach.

The proposed procedure in this paper is a utilization of the Kalman filter (KF) for the estimation and evacuation of the noise artifacts in ECG signal. The projected IIR filter methods based on the frequency selective components [6]. The state space model is coordinated with Kalman filter so as to approximate the state variables. The proposed technique recommends an appropriate way for estimation of the noise artifacts of an ECG signal, and is contrasted with the IIR filter methods which basis on the performance parameters.

2 Methodologies

2.1 Kalman Filter

An adaptive Kalman Filter (KF) is a recursive prescient filter that depends on the state model and time varying recursive algorithms. An ECG signals complexes that relates from back to back heartbeats are fundamentally the same yet not indistinguishable. However, while the recording of ECG, the signal is defiled due to some noise interference. An adaptive Kalman filter appraises the state of a dynamic system. This dynamic system can be contaminated by noise. The Kalman filter utilizes estimations to enhance the estimated state [7].

The KF method consists of the prediction and correction of states of the system.

$$x_{t+1} = x_t + v_t \tag{1}$$

$$Y_{t+1} = x_{t+1} + w_{t+1} \tag{2}$$

where x_{t+1} is the state input of the system, v_t is the process noise of the system, Y_{t+1} is the measurement output of the system and w_{t+1} is the measurement noise such as noise artifacts.

The prediction is an initial work of the Kalman filter. The predict state or prior state is intended by neglecting the noise of the system. In linear case state vector equation can be represented as:

$$X(t) = F \cdot x(t) + n(t) \tag{3}$$

$$X(t) = F \cdot x(t) \tag{4}$$

where, F is the dynamic grid and is consistent, state vector x(t) and dynamic interference(t) of the system.

The genuine predicted state is a linear combination of the primary state $x^-(t_0)$ From Eq. (3) and (4)

$$x^-(t) = A_0^t x^-(t_0) \tag{5}$$

where, A_0^t is called the conversion matrix, which transform the primary state $x^-(t_0)$ to its equivalent $x^-(t)$ at point t.

Covariance matrix $P^-(t_i)$ of the predicted state vector is attained with the law of error transmission,

$$P^-(t_i) = A \cdot P(t_{i-1}) \cdot A^T + Q \tag{6}$$

where, covariance matrix of the noise Q is a utility of time.

In a correction process we obtained the improved predicted state with observation form at time t_i, thus the posteriori state has form,

$$x^+(t_i) = x^-(t_i) + \Delta x(t_i) \tag{7}$$

And covariance matrix,

$$P^+(t_i) = P^-(t_i) + \Delta P(t_i) \tag{8}$$

$$\Delta x(t_i) = K(t_i) \cdot [l(t_i) - l^-(t_i)] \tag{9}$$

$$K(t) = P^{-}H^{T}\left(HP^{-}H^{T} + R(t_i)\right)^{-1} \tag{10}$$

where, K is called the gain matrix. The difference $[l(t_i) - l^{-}(t_i)]$ is identified the extent residual. It reproduces the inconsistency between the predicted measurement and actual extent. At the end corrected state is received by,

$$x^{+}(t_i) = x^{-}(t_i) + \Delta x(t_i) \tag{11}$$

The Kalman Filter approach uses to predict and remove the noise artifacts from ECG signal. Though, the equation given in (11) is repeated over input signal. Equation (11) is updated for the Kalman filter.

2.2 *Butterworth Filter*

Butterworth filters are having an attribute of maximally level recurrence response and no ripples in the pass band. It moves of nears zero in the stop band [8]. Its reaction inclines off directly toward negative infinity on Bode plot. For example, other filter types which have non-monotonic swell in the pass band or stop band, these filters are having a monotonically changing size capacity with ω. The initial 2n-1 subordinates for the force capacity as for recurrence are zero. Thus it is conceivable to determine the formula for frequency response,

$$|H(j\omega)| = \frac{1}{\sqrt{1 + (\omega + \omega c)^{2n}}} \tag{12}$$

3 Simulations and Result

The proposed approaches implemented in Matlab version 2009. To study the performance of the planned method numerous standard data sets have been taken from physio.net, including the MIT-BIH database. This ECG signal corrupted with some noise artifacts and corrupted ECG signal passes through filter and get noise free signal (Figs. 1 and 2).

(a) Graphical representation of Kalman filter method

Figure 3 illustrates the Kalman filter response of ECG signal and blue line shows the true response of the filter and red shows the filtered response of the signal.

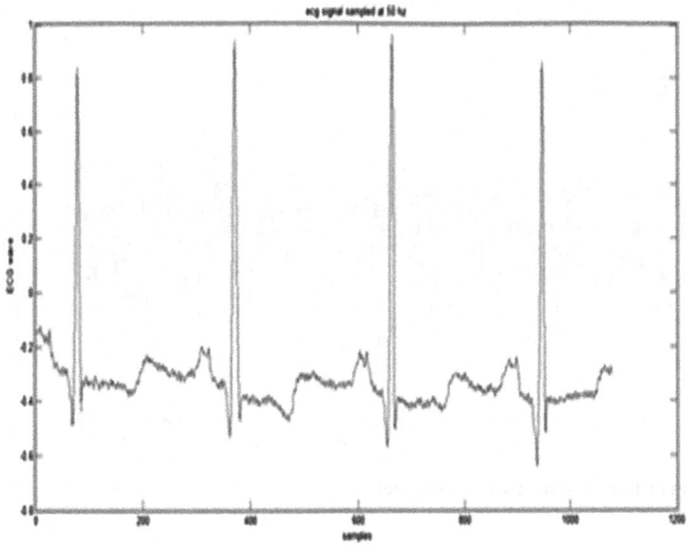

Fig. 1 Typical ECG signal

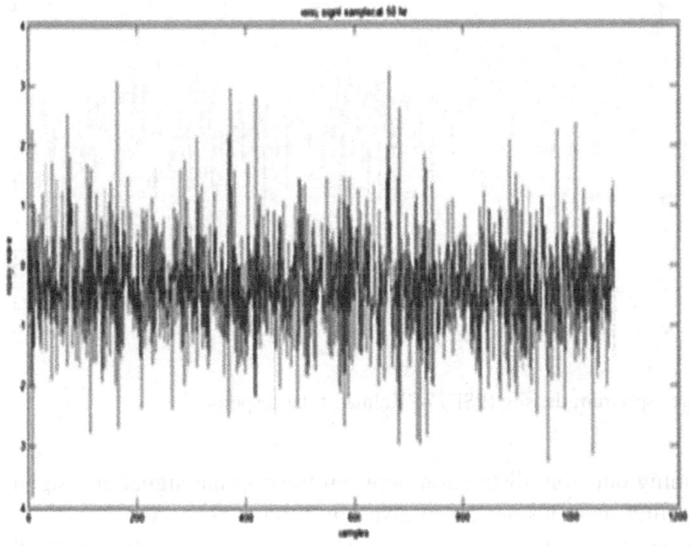

Fig. 2 Noisy ECG signal

(b) Graphical representation of Butterworth filter method

Figures 4 and 6 illustrate the power spectrum density of filtered ECG signal. To compare PSD graph of both the filter and chose best one on the basis of numeric values of graph. The power spectrum density (PSD) and signal to noise ratio

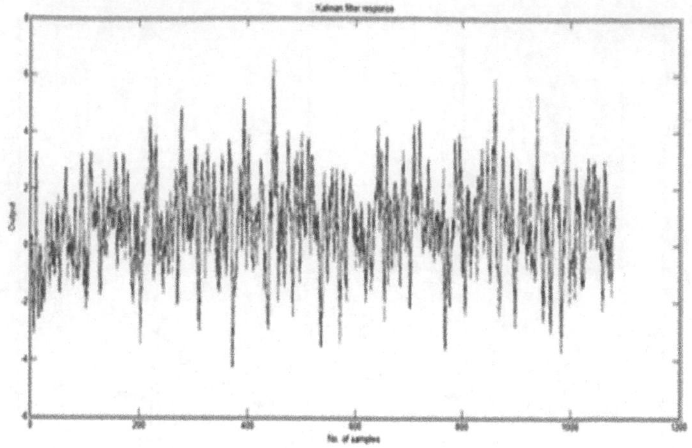

Fig. 3 Kalman filter response of ECG signal

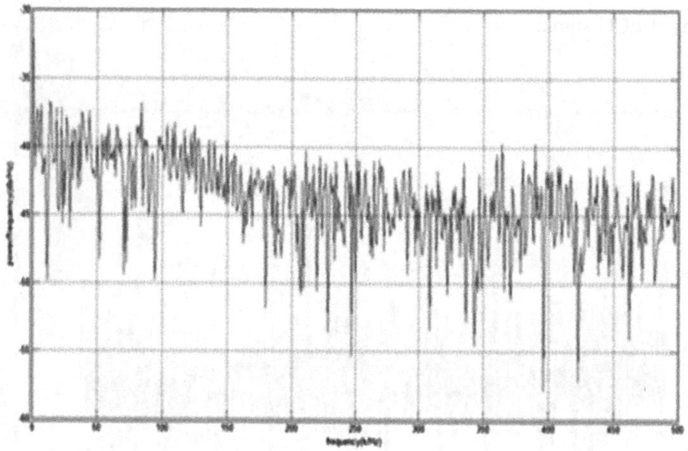

Fig. 4 Power spectrum density (PSD) of Kalman filter response

(SNR) coming out from distinction between the original signal and signal obtained from each filter methods are given away in Table 1 (Fig. 5).

As can be seen from the table over, the level of contortion is negligible for the Kalman filter technique. The Butterworth filter approach is highly dependent on the filter order and patient's heart rate. Kalman Filter (KF) approach shows better results for denoising ECG signal. Its SNR and PSD is high as compare to Butterworth filter thereby high SNR enhanced the performance of signal as well as PSD of the filtered signal referred to as frequency domain analysis or spectral density estimation. PSD is basically uses for decomposing or denoising a complex signal.

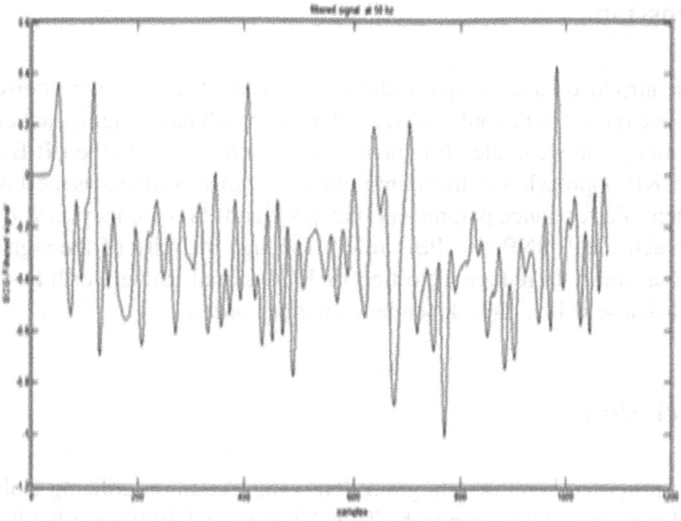

Fig. 5 Filtered ECG signal using Butterworth filter

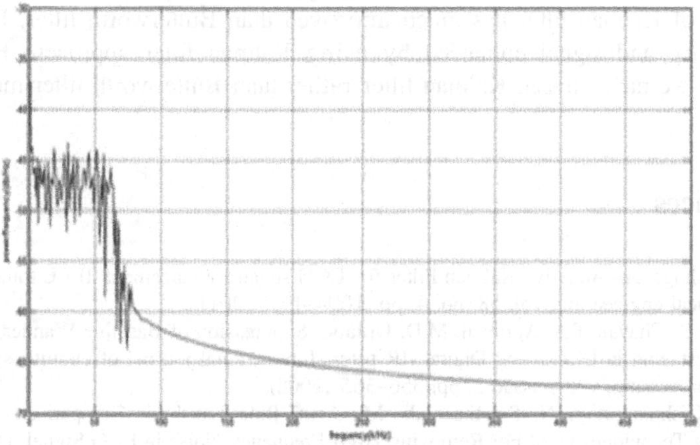

Fig. 6 Power spectrum density (PSD) of filter Butterworth signal

Table 1 Comparison between performance parameters of Kalman and Butterworth filter

Parameters	Original ECG signal	Kalman filter response	Butterworth filter response
Signal to ratio (SNR)	−17.4807	26.6035	5.5091
Power spectral density (PSD)	−34.5891	−30.1931	−34.9085

4 Discussion

The paper introduced a technique utilized as a part of present time noise artifacts removal. As exposed in the outcomes, the KF approach had insignificant contortion, especially in the QRS complex fragment, when contrasted with the IIR Butterworth filter. The KF approach for the elimination of noise artifacts is best for signal enhancement. Performance parameters like SNR and PSD are increased by Kalman filter approach. High SNR and PSD refers for high stability of the signal So, KF approach has ability to reduce distortion of ECG signal. Butterworth has less SNR and PSD because it is highly dependent on filter order.

5 Conclusion

This paper proposed Electrocardiogram denoising execution utilizing Kalman filter and IIR Butterworth filter approach. This Kalman and Butterworth filter method removes noise artifacts and KF approach has great denoising ability. Simulation Results recover the denoising execution of both Kalman Filter and Butterworth filters are completely different because the performance parameters such as SNR and PSD of Kalman filter has much improved than Butterworth filter. Denoising performance and signal enhanced by using Kalman filter approach. For better execution we have chosen Kalman filter rather than Butterworth filter method.

References

1. Rik Vullings: An Adaptive Kalman Filter for ECG signal Enhancement. IEEE transactions on biomedical engineering, vol. 58, no. 4, pp. 1094–1103 (2011).
2. Mahesh S. Chavan, RA. Agrawal, M.D. Uplane: Suppression of Baseline Wander and Power line interfacing in ECG using Digital IIR filter. International journal of circuits, systems and signal processing, vol. 2 issue 2, pp. 356–365 (2008).
3. Priya Krishnamurthy, N. Swethaanjali, M. Arthi Bala Laxshmi: Comparison of Various Filtering Techniques Used For Removing High Frequency Noise in ECG Signal. International journal of students research in technology & management, Vol. 3 no. 1, pp. 211–215 (2015).
4. Nidhi Rastogi, Rajesh Mehra: Analysis of Butterworth and Chebyshev Filters for ECG Denoising Using Wavelets. IOSR Journal of Electronics and Communication Engineering (IOSR-JECE), vol. 6, issue 6, pp. 37–44, (2013).
5. Lukas Smital: Adaptive Wavelet Weiner Filtering of ECG signal. IEEE transactions on biomedical engineering, vol. 60 no. 2, pp. 437–445 (2013).
6. Omid Sayadi: ECG Denoising and Compression Using a Modified Extended Kalman Filter Structure. IEEE transaction on biomedical engineering, vol. 55 no. 9, pp. 2240–2248 (2008).
7. Kohler, B.-U, Henning, C., Orglmeister, R.: The Principles of software *QRS* detection. IEEE Engineering in Medicine and Biology Magazine, vol. 21 issue 1, pp. 42–57 (2002).
8. Mohan M. Kumar: Simulating Motion Artifacts in ECG using Wiener and Adaptive Filter structure. International journal of emerging technology and advanced engineering, vol. 2 issue 2, pp. 237–260 (2012).

Bayesian Approach for Automotive Vehicle Data Analysis

Heena Timani, Mayuri Pandya and Mansi Joshi

Abstract Streaming network data can be analyzed by advance machine data methods. Machine data methods are ideal for large scale and sensor concentrated applications. Prediction analytics can be used to support proactive complex event processing while probabilistic graphical model can be extensively used to ascertain data transmitted by sensors. The structure of probabilistic graphical models encompasses variety of different types of models and range of methods relating to them. In this paper, real time sensor (OBD ha-II) device data has been used from telematics competition organized by kaggle.com for driver signature. This device is highly equipped to extract sensors related information such as Accelerometer, Gyroscope, GPS, and Magnetometer. Data cleaning, pre-processing, and integration techniques are performed on data obtained from OBD-II device. We have performed various classification algorithms on sensor data using data mining and machine learning open source tool "WEKA 3.7.10" and have identified that Bayes Net classification technique generates best results.

Keywords Probabilistic graphical model · Machine learning · OBD-II
Sensor data analysis

H. Timani (✉) · M. Joshi
School of Computer Studies, Ahmedabad University, Ahmedabad, India
e-mail: heena.timani@ahduni.edu.in

M. Joshi
e-mail: mansi.joshi@ahduni.edu.in

M. Pandya
Maharaja Krishnakumarsinhji, Bhavnagar University, Bhavnagar, India
e-mail: mayuri.dave@rediffmail.com

D.K. Mishra et al. (eds.), *Information and Communication Technology for Sustainable Development*, Lecture Notes in Networks and Systems 10,
https://doi.org/10.1007/978-981-10-3920-1_33

1 Introduction

Internet of Things (IoT) is one of the forces to push different cycles of innovation and improvement between Data Analytics, Big Data, Cloud Computing, and Mobile Computing. There are various applications of IoT in automobile where built in sensors are used to alert the driver when pressure of the tyre is low. Built-in sensors on equipment's present in the power plant transmit real time data and thereby enables for better transmission planning and load balancing. In oil and gas industry, it can help in planning better drilling, track cracks in gas pipelines. IoT will lead to better predictive maintenance in the manufacturing and utilities and this will in turn lead to better control, track, monitor or back-up of the process. Even a small percentage improvement in machine performance can significantly benefit the company bottom line. IoT in some ways is going to make our machines more brilliant and reactive [1].

Combination of Bayesian network and data mining algorithms encode both qualitative and quantitative knowledge from data. It merges different information data facts and provides all gathered knowledge into a distinct form of representation. In situation of uncertainty, this combination provides crucial information. From the graphical representation of Bayesian network, we can find conditional probability and causation of different attributes. From this, we can identify dependent and independent attributes from the data. It is very useful for model building [2].

IoT starts with embedding advanced sensors in machines and collecting the data for advanced analytics. As we start receiving data from the sensors, one important aspect that needs all the focus is the data transmitted correct or erroneous. To validate the data quality is a major issue. While dealing with uncertainty the most commonly used methods for modeling uncertainty is Probabilistic Graphical Models. IoT embeds various types of sensors and data gathered from these sensors are multi-type data. Sometimes, it may contain sequential series of information. It's not a good practice to deal with multi-variate data independently as we may fail to notice important information from the data. In this paper, sensor data are obtained from https://www.kaggle.com. Various machine learning algorithms of classification such as Bayes Net, Gaussian Process, and clustering algorithms EM, K means, etc. have been applied to get knowledge from the data of different trips. WEKA 3.7.10 tool is used for knowledge discovery from the data [3].

2 Methodology

For analytic purpose of conceptual map, X axis and Y axis are shown in Fig. 1 where X axis explains the association and correlation exists in the data among attributes, it also represents the succession of forecast optimization and replication. Model measurement is explained by Y axis, it also represents the data related to artificial

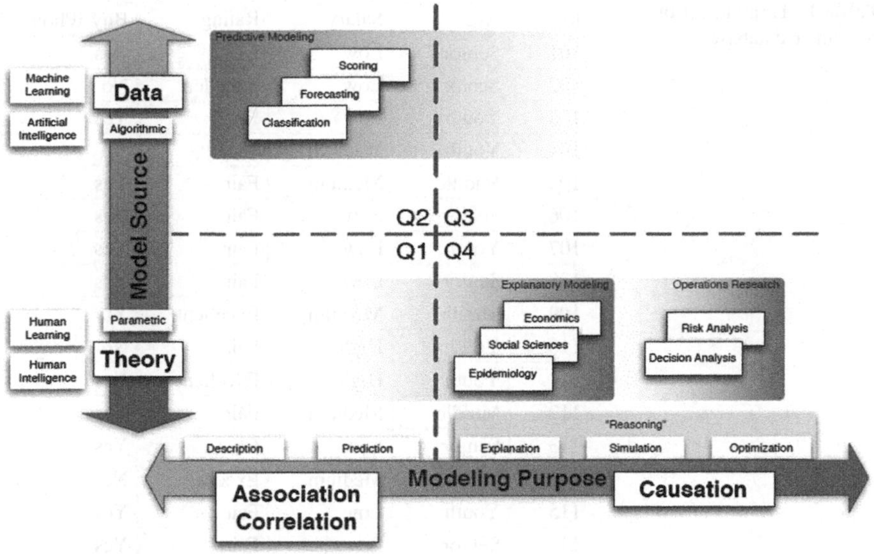

Fig. 1 Conceptual map of analytic modeling [4]

intelligent, machine learning, and human intelligence. Various data mining algorithms can be applied to extract knowledge of parameters from the sensor data [4].

A set $G = \{V, E\}$ represents random variable related to graphical structure of probabilistic model and conditional probability. The probabilistic graphical model is required to extract information from the data collected from sensors. Probabilistic graphical model can be extensively used in IoT projects to ascertain data transmitted by the sensors [5]. This model portrays the joint probability distribution over a set of random variables. The term "Probabilistic Graphical Model" emphasis on three aspects: Characteristics of the input data, reliability on Bayesian conditioning for providing a support to upgrade the information and differentiation of evidential with causal modes of reasoning [6].

For Bayesian learning, some prior structures are required in order to get good results. The structures of Bayesian network depends on how the attributes are correlated. These graphical structures are used to represent knowledge about an uncertain domain. Bayesian network is made of nodes that represent conditional probabilities and an arc represents the dependencies of each node. Conditional dependencies in the graph can be obtained by various computational method involved in probability theory.

We take an example for prediction of a class label using a Naive Bayes classifier: The training data are in class label training instances from the Consumer database. Table 1 represents a training set T_1 of class labeled instances randomly selected from the Consumer database.

Table 1 Training set of consumer database

Id	Age	Salary	Rating	Buy iPhone
101	Senior	Low	Fair	Yes
102	Senior	Low	Excellent	No
103	Youth	Low	Fair	Yes
104	Youth	Medium	Fair	No
105	Middle	Medium	Fair	Yes
106	Senior	Low	Fair	Yes
107	Youth	Low	Fair	Yes
108	Senior	Low	Fair	Yes
109	Middle	Medium	Excellent	Yes
110	Youth	High	Fair	No
111	Youth	High	Excellent	Yes
112	Middle	Medium	Fair	Yes
113	Senior	Low	Fair	Yes
114	Middle	Medium	Excellent	No
115	Youth	Low	Fair	Yes
116	Senior	Low	Fair	Yes
117	Middle	Medium	Poor	No
118	Youth	Medium	Good	Yes
119	Senior	Medium	Poor	No
120	Middle	Medium	Excellent	Yes

Table 1 shows the value of each attribute is discrete. We can generalize the continuous variable. The table shows three age groups buying an iPhone and it has two discrete values—Yes and No. There are fourteen instances of Yes and six instances of No. A root node N is created for the instances T_1. The splitting criteria of instances can be obtained from information gain of every attribute and it can calculated using

$$\text{Information } T_1 = -\sum_{i=1}^{n} p_i \log_2(p_i) \tag{1}$$

where p_i can be estimated by $|c_i, T_1|/|T_1|$. Here, c_i represents Yes and No classes.

To obtain average amount of information of class Yes and No, a logarithmic function of base 2 is used. Each class represents information of the preposition. This information can be considered as entropy T_1. For the given example, we can classify an instance in T_1 using

$$\text{Information } (T_1) = -(14/20)\log_2(14/20) - (6/20)\log_2(6/20) = 0.881 \textit{ bits} \tag{2}$$

requirement for each attribute.

Information of each attribute can be calculated using the above formula. We can find the information gain of the credit rating using various probabilistic classification methods. We can also find prior probability of each attribute and estimate the

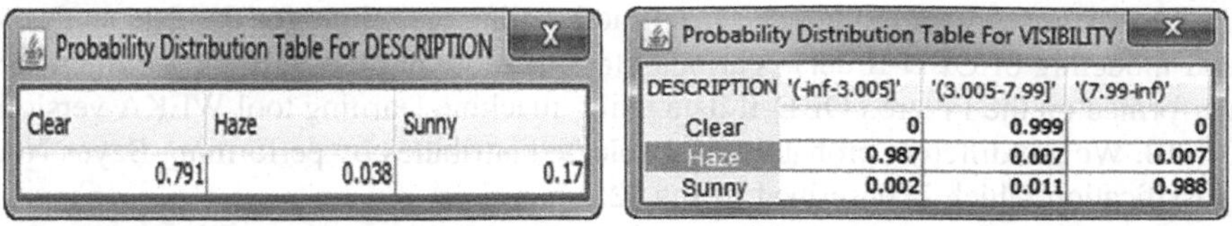

Probability Distribution Table For DESCRIPTION

Clear	Haze	Sunny
0.791	0.038	0.17

Probability Distribution Table For VISIBILITY

DESCRIPTION	'(-inf-3.005]'	'(3.005-7.99]'	'(7.99-inf)'
Clear	0	0.999	0
Haze	0.987	0.007	0.007
Sunny	0.002	0.011	0.988

Fig. 2 Probability of various attributes generated by Bayes Net

likelihood function. Classification refers to the task of predicting a class label for a given unlabeled attribute. Bayesian classifier uses Bayes theorem to predict class label and estimate joint priority density function for each class.

Therefore, the classification obtained by Naive Bayes predicts buys iPhone = yes for instance A1. In this paper, for modeling and data analysis purpose, high quality of data is required which is obtained from OBD-II sensor data through https://www.kaggle.com [7]. Kaggle organized a competition to use the telematics data for identifying the driver's score in terms of probability. The following steps are taken for analyzing and modeling the OBD-II data:-

(a) Data cleaning
(b) Data visualization
(c) Attribute selection
(d) Training, testing and 10-fold cross validation of data
(e) Averaging the data.

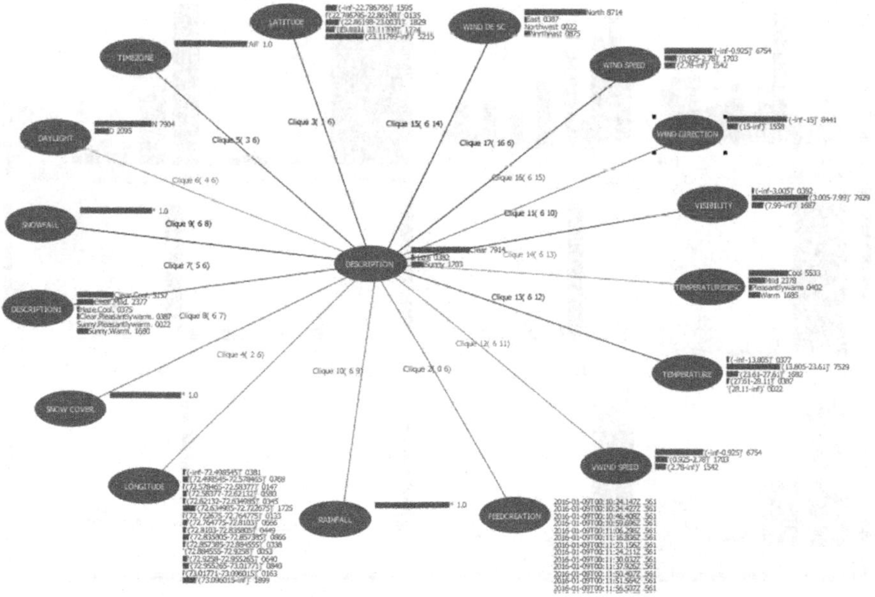

Fig. 3 Attributes involved in the data with its probability and cliques

We have used data mining and machine learning technique for the data analysis and modeling of OBD-II data. Various classification and clustering algorithms are performed on the Filtered OBD-II data using machine learning tool WEKA version 3.7.10. We got different probability for selected attributes by performing Bayes Net classification which is described in Fig. 2.

Every clique has received message from all neighbors. We can extract posterior probability of any variable X in a clique that contains X. The posterior probabilities of unobserved variables can be evaluated. Different queries of computation sharing are permitted while pruning of irrelevant variables is not permitted [8] (Fig. 3).

3 Data Analysis and Results

Some of the attributes of sensor data obtained of sensor data are GPS latitude and longitude, GPS speed, temperature, sky description, visibility, wind speed, etc. (Fig. 4).

We have performed an averaging method for generating the below statistics. Training and testing has been performed several times for more accurate results using 10-fold cross validation approach [9]. Using different classifications, we came to a conclusion that Bayes Net classifier generates best results among the other classifiers [10] (Table 2).

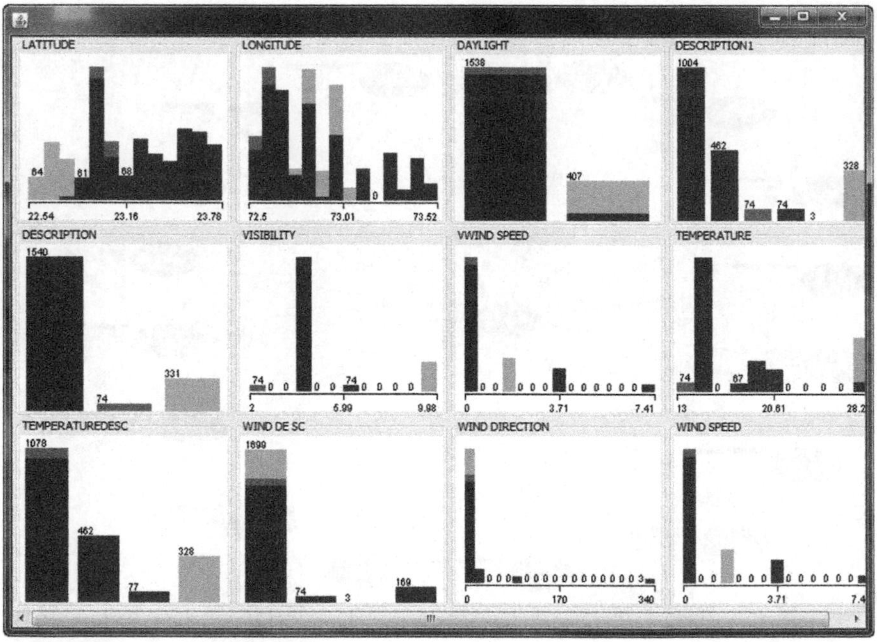

Fig. 4 Data visualization of attributes with values of training data set

Table 2 Comparison of Bayesian classifiers

Comparison of various Bayesian classifiers (Using open source Weka software 3.7.10)					
10-fold cross validation	NaiveBayes	Bayes Net	A1DE -F 1 -M 1.0	J48 -C 0.25 -M 2	Decision table -X 1 -S
Kappa statistic	0.9955	1	0.9992	0.9955	1
Mean absolute error	0.001	0	0.0011	0.0013	0.0043
Root mean squared error	0.0321	0	0	0.032	0.0153
ROC area	0.997	1	1	0.999	1
Time taken to build model (in s)	0.02	0.01	0.02	0.04	0.45

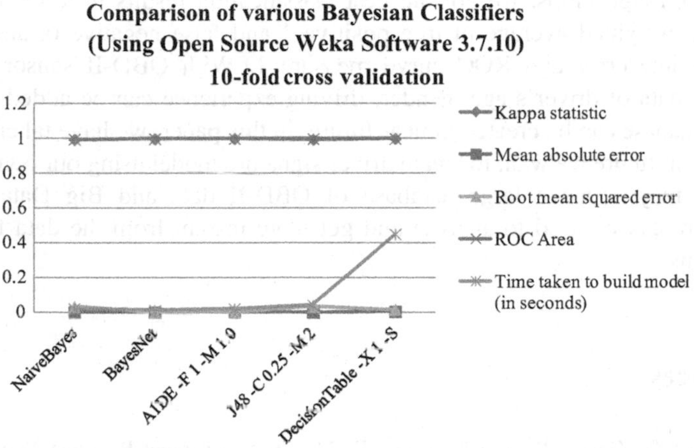

Fig. 5 Bayesian classifiers comparison

```
=== Detailed Accuracy By Class ===

                 TP Rate  FP Rate  Precision  Recall   F-Measure  MCC      ROC Area  PRC Area  Class
                 1.000    0.000    1.000      1.000    1.000      1.000    1.000     1.000     Clear
                 1.000    0.000    1.000      1.000    1.000      1.000    1.000     1.000     Haze
                 1.000    0.000    1.000      1.000    1.000      1.000    1.000     1.000     Sunny
Weighted Avg.    1.000    0.000    1.000      1.000    1.000      1.000    1.000     1.000

=== Confusion Matrix ===

     a     b     c   <-- classified as
 15400     0     0 |     a = Clear
     0   740     0 |     b = Haze
     0     0  3310 |     c = Sunny
```

<= 100% classification

Fig. 6 Bayes Net classifier result

From the results, the Bayes Net classifier takes 100% coverage of cases at 0.95% level of significance, 19450 numbers of instances are covered and weighted average of all the classes is 1. So, this classifier gives 100% classification of data as shown in Fig. 6. Detailed accuracy obtained by Bayes Net is explained in Fig. 5.

4 Conclusion and Future Work

We have applied different Bayesian classification and decision trees related machine learning algorithms to get knowledge about inferential probability. We have performed different data mining, machine learning classification and decision tree algorithms using open source tool WEKA version 3.7.10. Using Bayes Net classification algorithms, we got the best classification results in terms of probability with weighted average of true positive 1 and false negative 0, almost zero mean absolute error and ROC curve measure 1. With OBD-II sensor data, an additional data of driver's age, gender, driving experience can be added and augmented database can be created in near future. In this paper, we have taken the data from web, in future we want to create driver signature model using our own OBD-II data. We may create a large database of OBD-II data and Big Data Hadoop architecture can be used to analyze and get more insight from the data for future applications.

References

1. Haussler, D., Geiger, D., & Chickering, D. M. (1995). Learning Bayesian Networks: The Combination of Knowledge and Statistical Data. Boston: Academic Publishers.
2. Pearl, J. (1988). Probabilistic Reasoning in Intelligent Systems: Networks of Plausible Inference. San Francisco: Morgan Kaufmann Series in Representation and Reasoning.
3. Seewald, A., & Scuse, D. (2013). WEKA Manual for Version 3-7-10. Hamilton, New Zealand: The University of Waikato.
4. Conrady, S., & Jouffe, L. (2015). Bayesian Networks and BayesiaLab: A Practical Introduction for Researchers. Franklin: Bayesia USA.
5. Neapolitan, R. E. (2004). Learning Probabilistic Graphical Models. Prentice Hall Series in Artificial Intelligence.
6. B., & Nicholson, A. E. (2011). Computer Science and Data Analysis Series—Bayesian Artificial Intelligence. London: Chapman & Hall/CRC.
7. Kaggle Inc. (n.d.). Driver Telematics Analysis. Retrieved from Kaggle: https://www.kaggle.com/c/axa-driver-telematics-analysis.
8. Korb, K. B., & Nicholson, A. E. (2011). Bayesian Artificial Intelligence Second Edition. Boca Raton, Florida: CRC Press.
9. Timani, H., Shah, A., & Panchal, D. (2014). Knowledge discovery from music metadata using semantic web and open linked data. Proceedings of International Conference on "Emerging Research in Computing, Information, Communication and Applications". ERCICA.
10. Wijayatunga, W. (2007). Statistical Analysis and Application of Naive Probabilistic Graphical Model Classifier.

Design of Single Supply and Current Mirror-Based Level Shifter with Stacking Technique in CMOS Technology

Kamni Patkar and Shyam Akashe

Abstract The paper presents a new design for low power application of single supply and current mirror-based level shifter which is a 45 nm CMOS technique. Those recommended circuits use the benefits for Stacking technique with smaller leakage current and reduction in leakage power. The circuit is deliberate using 45 nm CMOS method. The modified stacking structure is more appropriate to system for high voltage supply. In such SoC's, level shifters play an important part in translating the signals from one voltage level to another. Single supply level shifter has been modified with two additional NMOS transistors. Another circuit, current mirror level shifter (CMLS) has been modified with three additional NMOS transistors. Performances of the proposed level shifters are compared in terms of power, noise, and leakage parameters. For single supply and current mirror-based level shifter, Supply voltage V_{ddH} is initialized as 0.7 V and V_{ddL} is initialized as 0.2 V.

Keywords CMOS · Level shifter · Single supply · Current mirror
Power · Noise

1 Introduction

With the increasing require of handheld strategy like cell phone, personal note books multimedia devices, and so on. Low power utilization has begun to be main design thought for VLSI (very-large-scale integration) circuits and system. Power utilization is essential because of the expanded bundling and cooling overheads as well as possible dependability problems. Therefore, the primary configuration goes for VLSI (very-large-scale integration) designer is to meet presentation supplies

K. Patkar (✉) · S. Akashe
ITM University, Gwalior, Madhya Pradesh, India
e-mail: patkar.kamni@gmail.com

S. Akashe
e-mail: shyam.akashe@itmuniversity.ac.in

D.K. Mishra et al. (eds.), *Information and Communication Technology for Sustainable Development*, Lecture Notes in Networks and Systems 10,
https://doi.org/10.1007/978-981-10-3920-1_34

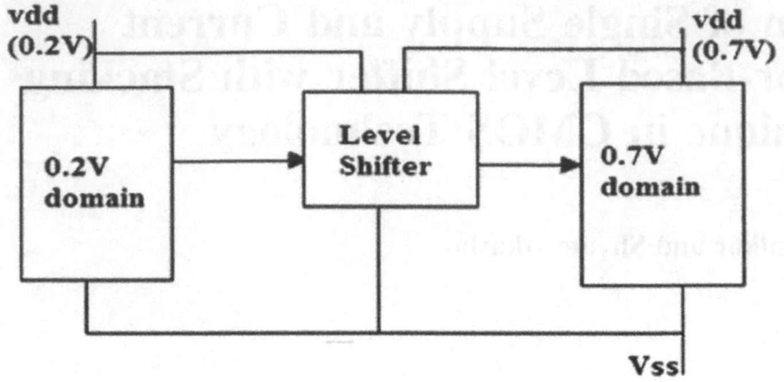

Fig. 1 Block diagram of level shifter cells

within power resources, delay, and area [1, 2]. In system on chip propose, various elements like analog, digital, passive elements be fabricated on a particular chip and requirements multiple voltages to acquire best presentation. Level Shifter unit is utilized to move a signal voltage arrangement from one voltage space to one more voltage space. This is required while the chip is working on various voltage spaces. A signal in one voltage space may have a voltage arrangement which is diverse to the signal in one more voltage spaces [3]. These variations in the voltage series can possibly undependable performance of the target area. Consequently Level shifter units are embedded in the voltage area crossings. Leakage power relies on upon the overall number of transistors and their operating situation in spite of their switching motion. The leakage power of CMOS circuits can be considered through the leakage current, while the threshold voltage and the source voltage's are not consider as similar voltages, transistor is at standby mode except here the outflow of current starts from drain to ground. With the advancement of innovation in leakage, power has turn out to be considerable component of entirety power dissipation [4]. The important part of the applied power is obsessive for the duration of process in the variety of leakage (Fig. 1).

Section 2 describes the circuit process of single supply and current mirror level shifter. Section 3 describes the modified single supply level shifters and current mirror level shifter and describe the operation of both the level shifters, Sect. 4 describes the imitation outcome, discussion and relative study the single supply, current mirror and modified level shifters, and finally the Conclusions have been drawn in Sect. 5.

2 Single Supply and Current Mirror-Based Level Shifters

Different outline for level shifters have been accounted for in literature with single and dual supply. The single supply level shifter licenses correspondence in the middle of parts excluding any additional supply pin. Single supply is appeared in

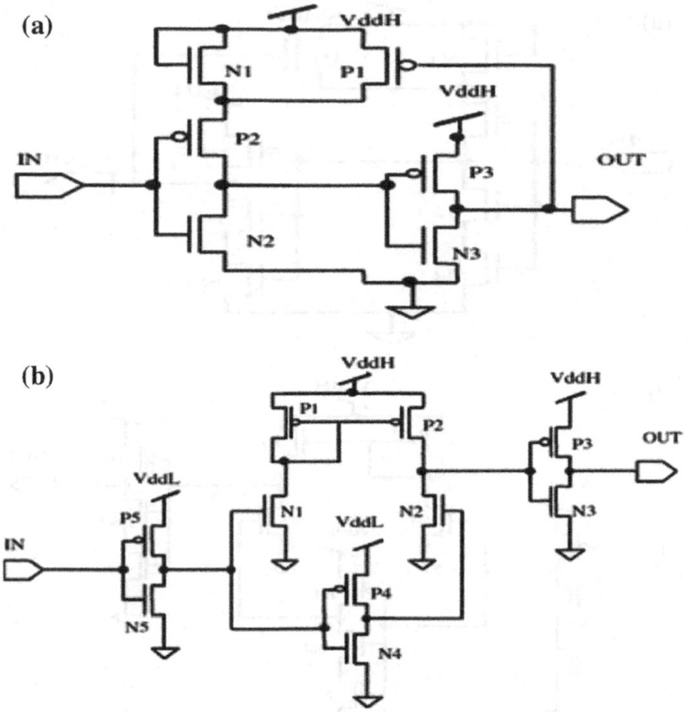

Fig. 2 Level shifter circuits **a** single supply **b** current mirror

Fig. 2a. Single supply level shifters have central focuses more than twofold supply as far as pin ascertain, blockage in controlling and general cost of the framework. Another point of interest of single supply is adaptable arrangement and steering in considerable configuration. Single supply level shifters dissipate lifted leakage power on account of ascend in leakage current when source supply level is lesser or V_{ddH} is raised than information supply level [5]. A current mirror peruses a current entering in a read-hub and mirror this current (with a sensible build variable) to a yield hub. Current mirror is appeared in Fig. 2b. The current mirror level shifter can switch an unlimited Sub edge level to above threshold subsequent to a high drain-to-source voltage of PMOS transistors supports the assembling of a consistent current mirror, which gives a successful ON–OFF current relationship on the yield hub. Lifted measures of calm current come to pass while the information voltage is sub threshold. This power consumption utilization compels the use of the current mirror level shifter. Input is connected to N1 and an inverted input is connected to the N2. MOSFETS P1 and P2 form a current mirror. The drain node power of N1 is mirrored at gate of P2.

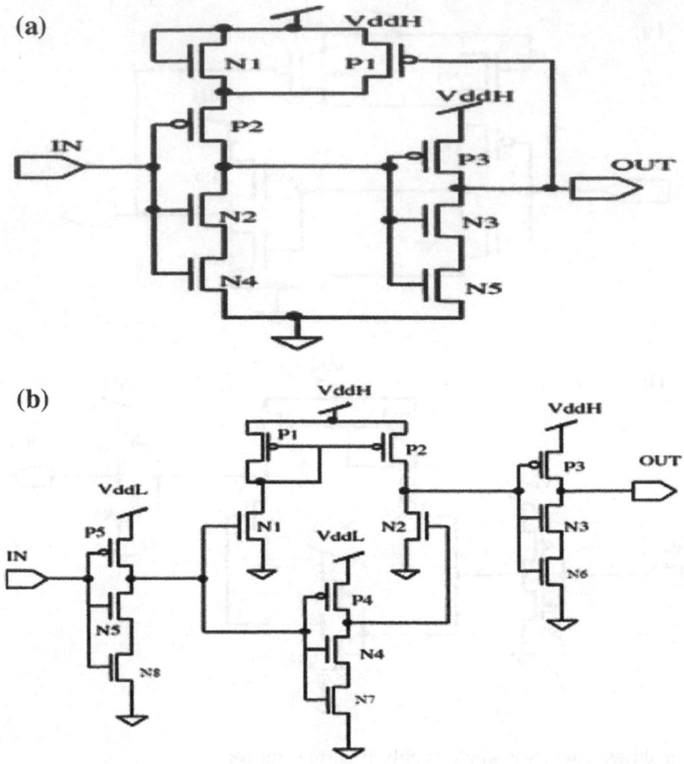

Fig. 3 **a** Single supply level shifter with stacking technique, **b** Current mirror-based level shifter with stacking technique

3 Modified Level Shifters

In present work, changes have been modified in existing level converter circuit's single supply and current mirror for improvement in average power, noise, and leakage parameters. Figure 2a illustrates modified single supply level shifter with stacking technique using two additional NMOS transistors. Current mirror-based level shifter with stacking uses three additional NMOS transistors as shown in Fig. 2b. Supply voltage V_{ddH} (high supply voltage) is initialized as 0.7 V and V_{ddL} (low supply voltage) is initialized as 0.2 V (Fig. 3).

4 Simulation Results and Discussion

The single supply level shifters and current mirror-based level shifters scheme are represented with their timing waveforms are shown in Fig. 4a and b for V_{ddH} = 0.7 V and V_{ddL} = 0.2 V. As discussed, while input IN is high, so output OUT is

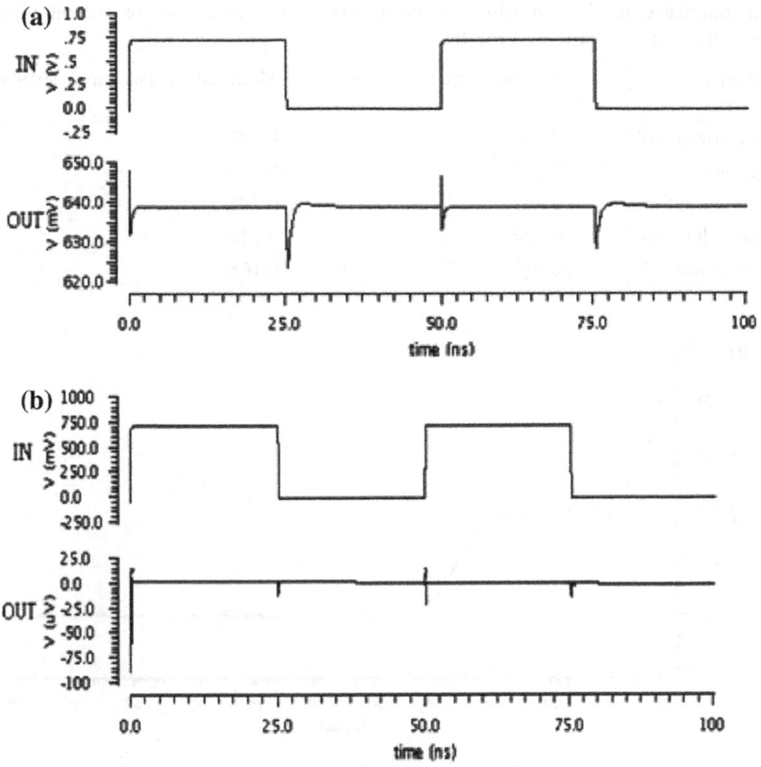

Fig. 4 Timing waveform of **a** single supply level shifter, **b** current mirror level shifter results with V_{DDL} = 0.2 V and V_{DDH} = 0.7 V

high and also while input IN is low, so output OUT is low in single supply level shifter. In case of current mirror-based level shifter, when input IN goes high, the output OUT also goes high. But when input IN goes low, there is distortion within the output OUT as shown in the figure. Due to this, the output signal OUT signifies some delay while shoots to high.

Modified level shifter circuits (Fig. 3a, b) with stacking technique have been designed and simulated in 45 nm technology with the help of cadence tool. Table 1

Table 1 Simulated result comparison between performance parameter of single supply and modified single supply level shifter

Performance parameters	Single supply level shifter	Modified single supply level shifter
Average power (nW)	18.89	7.849
Average noise (fV)	3.773	0.618
Leakage current (pA)	26.66	18.53
Leakage voltage (V)	2.657	0.438
Leakage power (pW)	11.98	7.086

Table 2 Simulated result comparison between performance parameter of current mirror and modified current mirror-based level shifter

Performance parameters	Current mirror level shifter	Modified current mirror level shifter
Average power (nW)	0.745	0.481
Average noise (fV)	0.212	0.140
Leakage current (pA)	1.467	0.798
Leakage voltage (V)	0.434	0.249
Leakage power (pW)	0.312	0.168

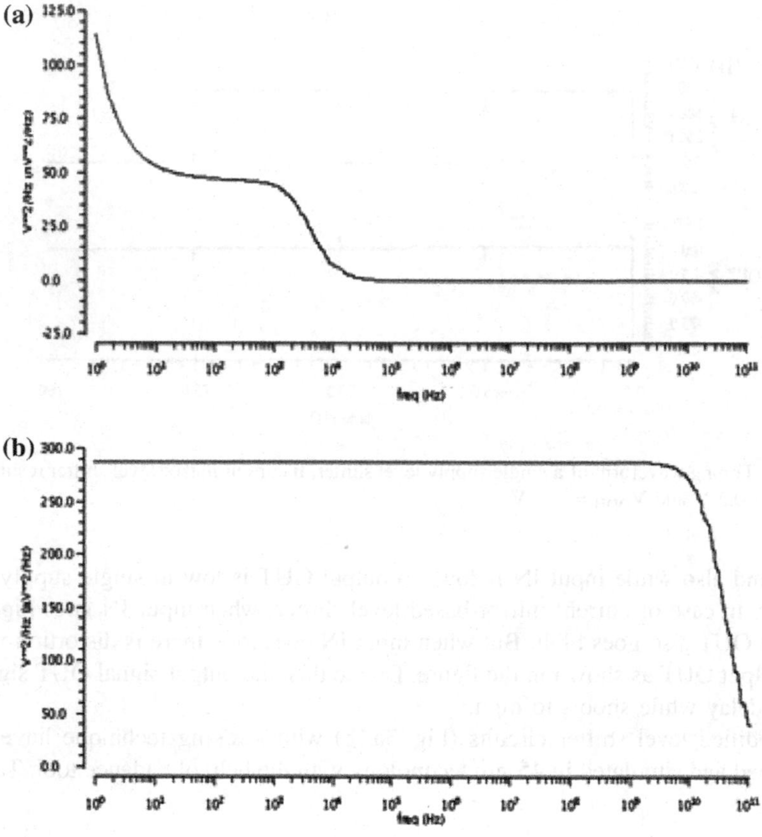

Fig. 5 Noise figure of **a** single supply level shifter, **b** current mirror level shifter

shows the results comparison between performance parameter of single supply and modified single supply level shifter and Table 2 shows results comparison between performance parameter of current mirror and proposed current mirror-based level shifter. Modified single supply level shifter gives average power of 7.849 nW as compared to 18.89 nW and average noise of 0.618 fV as compared to 3.773 fV with

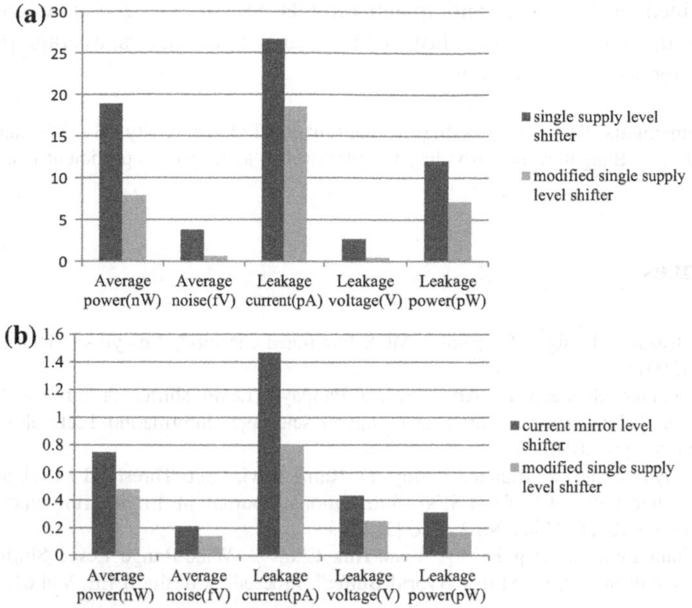

Fig. 6 Comparative analysis of **a** single supply and modified single supply level shifter, **b** current mirror and modified current mirror-based level shifter with stacking technique

existing single supply circuit. Modified current mirror-based level shifter gives average power of 0.481 nW compared to 0.745 nW and average noise of 0.140 fV as compared to 0.212 fV with existing circuit. Results show that average power and average noise has been reduced in modified circuits with application of stacking technique. Leakage parameters of existing and proposed circuits also have been obtained and shown in Tables 1 and 2. Results show that two modified circuit's shows improved performance in terms of average power and noise with a little conciliation in leakage parameter. Figure 3a shows correlation between execution parameter of single supply and modified single supply level shifter, and Fig. 3b indicates current reflect and modified current mirror-based level shifter (Figs. 5 and 6).

5 Conclusion

In current paper two circuits of level shifters namely modified single supply and modified current mirror have been presented. Modified single supply level shifter gives power of 7.849 nW as compared to 18.89 nW and average noise of 0.618 fV as compared to 3.773 fV for single supply circuit. Second Proposed circuit's current mirror shows average power of 0.481 nW compared to 0.745 nW and average noise of 0.140 fV as compared to 0.212 fV for current mirror circuit. Results are also obtained on behalf of anticipated circuits and it has been practical proportionally to facilitate with small allowance in leakage parameters, average power and noise has reduced considerably.

Acknowledgements The author is highly thankful to ITM University, Gwalior and Cadence Virtuoso Design, Bangalore for providing the platform so as to get the proficient result.

References

1. Behzad Razavi, "Design of Analog CMOS Integrated Circuits", Newyork: McGraw-Hill, 2nd edition (2001).
2. Manoj Kumar, Sandeep K. Arya, Sujata Panday, "Level shifter design for low power applications", International journal of computer science & information Technology (IJCSIT) Vol.2 No.5, Oct-(2010).
3. J. Chaitanya Verma, R. Ramana Reddy, D. Rama Devi, "Sub Threshold Level Shifters and Level Shifter with LEC for LSI's", International Journal of Engineering and Advanced Technology (IJEAT) Vol.4 No.2, Dec-(2014).
4. Shien-Chun Luo, Ching-ji Huang, Yuan-Hua Chu, "A Wide-Range Level Shifter Using a Modified Wilson Current Mirror Hybrid Buffer", Proceeding of the IEEE, Vol.61, No.6, Jun-(2014).
5. Lanuzza M., Corsonello P., Perri S., "Low Power Level Shifter for Multi-Supply Voltage Designs", IEEE Proceeding on Trans. Circuits System, Vol.59, No.12, pp. 922–926, (2012).

Privacy Preserve Hadoop (PPH)—An Implementation of BIG DATA Security by Hadoop with Encrypted HDFS

Prashant Johri, Sanchita Arora and Mithun Kumar

Abstract As data is growing exponentially than linearly, the rising abuse of large data set emphasizes the need to preserve and protect the Data. Hadoop, a big data solution, has increasingly become popular and adopted by most of the trades. However, Hadoop by default does not contain any security mechanism. Though, it does not support data encryption which makes data privacy and security becomes a cardinal concern. The generally extensively compliant methodology of preservation and protection of data is through cryptography algorithms which is computationally intensive. Exploiting cryptography with apportioning the processing with MapReduce framework will improve the security of Hadoop. This paper presents two applications which disseminate the cryptographic process among MapReduce jobs. The first application will handles encryption of an input file that is resides in HDFS and second application will handle decryption of encrypted input file. Our experimental results show the comparison between the two cryptographic algorithms.

Keywords Hadoop · Data security · Cryptography · Encrypted HDFS MapReduce

1 Introduction

1.1 Big Data—A Imminent Frontier of Growth, Potency and Competition

The encroachment in tools and techniques has given birth to the Big Data which desire more mature data processing and storage systems. Gartner and company in 2010 have entitled "Information will be the oil of twenty-first century and analytics

P. Johri (✉) · S. Arora · M. Kumar
School of Computer Science and Engineering, Galgotias University, Greater Noida, India
e-mail: johri.prashant@gmail.com

D.K. Mishra et al. (eds.), *Information and Communication Technology for Sustainable Development*, Lecture Notes in Networks and Systems 10,
https://doi.org/10.1007/978-981-10-3920-1_35

will be the combustion engine" and this is so true here because in today's world data is entirety and entirety is data. Beneath the unstable escalation of data in digital world big data is used to depict the substantial amount of data. Additionally big data gives new visions and understanding to business by altering the way of turns out the new value and understand the deeply knowledge of hidden values, from large datasets. Moreover, big data has other important characteristics other than volume such as velocity, variety, value, veracity and complexity. At present when the influence of big data has been recognized by almost all trades and industries, people still have different perspective on its definition. In general, big data is the amount of data that could not be reasonable to stored, processed and managed with existing tools. Big data usually includes piles of unstructured data that requires more real time analysis. Although big data distended new opportunities and advantages for realizing new benefits and values from large amount of data but at the same time incurs new challenges such as how to integrate, store, manage and process such immense amount of data.

1.2 Hadoop—A Big Data Solution

Hadoop [1] is a data and I/O intensive general purpose open source framework which consists of two prominent parts, MapReduce model, which further consist of two logical functions Mapper and Reducer, and Hadoop Distributed File System (HDFS) [2]. Its distributed file system facilitates storing large data sets with high fault tolerance and throughput. While MapReduce is a programming model which assigns tasks to each of nodes in the cluster and process data. There are two types of nodes in the Hadoop, Namenode and Datanode which are responsible for file locations and replications and block creation and takes orders from Namenode, respectively [1]. Namenode is a master process and runs only on a single node while Datanode is a slave process and runs on all nodes in the clusters. In terms of MapReduce there are two daemons Job-Tracker and Task-Tracker which are master and slave, respectively. Job-Tracker manages MapReduce jobs, task failures, restarts job on another node and speculative execution. Task-Tracker creates individual map and reduce tasks and report task status to Job-tracker.

1.3 MapReduce—A Parallel Programming Model

MapReduce [3] is an essence part of Hadoop. It is a programming model which is based on the Lisp commands, Map and Reduce, proposed by Google in 2004. Employment of this model is generally used for processing of elevated datasets. User stipulates a map function that processes data in terms of key-value pair and generates a set of intermediate key-value pairs as a result. On the other hand a reduce function merges all the intermediate values coupled with the same

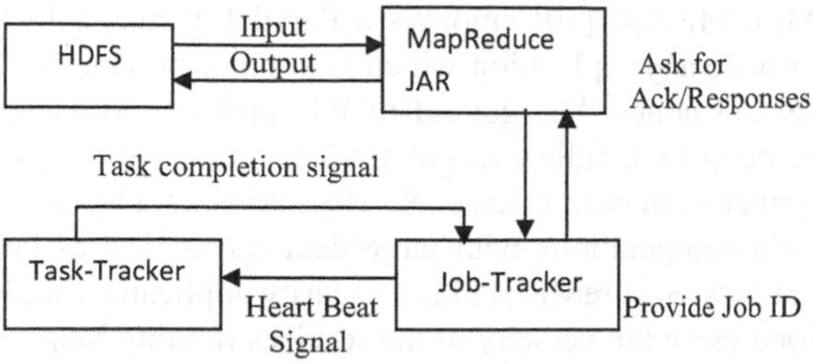

Fig. 1 Working of MapReduce

intermediate key, generated by map function. MapReduce jobs are parallel in nature and executed on a large clusters of commodity machines. Hadoop run-time takes care of details of partitioning the input data, scheduling the execution among the nodes in cluster, handling nodes failures and managing the inter-node communication [3] (Fig. 1).

2 Related Work

There have been a huge number of file systems which is cryptographic-based and contemplate data security and confidentiality. General Parallel File System (GPFS), currently known as IBM Spectrum Scale [4], is a high performance clustered file system which can be deployed in shared nothing and shared disk distributed modes. In comparison with HDFS, it is less fault-tolerant because it does not replicate data over different nodes. So data can be temporarily lost which is more serious event. iRODS [5] is highly customizable system developed by DICE (Data Intensive Cyber Environments) research group. It has two main components which are iCat server and several iRODS server which is responsible for management metadata and handles queries upon these metadata and storage resources, respectively. These two components are forms a zone. In comparison to other distributed systems, iRODS relies on the storage resources such as local file system; it does not deploy its own file system. Ceph [6] is an open source distributed file system like HDFS. But unlike HDFS, to certify scalability, Ceph supports a distributed dynamic metadata management using Meta Data Cluster (MDS) and stores data and its metadata in object storage devices. Ceph and HDFS are somewhat same but performance ratio of Ceph is less as compared to HDFS. GlusterFS is another distributed open source file system which is based on client server design which makes it different from other distributed file system. GlusterFS stores data and metadata in several devices attached to different servers [7]. The set of devices in which data and metadata is stored called volume and in volume data and metadata is distributed across several devices. Lustre [8] is a centralized distributed file system available for Linux. It is different from other DFSs in the sense that it does not provide any replication of data and metadata instead it relies on independent software. The Message Passing Interface (MPI) [9] is a tool which is very valuable when handling parallel computing. MPISec is an implementation of cryptographic services via

MPI. MPISec [10] employs a Parallel Virtual File System (PVFS) and does not provide any replication which is a major bottleneck. Hadoop replicates data across several nodes thus get rid of this problem. Biodoop [11], a framework built on Hadoop to handle computationally intensive task of bioinformatics and handles immense amount of data. So Hadoop is an adaptable framework for nuisance with light computations with large datasets as well as heavy computations with small datasets. SecureMR [12] is a security application implemented using Hadoop which focuses on the veracity of the services running MapReduce. The application used to analyzes the activities of datanodes to assure the virtue of services running on Hadoop. But rather than focus on integrity, it does not focus on confidentiality. The cryptographic distributed file system is an excellent solution but it does not fit well with MapReduce model because if cryptography is implemented at the file system level it surpasses the advantages of MapReduce Model. This is the grounds that our solution is implemented at application level.

3 Proposed Work

3.1 Hadoop Application Behaviour

When Hadoop application starts up, the Namenode situates which node, on its local storage, have the data replicated. The input file is divided into many chunks which are termed as Input Splits. Then all splits are passed individually to application's mapper class. Mapper sends each record in split at a time to map function. All the output from map function is stored in Output Buffer.

The reducer class is responsible for shuffle/sort and merges the each split that is executed by mapper. We can set the number of reducer that reduces the output splits given by mapper and hence the final output is produced in number of files depend upon number of reducer.

3.2 Privacy Preserve Hadoop Application Behaviour

In our experiment we use the password-based cryptography in order to certify that the keys will be the same for every mapper class without using key management.

The first application will consist of a driver, mapper and reducer class to implement the encryption process. The driver class has the configuration methods needed to run the mapper and reducer. The mapper class will run a stagnant block of code that will set up the encryption key and convert the given input file to cipher stream. The same map function will be called for each chunk of input and give the output in the form of intermediate key-value pair. Reducer on the other hand reduces the intermediate key-value pairs generated by the map function and gives a single file.

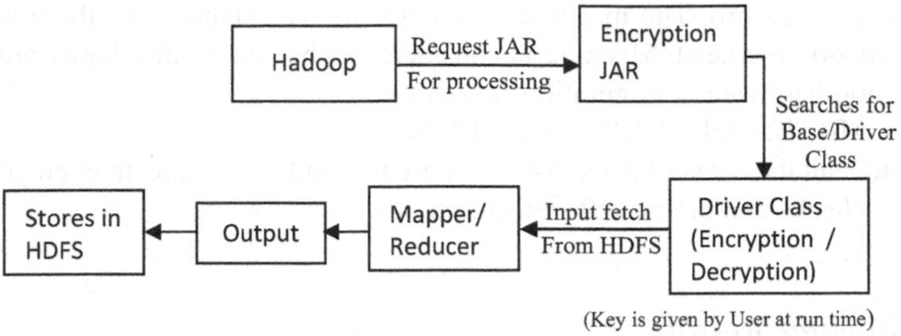

Fig. 2 Cryptography algorithm working

The second application will consist of driver, mapper and reducer class to implement the decryption process. The driver will act as same that of encryption driver class. The mapper will run a static block of code that set up the decryption key and ciphers for decryption. The input file will be an encrypted single file that is a result generated by a first application. The map function will go through each line and decrypt using the ciphers setup in the stagnant block in the map function. The end result will be plaintext that is same as the original input (Fig. 2).

4 Experiments

4.1 Experimental Design

The experiment is intended to exhibit the difference in runtime between cryptography algorithms in Hadoop framework environment. We used password-based encryption in order to ensure that key would be the same during every check without the need of key management. For cryptography algorithms with Hadoop environment check, AES and DES algorithm of cryptography and five different file sizes were used. For algorithm, two programs were created; one for encrypt the data and another for decrypt the data. The checks consist of running each program for every file size. By experiment we want to show the overall runtime of different cryptographic algorithms with five different datasets by distributing the algorithm in MapReduce Framework.

4.2 Experimental Setup

Machine with Single node Hadoop framework Configuration: Sony E Series VPCEH18FG brand computer with a 2.3 GHz, 2 core, Intel i5-2410 M processor, 4 GB RAM and 500 GB HDD, running Ubuntu 14.04.3 (64 bit version) on

VMware workstation 12 pro. The machine is functioning as a standalone; there is no cluster or network overhead. Machine is considered to be commodity hardware due to its common hardware configuration and price.

File Sizes: 1 MB, 128 MB, 1 GB, 2 GB, 4 GB.

Data Collected: all the data collected for use in figures and graphs are an average of experimental checks and are articulated in seconds.

4.3 Experimental Results

Figure 3 shows the average runtime of AES cryptographic algorithm running on a Machine with Single node Hadoop framework Configuration. The runtime increases exponentially when data size increases. The algorithm's behaviour is undoubtedly demonstrated even with a limited number of experiment runs and as a file sizes get larger, the performance becomes more and more clear. Table 1 lists the data that is used for Fig. 3.

Figure 4 shows the DES cryptographic algorithm running on a same machine used in previous. The runtime increases as the file size increases. The figure shows that the Hadoop setup is an idyllic solution for large files.

Figure 5 compares the performance of AES and DES encryption runtime on machine with single node Hadoop environment. DES outperforms the AES for 1 MB or smaller files. The single node Hadoop environment extensively enhances the performance of cryptographic algorithms even multi node setup enhances the performance more uncompromisingly because of more computational power. The performance could be gained from the single node to multimode as the file sizes increases. Table 2 lists the data sizes with its run-time DES on single node Hadoop environment.

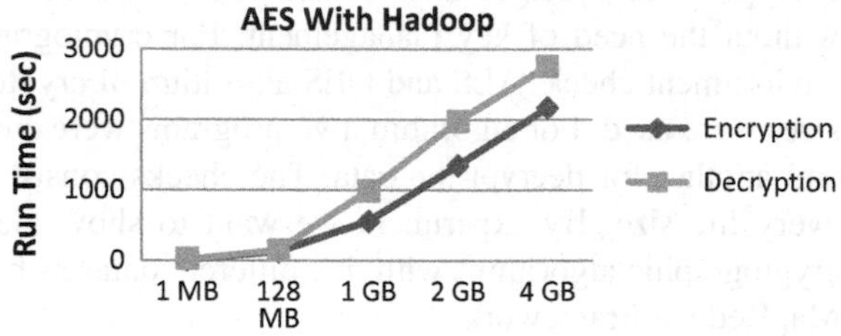

Fig. 3 AES encryption/decryption runtime of different datasets on machine with single node Hadoop environment

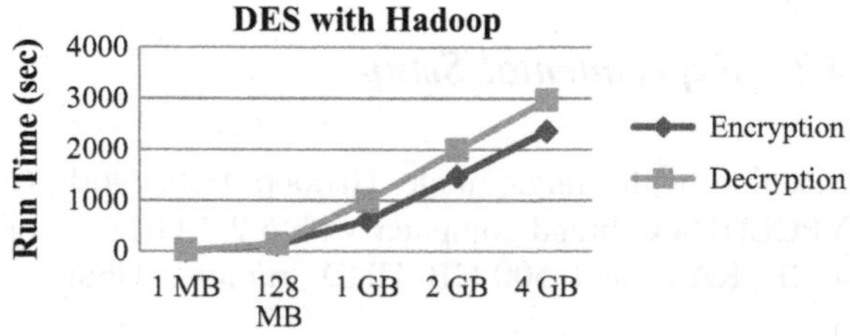

Fig. 4 DES encryption/decryption runtime of different datasets on machine with single node Hadoop environment

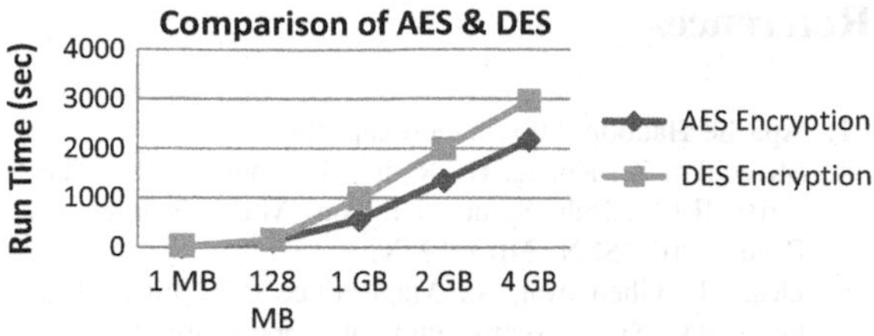

Fig. 5 Comparison of encryption runtime of AES and DES in sec

Table 1 Average runtime of AES encryption/decryption on single node Hadoop environment in sec

File size	Encryption time	Decryption time
1 MB	10.063	15.039
128 MB	127.03	150.043
1 GB	560.068	992.097
2 GB	1349.061	1978.049
4 GB	2167.031	2760.066

Table 2 Average runtime of DES encryption/decryption on single node Hadoop environment in sec

File size	Encryption time	Decryption time
1 MB	9.067	14.013
128 MB	128.07	152.045
1 GB	600.054	995.086
2 GB	1467.098	1998.089
4 GB	2361.041	2970.088

5 Conclusion and Future Work

The results from the experiments show Encryption/Decryption runtime of AES and DES. However, AES is outperformed the DES on single node Hadoop environment as expected. Even Multi node set up of Hadoop significantly improves the runtime more drastically. This solution is optimal for organizations with more sensitive data and as well as data sixe is large. Because Hadoop does not require costly hardware, so every company can easily implement it. As data is encrypted in HDFS by our solution, so organization does not worry about where the data is resides. Privacy Preserve Hadoop provides an option to securely manage and access the data on Hadoop that is scalable and cost effective. We have used the password-based encryption to decrease the overhead of key management. In future we could implement the key management server different from HDFS which will manage only keys. We could also implement multi node Hadoop setup with larger files and more cryptography algorithms.

References

1. Apache Hadoop. http://hadoop.apache.org/.
2. Shvachko, K., Kuang, H., Radia, S., Chansler, R.: The Hadoop Distributed File System. In: 2010 IEEE 26th Symposium on Mass Storage Systems and Technologies (MSST). Page: 1–10, ISSN: 2160–195X.
3. Dean, J., Ghemawat, S.: MapReduce: Simplified Data Processing on Large Clusters. In: OSDI'04: Sixth Symposium on Operating System Design and Implementation, San Francisco, CA, December, 2004.
4. Schmuck, F., Haskin, R.: GPFS: A Shared-Disk File System for Large Computing Clusters. In: Proceedings of the FAST'02 Conference on File and Storage Technologies. Monterey, California, USA: USENIX. pp. 231–244. ISBN 1-880446-03-0. Retrieved 2008-01-18.
5. Fortner, B., Ahalt, S., Coposky, J., Fecho, K., Heinzel, S., Krishnamurthey, A., Moore, A., Rajasekar, A., Schmitt, C., P., Schroeder, W.: Control Your Data iRODS: integrated Rule-Oriented System. In: Morgan and Claypool Publishers (2010), Volume 2, No. 2, The RENCI White Paper 2014.
6. Weil, S.,A., Brandt, S.,A., Miller, E.,L., Long, D.,D.,E., Maltzahn, C.: Ceph: A scalable, high-performance distributed file system. In: Proceedings of the 7th Symposium on Operating Systems Design and Implementation (OSDI), 2006 307–320.
7. Gluster: An Introduction to Gluster Architecture (2011).
8. Lustre: A scalable, high-performance file system. Cluster File Systems Inc. white paper, version 1.0 (Nov 2002).
9. LeBlanc, T., Subhlok, J., Gabriel, E.: A High-Level Interpreted MPI Library for Parallel Computing in Volunteer Environments. In: Proceedings of the 2010 10th IEEE/ACM International Conference on Cluster, Cloud and Grid Computing (CCGRID'10). IEEE Computer Society, Washington, DC, USA, 673–678.
10. Prabhakar, R., Patrick, C., Kandemir, M.: MPISec I/O: Providing Data Confidentiality in MPI-I/O, 9th IEEE/ACM International Symposium on Cluster Computing and the Grid, vol., no., pp. 388–395, 18–21 May 2009.
11. Leo, S., Santoni, F., Zanetti, G.: Biodoop: Bioinformatics on Hadoop. In: ICPPW, International Conference on Parallel Processing Workshops, 2009. pp. 415–422.
12. Wei, W., Du, J., Yu, T., Gu, X.: SecureMR: A Service Integrity Assurance Framework for MapReduce. Annual Computer Security Applications Conference, 2009., vol., no., pp. 73–82, 7–11 Dec. 2009.

Design a Circular Slot Patch Antenna with Dual Band Frequency

Karishma Patkar, Neelesh Dixt and Anshul Agrawal

Abstract Locale of research work indicates some basic concepts related to patch antenna. The techniques for increasing bandwidth of circular patch antenna are explained with other parameters. Patch antenna is basically used for wireless communication systems. Design a circular slot patch antenna of dual band frequency. Each type of antenna is good in their properties and usage. Antennas are those backbones also almost all that in the wireless communication without which the world could have not arrived at in this period of technology. The proposed micro-strip patch antenna has FR4 lossy as a dielectric substrate with thickness of 1.6 mm and relative permittivity ε_r is 4.3. The simulation results of directivity, gain, and return loss of designed patch antenna are determined successfully. It is designed the dual band frequency having a return loss −30 dB at 1.5 GHz and second one is −40 dB at 2.5 GHz, analyzed in CST software.

Keywords Circular micro-strip patch antenna · Dielectric substrate
Gain · Directivity · Return loss · CST software

1 Introduction

Micro-strip Antennas are accepted in the starting of 1970s [1]. Antenna would an irreplaceable and only modern culture, serving as the link between man and his location extending to the external space. Antennas need been around for more than a century now, Also appear will bring an infinite variety, all operating in the same fundamental principles for electromagnetic. Micro-strip Patch antenna is diminutive of antenna implemented for wireless solicitation due to their numerous benefits such as short profile, minimum weight, and relaxed assemble. Micro-strip patch antenna comprises limited restrictions such as low profile, minimum gain, deprived polarization, and minimum proficiency. The micro-strip patch antenna entails in

K. Patkar (✉) · N. Dixt · A. Agrawal
ITM University, Gwalior, Madhya Pradesh, India
e-mail: patkar.karishma21@gmail.com

D.K. Mishra et al. (eds.), *Information and Communication Technology for Sustainable Development*, Lecture Notes in Networks and Systems 10,
https://doi.org/10.1007/978-981-10-3920-1_36

accompanying substantial which are copper and gold. Between the published micro-strip patch antenna of numerous contours rectangular, circular and triangular patch antenna are accessible.

An enormous of micro-strip patches use in wireless solicitation has been established. In assessment to patch features, the antennas having slot alignment establish superior appearances comprising broader bandwidth, few conductor losses and enhanced isolation.

A feed procedure is a route to provide radio effect within antenna assembly. Conducting and non-conducting substantial is utilized for antenna strategy. In conducting scheme, the feed line is unswervingly related to the power of RF like as micro-strip and coaxial line. In assessment to the micro-strip patch, the conducting strip has lesser magnitude. A plane erection is acquired if the feed is engraved on equivalent substrate [1].

The parameters of the antenna are bandwidth, emission pattern and directivity is resolute with CST software and is used to propose a high frequency array of the antenna. This software is more valuable for regularity application. Electromagnetic analysis is performed using CST MICROWAVE STUDIO.

2 Antenna Designing

Structure of recommend circular micro-strip patch antenna is represented in Fig. 1 where radius is calculated by

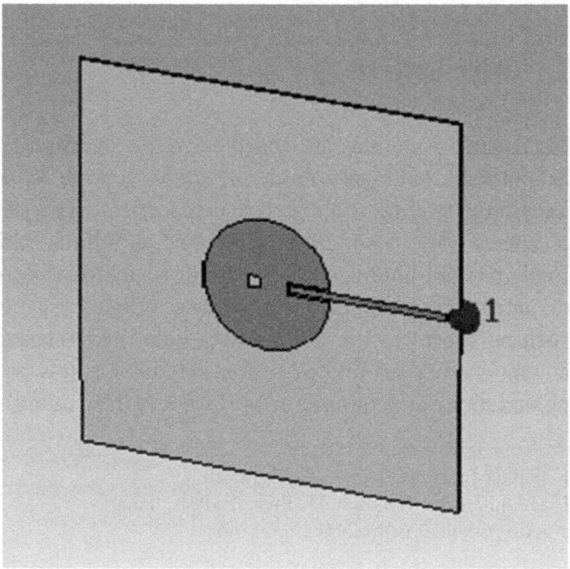

Fig. 1 Design of circular micro-strip patch antenna

$$a = \frac{F}{\left\{1 + \frac{2h}{\pi \varepsilon_r F}\left[ln\left(\frac{\pi F}{2h}\right) + 1.7726\right]\right\}^{1/2}} \quad (1)$$

$$F = \frac{8.791 \times 10^9}{f_r \sqrt{\epsilon_r}} \quad (2)$$

where,

a Radius of circular patch antenna
F .
ε_r Permittivity of antenna
h Thickness of the substrate.

3 Simulation Results

This circular micro-strip patch antenna was simulated on CST Software. The radiations were measured with the help of CST software in far field. The RF signals were set of connections form 1–3 GHz, simulated adaptively. The simulation results are shown in figure; in Fig. 2 demonstrates the return loss against frequency range in 1–3 GHz for the proposed dual band circular patch antenna at different frequencies. The response of this design of first return loss is −30 dB at the resonant frequency 1.489 GHz and second return loss is −40 dB at the resonant frequency. Circular patch antenna has dual band characteristics, showing at frequencies 1.5 and 2.5 GHz. Figure 3 shows the ratio for circular patch antenna (Figs. 4, 6).

Gain and radiation pattern with dual band frequency are shown in Fig. 3 or 5, respectively, for circular patch antenna. From Fig. 5 a maximum gain of 2.87 dB is reported (Fig. 7).

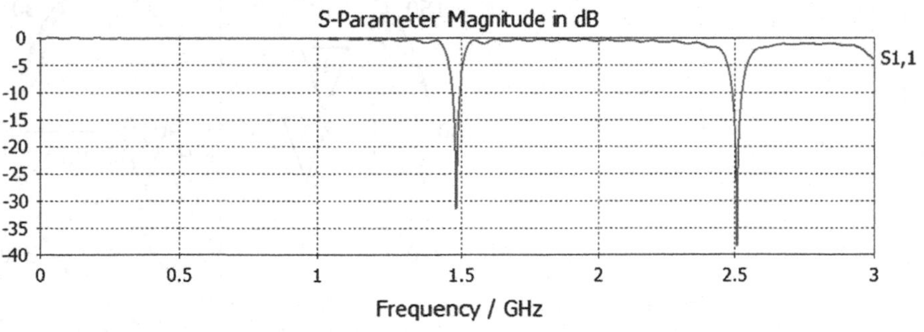

Fig. 2 Return loss simulation

Fig. 3 Radiation pattern Simulation

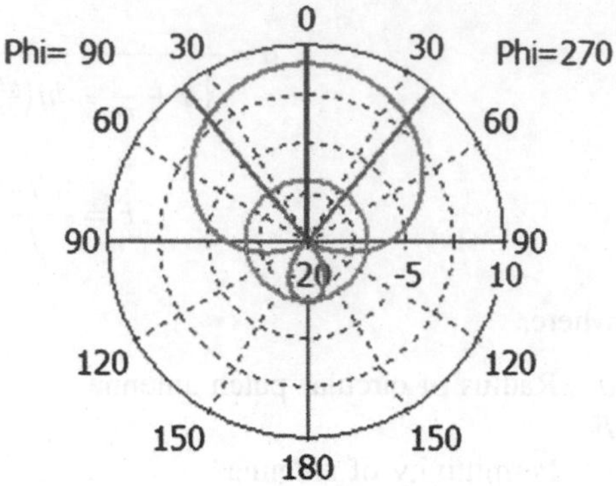

Fig. 4 Smith chart simulation

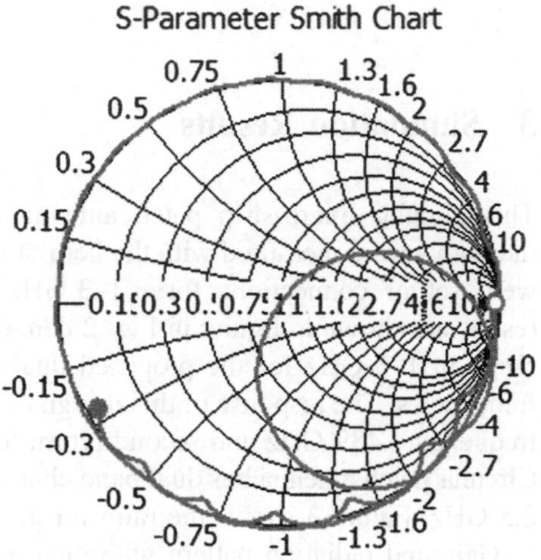

Fig. 5 Polar plot simulation

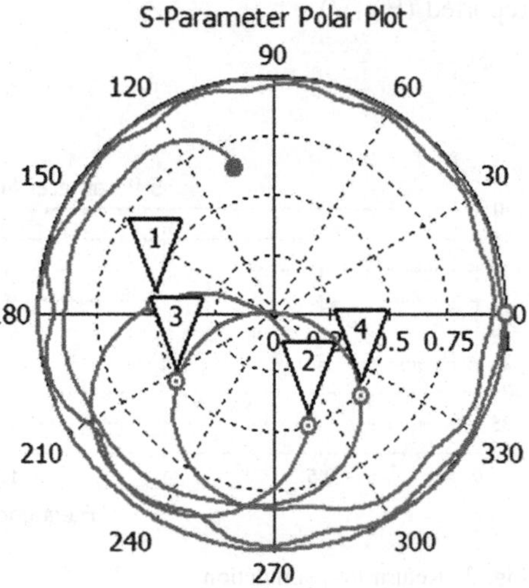

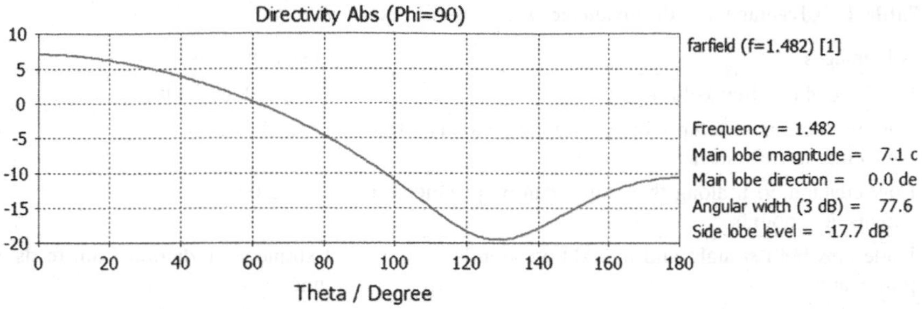

Fig. 6 Cartesian plot simulation

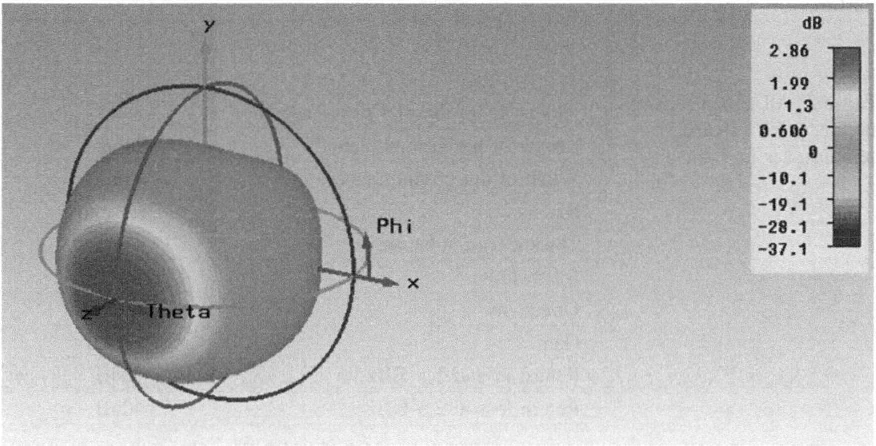

Fig. 7 Gain of circular patch antenna with dual band frequency

4 Conclusion

The design of circular patch antenna with dual band frequency is presented in this paper, first return loss is −30 dB at 1.5 GHz and second one is −4 dB at 2.5 GHz and results were analyzed (Table 2). Then the directivity of this antenna is 7.05 dBi and gain is 2.87 dB in CST studio software. The results confirm good performance of the dual bands antenna design as shown in Fig. 2. Antenna parameters such as return loss, gain, and directivity are calculated with good results. In this paper enhancement in return loss in large amount this will give the maximum output.

Table 1 Advantage and disadvantage of micro-strip patch antenna

Advantages	Disadvantages
Light weight & low volume	Narrow bandwidth
Low profile planar setup which can be effectively made conformal to host surface	Low efficiency
Low creation cost, along these lines can be produced in expansive amounts	Low gain
Underpins both straight and in addition round polarization.	Extraneous radiation from feeds & junctions
Can be effortlessly incorporated with microwave integrated circuits (MIC's)	Poor end fire radiator except tapered slot antennas
Fit for double and triple recurrence operations	Low power handling capacity
Mechanically vigorous when mounted on unbending surfaces	Surface wave excitation

Table 2 All dimensions of circular patch antenna is demonstrates in Table 1

Parameter for circular patch antenna	Values
Length of the ground plane	80 mm
Width of the ground plane	80 mm
Radius	28.5 mm
Thickness of substrate	1.6 mm
Bandwidth	0.018 GHz
Directivity	7.056 dBi
Gain	2.85 dB
Return loss at 1.5 GHz	−30dB
Return loss at 2.5 GHz	−40dB

References

1. Balanis, Constantine, "Antenna Theory Analysis and Design", John Wiley & Sons Ltd (2005).
2. NehaParmar, Manish Saxena, KrishnkantNayak, "Review of Micro strip patch antenna for WLAN and Wimax application", International Journal of Scientific & Engineering Research, pp. 168–171, January (2014).
3. Saurabh Jain, Vinod Kumar Singh, Shahanaz Ayub, "Design of slotted micro strip patch antenna having high efficiency and gain'', National conference on synergetic trends in engineering and technology, pp. 21–26, November (2014).
4. Sanjay Singh, AmitSaini, Kishor Chandra Arya, RachnaArya, "Design of a Tri Band H-Shaped Micro strip Patch Antenna for L, C and X-Band Application", IJARCCE Trans., vol. 4, pp. 9–11, June (2015).
5. Rahul Tiwari, Dr.SeemaVerma,"Inverted l-slot Wideband Rectangular Micro strip Patch Antenna", International Journal of Advanced Technology & Engineering Research, vol 4, pp. 118–123, January (2014).

A Methodology to Segment Retinal Vessels Using Region-Based Features

Vinita P. Gangraj and Gajanan K. Birajdar

Abstract Analysis of the retinal blood vessels has become remarkable area of research in biomedical field. This paper presents fundus image blood vessel segmentation approach using region-based features. In the pre-processing phase, the input fundus image is segmented as major vessel and minor vessel region. Further, to enhance segmentation accuracy region-based features are extracted from minor vessels by applying morphological operations. Fuzzy entropy measure is used to select the relevant features and for classification, a k-NN classifier is employed. The proposed algorithm is evaluated using two openly available data sets DRIVE and CHASE_DB1. The method presented is independent of training samples and achieves 96.75% of classification accuracy.

Keywords Diabetic retinopathy · Morphology · Region based feature Fundus images · Fuzzy entropy measure · k-NN classifier

1 Introduction

Diabetic eye disease is the leading pathological cause of blindness in adults. Abnormal blood glucose level is the main cause of diabetic eye disease, that upsurges permeability of the vessel which later, leads to retinal rupture. Diseased patients reckons no symptoms until loss of optical vision. Study of the blood vessels from fundus images of retina has been extensively used by the medical masses for detecting complications instigated owing arteriosclerosis, hypertension, glaucoma, cardiovascular disease, diabetic retinopathy (DR) and stroke [13]. Blood vessel segmentation is essential to successfully detect the optical diseases that manifests in eye.

V.P. Gangraj (✉) · G.K. Birajdar
Department of Electronics & Telecommunication, Pillai HOC College of Engineering & Technology, Raigad 410206, Maharashtra, India
e-mail: vinitagangraj@gmail.com

G.K. Birajdar
e-mail: gajanan123@gmail.com

D.K. Mishra et al. (eds.), *Information and Communication Technology for Sustainable Development*, Lecture Notes in Networks and Systems 10,
https://doi.org/10.1007/978-981-10-3920-1_37

This paper presents a two phase methodology for segmentation of blood vessel. In the first phase, the total count of pixels relayed for classification are distinctly reduced by extracting the major essential vessels. In the second phase, the minor part of the vessel sub-image is analysed instead of classifying all the vessel pixels. This radically reduces the intricacies of segmentation time. For classification of vessel and non-vessel pixel in vessel sub-image, 50-dimensional feature vector is extracted using region-based morphological descriptors. Fuzzy entropy measure-based feature selection technique is used to choose important features. Finally, k-NN classifier is used for vessel and non-vessel pixel classification. This technique is computationally less complex and is independent of training samples. The proposed approach achieves 96.75% classification accuracy.

The organization of the paper is formulated as follows: Sect. 2 reflects literature survey of existing algorithms. Proposed algorithm is described in Sect. 3 of the paper. In Sect. 4 feature extraction is explained. Section 5 describe the experimental results. Lastly, Sect. 6 portray the conclusion of the paper.

2 Literature Survey

The complication in diabetes causes Diabetic Retinopathy (DR). The dispute in automatic segmentation of blood vessels from fundus image of a retina has gained serious deliberation since decade. This attention has classified the segmentation approach as unsupervised and supervised methods. The unsupervised algorithm involves model-based technique, morphological transformations, matched filter and tracking of blood vessel. The unsupervised algorithms are computationally sensitive to retinal abnormalities. The supervised approach classifies the pixels as vessels and non-vessels thus partitioning the image by using K-NN, GMM, SVM, boosting and bagging strategies.

The method presented in [2] extracts the centerline pixel using fuzzy segmentation algorithm. In [7], the segmentation of retinal vessels is modelled by considering the multi-concavity in healthy as well as pathological eye, but the algorithm requires high computational time. In colour retinal image, a hybrid technology for successful segmentation of several oriented vessels is recommended in [11]. Image of the blood vessel is upgraded and surrounding noise is quenched with complex Gabor filter. The major limitation is associated with detection of tiny vessels that are likely affected to any shift in intensity. In [12], probabilistic neural network (PNN), support vector machine (SVM) classifiers and Bayesian classifiers are compared where SVM results in highest classification accuracy, but it increases the computational cost of the system.

A supervised method is proposed in [4] which utilizes fusion of bagged and boosted decision trees. The method discussed in [5] extracts 7-D feature set per pixel. A 7-dimensional set per pixel is figured by morphological linear operators, multiple scale oriented Gabor filters, and by measuring line strengths. The segmentation is performed using GMM classifier. Table 1 summarizes all the existing supervised segmentation approaches.

Table 1 Summary of various supervised blood vessel segmentation techniques

Method	Technique	Extracted features	Classifier	Accuracy
[7]	Multi-concavity modelling approach	ZLine, ZDiff, ZLocal	–	DRIVE: 94.72% STARE: 95.67%
[11]	Gabor filter	Directional features	NN	–
[5]	Top hat reconstruction	Morphological operator features	GMM	95.79%
[4]	Classifier ensemble	9-D	200 decision trees	DRIVE: 94.80% STARE: 95.34% CHASE_DB1: 94.69%
[12]	Machine learning techniques	Radius, diameter, area, arc length, centre angle and half area	PNN, Bayesian, SVM	95.38%
[2]	Fuzzy vessel segmentation	Vessel centerline detection	–	DRIVE-95.13% STARE-95.37%
[3]	2-D gabor filter	Region based	SVM	87.3%

3 Proposed Algorithm

In this paper, the algorithm proposes a technique to segment the blood vessels in fundus image of a retina. Figure 1 depicts blood vessel segmentation algorithm. It consist of following steps:

1. *Pre-processing*: To remove the noise using gaussian filter, first step involves pre-processing of the fundus image.
2. *Major blood vessel extraction*: In the first stage, major blood vessels from the retinal fundus image are extracted using high pass filtering and morphological operations. Secondly, minor blood vessels are segmented using following steps.
3. *Region-based feature extraction*: The sub-image (image after extraction of major blood vessel) is segmented into four sub-regions. Various region-based features are extracted using these four sub-regions. Total of 50 features are extracted.
4. *Feature selection*: Fuzzy entropy measure is employed to identify and select important features from the input feature space. Removing the redundant features speeds up the classification accuracy by reducing the feature vector size. The resultant feature vector size after applying feature selection is 30.
5. *k-NN classifier*: Finally, k-NN classifier is used to classify the pixel as blood vessel and non-blood vessel.

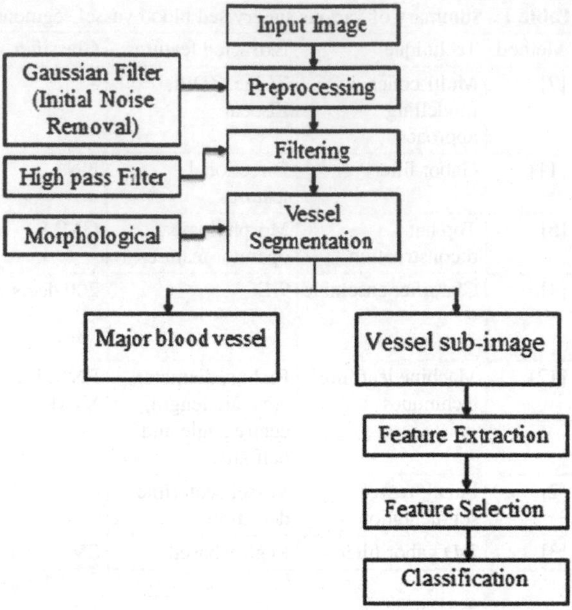

Fig. 1 Block diagram of blood vessel segmentation method

4 Feature Extraction

Vessels apportionment is dependent on classification of pixels as vessels or non-vessels, significantly, on the basis of features extracted. In our methodology, a 50-dimensional feature vector set consisting of the vessel graph is obtained. The feature set is extracted by analysing the orientation of the gradient vector field and by applying morphological operations. The description of the features extracted are:

1. *Area (f1)*: After measuring the diameter, this feature is measured as $\text{Area} = \pi\,(Diameter/2)^2$
2. *Centroid (f2)*: It is first point of center of an image, its co-ordinate is the mathematical average of co-ordinates of all the points on the image.
3. *Eccentricity (f3)*: It specifies rotatory of a region until a bounding box is fixed in the image to discriminate between false vessels and true vessels.
4. *Equivalent Diameter (f4)*: It is twice of the average radius measured for every pixel of an image.
5. *Extent (f5)*: It is the ratio of total pixels covered in the specified region of interest to total pixels in the total bounding box [3].
6. *Major axis length (f6)*: The feature reflect elongated true vessels.
7. *Minor axis length (f7)*: It specifies reduced values for the pseudo vessels.

8. *Orientation (f8)*: If ϕ(i) is the angle of element vector of curve in image Img(i). Then orientation is defined as [8]

$$\phi = E(\phi) = \sum \phi(i) prob(\phi(i)) \tag{1}$$

9. *Perimeter (f9)*: The length of total boundary of a bounding box to determine the outer limit of an area feature.

5 Experimental Results

To calculate the system performance two data sets are used: (1) CHASE and (2) DRIVE. DRIVE [14] data set consists of 40 JPEG compressed digital retinal images. CHASE DB1 [1] data set consist of 28 images with 30° field of view. The data set corresponds to 14 children with two images per child (one image per eye). Segmented blood vessel is evaluated, and is examined by computation of sensitivity (Sens), specificity (Specf) and accuracy (Acc) as expressed in equations (2)–(4),

$$Acc = \frac{(TruePositive + TrueNegative)}{(TruePositive + TrueNegative + FalsePositive + FalseNegative)} \tag{2}$$

$$Sens = \frac{(TruePositive)}{(TruePositive + FalseNegative)} \tag{3}$$

$$Specf = \frac{(TrueNegative)}{(TrueNegative + FalsePositive)} \tag{4}$$

True positive point towards pixels clearly determined as vessel pixels, whereas, clearly identified non-vessel pixels are represented by true negative. False positive point towards the vessel pixels determined as non-vessel pixels, whereas, the non-vessel pixels identified as vessel pixels is represented by false negative.

Feature selection has a crucial role in biomedical imaging application. Using k-NN classifier, a fuzzy entropy-based measure is exercised on the 20 images which belong to DRIVE train database, to select the most prominent and discriminating features. The probabilistic measure of fuzzy entropy:

$$H_1A = -Sum \sum_{j=1}^{n} (\mu_A(X_j) log \mu_A(X_j) + (1 - \mu_A(X_j)) log(1 - \mu_A(X_j))) \tag{5}$$

where $\mu_A(X_j)$ are fuzzy values and H is entropy of the features [9]. These fuzzy entropy measures are used for selection of the relevant features which simplifies the structure.

The training set, consist of a class of candidates whose feature set, and the result of classification (C1 or C2 : vessel or non-vessel) are already calculated

$$S_T = (F^n, C_k^n)|n = 1, \ldots, N; \tag{6}$$

In the feature extraction and selection a 30-D feature vector is used for classification:

$$F(x, y) = (f_1(x, y), \ldots, f_{30}(x, y)) \tag{7}$$

As a result of classification, each candidate pixel is assigned a class as vessel or non-vessel when its pixel representation is known.

As explained in the Sect. 3, the input fundus image is low pass filtered by gaussian filtering to remove the irrelevant features. In the later stages, all the operations and feature extraction is performed using green channel image. Since, green channel of the colour image differentiate the best between dark vessels and bright surrounding, taking advantage of this, its intensity is reserved for subsequent processing. Altogether, 50 region based features are extracted from the four sub-regions. Fuzzy entropy- based feature selection measure reduces the feature dimension to 30. Finally, this feature vector is applied to k-NN classifier as the input for classification. For training the classifier 70% of the total images are applied and remaining 30% are used for testing.

Figure 2 shows input fundus images where as its contrast enhanced and Gaussian filtering output is depicted in Fig. 3. Green channel is extracted (Fig. 4) for further processing and high pass filtering operation is performed. Then the conversion of the high pass filtered image into binary image is viewed in Fig. 5.

Major vessels from the high pass filtered binary image are extracted and presented in Fig. 6. To obtain remaining minor vessels, the algorithm presented in Sect. 3 is applied. Figure 7 shows the minor blood vessel extracted image.

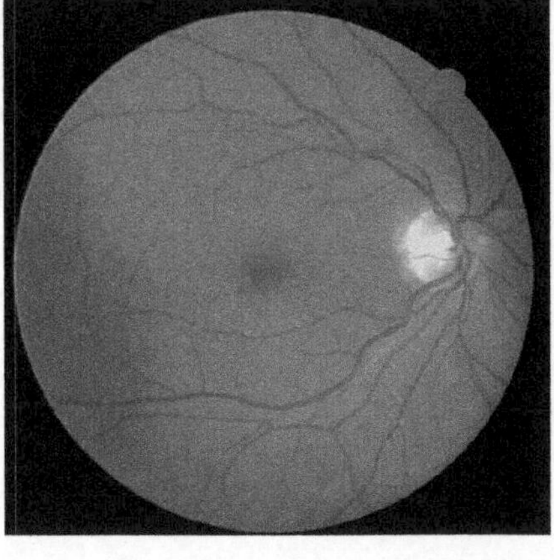

Fig. 2 Input fundus image

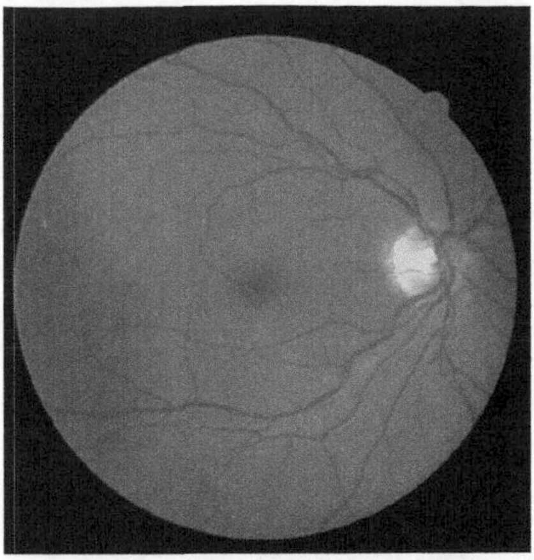

Fig. 3 Contrast adjusted fundus image

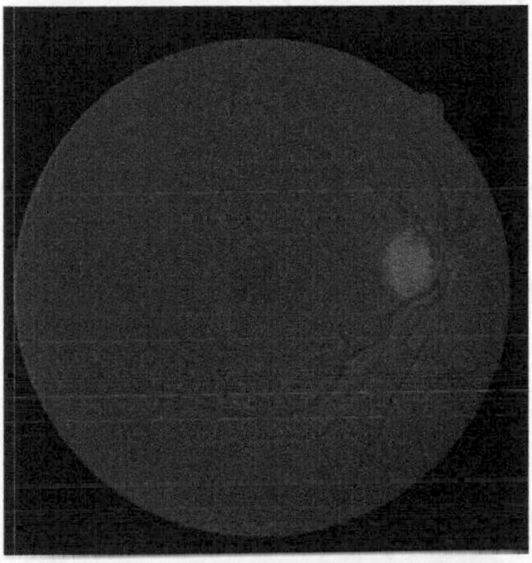

Fig. 4 Green channel image

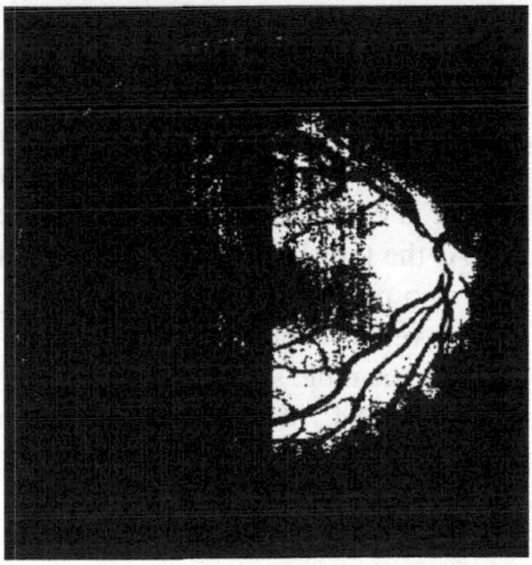

Fig. 5 Binary high pass filtered image

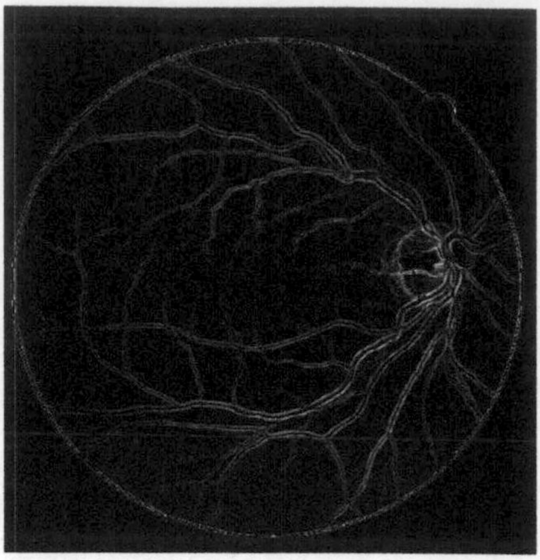

Fig. 6 Extraction of Major vessel

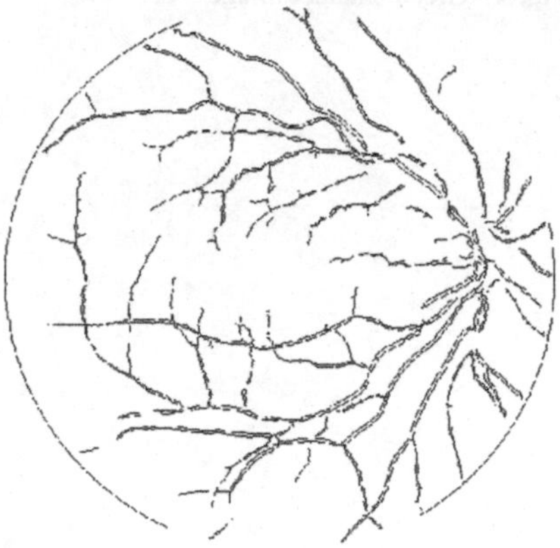

Fig. 7 Extraction of minor vessel

Table 2 presents the segmentation performance of the approach for DRIVE and CHASE_DB1 database. DRIVE database achieves 96.75% accuracy, whereas CHASE_DB1 database resulted in 96.42% accuracy. To analyse the effect of fuzzy entropy-based feature selection, by varying the threshold T accuracy for DRIVE dataset is computed. Table 3 shows the effect of feature selection on DRIVE data set. As the threshold (T) is lowered, resulting in reduction of feature vector dimension also reduces sensitivity, specificity and accuracy. For example when $T = 0.8$ feature vector dimension reduces to 39 resulting in 92.73% classification accuracy. Another example shows that when $T = 0.7$, feature vector dimension reduces to 35 resulting in 84.70% classification accuracy.

Table 2 Segmentation performance (%) for DRIVE and CHASE data sets

Database	Sensitivity	Specificity	Accuracy
DRIVE	96.38	97.12	96.75
CHASE DB1	95.89	96.94	96.42

Table 3 Effect of feature selection on DRIVE data set

T	Feature dimension	Sensitivity (%)	Specificity (%)	Accuracy (%)
0.9	45	94.98	95.17	95.10
0.8	39	92.47	92.98	92.73
0.7	35	84.12	85.27	84.70
0.6	29	74.53	73.68	74.11
0.5	24	68.18	67.87	68.10

Table 4 Performance of system by cross-training

Training	Testing	Sensitivity (%)	Specificity (%)	Accuracy (%)
DRIVE	CHASE_DB1	93.98	94.05	94.10
CHASE_DB1	DRIVE	93.42	94.28	93.85

Table 5 Comparison of the proposed method

Method	Database	Accuracy (%)
Maruthusivarani [10]	DRIVE	88%
Carmen Holbura [6]	DRIVE	94%
Waheed [3]	AFIO, DRIVE and STARE	87.3%
Wankhede [15]	DRIVE	96.26%
Akhavan and Faez [2]	DRIVE, STARE	95.13%, 95.37%
Proposed method	DRIVE, CHASE DB1	96.75%, 96.42%

Additional experiments are performed to testify the robustness of proposed model against dependency on training samples. Table 4 demonstrate the dependency of proposed segmentation algorithm by exchanging the training and the testing data sets, then analysing the vessel segmentation accuracy. Average classification segmentation accuracy obtained is 94%. The proposed segmentation system excel most of the existing system as viewed in Table 5.

6 Conclusion

Supervised approach using region based features is presented to spot vessels in retinal images. Fuzzy entropy based feature selection lowers the dimensionality hence reducing the training and testing time of classifier. By visually inspecting the result of the event and correlating it with the current methodologies, it is concluded that classification under such prominent region based morphological feature and their precise selection leads to more accurate and upgraded results compared to that of existing methods. The algorithm is of great use to the ophthalmologist as it can prove beneficial for detailed and appreciative diagnosis of diseases.

References

1. K.U. research, chase db1, online available: http://blogs.kingston.ac.uk/retinal/chasedb1 (January 2011)
2. Akhavan, R., Faez, K.: A novel retinal blood vessel segmentation algorithm using fuzzy segmentation. International Journal of Electrical and Computer Engineering (IJECE) 04(04), 561–572 (August 2014)
3. Awan, A.W., Awan, Z.W., Akram, M.U.: A robust algorithm for segmentation of blood vessels in the presence of lesions in retinal fundus images. In: 2015 IEEE International Conference on Imaging Systems and Techniques (IST). pp. 1–6 (Sept 2015)
4. Fraz, M.M., Remagnino, P., Hoppe, A., Uyyanonvara, B., Rudnicka, A.R., Owen, C.G., Barman, S.A.: An ensemble classification-based approach applied to retinal blood vessel segmentation. IEEE Transactions on Biomedical Engineering 59(9), 2538–2548 (Sept 2012)
5. Fraz, M.M., Remagnino, P., Hoppe, A., Velastin, S., Uyyanonvara, B., Barman, S.A.: A supervised method for retinal blood vessel segmentation using line strength, multiscale gabor and morphological features. In: Signal and Image Processing Applications (ICSIPA), 2011 IEEE International Conference on. pp. 410–415 (Nov 2011).
6. Holbura, C., Gordan, M., Vlaicu, A., Stoian, I., Capatana, D.: Retinal vessels segmentation using supervised classifiers decisions fusion. In: Automation Quality and Testing Robotics (AQTR), 2012 IEEE International Conference on. pp. 185–190 (May 2012)
7. Lam, B.S.Y., Gao, Y., Liew, A.W.C.: General retinal vessel segmentation using regularization-based multiconcavity modeling. IEEE Transactions on Medical Imaging 29(7), 1369–1381 (July 2010)
8. LUO, D.: pattern Recognition and Image Processing. Woodhead publishing limited, UK (1998)
9. Luukka, P.: Feature selection using fuzzy entropy measures with similarity classifier 38(4), 4600–4607 (April 2011)
10. Maruthusivarani, M., Ramakrishnan, T., Santhi, D., Muthukkutti, K.: Comparison of automatic blood vessel segmentation methods in retinal images. In: Emerging Trends in VLSI, Embedded System, Nano Electronics and Telecommunication System (ICEVENT), 2013 International Conference on. pp. 1–4 (Jan 2013)
11. P. C. Siddalingaswamy, K.G.P.: Automatic detection of multiple oriented blood vessels in retinal images. Biomedical Science and Engineering 3(1), 101–107 (January 2010)
12. Priya, R., P. Aruna: Diagnosis of diabetic retinopathy using machine learning techniques. ICTACT Journal on soft Computing 03(04), 563–575 (July 2013)
13. Roychowdhury, S., Koozekanani, D.D., Parhi, K.K.: Blood vessel segmentation of fundus images by major vessel extraction and subimage classification. IEEE Journal of Biomedical and Health Informatics 19(3), 1118–1128 (May 2015)

14. Staal, J., Abramoff, M.D., Niemeijer, M., Viergever, M.A., van Ginneken, B.: Ridge-based vessel segmentation in color images of the retina. IEEE Transactions on Medical Imaging 23(4), 501–509 (April 2004)
15. Wankhede, P.R., Khanchandani, K.B.: Retinal blood vessel segmentation using graph cut analysis. In: Industrial Instrumentation and Control (ICIC), 2015 International Conference on. pp. 1429–1432 (May 2015)

Recommendation System for Learning Management System

Phalit Mehta and Kriti Saroha

Abstract Education field is core area of recent researches in electronic technology. There are various aspects which are included in the researches for analysis and evaluation like, learners, learning management system, evaluators, etc. Taking one aspect into consideration, i.e., learning management system, there are very few techniques available for the evaluation of learning management system. The approach using the learner's behavior and teaching evaluation for the evaluation and recommendation of Learning Management System has already been proposed. The paper presents an implementation and the results of the proposed approach. Apart from the results, the paper also discusses further improvements for the learning management system.

Keywords Learning behavior · Learning management system
Learning style · Online learning · Teaching evaluation

1 Introduction

In the age of electronic technology, it is not so difficult to learn even if we are far away from the teaching source. This far away learning can be known as distance learning. The learning is helpful for learners who live far from school or university or education ground or even at distant places and also learners with disability. Online learning through Learning Management System (LMS) is the recent way to carry out the distance learning for various learners [4]. Online learning is also helpful for the learners with regular courses as it is the best way to distribute the learning content for the learner as well as efficient way to judge the behavior of the learners.

P. Mehta (✉) · K. Saroha
School of Information Technology, Centre for Development of Advanced Computing, Noida, India
e-mail: phalitm@gmail.com

K. Saroha
e-mail: kritisaroha@cdac.in

D.K. Mishra et al. (eds.), *Information and Communication Technology for Sustainable Development*, Lecture Notes in Networks and Systems 10,
https://doi.org/10.1007/978-981-10-3920-1_38

Further, there has to be a mechanism to evaluate the performance of the learners as well as the LMS for the improvement of both.

There are two main aspects to evaluate learning management system namely, learning behavior and learning style and thus, these are taken into consideration for the evaluation of the system. Apart from this, it is beneficial for the evaluator to evaluate the learners of different learning behavior and different learning styles. LMS also helps to enhance the learning capabilities of the learners on the basis of learner's style [1]. Various techniques may be applied for the evaluation of learning management system.

Numerous researches have been accomplished in the online learning environment in the past. Lingyan Wang et al. [8] proposed an approach in which they collected and examined learners' behavior data. Then, the learning behavior and learning effect is combined and considered for decision tree model. The evaluation of decision tree model includes data collection, data analysis, and finally the evaluation. Sabine Graf et al. [11] focused on developing the online learning by web mining. Web mining handles the web log files for the technique. This technique includes data preprocessing, data analysis and association rule finding and grouping of these finding. It also takes care of hit analysis in terms of access hits and unique user hits. Then association rules on the hits is being carried out for the unseen learners' access patterns. Mohammad Hassan Falakmasir et al. [9] studies the learners' usage logs with some established tools and Moodle, which are refined to test web server logs and to fulfil the teaching requirements. The author used the log details of all learners' activities and then used association rule mining, classification, clustering, pattern analysis and some other statistical methods for the evaluation of the learners. Supriya Solaskar et al. [12] used a fuzzy inference system to give recommendations for a learning management system. Their recommendations are based on learners' styles and teacher's evaluation. The main motive is to improve the existing LMS that are best for learner's style. Authors have implemented Felder and Silverman for the learning styles of learners which are explained based on questionnaire [8]. Fuzzy Inference System takes the Learning styles and Teacher Evaluation as input in which Learning style is based on questionnaire [5] and Teacher's Evaluation is based on four classifications, i.e., fail, average, good and excellent. After examination and determination of learning management system, authors suggested some recommendations for the system.

This paper focuses on the evaluation of learning management system and its implementation and discusses the result. The paper analyzes the log details of a group of 115 students of first-year of undergraduate engineering major of the University of Genoa. The study over a simulation environment named Deeds [6] (Digital Electronics Education and Design Suite), which is used for LMS in digital electronics is also included in the dataset. In the dataset, we have the learners' time series of activities during six sessions of laboratory sessions of the course of digital electronics where each session includes 13 features. It is proposed to evaluate the system with respect to learners who are the key factors in the evaluation process. The teaching evaluation, which includes the marks given to learners by the educator for each session are also included. Evaluation and recommendation for LMS would be effective if learners' access behavior is also examined for the study.

The paper is organized as follows. Section 2 presents the methodology to evaluate the system and Sect. 3 explains the implementation of methodology. The results are presented in Sect. 4. Finally, Sect. 5 discusses future work and conclusion followed by references.

2 Methodology

Learning Management System has various aspects on which the system can be evaluated. In the existing approaches, main focus is given to the evaluation of learners, who are accessing the Learning Management System but we are focusing on the evaluation of the system, which is important for the learners. In the proposed approach, shown in Fig. 1, the learner's behavior is considered to identify various learners' styles. There are various models which can be used to determine learner's styles. FSLSM Model [2, 3] is referred in the approach. Moreover, some other learner's style are also identified based on the learner's behavior.

The various phases of the proposed model are explained below with the dataset in consideration.

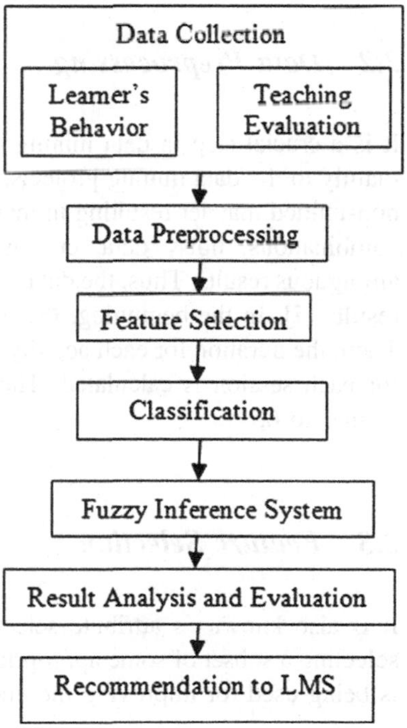

Fig. 1 Methodology

2.1 Data Collection

It is the method of collecting and calculating information in a standardized way. In this phase, the dataset for a group of 115 students of the University of Genoa is collected. The learners' behavior and teaching evaluation is also included for the study.

Learner's Behavior. Learner's behavior is based on the access log details of the learners using the LMS. This can be stored by the system for further analysis. The data set contains the learners' activities during six sessions. Each session contains 13 features namely, session, student_Id, exercise, activity, start_time, end_time, idle_time, mouse_wheel, mouse_wheel_click, mouse_click_left, mouse_click_-right, mouse_movement, keystroke. Moreover, activity features includes, Study_Es_# of session_# of exercise, Deeds_Es_# of session_# of exercise, Deeds_Es, Deeds, TextEditor_Es_# of session_# of exercise, TextEditor_Es, TextEditor, Diagram, Properties, Study_Materials, FSM_Es_# of session_# of exercise, FSM_Related, Aulaweb, Blank, Other. As discussed, the dataset contains the log details of the learners which can be used to identify the behavior as well as style of the learners.

Teaching Evaluation. Teaching Evaluation is the grades or marks given to the learners by the teacher of their respective course. The marks of individual session as well as total marks of the specific semester are considered for evaluation of the system.

2.2 Data Preprocessing

It is a crucial step in data mining process. It shows the "garbage in-garbage out" mainly to the data mining projects. In the previous phase, the data is gathered in a non-unified manner resulting in many problems like missing values, irrelevant data combinations, noisy data, etc. Analysis of these types of data will produce ambiguous results. Thus, the data is preprocessed to get the relevant and appropriate results [7]. In the beginning, the various activities converted to unified activities. Then, the duration for each activity used the sum of all the activities of each student for each session is calculated. The work is carried out for each session, ranging from 1 to 6.

2.3 Feature Selection

It is also known as attribute selection. Attribute subset selection is a method of selecting a subset of some appropriate and significant features for use in study. This is being used for improving the understanding and observations of the study.

```
Ranked attributes:
 0.089319272    7 fsm
 0.045031411    2 study
 0.03805518     4 texteditor
 0.033145178    6 properties
 0.029544935    1 student_id
 0.018831927    3 deeds
 0.010878562   10 other
 0.007581719    8 aulaweb
 0.002339253    5 diagram
 0.000000215   11 idletime
-0.0002187      9 blank

Selected attributes: 7,2,4,6,1,3,10,8,5,11,9 : 11
```

Fig. 2 Feature selection

Table 1 Classification of learning style

Learning style	Learning behavior
Intuitive	Study
Sensitive	Deeds, texteditor, properties, fsm
Visual	Diagram
Verbal	Study

In the approach, the ReliefFAttributeEval Algorithm and Ranker method is being used for feature selection and evaluation. The set of ranked features are shown in Fig. 2. From the set of ranked features available, the following 9 features are selected namely, fsm, study, texteditor, properties, student_id, deeds, other, aulaweb, diagram.

2.4 Classification

In this phase, the learners' behavior is classified in order to understand the various learning style. Moreover, the teaching evaluation is also categorized for the learners. Learner's style is the style which learners opt at the time of learning the content of LMS. There can be different styles of the learners which can be analyzed. From the features selected, four learning style are used for evaluating the LMS namely, intuitive, sensitive, visual, and verbal.

The selection of the features is very tedious task as the features which are best suited with the learning behavior need to be selected so that the system can be analyzed and evaluated for the recommendation. In the study, nine features were selected from which the following seven features have been classified for the learning style as shown in Table 1.

Apart from this, the teaching evaluation is also used to analyze the overall performance of the learners which can be divided in four categories as shown in Table 2.

3 Implementation

Fuzzy Inference System is used for the implementation of the proposed approach and Mamdani model is used for each session. With two input variables and one output variables, learning style and teaching evaluation are used for input variables and the recommendation is considered as output variables as shown in Fig. 3.

The learning behavior of the learners for the different sessions is used to identify the learning style and is evaluated for each session separately. Table 3 gives the classification of learning style for one session. The output variable would generate some recommendation for the LMS. The recommendation is based on learning style as well as teaching evaluation. The proposed recommendations are as shown in Table 4. Fuzzy Inference System is used with Triangular Membership function for the inputs and output of the system. According to the Triangular Membership function, the functions can be defined for the two input variables and one output variable as given by Eq. (1).

Table 2 Teaching evaluation

Input	Range	Output
Teaching evaluation	79–100	Excellent
	59–80	Good
	39–60	Average
	<40	Fail

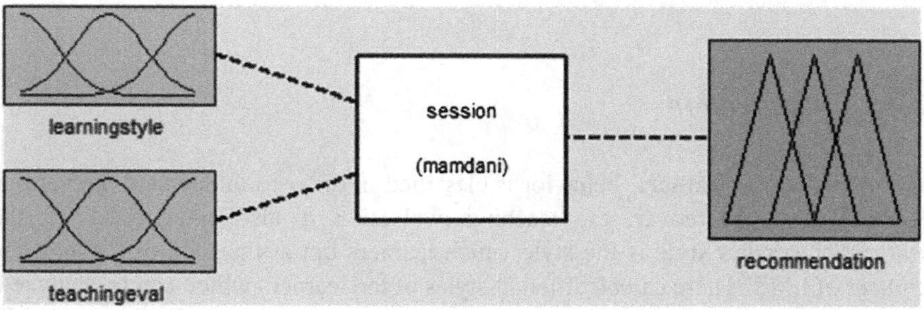

Fig. 3 Fuzzy inference system

Table 3 Learning style

Input	Range	Output
Learning style	<1177	Intuitive
	<4913	Sensitive
	<2203	Visual
	<1177	Verbal

$$f(x; a, b, c) = \max(\min(\frac{x-a}{b-a}, \frac{c-x}{c-a}), 0) \quad (1)$$

The learning styles shown in Fig. 4 are suggestive that there is some overlap existing between the learning styles. The overlapping of the learning styles can be described as shown in Fig. 5. The function for teaching evaluation is shown in Fig. 6.

After the implementation of membership function for the input and output variables, rules are defined for the function [10]. Rules can be generated by means

Table 4 Recommendation

Input	Range	Output
Recommendation	70–100	Accepted
	45–75	Give more exercise
	25–50	Add more content
	<30	Replace existing content

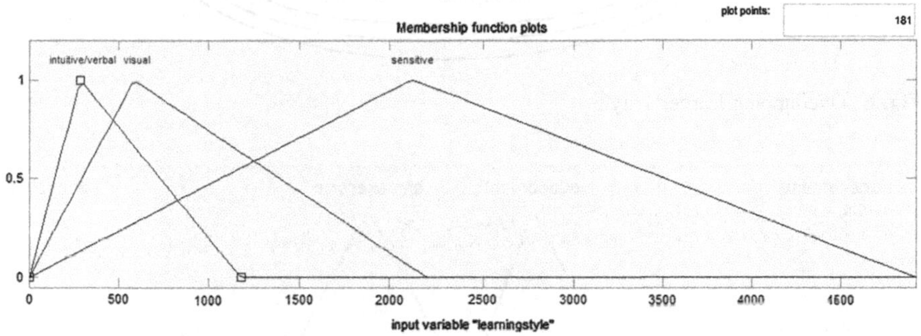

Fig. 4 Learning style

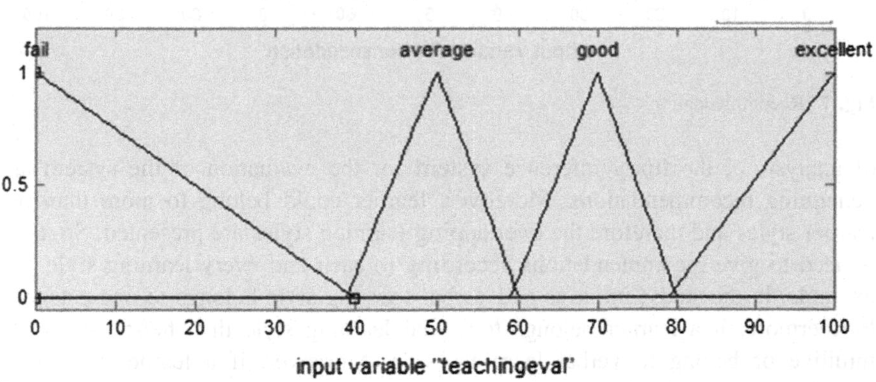

Fig. 5 Teaching evaluation

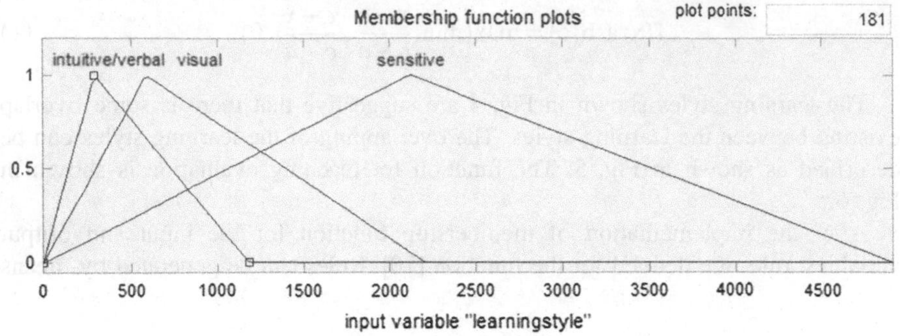

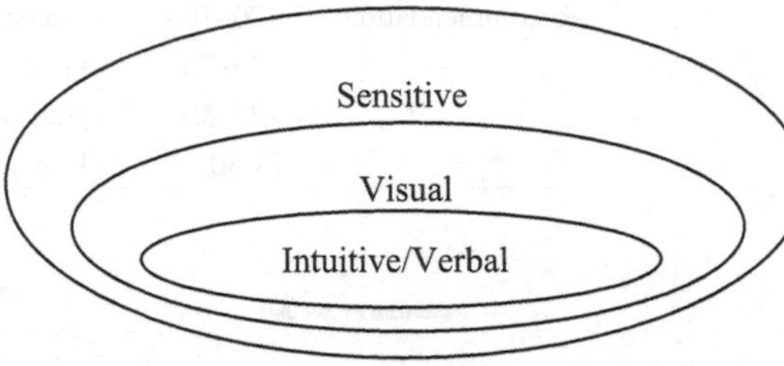

Fig. 6 Overlapping learner's style

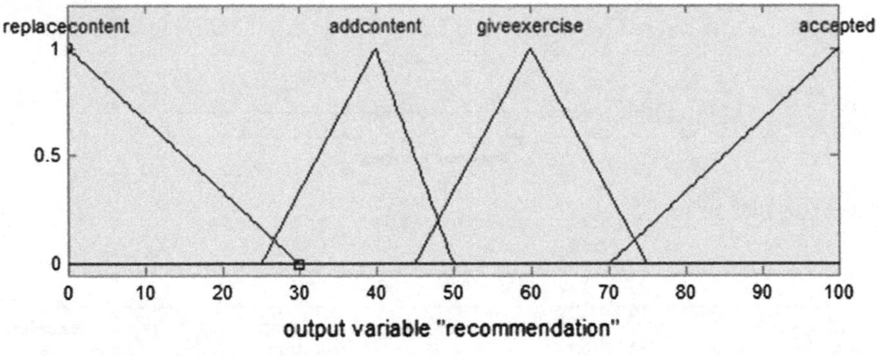

Fig. 7 Recommendation

of analysis of the fuzzy inference system for the evaluation of the system and generating recommendations. Moreover, learner could belong to more than one learner styles and therefore the overlapping learning styles are presented. So, there is need to give recommendations according to each and every learning style. For example, in the study, intuitive and verbal learning style belongs to one category. Furthermore, if a learner belongs to visual learning style then he/she would be intuitive or belong to verbal learning style. Moreover, if a learner has visual learning style then he/she might have learning style as sensitive, as shown in Fig. 5. The function for recommendation is shown in Fig. 7.

4 Result

After the implementation, the results can be evaluated by the rules defining the membership functions of the Fuzzy Inference System. Total 12 rules for the two input variables have been defined. On the basis of the two input variables and the rules, the fuzzy inference system gives the output result, as shown in Fig. 8.

The evaluation is based on rules defined in the membership functions. Some of the rules are, if learning style is intuitive and the teaching evaluation is excellent then the content of LMS is accepted without modification, if learning style is sensitive and teaching evaluation is good then recommendation is generated to give more exercise, if learning style is visual and teaching evaluation is average then recommendation is to add more content, if learning style is verbal and teaching evaluation is fail then recommendation is to replace the existing content. On the basis of these rules, the recommendations for LMS are shown in Table 5.

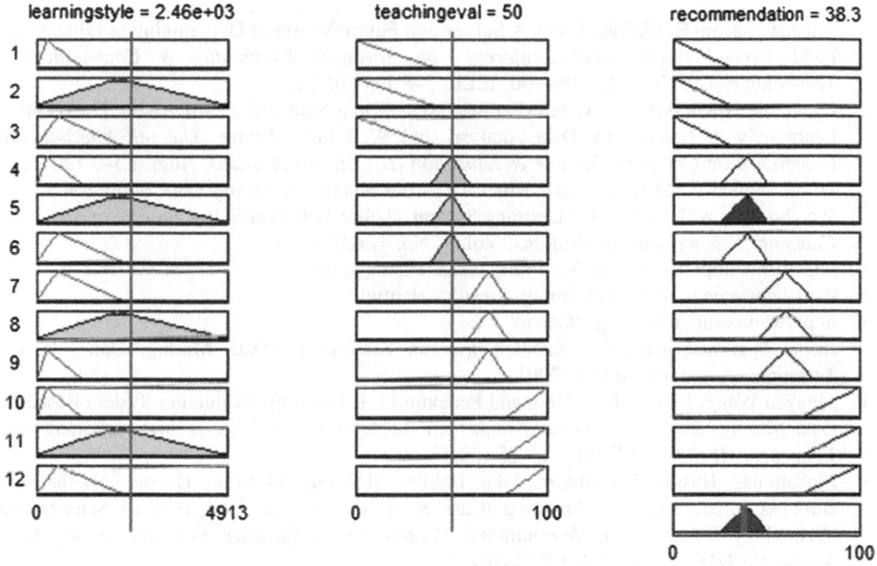

Fig. 8 Evaluation of FIS

Table 5 Recommendation to LMS

Teaching evaluation				
Learning style	Excellent	Good	Average	Fail
Intuitive	Accepted	Give more exercise	Add more content	Replace existing content
Sensitive	Accepted	Give more exercise	Add more content	Replace existing content
Visual	Accepted	Give more exercise	Add more content	Replace existing content
Verbal	Accepted	Give more exercise	Add more content	Replace existing content

5 Conclusion

The approach targets on the learners' behavior and teaching evaluation. Further, learners' behavior is analyzed for identifying the learning style. As we get the learning style, it is combined with teaching style for the evaluation and recommendation. The model is expected to improve the evaluation of Learning Management System and thus giving the relevant recommendations. So, it would be a benefit to the learners to get well defined and well evaluated learning content which can improve teaching evaluation and would help to understand the various learning styles. In future, more learning styles would be explored to be included by analyzing various learners' behavior and then recommendations would be generated for new learning styles.

References

1. Barahate Sachin R, Shelake Vijay, A Survey and Future Vision of Data mining in Educational Field, *Second International Conference on Advanced Computing & Communication Technologies (ACCT 2012)*, 96–100, IEEE, 7–8 Jan (2012)
2. Chakarida Nukoolkit, Praewphan Chansripiboon, Satita Sopitsirikul, Improving University e Learning with Exploratory Data Analysis and Web Log Mining, *The 6th International Conference on Computer Science & Education (ICCSE 2011)*, IEEE, August 3–5 (2011)
3. Dr. Mohamed F. AlAjmi, Shakir Khan, Dr. Arun Sharma, Studying Data Mining and Data Warehousing with Different E-Learning System, *(IJACSA) International Journal of Advanced Computer Science and Applications*, Vol. 4, No. 1. (2013)
4. https://en.wikipedia.org/wiki/Learning_management_system
5. https://www.engr.ncsu.edu/learningstyles/ilsweb.html
6. http://www.esng.dibe.unige.it/deeds/
7. Jiawei Han and Micheline Kamber, Morgan Kaufmann, "Data Mining: Concepts and Techniques", Second Edition (2001)
8. Lingyan Wang, Jian Li, Lulu Ding and Pengkun Li, E-Learning Evaluation System Based on Data Mining, *2nd International Symposium on Information Engineering and Electronic Commerce (IEEC)*, IEEE, July 23–25 (2010)
9. Mohammad Hassan Falakmasir, Jafar Habibi, Shahrouz Moaven, Hassan Abolhassani, Business Intelligence in E-Learning (Case Study on the Iran University of Science and Technology DataSet), *2nd International Conference on Software Engineering and Data Mining (SEDM)*, 473–477, IEEE (2010)
10. Phalit Mehta, Kriti Saroha, Analysis and evaluation of learning management system using data mining techniques, *Fifth International Conference on Recent Trends in Information Technology*, 2016, Chennai, IEEE. [unpublished]
11. Sabine Graf, Silvia Rita Viola, Kinshuk, Tommaso Leo, Representative characteristics of Felder-Silverman Learning Styles: An empirical Model, *IADIS International Conference on Cognition and Exploratory Learning in Digital Age*, p 235, IADIS Press, Barcelona, Spain (2006)
12. Supriya Solaskar, Anuradha. G, Dr. A. K. Sen, Improving Learning Management System using Data Mining, *International Journal of Engineering Research & Technology (IJERT)*, Vol. 4 Issue 02, February (2015)

Classification of Hematomas in Brain CT Images Using Support Vector Machine

Devesh Kumar Srivastava, Bhavna Sharma and Ayush Singh

Abstract Hematoma is caused due to traumatic brain injuries. Automatic detection and classification system can assist the doctors for analyzing the brain images. This paper classifies the three types of hematomas in brain CT scan images using Support vector machine (SVM). The SVM has been simulated and trained according to the dataset. Trained SVM classifiers performances were compared on the basis of parameters, i.e., classification accuracy, mean square error, training time, and testing time. The classification process depends on the training dataset and results are based on simulation of the classifiers.

Keywords CT scan · Hematoma · Data mining · Support vector machine

1 Introduction

Brain Hematoma is caused due to traumatic injury in brain and blood is collected inside the brain [1]. Three types of brain hematomas are Epidural Hematoma (EDH), Subdural Hematoma (SDH), and Intracerebral Hematoma (ICH). Hematomas are hyper dense in nature so they appear brighter than the other brain tissues. EDH involve bleeding between the skull and the dura matter of the brain and has a biconvex shape while a SDH is caused between dura and subarchnoid layer of the brain having a crescent shape. ICH is inside the brain having a circular shape.

D.K. Srivastava (✉) · A. Singh
SCIT, Manipal University Jaipur, Jaipur, Rajasthan, India
e-mail: devesh988@yahoo.com

A. Singh
e-mail: ayushsingh@gmail.com

B. Sharma
Apex Institute of Engineering & Technology, Jaipur, India
e-mail: bhavnapareek@rediffmail.com

D.K. Mishra et al. (eds.), *Information and Communication Technology for Sustainable Development*, Lecture Notes in Networks and Systems 10,
https://doi.org/10.1007/978-981-10-3920-1_39

The CT scan is among various available brain scanning techniques, it is preferred in comparison to the other techniques because of its extensive availability, little cost, fast scanning and superior contrast [2]. Classification task is the most frequently encountered problem in medical image analysis. In these tasks an object should be allocated into a particular group or class depending upon the observed properties or attributes called features related to that object. Many classifiers are available, developed, and successful in medical imaging applications but all are having different initial parameter that varies according to the applications [3]. The accuracy of classification methods must be high because the diagnosis and treatment is based on this categorization. This paper focuses on SVM classifier for classifying the hematomas in brain CT images. There are varieties of other approaches available for such an application [4]. There are no clear rules or procedures that can be followed to choose the best classifier for this specific problem [5]. We compared performances of classifiers on the basis of parameters, i.e., classification rate, mean square error, training, and testing time, etc. The classification process depends on the training dataset and results are based on simulation of the classifiers.

2 Problem Statement

The aim of the research paper is classification of brain Hematomas by extracting features from brain CT images. Figure 1 shows axial slice of brain CT scan images for normal and all three kinds of hematomas which are Subdural (SDH), Intracerebral (ICH) and Epidural (EDH) Hematomas. The organization of the research paper consists of introduction and problem statement given in Sects. 1 and 2, respectively, and Sect. 3 describes methodology for brain CT image analysis. Section 4 describes the simulation and training of two SVM classifiers (including one on one and one versus all) and interprets the inference. Finally, we get to the conclusion of the paper and discuss any possibility of future work.

Fig. 1 Normal image and EDH image

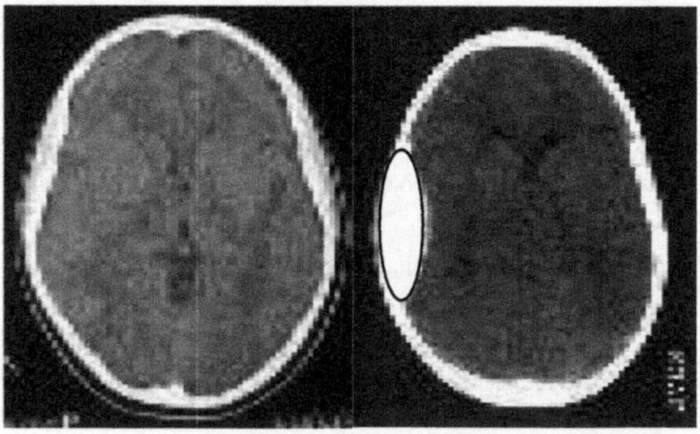

3 Methodology

The Methodology for analyzing the brain CT Image is sectioned into several steps which are shown in Fig. 2. In the initial step image is acquired from CT scan machine that is converted into suitable format. We collected images of brain CT scan in DICOMM and convert it into jpeg format. The images are from same CT scan machine and having resolution of 512 * 512 with same window width and window level. Second step is preprocessing of the original image to eliminate preexisting noise from that image. The region of the brain is obtained through extraction to eliminate the artifacts from that image. The extracted region contains only the brain portion, which is further processed in analysis steps. Third step segments the image automatically into regions. A new hybrid method [6] based on peak detection and K-means clustering is used, that automatically segments the preprocessed image which is shown in Fig. 3.

In the fourth step statistical features [7], intensity features [8], shape features [9] and texture features [10] are obtained through extraction from the segmented image. Features are thus obtained from each tested image. Large number of features would be more informative but due to the complexity, computational cost, increase in generalization error, feature selection [11] methods are used that bring out the most informative features. We used backward feature selection method in which we started with all feature vectors and then remove the features one by one, at each step, until any further removal of the feature increases the error (misclassification rate) significantly. After rigorous checking and many trials, finally 12 features are selected these are given in Table 1. Area of the segmented region is 3055 pixels that is important in calculating the size of hematomas and other important features but not used as an independent feature. The last step, classification task which is organization the data as dissimilar groups by referring these features comprises of phase of testing and training. In phase of training, features and highlights the image

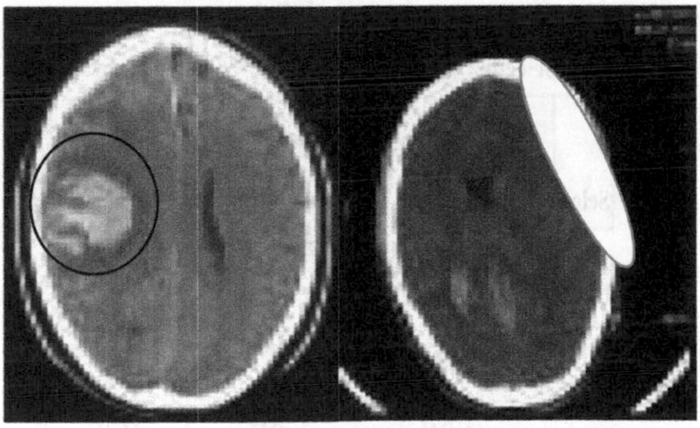

Fig. 2 ICH image and SDH images

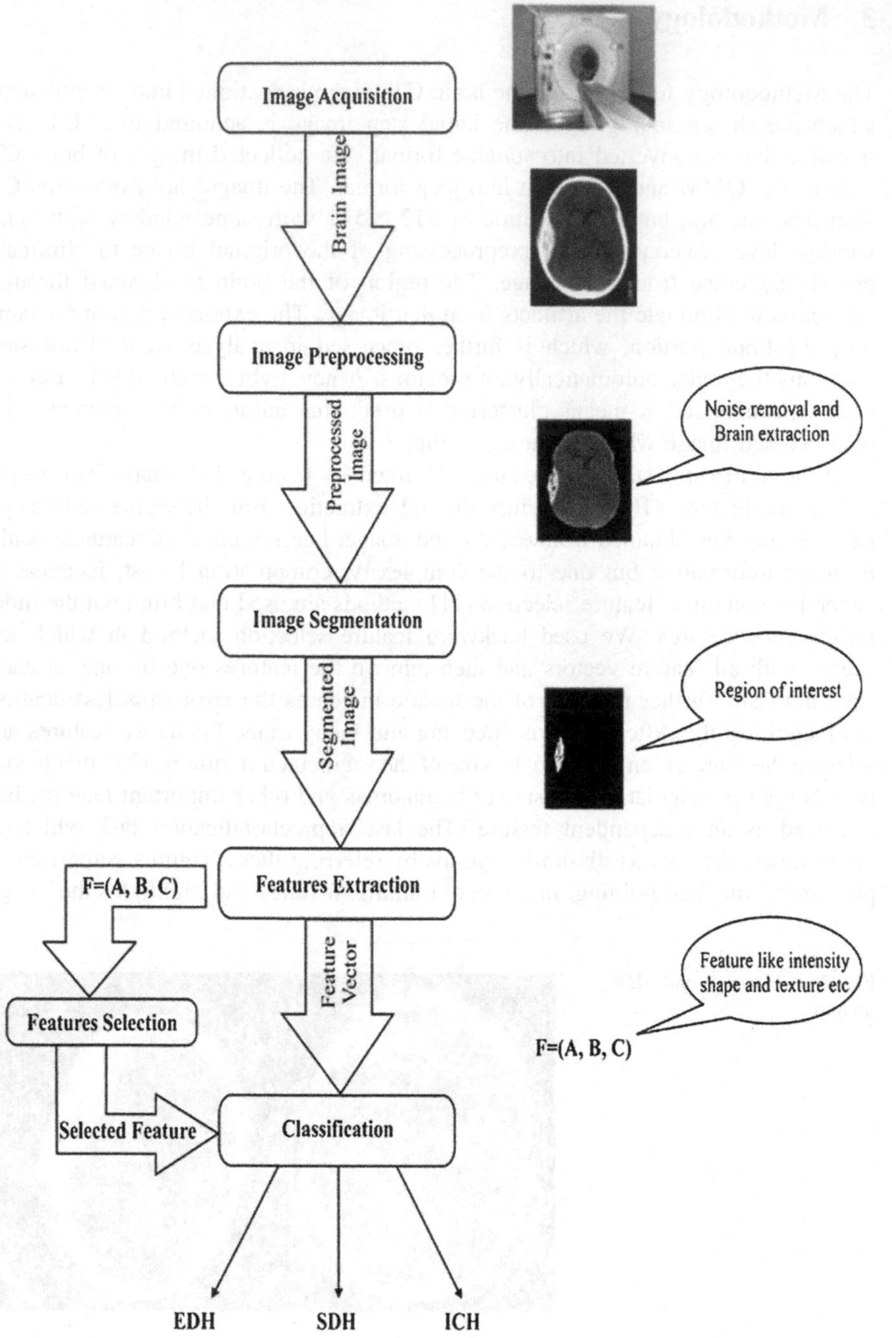

Fig. 3 Steps in brain CT image analysis for classification of hematomas

Table 1 Extracted features and their values obtained from the segmented image

S. no.	Feature	Value (in pixels)
1	Circularity	0.0830
2	Major axis	235.18
3	Minor axis	37.34
4	Eccentricity	0.9834
5	Solidity	0.3245
6	Extent	0.1008
7	Convex hull area	10564
8	Adjacent to skull boundary	Yes (1)
9	Ratio of bounding box width to minor axis	0.2534
10	Energy (ASM)	0.9621
11	Entropy	0.1188
12	Contrast	0.1559

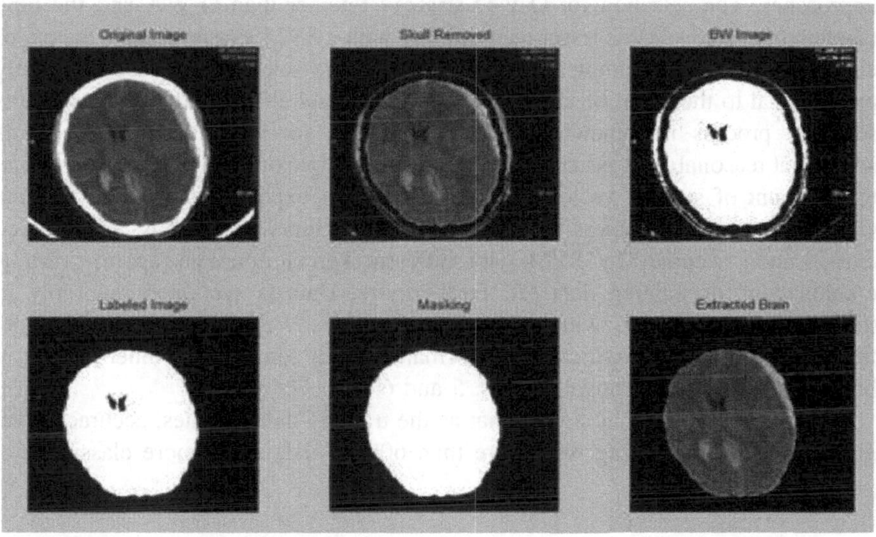

Fig. 4 Preprocessed image

must be presented to the network and creation of an exclusive description of every class is done. In the phase of testing, those features and highlights are employed for classification of the images in different categories. Multi-class SVM (one-against-one (O-V-O) and one-against-all (O-V-A)) is already simulated and trained [12] for the training samples that are tested by the system. Performances of each classifier are evaluated according to the performance parameters (Fig. 4).

4 Experimental Results

Every method is implemented on MATLAB v7.1 running on Windows 7. Computer possesses an Intel i3 Processor with 4 GB of RAM and a NVIDIA GeForce 2 GB Graphics Driver. The Sample Data comprises of 150 Brain CT images. The Quality of every image is similar but bears dissimilar resolutions. First 50 images for a brain with no Hematoma and other 100 images have Hematomas in the brain of several types, forming dissimilar shapes at random locations. We selected 12 features among several features from segmented images. These 12 features from each image with a sample size of 100 are given to classifiers as input patterns and three outputs are obtained for each type (EDH, SDH, and ICH). The sample images are randomly sectioned into three sets of validation, training and testing.

SVM [13] is supervised binary classifier capable of giving high performance in classification of patterns. Although SVMs are designed to do only two-class classification, they have shown consistent performance when applied to multi-class classification problems, using one-against-one (O-V-O) or one-against-all (O-V-A) approaches. The selection for O-V-O strategy tougher than O-V-A and the time taken to train $c(c - 1)/2$ is lesser than training with c SVM's (c represents the no. of classes). The overall complexity for SVM's results selection process is directly proportional to the count of all supporting vectors and although the SVM's results selection process is somewhat more complicated for a multi-class case, it is somewhat reasonable to assume that the overall complexity is directly proportional to the count of support vectors. But somehow, our experiments showed that the O-V-O strategy assumes less number of support vectors from O-V-A. We checked classification accuracy by SVM with different kernel functions against various combinations of training data set. Furthermore, Data is sectioned randomly in training and testing sets with different proportions. We found that RBF kernel function gives better classification performance when compared to other kernels in both types of SVM as shown in Figs. 5 and 6.

It is clear from both the figures that as the training data elevates, accuracy level also increases. So training with more than 60% of data gives more classification

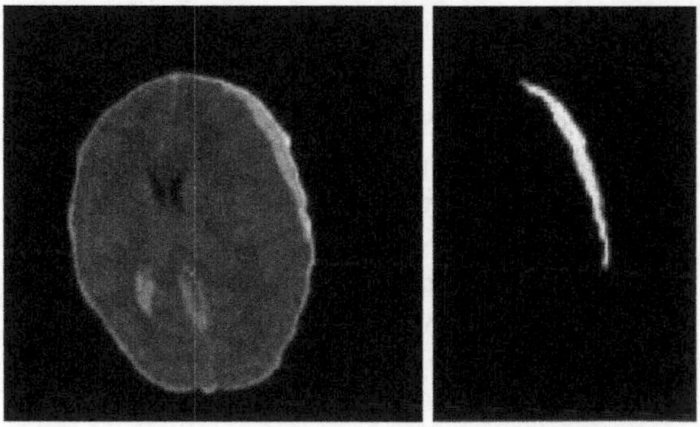

Fig. 5 Original image and segmented image

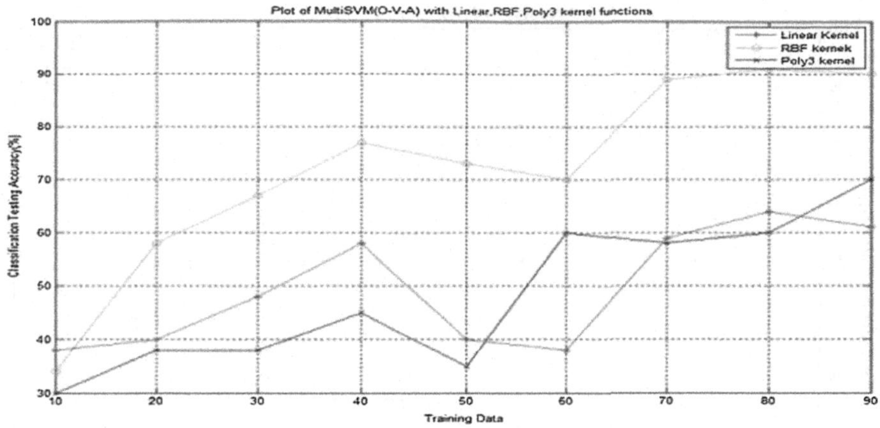

Fig. 6 Classification accuracy in O-V-A multiclass SVM with linear, RBF and polynomial kernel functions with varying number of data set size for training and remaining data for testing chosen randomly

accuracy. RBF kernel function qualifies among other kernels in both SVMs and accuracy of classification is better in O-V-O SVM than O-V-A SVM.

Randomly segmenting data into training and testing set and running it many times gives different accuracy levels. Moreover, this random segmentation might result in training or test datasets with dissimilar portions of outputs in dissimilar classes.

An improvement to the random sampling is cross validation. In this method, every record is used only once for testing and a number of times for testing. If data is partitioned in two equal sized subsets, one of the subset for training and the other for testing then the roles of the two subsets are swapped. This approach is called a twofold cross validation. The total error is obtained by summing the errors for both the runs. In this paper we used ten fold cross validation where data set is segmented into ten folds randomly and the classifier is tested with one fold and trained with remaining folds. This process repeated 10 times such that data is tested with each fold. Finally average of training, testing and overall accuracy is calculated (Fig. 7).

In SVM classifier applying the ten fold cross validation for both one-against-one (O-V-O) and one-against-all (O-V-A) and the training and testing accuracy are measured [14]. In SVM RBF kernel function is used as it is more reliable than other kernel functions. We assumed a RBF kernel in collaboration with the ten fold cross validation method to estimate the optimal parameters γ and C. Values of (C, γ) are applied using Hit and Trial method and the value having the best accuracy for cross validation is selected. Using Hit and Trial method exponentially sequentially of γ and C is a practical and optimal method for identifying best parameters. After rigorous checking and running several times we choose (C, γ) as (1.0, 0.5).

After selecting suitable parameters the testing accuracy on ten folds are calculated and the results proved that the O-V-O SVM gives the average accuracy of

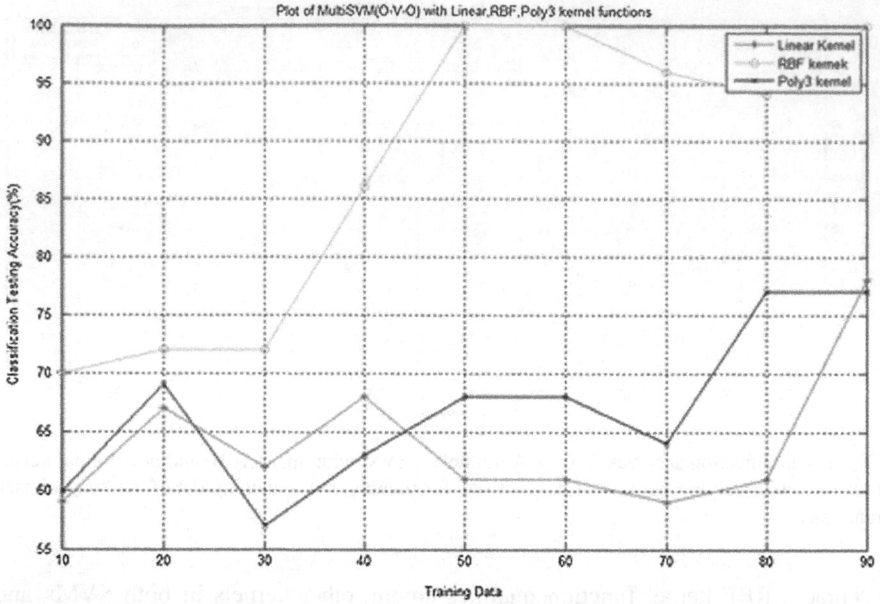

Fig. 7 Classification accuracy in O-V-O multiclass SVM with linear, RBF and polynomial kernel functions with varying number of data set size for training and remaining data for testing chosen randomly

97% which is greater than the O-V-A SVM having 88% of average training accuracy.

Applying ten fold cross validation on each classifier testing accuracy varies with each fold. This may be due to the fact that features are randomly partitioned in each fold so calculation of average accuracy is necessary. O-V-O multi-class SVM gives average accuracy of 97% and O-V-A multi-class SVM gives average accuracy of 88%.

The important factor in simulating these classifiers are training time and setting the right initial parameters which is again a time consuming task. We ignored here parameters selection time. The training time and testing time for classifiers after setting the initial parameters are shown in Table 2. This is the time of testing one fold and training given by nine folds. There is not much variation in training time and testing time for each fold so we have not shown the time for each fold. Average training and testing time are shown in Table 2, we conclude that training time for O-V-O multiclass SVM classifier is lesser than O-V-A multiclass SVM.

Table 2 Comparison of testing and training time of SVM classifiers

Classifiers	Training time	Testing time
SVM (O-V-O)	0.015600	0.00015
SVM (O-V-A)	0.043300	0.0013

We also compared SVM with other classifiers like MLP and RBF networks [15]. We draw some conclusion like SVM scales well and constructing and setting initial parameters are simpler than MLP and RBFN. Secondly RBFN and SVM require less trial and error and thus, less time and effort for better classification rate than MLP. Overall performance of (O-V-O) multi SVM classifier outperforms others in terms of complexity, lesser number of training samples, faster training and testing speeds and reproducibility of results. The second choice is RBFN because of lesser number of training parameters and faster testing speed. MLP is not a well-controlled learning machine and does not give accurate classification rate.

Place of Interest also plays an important role in every surgical decision and its procedure. If a temporal hematoma is already large or is regularly expanding in size, then it might lead to a faster deterioration and herniation. EDH present in posterior fossa, generally demands a quick evacuation as there is only some limited space which is available if compared to the Epidural Hematoma on frontal segment. In this system for finding the location of the hematomas we have divided the image in four segments and find out the pixels in the ROI belongs to which segments and at what ratio.

We only considered four lobes of the brain and set rules according to expert knowledge to find the location of the hematomas. The first segment is right frontal lobe, second segment is left frontal lobe and third and fourth segment belongs to parietal and temporal lobe and if hematoma is in third and fourth segment than it is in posteria fossa. This whole procedure takes 0.5 s on an average as the calculation that how many pixels belongs to which segments needs labeling and counting the pixels in each segment. Here we have found the approximate location of the hematomas but for finding the exact location for example near the ventricles, syri, CSF, etc., more segments are needed, more segmentation of the brain axial slice should be done and more knowledge should be incorporated into the system.

For calculating the size of hematomas, area of region feature extracted from segmented image of Fig. 8 is considered. As the radiologist are concerned with not only the size of hematoma region but also the width of the region. We have calculated the width of EDH and ICH from the minor axis of the fitted ellipse on the region and multiplying with the size of pixel. Due to SDH shape, fitted ellipse minor axis has almost double than the region so width of SDH region is calculated by dividing minor axis by 2 and multiplying with pixel size.

The Size of the Pixel is estimated from field of view (FOV) and by the matrix size, for example if CT FOV is d and the size of matrix is M, then generally for a head scan having an FOV of 10 cm with a matrix of exactly 512 pixels, Size of the Pixel is 0.2 mm. Pixel size = d (in mm)/M = 100/512 = 0.2 mm

Practically, the size of the pixel is smaller than the theoretical values due to resolution limitation imposed because of the finitely small size of every focal spot and x-ray detections. Axial resolution of scan plane may be enhanced with the help of operation in a greater resolution mode applying a larger matrix size or a smaller FOV. For the image in Fig. 9, Area of the region is 3055 pixels and minor axis of fitted ellipse is 38 pixels than the size and thickness of the SDH is calculated as

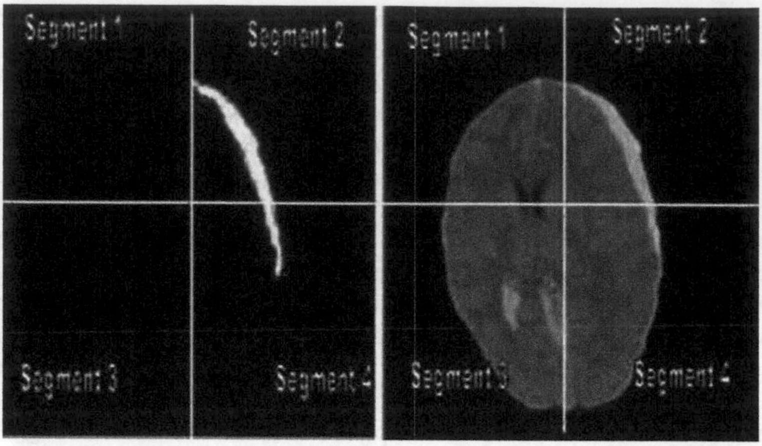

Fig. 8 Division of image into four equal parts. **a** Original which shows that SDH is located in *Left*-Frontal-Parietal lobe. **b** Percentage of pixels in each segment in segmented image

$$\text{Size of SDH} = (\text{Area}) * \text{Pixel size} = 3055 * 0.2 = 611\ \text{mm}^2$$

$$\text{Thickness of SDH is} = (\text{Minor axis})/2 * \text{Pixel Size} = 17 * 0.2 = 3.4\ \text{mm}$$

This is moderate size SDH and surgical evacuation via craniotomy is often considered in patients with a SDH thicker than 10 mm. There is no problem in recommending surgery in the large size hematomas and recommending medicines and therapies in smaller size hematomas. The main problems are with moderate size hematomas that requires expert knowledge and other parameters like age, Glasgow comma scale. The brain image analysis and recommendation for diagnosis may vary with radiologist.

5 Conclusion

In this paper SVM classifiers are proposed to classify the hematoma types in brain CT images. The input dataset (features) to train the classifiers are obtained from segmented CT scanned images. Constructing and setting initial parameters in SVM classifier is simpler than other classifiers. SVM require less trial and error and thus, less time and effort for getting better classification rate than the other classifiers. Overall performance of SVM (O-V-O) classifier outperforms others in terms of complexity, lesser number of training samples, lesser time, effort and high classification accuracy.

Our Future work is to find out the exact location and exact volume of hematomas by adding more expert knowledge in the system. However, the ultimate future goal of this ongoing research is to provide an everyday clinical tool that will assist clinicians in the diagnosis of brain hematomas. To achieve this, system must undergo exhaustive clinical trials to ensure its proper functionality.

References

1. Dhawan, A. P., Huang, H. K., Kim D. S.: Principles and Advanced Methods in Medical Imaging and Image Analysis. World Scientific. (2008)
2. Seletchi, D., Duliu, O. G.: Image processing and data analysis in computed tomography. Roman Journal of Physics, vol. 52, no. 5, pp. 667–675. (2007)
3. Burges, C. J. C.: A Tutorial on Support Vector Machines for Pattern Recognition. Data Mining and Knowledge Discovery, vol. 2, No. 2, pp. 121–167. (2002)
4. Loncaric, C. S.: Rule-based labeling of ct head image. Proceedings of the sixth Conference on Artificial Intelligence in Medicine in Europe, pp. 453–456. (1997)
5. Zhang, W. Z., Wang, X. Z.: Extraction and classification for human brain CT images. In Proceedings of Sixth International Conference on Machine Learning and Cybernetics, pp. 1155–1156. IEEE (2007)
6. Sharma, B., Venugopalan, K.: Automatic segmentation of brain CT scan images. International Journal of Computer Applications, vol. 40, no. 10, pp. 1–4. (2012)
7. Mark, N., Aguado, A. S.: Feature Extraction & Image Processing. Academic Press. (2008)
8. Yang, M., Kidiyo K., Joseph, R.: A survey of shape feature extraction techniques. Pattern Recognition, pp. 43–90. (2008)
9. Haralick, R. M., Karthikeyan, S., Dinstein, I. H.: Textural features for image classification. IEEE Transactions on Systems, Man and Cybernetics, vol. 6, pp. 610–621. (1973)
10. Kohavi R., John, G. H.: Wrappers for feature subset selection. Artificial Intelligence, vol. 97, no. 1, pp. 273–324. (1994)
11. Gong, T., Li, S., Wang, J., Tan, C. L., Pang, B. C., Lim, C. T., Zhang, Z. Automatic labeling and classification of brain CT images. 18th IEEE International Conference on Image Processing (ICIP), pp. 1581–1584. (2011)
12. Vapnik: Support-Vector Networks. Machine Learning, pp. 273–297. (1999)
13. Suykens, J. A. K., Gestel, T. V., Brabanter, J. D., Moor, B. D., Vandewalle J.: Least Squares Support Vector Machine. World Scientific Publishing Co., Singapore. (2002)
14. Chang C., Lin, C.: LIBSVM: a library for support vector machines. ACM Transactions on Intelligent Systems and Technology, Vol. 2, No. 27, pp. 127. (2011). Software available at http://www.csie.ntu.edu.tw/~cjlin/libsvm
15. Sharma B., Venugopalan K.: Comparative Analysis of MLP and RBF Neural Networks for Classification of Brain CT Images. International Journal of Advanced Engineering & Computing Technologies (Special Edition), pp. 44–48. (2013)
16. Sharma, B., Venugopalan, K.: Classification of hematomas in brain CT images using neural network. In proceedings of IEEE Conference on Issues and Challenges in Intelligent Computing Techniques, pp. 41–46. IEEE (2012)

Secured Neighbour Discovery Using Trilateration in Mobile Sensor Networks

Jeril Kuriakose, Ravindar Yadav, Devesh Kumar Srivastava and V. Amruth

Abstract In a wireless network, the initial step after deployment of a network is identifying the nodes neighbours. Neighbour discovery is the building block of a network applications and protocols and its responsibility is to identify the one hop neighbours or the nodes that are in the direct communicational range. A minor vulnerability in the neighbour discovery can lead to severe disruptions in the functionalities of the network. In this paper we propose a novel technique to identify the adversary nodes that disrupt the networks functionalities. We have modified the trilateration technique to identify the adversary node. Our security mechanism has been carried out along with the localization procedure without causing any additional overhead. Our technique can achieve successful neighbour discovery even in the presence of cheating nodes. We have also identified the probability of detecting malicious nodes for two different scenarios.

1 Introduction

The advancements in wireless sensor networks (WSNs) and mobile ad hoc networks (MANETs) have increased the rise of miniaturized gadgets such as sensors, smartphones, and personal digital assistants. WSN is a broad area in wireless networks where wireless motes transfer small amount of data mostly sensed data to a central server. These sensed data mostly contain information about the environment such as temperature, humidity, and seismic wave's data. The nodes in the network are required to configure themselves during deployment as there is no communicational infrastructure between them. The chances of knowing a node's

J. Kuriakose (✉)
St. John College of Engineering and Technology, Palghar, India
e-mail: jeril@muj.manipal.edu

J. Kuriakose · R. Yadav · D.K. Srivastava
Manipal University Jaipur, Jaipur, India

V. Amruth
Maharaja Institute of Technology, Mysore, India

D.K. Mishra et al. (eds.), *Information and Communication Technology for Sustainable Development*, Lecture Notes in Networks and Systems 10,
https://doi.org/10.1007/978-981-10-3920-1_40

neighbour is very less, and all the nodes needs to communicate with each other to comprehend about its neighbouring nodes. The neighbouring nodes information is required during routing [1], for efficient working of the topology control algorithms [2], and by medium access control protocols [3].

Neighbour discovery is the preliminary step after network deployment. It can be categorised into two types such as deterministic neighbour discovery and randomised neighbour discovery. A predetermined transmission schedule is known in prior to the all the nodes in the network and each node transmits its information at these intervals, thus identifying the neighbour in a deterministic manner. Whereas; in randomized neighbour discovery, nodes transmits its information at random intervals and tries to identify its neighbour with a high probability. Increased frequency of transmission time can reduce induce guaranteed neighbour discovery. Various studies have been carried out in neighbourhood discovery [4–6], and the research is mostly focussed on the following necessities:

1. Energy efficiency by increasing the wake-up interval.
2. Time taken to provide guaranteed neighbour discovery.
3. Synchronization with respect to the battery life.

One better way to overcome the replay discrepancies is to use distance bounding [7] based approach. Distance bounding technique uses the time of flight concept to identify whether a node is a one-hop neighbour or not. The transmission time of a two-hop neighbour is slightly greater than the transmission time of one-hop neighbour, and this can be used to provide some amount of security during the neighbour discovery. Proximity alone cannot be a solution to identify the communication neighbours [8]. Adversary node can reduce the coverage range or the proximity range of one node by inducing the obstacles in the line of sight.

In this paper, we address the above mentioned problems by providing a novel approach towards neighbour discovery. The location information sent along with the beacon frame will be used with trilateration technique [9] to identify the one-hop neighbour. An additional security mechanism has also been introduced to reduce the effect of adversarial node. Trilateration technique uses the geometry of spheres to identify the point of intersection of three spheres. We have exploited this to identify the distance between the sender and the receiver. In order to overcome relay attacks such as wormhole attack and black hole attack, we have introduced few conflicting sets to recognise the abnormalities during message exchange which is then used to filter out the suspicious locators. Previous work related to wormhole detection [10] during neighbourhood discovery was with the help of graph-based approach. This particular approach cannot be applied to mobile network due to the irregularity in the network and will also lead to incorrect wormhole detection.

Trilateration technique works with the help of beacon frames. Each beacon frame contains the location information about the node which is sending the packet along with some additional security parameters. The location information is known to the node in prior and it is either carried out during deployment in case of a static node or network or by using GPS [11] in case of a mobile node or network.

GPS is generally not preferred in the network density is high, and also if the network is in an indoor environment. Existing solutions [12] provided secure neighbour discovery with an additional overhead, whereas in our paper we were able to provide it without any additional overhead.

Paper organization: Sects. 2 and 3 demonstrates the system and attack model respectively, Sect. 3.1 shows the working of our neighbour discovery scheme. Section 4 computes the modified trilateration technique, and Sect. 5 evaluates the performance of our system. Section 6 concludes the paper.

2 System Model

Our system consist of the following:

i. A method M which is used to determine the type (genuine or adversary) of node, location and the type channel or medium used for communication.
ii. Two sets of protocols P_1 and P_2 to identify the adversary node in the network.

Each node in the network is equipped with a unique id. For our work we used the MAC address as the unique id of the node. Two nodes are considered to be neighbours if received signal strength of exceeds a given decoding threshold value. When a collision happens a practical recovery of packet at the destination node is not possible. For collision detection in our work we used the technique proposed by [13]. Collision detection plays a vital role during neighbour discovery. The impacts of collision detection is also shown in our paper.

The initial location of the mobile nodes are arbitrarily chosen. Each mobile node can be in any one of the two states: static state where the node is not moving and mobile state where the node is moving from one location to another location. Each state can have any one of the following three modes:

i. Transmission mode—sending a neighbour discovery message or advertising its location.
ii. Listening mode—listening for neighbour discovery message. On receiving a message it updates its local topology table.
iii. Energy-saving mode—going to a sleep mode when not in use.

Each node in the network transmits and listens are some specified time interval. All nodes cannot transmit at the same time or listen at the same time. In order to overcome this we have used birthday-listen-and-transmit (BLT) to determine the transmission and listening time. Each node selects a transmission mode with a probability P_T, and each node listens with a probability P_L, and chooses its energy-saving mode with a probability of $1-P_T-P_L$. Figure 1 shows the three modes during neighbour discovery.

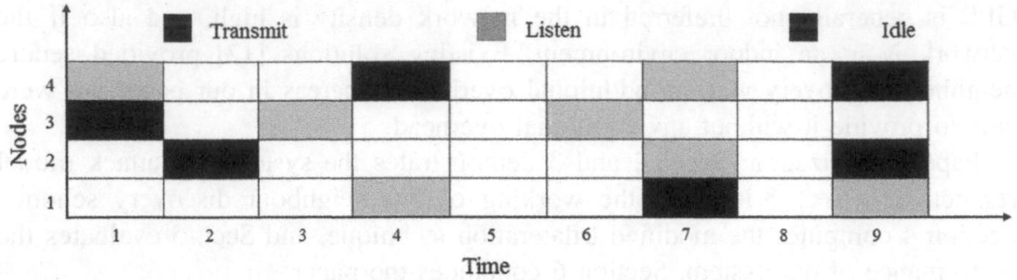

Fig. 1 Modes during neighbour discovery while using BLT

3 Neighbour Discovery System

Our neighbour discovery scheme extends the system of [14]. During an active mode, the nodes are endlessly engaged in transmitting and receiving information. During a passive mode, the node listens only for a fraction of time and goes to its idle state. This is normally done to improve the energy efficiency of the device. For providing the security during neighbour discovery we use trilateration technique [15]. Our security scheme is based on location-based protocols, where the security and neighbour discovery is carried out with the help of beacons. We have added the zero knowledge proof protocols along with the location-based protocols.

In our work we have modified the traditional trilateration techniques and added an additional security parameter to provide secured localisation and neighbour discovery. Figure 5 shows the modified technique and the modified location identification is as follows:

1. The node that wants to find its neighbours and its location (new node) sends a location identification request & neighbour discovery request (LIReq) to the one hop neighbours.
2. The neighbouring node sends the location identification reply (LIRep) along with the range measurement, time stamp, and the location coordinate [range, (location coordinate)].
3. New node sends its newly identified location coordinates (XU; YU) along with the time stamp to the neighbouring nodes that assisted in its localization.
4. The assisted neighbouring nodes shares the location coordinates of the new node (XU; YU) with the other neighbouring nodes of the new node.
5. The other neighbouring nodes of the new node replies whether the shared location coordinates is valid or not with a node discovery & location identification response LIRes (0 or 1) message, '0' indicates that the shared location coordinates is false and '1' indicates that the shared location coordinates is true. In our work we have given a threshold value of ± 1 m in the verification process.
6. The assisted neighbouring nodes sends the honesty scores HS (+1 or -1) message to the new node, ' +1' if the new node is send the correct location reference and '-1' if the new node sends a false location reference (Fig. 2).

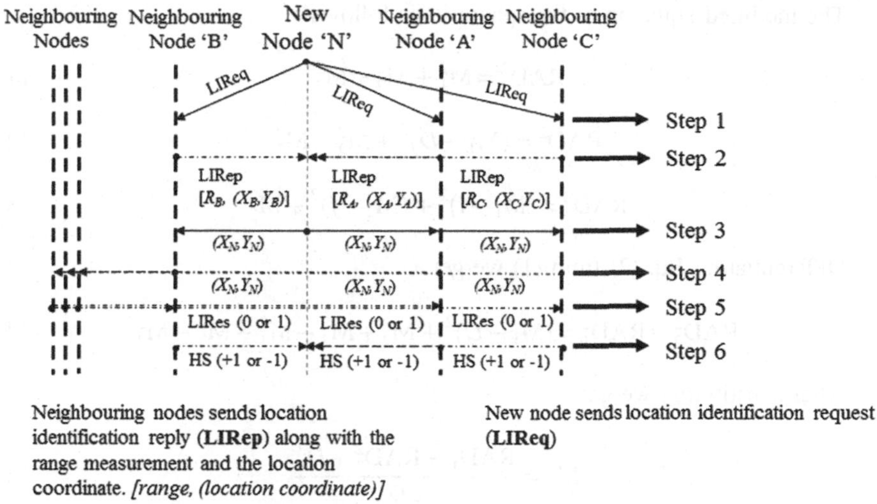

Fig. 2 Modified location identification process

3.1 *Honesty Scores*

During deployment the beacon nodes are given an honesty score of 10 points. This is given to distinguish the normal neighbouring node from the beacon nodes. Normally after assist in location the nodes that assist a new node in localization is given a point. When a vulnerable node tries to manipulate the location reference, the neighbouring nodes identifies it and reduces the honesty score by 1 point. Because of the honesty scores an adversary node cannot survive in a network for more time.

4 Modified Trilateration Technique

The geometry of spheres is used by Trilateration technique to identify the location of a new node. It uses three range measurements to identify the location of the new node. Generally a new node sends a location request to the neighboring nodes or the neighboring beacon nodes. The neighboring beacon nodes or nodes reply with the range measurements to the new node. The new node identifies its location as follows:

General equation of sphere is as follows:

$$\sum_{k=1}^{3} (M_k - S_k)^2 = \mathrm{RAD}^2$$

The modified equation of the sphere is as follows:

$$\mathrm{RAD}_1^2 = \mathrm{M}_1^2 + \mathrm{M}_2^2 + \mathrm{M}_3^2 \tag{1}$$

$$\mathrm{RAD}_2^2 = (\mathrm{M}_1 - D)^2 + \mathrm{M}_2^2 + \mathrm{M}_3^2 \tag{2}$$

$$\mathrm{RAD}_3^2 = (\mathrm{M}_1 - i)^2 + (\mathrm{M}_2 - j)^2 + \mathrm{M}_3^2 \tag{3}$$

Differentiating Eq. (2) from (1) we get,

$$\mathrm{RAD}_2^2 - \mathrm{RAD}_1^2 = (\mathrm{M}_1 - D)^2 + \mathrm{M}_2^2 + \mathrm{M}_3^2 - \mathrm{M}_1^2 - \mathrm{M}_2^2 - \mathrm{M}_3^2 \tag{4}$$

After substituting we get,

$$\mathrm{M}_1 = \frac{\mathrm{RAD}_1^2 - \mathrm{RAD}_2^2 + D^2}{2D} \tag{5}$$

The following equation state that two spheres intersect at two different points,

$$D - \mathrm{M}_1 < \mathrm{M}_2 < D + \mathrm{M}_1 \tag{6}$$

Differentiating Eq. (5) from (1) we can find out,

$$\mathrm{RAD}_1^2 = \left(\frac{\mathrm{RAD}_1^2 - \mathrm{RAD}_2^2 + D^2}{2D}\right)^2 + \mathrm{M}_2^2 + \mathrm{M}_3^2 \tag{7}$$

After substituting we get the point of intersection,

$$\mathrm{M}_2^2 + \mathrm{M}_3^2 = \mathrm{RAD}_1^2 - \frac{(\mathrm{RAD}_1^2 - \mathrm{RAD}_2^2 + D^2)^2}{4D^2} \tag{8}$$

Modifying Eq. (1) with (3) we get,

$$\mathrm{RAD}_3^2 = (\mathrm{M}_1 - i)^2 + (\mathrm{M}_2 - j)^2 + \mathrm{RAD}_1^2 - \mathrm{M}_1^2 - \mathrm{M}_2^2 \tag{9}$$

$$\begin{aligned} \mathrm{M}_2 &= \frac{\mathrm{RAD}_1^2 - \mathrm{RAD}_2^2 - \mathrm{M}_1^2 + \left(\mathrm{M}_1^2 - i\right)^2 + j^2}{2j} = \frac{\mathrm{RAD}_1^2 - \mathrm{RAD}_2^2 + i^2 + j^2}{2j} \\ \mathrm{M}_2 &= \frac{i}{j}\mathrm{RAD}_1 \end{aligned} \tag{10}$$

The values of M_1 and M_2 can be obtained from Eqs. (5) and (10). M_3 can found by substituting the values of M_1 and M_2 in Eq. (1).

$$M_3 = \pm\sqrt{RAD_1^2 - M_1^2 - M_2^2}$$

After identifying the location of a new node, the new node proves its trustworthiness to the neighbouring nodes by advertising its location information. The neighbouring nodes identifies the trustworthiness by again performing the trilateration technique with nodes other than the new node. If the node is found to be a genuine node then all the neighbouring nodes of the new nodes update its neighbour table.

5 Performance Evaluation

We have done our simulation in a 20 m × 20 m space. The beacon nodes were uniformly distributed across the space, and are all static. The location coordinates of the beacon nodes are manually configured with the help of a GPS. The beacon were deployed in such a way that each beacon node's coverage area overlaps a minimum of three beacon nodes coverage areas. Initially a single node was made to enter the network and trilateration technique was carried out by the node to identify its location. Slowly the density of the network was increased and the time taken for the trilateration technique was taken. Later cheating nodes were introduced in the network and the time taken for the trilateration technique were noted. Figure 5 shows the time taken for the trilateration technique in the presence of cheating nodes. The time taken for the trilateration technique in the presence and absence of beacon nodes were both same. Figure 6 shows the mean localization error in the presence of cheating nodes (Figs. 3 and 4).

In order to check the efficiency of our algorithm we have made two scenarios. In the first scenario 7 new nodes were introduced in the network for every 15 min. Among the 7 nodes, 2 nodes were malicious and our algorithm tried to detect the malicious nodes. In this scenario no nodes were allowed to leave the network, and Fig. 7 shows the actual number of cheating nodes and the detected nodes.

For our second scenario we introduced nodes randomly in the network, but with a concentrated threshold of 9 nodes per 25 min, and among them a maximum of 5 cheating nodes were introduced. Figure 5 shows the detected cheating nodes with respect to the actual cheating nodes. We have also checked the probability of identifying the cheating node with respect to the total number of anchor nodes and Fig. 6 summarises the results.

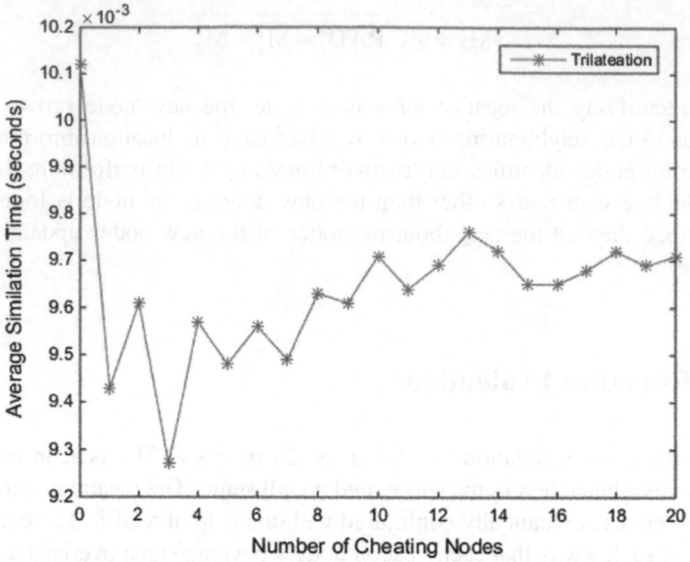

Fig. 3 Average simulation time

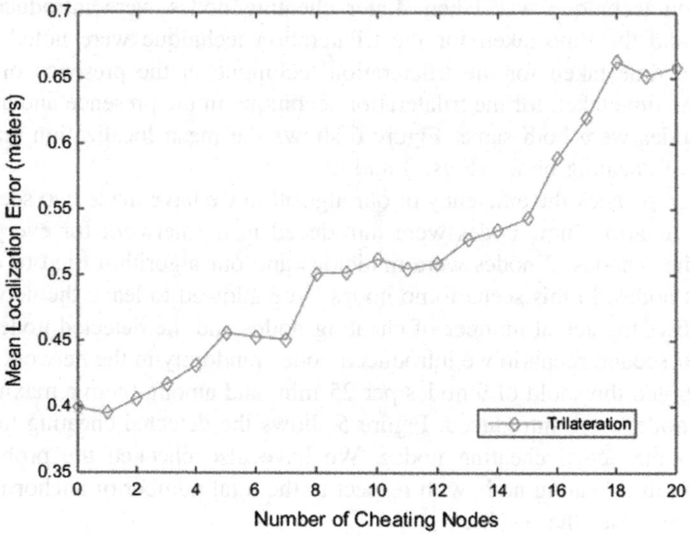

Fig. 4 Mean localization error

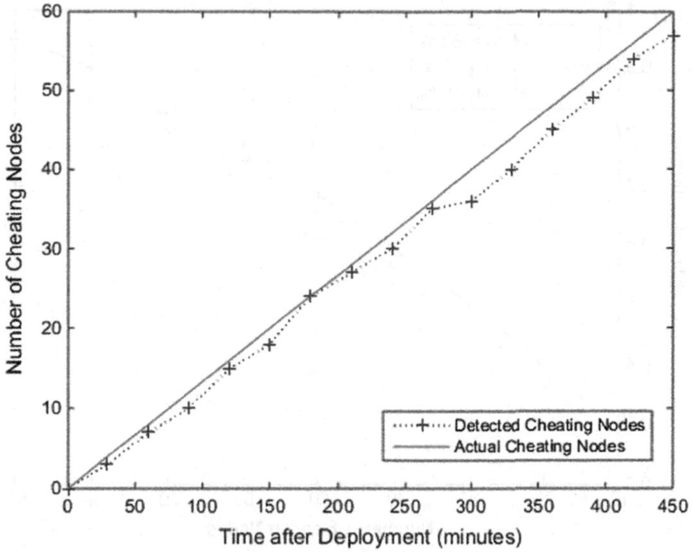

Fig. 5 Scenario 1 results

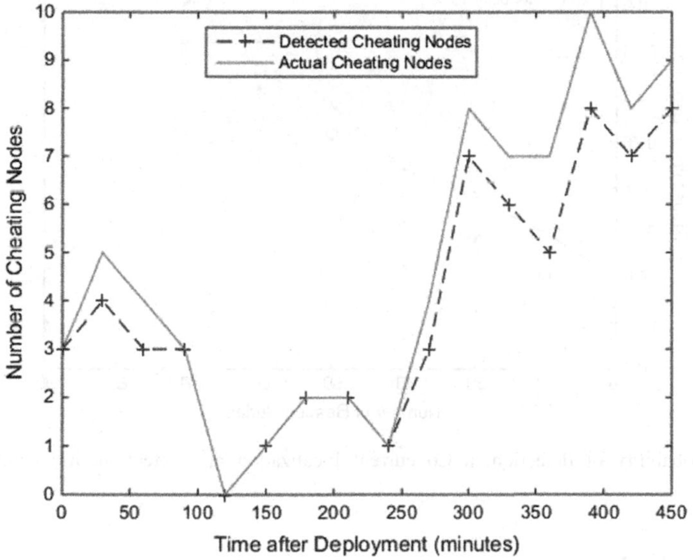

Fig. 6 Scenario 2 results

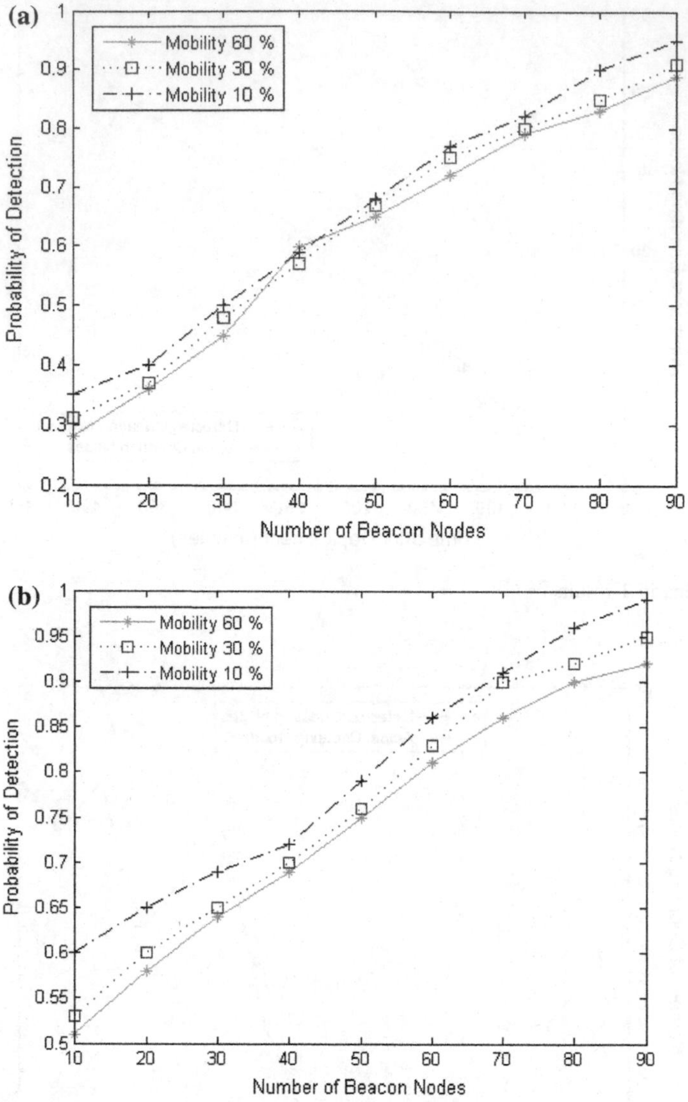

Fig. 7 Probability of detection. **a** Concurrent localization and detection, **b** localization later detection

6 Conclusions

We have presented a novel technique to detect the cheating nodes during neighbour discovery. Our scheme does cause any additional overhead, since the security step is carried out along with the normal localization technique. A cheating node cannot survive in the network for more time. We were able to detect the cheating nodes with an accuracy of around 85%. Time complexity of our work depends on the network scenario and space, and it is not depended on the density of the network. The neighbour table of each node is updated only if the neighbouring node is found to be non-malicious. We compared our scheme in two different scenarios, from scenario 2 we can understand that the identification of malicious nodes depends on the mobility of the nodes. From time 100 to time 250, we were able to identify the cheating nodes with a high probability and this was because of the low mobility of the network. Furthermore, our technique can be modified for deterministic scenario also.

References

1. Johnson, David B., and David A. Maltz. "Dynamic source routing in ad hoc wireless networks." *Mobile computing*. Springer US, 1996. 153–181.
2. Li, Li Erran, et al. "A cone-based distributed topology-control algorithm for wireless multi-hop networks." *Networking, IEEE/ACM Transactions on* 13.1 (2005): 147–159.
3. Bao, Lichun, and J. J. Garcia-Luna-Aceves. "A new approach to channel access scheduling for ad hoc networks." *Proceedings of the 7th annual international conference on Mobile computing and networking*. ACM, 2001.
4. Bakht, Mehedi, Matt Trower, and Robin Hilary Kravets. "Searchlight: won't you be my neighbor?." *Proceedings of the 18th annual international conference on Mobile computing and networking*. ACM, 2012.
5. Dutta, Prabal, and David Culler. "Practical asynchronous neighbor discovery and rendezvous for mobile sensing applications." *Proceedings of the 6th ACM conference on Embedded network sensor systems*. ACM, 2008.
6. Kandhalu, Arvind, Karthik Lakshmanan, and Ragunathan Raj Rajkumar. "U-connect: a low-latency energy-efficient asynchronous neighbor discovery protocol." *Proceedings of the 9th ACM/IEEE International Conference on Information Processing in Sensor Networks*. ACM, 2010.
7. Kuriakose, Jeril, et al. "A review on localization in wireless sensor networks." *Advances in signal processing and intelligent recognition systems*. Springer International Publishing, 2014. 599–610.
8. Papadimitratos, Panos, et al. "Secure neighborhood discovery: a fundamental element for mobile ad hoc networking." *Communications Magazine, IEEE* 46.2 (2008): 132–139.
9. Kuriakose, Jeril, et al. "A review on mobile sensor localization." *Security in Computing and Communications*. Springer Berlin Heidelberg, 2014. 30–44.
10. Amruth, V., et al. "Attacks that Downturn the Performance of Wireless Networks." *Computing Communication Control and Automation (ICCUBEA), 2015 International Conference on*. IEEE, 2015.
11. Taylor, Michael AP, Jeremy E. Woolley, and Rocco Zito. "Integration of the global positioning system and geographical information systems for traffic congestion studies." *Transportation Research Part C: Emerging Technologies* 8.1 (2000): 257–285.
12. Kuriakose, Jeril, V. Amruth, and R. Ihram Raju. "Secure Multipoint Relay Node Selection in Mobile Ad Hoc Networks." *Security in Computing and Communications*. Springer International Publishing, 2015. 402–411.
13. Jacquet, Philippe, et al. "Priority and collision detection with active signaling-the channel access mechanism of hiperlan." *Wireless Personal Communications* 4.1 (1997): 11–25.

14. Kuriakose, Jeril, and Sandeep Joshi. "A Comparative Review of Moveable Sensor Location Identification." *International Journal of Robotics Applications and Technologies (IJRAT)* 3.2 (2015): 20–37.
15. Thomas, Federico, and Lluís Ros. "Revisiting trilateration for robot localization." *Robotics, IEEE Transactions on* 21.1 (2005): 93–101.

MAICBR: A Multi-agent Intelligent Content-Based Recommendation System

Aarti Singh and Anu Sharma

Abstract This study aims at proposing an intelligent and adaptive mechanism deploying intelligent agents for solving new user and overspecialization problems that exist in Content Based Recommendation (CBR) systems. Since the system is designed using software agents (SAs), it ensures highly desired full automation in web recommendations. The proposed system has been evaluated and the results suggested that there is an improvement in positive feedback rate and the decrease in recommendation rate.

Keywords Content · Ontology · Overspecialization · New user problem
Recommendation · Semantic · Software agents

1 Introduction

Adaptive web sites that provide personalized experience to web users are very effective in reducing the surfing time and searching the relevant information from internet [1]. Many techniques are discussed in literature to provide recommendations to the user [2]. Some of the important techniques are CBR, Collaborative Filtering (CF), Case Based Filtering (CBF), hybrid, and intelligent recommendation techniques [3]. CBR considers the contents of items/web page for generating recommendations to the web user. Some of the important drawbacks of CBR are limited content analysis, new user problem, and overspecialization. Many researchers [4–6] have put their efforts in solving these problems in CBR by proposing hybrid and intelligent techniques. With the advent of semantic web, orientation of web has been changed to knowledge provider rather than information dissemination medium. The amalgamation of semantic web technologies with

A. Singh (✉) · A. Sharma
MMICT & BM, Maharishi Markandeshwer University, Haryana, India
e-mail: singh2208@gmail.com

A. Sharma
e-mail: anu@iasri.res.in

D.K. Mishra et al. (eds.), *Information and Communication Technology for Sustainable Development*, Lecture Notes in Networks and Systems 10,
https://doi.org/10.1007/978-981-10-3920-1_41

ontologies and SA seems to be an effective solution to solving inherent problems in CBR system.

This study aims at solving the new user problem and overspecialization problem efficiently and effectively by proposing a fully automated system through application of Multi-Agent Systems (MAS) and ontology [7]. Semantic enhancement of user's preference through domain ontology and semantic association discovery in user profile database is considered for better recommendations and solving new user problem. Diversification is provided by incorporating the group preferences with user profile. Further, the working of the proposed system relies on the availability of domain ontologies. So, an automatic construction mechanism for ontology creation through an agent-mediated automated process is given. The paper is organized in five sections. Section 1 deals with introductory concepts and Sect. 2 with the related work along with the motivation for undertaking the study. Section 3 describes the proposed recommendation system in detail. The flow of the information and algorithmic details of proposed SAs in the system has been described in Sect. 4. Empirical evaluation of the system is given in Sect. 5. Next section discusses the related work and motivation for the study.

2 Related Work

This section attempts to enlist some of the important work done by researchers in solving various concerned issues automatically or through SA. FilterBots had been used by [4] which implicitly assign rating to the contents of the new UseNet group documents. This approach helped in improving the coverage and accuracy of a CF system. A MAS for personalized information gathering, named PIA, has been proposed by [8]. The PIA system has many agents co-operating and coordinating with each other to perform various tasks like gathering, preprocessing, filtering, and presenting information to the user. These approaches might be extended by providing better recommendation approaches and adding inferential capabilities to the SA. Existing RS generates recommendations solely on the preferences of individual user and do not focus on expert's opinions and general user preferences. This aspect has been studied by [9]. They proposed fuzzy cognitive personalized agent which are capable of learning the preferences and making inferences automatically. But, this study does not provide the detailed mechanism for handling new user problem. A MAS approach has been proposed by [10] for personalized recommendation system. But, the system structure, algorithms, and implementation can be improved further. A MAS has been proposed by [11] for ontology-based RS. This approach aims at solving many problems prevailing in RS using agents for handling each of these. This approach may be further extended by developing more refined algorithms for filtering agents.

Overspecialization problem has been considered by [12]. User interest and domain ontology are automatically created by observing the user preference, demographic data, and by integrating information from multiple resources. But this approach does not use SA. Semantic reasoning approaches had been applied by [13] along with ontologies in knowledge-based RS. A user's interest ontology is created by extracting concepts from domain ontology based on user preferences and inferred hidden semantic associations. Recommendation phase processes this information with CBR approach for generating the recommendations. This approach may be further taken forward by applying it on collaborative and hybrid approaches and SA. Various approaches for RS along with their limitation have been surveyed by [14]. They have proposed a hybrid intelligent recommendation approach combining utility-based and knowledge-based filtering for calculating utility function automatically. This approach overcomes the problem of calculating utility function manually. But this study does not use sematic web technologies and SA. Knowledge integration tools have been used by [15] to generate a group profile recommendation model to solve the new user problem by generating recommendations using hierarchical group profile. But this approach does not use SA.

From the above literature review it is clear that researchers are putting their efforts toward providing efficient recommendation of web contents. However, approaches presented so far lack scalability. Considering the size of the web and rate of information/content generation, scalability, and automation is highly desired in recommendation systems. Researchers have advocated the usage of SA for this purpose. However, a limited number of agent-based applications are available due to the lack of global and domain ontologies. Construction of domain ontology is considered as a difficult task. So there is a need to create an automatic ontology construction mechanism for creating feasible MAS-based solution. Also, there is a need to solve new user and overspecialization problem in CBR using semantic web technologies. This study is undertaken with the aim to solve these issues. The next section describes the proposed system.

3 The Proposed Recommendation System

This study proposes a multi-agent RS framework, named MAICBR—Multi-Agent Intelligent Content-Based Recommendation System, consisting of three modules, namely, agent mediated Semantic Profile Enhancement Module (SPEM), agent mediated Automatic Ontology Construction Mechanism (AOCM), and Agent based Recommendation Engine (ARE). SPEM has two sub-modules namely ontology-based Semantic Group Profile Enhancement Module (SGPEM) and ontology-based individual User Profile Enhancement Module (UPEM). The high-level view of MAICBR framework is shown in Fig. 1. Each module in the framework is attached with SA that performs its job independently. Various SA cooperate with each other by passing and receiving messages. Figure 2 elaborates detailed architecture of MAICBR.

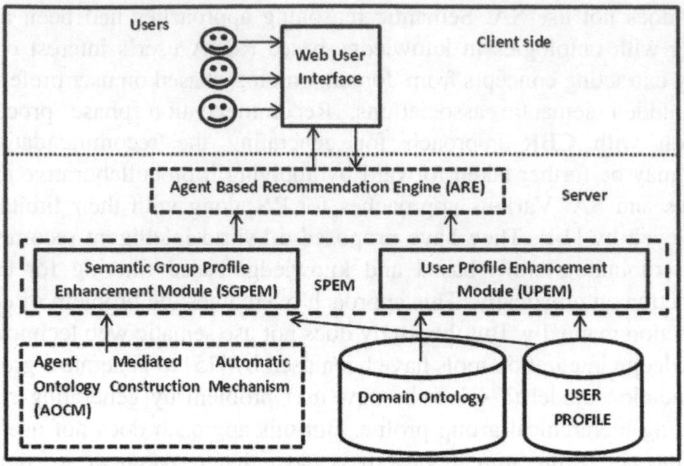

Fig. 1 High-level view of MAICBR framework

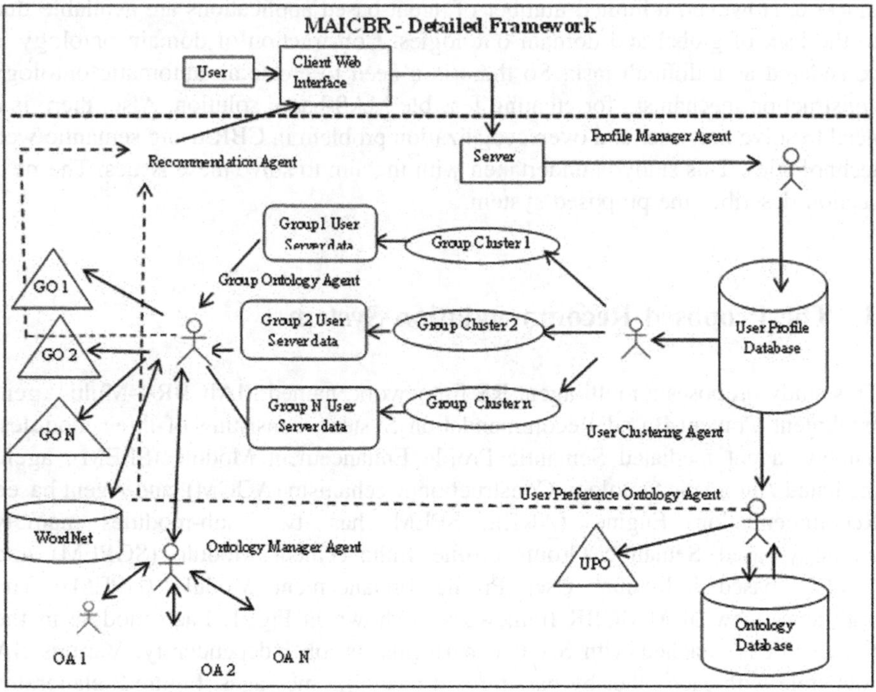

Fig. 2 Detailed architecture of MAICBR

(i) SPEM consist of four agents, namely, User Clustering Agent (UCA), Group Ontology Agent (GOA), Ontology Manager Agent (OMA), and User Preference Ontology Agent (UPOA). Functionality of each agent is as follows:

- UCA groups the user profile into various groups where all the members in the group share similar interests. This agent uses k-means clustering algorithm for making similar user groups [16].
- GOA monitors and processes the active sessions of all the users in a group to form ontology of items of interest for whole group termed as Group Ontology (GO).
- UPOA uses domain ontology and individual user profile to form User Preference Ontology (UPO).
- OMA is responsible for automatic semantic enhancement of top extracted concepts for each group. OMA advertises the concept to software agents named as Ontology Agents (OA) for relevant ontology. If any of the existing OA responded positively on availability of the ontology, then OMA sends a request for ontology schema. If none of the OA responded positively, then OMA sends a signal to GOA which generates the ontology using WordNet lexical database using the algorithm described in Fig. 9.

All these agents coordinate and interact with each other to generate UPO and GO.

(ii) ARE module consists of a Recommendation Agent (RA) which works in coordination with UPOA and GOA to generate recommendation. RA generates the two Recommendation List (RL) based on individual and group preferences. User may specify the explicit weightage to user and group preferences or otherwise equal weightage is given to both by RA. It also stores the implicit user feedback and readjusts the weights automatically. For recommendations based on group preferences, a negative feedback from user results in discarding the concepts from recommendation list in future for short term. Detailed working of MAICBR is discussed next.

Whenever a user starts searching something through web user interface, its profile is created in user profile database. UPEM uses spreading activation technique [17] to enhance the user profile from related concepts existing in domain ontology on query keywords. Consider a restaurant ontology (http://dbpedia.org/ontology/Restaurant) representing the various types of French Cuisine under subclass French Cuisine of class Cuisine in restaurant ontology. Initial user profile is expanded using sub-concepts from this ontology (which serves as the domain ontology) by SPEM. Concepts in the profile are matched with the concepts in domain ontology. All the suitable concepts are assigned an initial weight w and the further sub-concepts are assigned a weight w/2. A revised user profile is created as indicated in Fig. 3. These sub-concepts are randomly recommended to the user and user feedback is stored. Table 1 provides the initial and enhanced user profiles generated by UPEM. The remaining values in enhanced preferences column would also be inserted by matching the concepts with other domain ontologies.

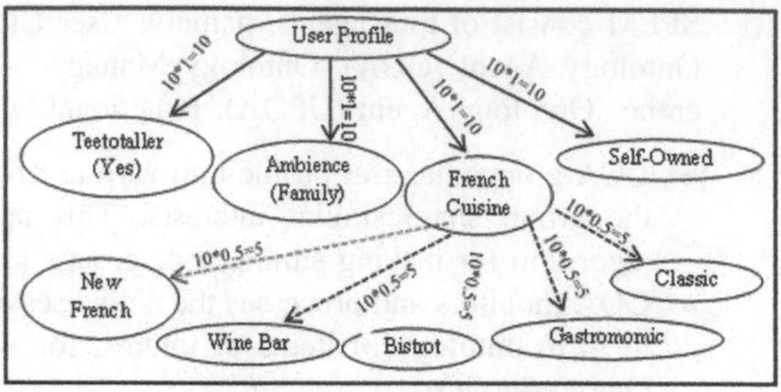

Fig. 3 Semantically enhanced user profile

Table 1 Initial and enhanced user profile

Preference	Initial preferences	Enhanced preferences
Teetotaller	Yes	Juice, Mojito
Cuisine	French Cuisine	New French Cuisine
Transport	Self-owned Car	Figo

Now, to deal with overspecialization problem, SPEM makes use of SGPEM module. SGPEM creates the groups of similar users using clustering. It also identifies aggregated preferred concepts in the group and creates a group ontology (GO) for each identified concept. GO is used for generating new recommendations that may be of interests to the user. This mechanism is used for handling over-specialization problem. The details of this mechanism are given here.

SGPEM uses UCA to cluster the users in different groups using k-means clustering algorithm. For each group, top trending concepts are identified from the group server log data. A group server log data consists of web server log data for session of each user in the pre-computed cluster. Suppose there are n clusters and each cluster has same number of users (m), then for each user important keywords have been identified from the web usage log data. A group session G_j is then the union of keywords by all users in that group.

Session: $(S)_i = \{kw_1, kw_2, kw_3 \ldots kw_m\}$ where S_i is the log data for a session for ith user.

Group: $G_j = \{S_1 \cup S_2 \cup S_3 \ldots \cup S_n\}$ where G_j is the group corresponding to cluster j and 1 ≤ i ≤ m and 1 ≤ j ≤ m. For each G_i, SGPEM prepares a term weight matrix where term is keyword (K_i) found and weight (w_i) is the frequency of the keyword.

$$\text{Term} - \text{weight matrix, } T_i = \begin{vmatrix} K_1 & w_1 \\ K_2 & w_2 \\ K_3 & w_n \end{vmatrix}.$$

Top five trending keywords are extracted from T_i. SGPEM requests for retrieval of domain ontology either from ontology database or by AOCM mechanism for the identified concepts. In case, the domain ontology is not provided by the above methods, and it is created automatically using lexical database WordNet

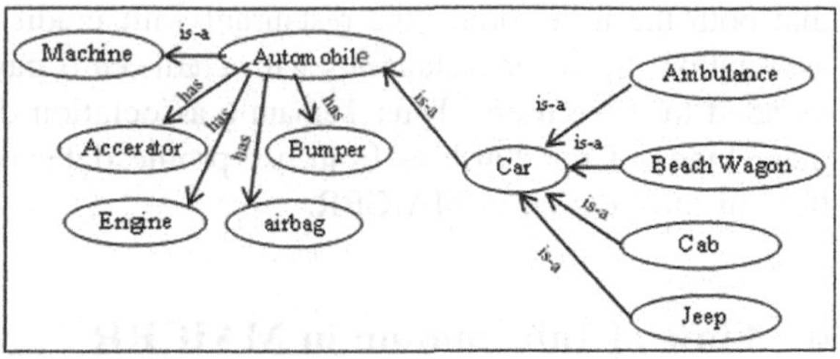

Fig. 4 Car ontology created by SGPEM using WordNet

(https://wordnet.princeton.edu/) by SGPEM. Let us suppose that most trending topic for a group cluster is *car* and there is no ontology available in database as well as provided by AOCM. So it is desired to create ontology. The concept *car* is passed to WordNet lexical database and the semantic information retrieved is used to create ontology for concept *car* by SGPEM as shown in Fig. 4.

Another approach used by MAICBR is inferring the association between the users through use of semantic association discovery to identify the potentially important but hidden relations between individual users and within the group from ontology. Semantic relationship between various classes in ontology is identified using the notion of Property Sequence (PS) [18]. A PS consists of various class instances in ontology linked together by common properties with first class instance known as origin and last as terminator. A PS is represented as finite sequence of properties $P_1, P_2, P_3 \ldots P_n$ where each P_i is a property defined in a RDF schema R_j as shown in Fig. 6. The interpretation of a property sequence S is the set of ordered sequences of tuples:

$$S^I = \{[(a_1, b_1), (a_2, b_2), \ldots (a_n, b_n)]\} | (a_i, b_i) \text{ is in } P_i^I$$
$$PS = \{P_1, P_2, P_3, P_4\}.$$

The node P_1 is called the origin of the sequence and P_4 is the terminus. The function NodesOfPS() returns the ordered set of nodes in a property sequence PS. The approach is based on the use of a query operator ρ (Rho).

ρ-path association: Two class instances are ρ-path associated in an ontology if it is possible to find a property sequence with a origin and terminus.

ρ-join association: A set of property sequences $PS_1, PS_2, PS_3 \ldots PS_n$ are called joined if they satisfy an interpretation I and

$$\cap NodesOfPS(PS_i) \neq 0 \; for \; i = 1 \ldots n$$

i.e., the sequences $PS_1, PS_2, PS_3 \ldots PS_n$ intersect at some node n.

As an example of the concept, let us consider two users. User 1 goes to restaurant 1 with credit card payment facility for lunch with friends and User 2 goes to same restaurant 1 for lunch. Using *ρ-path association*, it can be inferred that the user 1 is interested in restaurant 1, so there is an association between user 1 and restaurant 1 which are the instances of different classes in ontology. The relation that both the users prefer the restaurant with credit card facility is type of ρ-join association. So other restaurants with credit card payment facility may be recommended to these users. Thus, semantic association discovery is used for inferring these kinds of relationships from use profile database. Next section describes the flow of information in MAICBR.

4 Flow of Information in MAICBR

This section explains the interaction and coordination between various agents discussed in the proposed framework. The flow of information among agents is represented in Fig. 5.

- ***Step 1***: User profile information is collected from client machine and is passed to PMA.
- ***Step 2***: PMA stores the information in User Profile Database (UPD).
- ***Step 3 and 4***: UPOA sends a request to PMA for accessing the UPD. After the request is authenticated, PMA grants permission to UPOA for accessing UPD.
- ***Step 5 and 6***: UPOA uses the ontology database and user profile to create UPO.
- ***Step 7 and 8***: CA sends a request to PMA for accessing the UPD. After the request is authenticated, PMA grants permission to CA for accessing UPD. CA uses the UPD to create groups of users sharing similar interest.
- ***Step 9***: GOA access the web server log data for users in each group, preprocess it, and identifies the top N topics trending in that group.

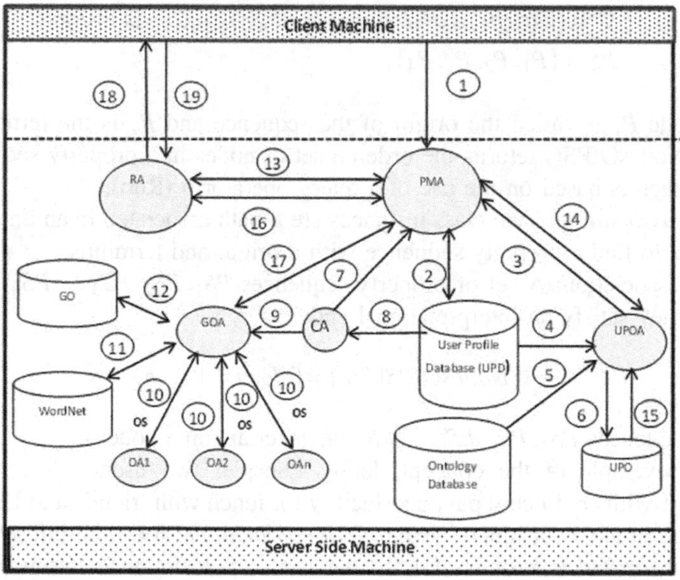

Fig. 5 Flow of information in MAICBR

- ***Step 10***: Once the top trending topics have been identified, GOA advertises the request for domain ontology for identified concepts to existing registered ontology agents. In case, the domain ontology is available with any one of the agents, and GOA requests for the OS for the found ontology.
- ***Step 11 and 12***: Otherwise, if none of the agents responded positively to the GOA's request, GOA uses the WordNet lexical database and automatically creates GO. Semantic enhancement of trending topic is done using this GO. GOA identifies the new topics of user interest and thus helps in providing a solution to overspecialization problem.
- ***Steps 13, 14, and 15***: RA sends a request to PMA for accessing the UPO. PMA authenticates the request and sends a request to UPOA for accessing UPO. PMA receives the UPO from UPOA and sends it to RA.

Fig. 6 Algorithm for UPOA

```
Input: Initial User Profile (U_i) and Ontology(O_i)
Output: Semantically Enhanced User Profile
{
For each user U_i in User Profile Database
C_e = explicitly specified concepts in U_i
O_F = Concept Found in Ontology

For each C_j in C_e
        WC_j = 1
        Add < C_j, WC_j > to U_i
        O_F =MapConcept (C_j, O_i)
        For each subConcept in O_F
                ChildWeight = C_j * 0.5
                Add < C_j, ChildWeight > to U_i
        End For
End For
End For
Return U_i
}
```

Fig. 7 Algorithm for OMA

```
Input: Concept Name, Ontology Agent ids
Output: Ontology Schema (OS), Agent id
Steps:
    Domain_Name = Concept Name
    For all the OA_I in Ontology Agents
        Send_Request_for_Ontology(Domain_Name, OA_I)
        R = Get_Response(OA_I)
        If (R > Yes) then
         Send_Request_for_Ontology_Schema(OA_I)
         Receive (OS, OA_I)
         Return(OS,OA_I)
         Exit;
        End if
  End For
End;
```

- ***Steps 16 and 17***: RA sends a request to PMA for accessing the GO. PMA authenticates the request and sends requests to GOA for accessing GO. PMA receives the GO from GOA and sends it to RA.

```
Input: User Profile, User Preference Ontology (UPO), Group
Ontology(GO)
Output: RL1 – Recommendation list based on enhanced user profile
        RL2 - Recommendation list based on group ontolgy
Steps:
      Send_request (UPO, UPOA) // sends a request to UPOA for UPO
      Receive (UPO, UPOA)      // Receive UPO from UPOA
      Send request (GO, GOA)   // sends request to GOA for GO
      Receive (GOA, GOA)       // receive GO from GOA
      For items in NewItemList
      {
         Compare the Item with enhanced user_profile
         If similar then
            Add item to RL1
         End if
         // Handling Overspecialization problem
         Compare the Item with GO
         If similar then
            Add item to RL2
         End if
      }
Return RL1, RL2
End;
```

Fig. 8 Algorithm for RA

```
Input: WLU_i – Web log usage data for each session in cluster C_i in C, Ontology
(O_i)
Output: Group Ontology (GO)
Steps:
  TC = null      // Term Weight Matrix for all clusters
 For each C_i in Cluster set C {
    TC_i = null // Term Weight matrix for C_i
    For each WP_i in WLU_i {
       Remove_Stop_Words(WP_i); // Procedure to remove stop words
       Stemming(WP_i);                     // Procedure to bring
       Calculate_Frequency_Words( WP_i) // Count of words in WP_i
       T_WPi=Extract_Top_Five_Words(WP_i) // Term-Weight Matrix for WP_i
       TC_i = TC_i + T_WPi //Store the top five words from the webpage to TC_i}}
  TC = Extract_Top_Five_Frequent_Words (TC_i)
  For each keyword (K_i in TC){
   If (Map (K_i, O_i) is true{
     GO = O_i;}
   Else if (get_Ontology(OMA) is true){
     GO = get_Ontology(OMA);}
  Else{
  GO = CreateOntology_WordNET); }
 Enhance the group profile with GO
Return GO
End
```

Fig. 9 Algorithm for GOA

- ***Steps 18 and 19***: Finally, RA uses UPO and GO to generate recommendations. It also stores the user feedback and updates GO and UPO. The algorithms for various agents involved in the framework are shown in Figs. 6, 7, 8, and 9.

5 Evaluation

For the purpose of evaluation of the proposed algorithm, we have selected the 100 web users who wish to know information about the famous and new restaurants. User profiles database have been prepared using explicit method of information collection with parameters (name, location, latitude, longitude, smoke, drink, ambience, transport, marital status, profession, religion, weight, date of birth). Restaurant database has been prepared with information collected on parameters (name, longitude, latitude, address, city, state, country, fax, zip, smoking, alcohol, price, ambience, area, other services, parking, timings, and credit card facilities). Google forms have been designed and circulated, and responses from 100 users have been collected.

Restaurant ontology (which servers as domain ontology) has been used to semantically enhance the user profile. These users are then clustered into five groups based on demographic data. For each of the groups, top five restaurants have been identified and these restaurants are recommended to users in the same cluster and the users in the other group. User feedback, for each of the recommendations, is recorded and stored for future use. It was observed that clustering-based recommendation with semantic profile enhancement resulted in better Feedback Rate (FBR) and decreased rate of recommendation as shown in Figs. 10 and 11.

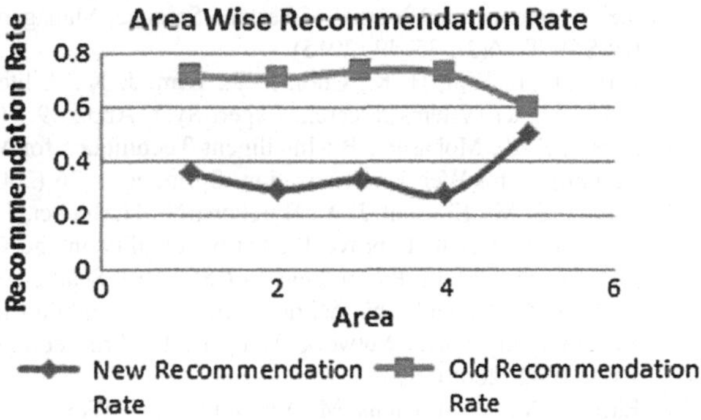

Fig. 10 Comparison of old and new recommendation rate

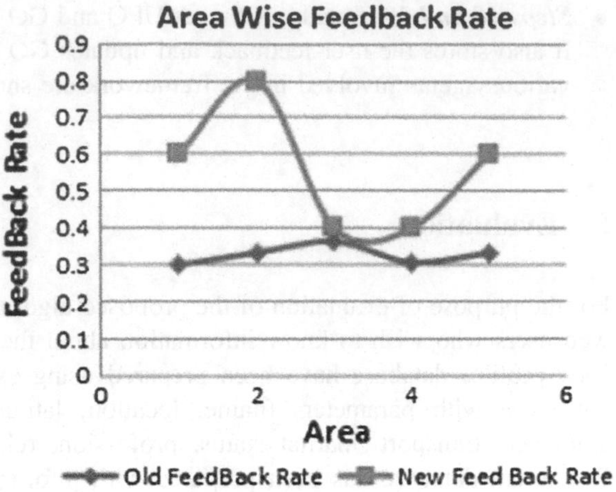

Fig. 11 Comparison of old and new feedback rate

6 Conclusion

This paper contributes a mechanism focusing on new user overspecialization problem existing in CBR by deploying software agents. Results indicated that this mechanism provides better user satisfaction with no additional manual operations and out performs the existing methods. The proposed mechanism has been implemented on a limited data set and its evaluation with real large dataset is still under process.

References

1. Singh, A., Sharma, A., Dey, N.: Semantics and agents oriented web personalization –state of the art. International Journal of Service Science, Management, Engineering, and Technology (IJSSMET), 6(2), 35–49 (2015)
2. Park, D. H., Kim, H. K., Choi, I. Y., Kim, J. K.: A literature review and classification of recommender systems research. Expert Syst. App., 39, 10059–10072 (2012)
3. Anand, S. S., Mobasher, B.: Intelligent Techniques for Web Personalization. In: Intelligent Techniques for Web Personalization, Springer, 1–36 (2005)
4. Sarwar, B. M., Konstan, J. A., Borchers, N., Herlocker, J., Miller, B. Miller, Riedl, J.: Using Filtering Agents to Improve Prediction Quality in the GroupLens Research Collaborative Filtering System, In: Proceedings in CSCW'98 Seattle, Washington, USA, 345–354 (1998)
5. Debnath, S., Ganguly, N., Mitra P.: Feature Weighting in Content Based Recommendation System Using Social Network Analysis, In: Proceedings in WWW 2008, Beijing, China, 1041–1042 (2008)
6. Eirinaki, M., Vazirgiannis, M., Varlamis, I.: SEWeP: Using Site Semantics and a Taxonomy to Enhance the Web Personalization Process, In: Proceedings in SIGKDD '03, Washington DC, USA, 99–108 (2003)

7. Singh, A. Juneja, D., Sharma, A. K.: Design of an intelligent and adaptive mapping mechanism for multiagent Interface. In: proceedings in International Conference on High Performance Architecture and Grid Computing (HPAGC'11), 373–384 (2011)
8. Albayrak, S., Wollny, S., Varone, N., Lommatzsch, A., Milosevic, D.: Agent Technology for Personalized Information Filtering: The PIA-System. In: ACM Symposium on Applied Computing, 54–59 (2005)
9. Miao, C, Yang Q., Fang H., Goh. A.: A cognitive approach for agent-based personalized recommendation. Knowl-Based Syst., 20(4), 397–405 (2007)
10. Huang, L., Dai, L., Wei, Y., Huang, M.: A personalized recommendation system based on multi-Agent. In: Proceedings in Second International Conference on Genetic and Evolutionary Computing, 223–226 (2008)
11. Pan, P., Wang, C., Horng, G., Cheng, S.: The Development of an ontology-based adaptive personalized recommender system, In: Proceedings in International Conference on Electronics and Information Engineering (ICEIE 2010), vol. 1, V176–V180 (2010)
12. Ge, J., Chen, Z., Peng, J., Li, T.: An ontology-based method for personalized recommendation. In: Proceedings in 11th IEEE Int. Conf. on Cognitive Informatics & Cognitive Computing, 522–526 (2012)
13. Blanco-Fernández, Y., López-Nores, M., Gil-Solla, A., Ramos-Cabrer, M., Pazos-Arias, J. J.: Exploring synergies between content-based filtering and spreading activation techniques in knowledge-based recommender systems. Inform Sciences, 181(21), 4823–4846 (2011)
14. Kahara, T., Haataja, K., Toivanen, P.: Towards more accurate and intelligent recommendation Systems. In: Proceedings in 13th International Conference on Intelligent Systems Design and Applications (ISDA), 165–171 (2013)
15. Maleszka, M., Mianowska, B., Nguyen, N. T.: A method for collaborative recommendation using knowledge integration tools and hierarchical structure of user profiles. Knowl-Based Syst., 47, 1–13 (2013)
16. Han, J., M. Kamber: Data Mining: Concepts and Techniques, 2nd edition, Morgan Kaufmann Publisher, ISBN 1-55860-901-6. (2006)
17. Rocha, C., Schawabe, D., Poggi, M.: A hybrid approach for searching in the semantic web. In: Proceedings in 13th International World Wide Web Conference (WWW-04), 74–84 (2004)
18. Anyanwu, K., Sheth A.: ρ-Queries: enabling querying for semantic associations on the semantic web. In: 12th International World Wide Web Conference (WWW-03), 115–125 (2003)

HMM-Based IDS for Attack Detection and Prevention in MANET

Priya Pathak, Ekta Chauhan, Sugandha Rathi and Siddhartha Kosti

Abstract MANETs are wireless networks which communicate without BS and centralized control nodes. Due to its mobile nature of nodes, topology of the network changes frequently. So it is most difficult to stimulate this network. The main task is to provide an efficient and effective routing in MANETs with limited resources. As MANET is an open medium, it is open to numerous attacks by the attackers. To avoid attacks, a good intrusion detection and prevention system is developed. This paper gives a brief survey about different IDS developed to protect attacks in MANET have been briefed. To strengthen the security of IDS, we propose a hidden Markova model-based IDS for MANET for preventing network from attacks. HMM implements learning on the nodes of the network. Based on this learning, the results show the best possible positions and probability of the attacker node.

Keywords MANET · IDS · Black hole · Gray hole · HMM

P. Pathak (✉)
MPCT, Banasthali, Gwalior, India
e-mail: shakhi.priya@gmail.com

E. Chauhan
ITM, Tonk, Gwalior, India
e-mail: ektachauhan81@gmail.com

S. Rathi
SRCEM, Gwalior, Rajasthan, India
e-mail: sugandharathi@gmail.com

S. Kosti
Indian Institute of Technology Kanpur, Kanpur, India
e-mail: sidmits05@gmail.com

D.K. Mishra et al. (eds.), *Information and Communication Technology for Sustainable Development*, Lecture Notes in Networks and Systems 10,
https://doi.org/10.1007/978-981-10-3920-1_42

1 Introduction

Wireless mobile communications are generally two types. Infrastructure wireless networks are the first type, where a mobile node communicates with the BS that is located within its range of transmission. MANETs consist of fixed or mobile nodes which can transmit and receive data, and nodes are associated without using secure central or infrastructure management. These nodes have self-arranging and self-maintaining capability. Two different nodes know which kind of communicating if they are within other's transmission range reach otherwise routers serve as the intermediate routers. Because of MANETs features, they are used in applications, for example, military conflict, medical emergency recovery, and human-induced disasters. The challenges which MANET has to face are time-varying in wireless links with nature. The nodes are allowed to transfer randomly because of dynamic topology of MANET, i.e., normally nodes and multi-hop changes erratically and quickly at random times, and may contain both unidirectional and bidirectional links. Dynamic update is required to identify new nodes in the network. Power conservation is also one of the factors because they have limited battery power. Some attacks may try to engage the mobile nodes unnecessarily, so that they keep on using their battery for early drainage. Because of lack of centralized management, it becomes problematic to the track attacks in MANET [1].

1.1 Attacks in MANET

MANET attacks can be divided into two types: active and passive attacks. In passive attack, they add unauthorized listening in network and information is transferred without any modification. In active attack, they extract knowledge and they permit data flow between various nodes [2].

The main three ad hoc layers that take part in the routing mechanism are physical layer, network layer, and MAC layer. In MANET all nodes behave like a router and forward packets, so it is easy for attacker to get into network. Main idea behind network layer attack is to place itself between the source and destination [2].

1.1.1 Black Hole Attack

In the black hole attack, the attacker creates vulnerabilities use in AODV routing discovery technique, DSR routing protocols [3]. It sends data from source to destination node, so that node with sequence number of highest destination than present destination sequence amount of node will reply and the destination sequence number is higher than the current destination number. Getting this false RREP packet source node will select path from this malicious node, assuming that

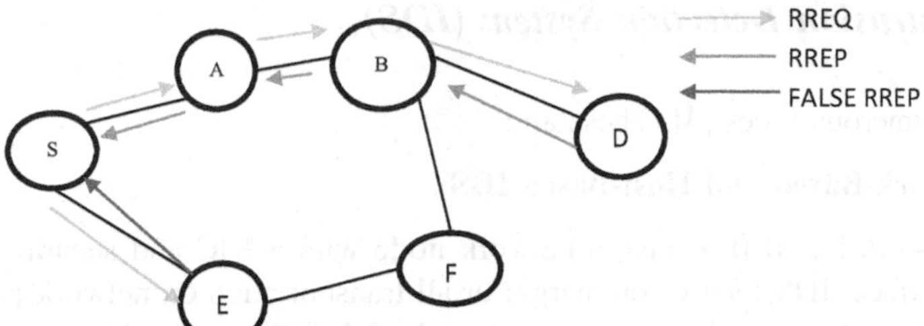

Fig. 1 *Black* hole attack

it is a fresh shortest path to destination. Then this malicious node can drop packets instead forwarding it. This attack type is known as a black hole attack (Fig. 1).

A source node "S" to the destination node "D" to the neighbors it sends RREQpacket. When node "E", malicious node, obtains RREQ request it sends RREPpacket advertising itself containing shortest route to the destination and it rejects RREP packet from legitimate route <S, A, B, D>. The S is a source node which starts packet forwarding via <S, E, F, B, D> route and node E, i.e., while passing via the packet was dropped by blackhole attacker.

There is a special difference between Gray hole attack and black hole attack. The attacker present in source middle in black hole attack. In this attack, the attention of data packets by advertising that they have the shortest route to the endpoint and when the data packet get attracted then they trap the data packet and drop it. While in gray hole, the there are selected data packets are drop or whether in statistical order. As we say that the data packets are dropped in a pattern or in any unique way or via any definite node [3] (Fig. 2).

Here the node E is an attacker node which drops the packet to the D node only and it sends packet via other nodes by generating a gray hole attack.

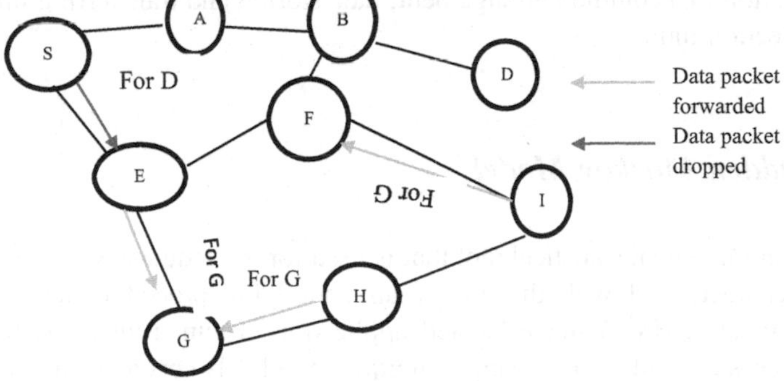

Fig. 2 *Gray* hole attack

1.2 Intrusion Detection System (IDS)

IDS is numerous types [4]. These are:

- **Network-Based and Host-Based IDS**

A network-based IDS have a network node with a NIC and separate management interface. IDS focuses on margin or all transportation on network part hence, it is located on network segment. In host-based IDS agents, the applications of software were connected on the workstations for monitored.

- **Active and Passive IDS**

The system is arranged for blocking the doubted attacks in evolution automatically without any interference with operator in active IDS. And in the passive IDS, the system is arranged only for monitoring and analyzing the activity of network traffic and attentive about possible attacks to an operator and weaknesses.

- **Knowledge Based and Behavior Based**

In knowledge-based IDS, the intrusion and attacks is recorded as an evidence in the database (DB) and the DB is mentioned when an attack or similarly the condition like attack is occurs so, each intrusion leaves a footprint and these footprints are denotes or used for identifying and stop such kinds of attacks in upcoming time. These kinds of footprints are known as signatures and will be used for identifying and preventing the attacks in future. In behavior-based IDS, the system's behavior is noticed and assumed intrusion if there is any non conformity via usual or projected behavior of the system.

Any set of activities which effort to cooperate with the confidentiality, availability or integrity of an element is called as Intrusion [5] and for the detection such kind of intrusions, the system is known as an IDS. Basic components are presented of an IDS, i.e., detection, set of information, and data response. The data collection elements are liable for collecting and preprocessing the data tasks, i.e., transferring of information to a common arrangement, data storing and transferring information to the detection unit.

1.3 Hidden Markov Model

HMM is an influential statistical tool that is used for the sequences of modeling that may be characterized with the aid of an underlying procedure an observation sequence creates. HMMs have located application in numerous areas focused on signal processing, and in processing of unique speech, also been useful with success to low stage NLP duties for example section-of-speech tagging, phrase chunking, and extracting goal know-how as of documents. Markov provided his identify to the mathematical thought of Markov approaches within the early twentieth century,

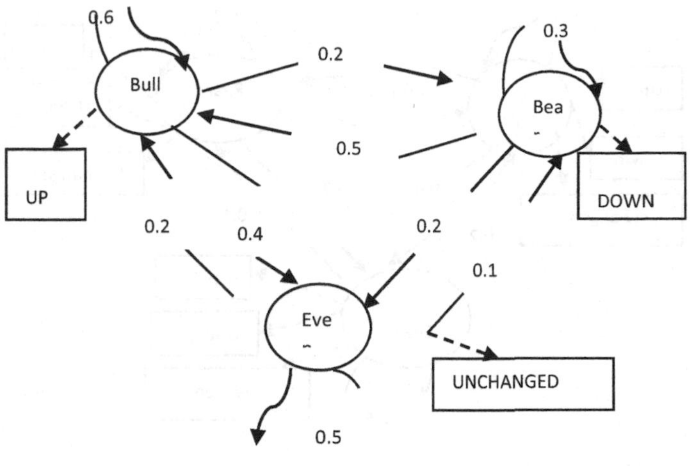

Fig. 3 Markov process example [12]

nevertheless it was Baum and his colleagues that developed HMMs speculation within the 1960s [6].

Markov Model, Fig. 3, depicts an illustration of a Markov system. The model provided defines a simply model for the inventory market index. Model is a finite state automaton, with probabilistic transitions between various states. Given an observations sequence, illustration: up-down-down are able to the readily confirm that state sequence that created these observations used to be: Bull-bear-bear, and the chance of the sequence is without difficulty the fabricated from the transitions, in this case 0.2 × 0.3 × 0.3.

Figure 4 shows an instance of how the earlier model may also be accelerated right into an HMM. Novel model now permits each statement symbols to be emitted from every state with a finite chance. This transformation types model much additional expressive.

Definition The HMM is a finite state machine variant consuming a suite of hidden states, H, an output alphabet, S, transition possibilities, M, output probabilities, N, and initial state chances, ɤ. The present state is not observable. Every state creates an output with a special likelihood (b). Most often states, H, and outputs, s, are understood, so an HMM is claimed to be a triple, (M, N, ɤ).

Hidden states $H = \{H_x\}, x = 1, \ldots, L$.

Transition probabilities M = lxy = P(hy) q +1, where (a) is conditional probability of a given a, q = 1,..., Q is time, and kx in K. Informally, A is the probability that the following state is ky given that the current state is kx.

Observations (symbols) $S = \{s_k\} = 1, \ldots, Z$.

Emission probabilities N = zxk = zx(s_k) = P(s_k|kx), sk = S. Informally, Z is the chance that the output is $\{s_k\}$ due to the fact the current state is kx. Preliminary state probabilities Π = px = P(kx at q = 1).

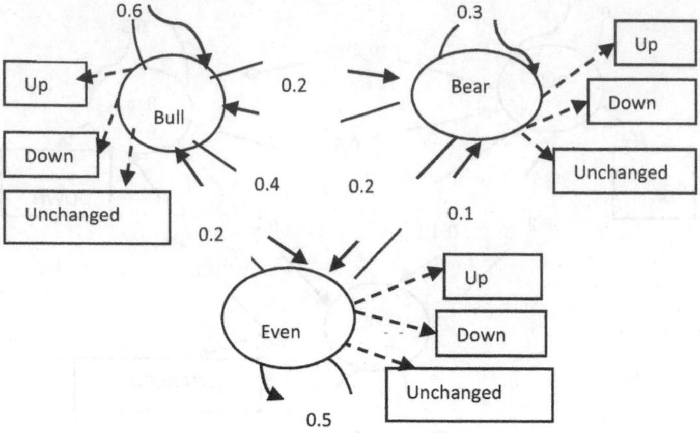

Fig. 4 Earlier Markov process example

Canonical problems

There are three canonical problems that are solved with **HMMs**:

i. Parameters of the model, compute likelihood of a designated output sequence. This concern is solved through the ahead and backward algorithms.
ii. Parameters of the model, to find the definitely states sequence which might have produced by specified output sequence. Resolved with the help of the Viterbi algorithm and posterior decoding.
iii. Output sequence, for finding state transition and output probabilities. Resolved using Baum–Welch algorithm.

2 Literature Survey

Modified DSR protocol: Searching and eliminating of particular choose attack of black hole in MANET proposed three steps of MDSR algorithm [7], The IDS nodes might be change to the immoral or indiscriminate mode after knowing that the occurrence of interrupting nodes. DOS Attack prevention: modifying finding the intrusion and elimination proposed DOS which generates fake alarm rates high and searching will be low [8]. AIDP reveals a high success price and very low false alarm expense with a cheap processing overhead. MANET security: GIDP proposed GIDP is an allowance to AIDP so the united technique of abnormality and knowledge-based ID [9] is a unique which takes benefit of these two techniques. It avoids various types of attacks, and it also have the capability to search new attacks and disturbing activities which increase performance humiliation. Intrusion prevention and detection system: Different MANET attacks proposed that CH uninterruptedly collects the parameter of network characteristic matrix and derived matrix and expresses the primary training summaries, blocks the infinite procedures of attacks in the MANETS but also have the probability to identify new unexpected attacks. The CH gathers parameters of the NCH and DM and conveys the main training summaries [10]. ID (Intrusion detection), Adaptive intrusion response, and Attack identification Intruder Identification [11].

3 Proposed Work

In our proposed work, we apply hidden Markov model for calculating reputation of vehicles, reputation is a one of the most important factors for communication; reputation value updates regularly on the basis of vehicles behavior either it pass the data to RSU or report for a particular condition. With the help of HMM, we calculate reputation. HMM have two defying properties:

1. Assume some time t observation of states is hidden observer.
2. Hidden observation state satisfy the Markov property.

In VANET scenario, we observe that vehicles have two states like either it malicious or trusted on the basis of these two state observations of these states are forward packet, drop packet, or send.

Case1: congestion

In case of congestion vehicles report RSU that due to overused of network bandwidth congestion occur so that packet drop happen, but if node is malicious it cannot be report RSU in this condition HMM find out the hidden state of vehicles and find out the malicious behavior execution of hmm in this case (Tables 1 and 2).

Case2: normally a node generate a packet within a limit if node generate more than packet than in this condition hmm execute to find the malicious behavior in case of denial of service attack.

In this condition, transition probability table is for vehicles transition or emission matrix for above scenario (Tables 3 and 4).

For above matrixes, we pass sequence (2, 1, 3).

Outcome of these input matrixes (Tables 5, 6, 7 and 8).

Table 1 States showing nodes

States	Safe	Malicious
Probability	0.8	0.2
	0.4	0.6

Table 2 Emission probability for vehicle

States\Observation	Send	Drop	Forward
Safe	0.1	0.4	0.5
Malicious	0.6	0.3	0.1

Table 3 Transition matrix

State/Observation	Safe	Malicious
Safe state	0.7	0.3
Malicious state	0.4	0.6

Table 4 Emission matrix

	X1	X2	X3
Safe state	0.1	0.4	0.5
Malicious	0.5	0.3	0.1

Table 5 New transaction matrix

State/Observation		
Safe state	0.4273	0.2105
Malicious state	0.5727	0.3629

Table 6 New emission matrix

	X1	X2	X3
Safe state	0.1008	0.2561	0.2615
Malicious	0.3093	0.1382	0.1322

Table 7 Forward path

1.000	0.2800	0.0232	0.0247	0.0195
0	0.1200	0.11040	0.0076	0.0144

Table 8 Backward path

0.0144	0.0345	0.1640	0.4000	0
0.0112	0.0395	0.0960	0.6000	1.000

Path Outcome sequence is:

1	2	1

4 Conclusion

Wireless technology is used greatly in today's generation and it is must and applicable where fixed infrastructure, that is, wired technology cannot be used. But with increased numbers of users, security issues arises so intrusion detection systems play a vital role in this situation. The infrastructure less ad hoc network provides greater flexibility but it has pros and cons with flexibility numbers of mobile devices that can connect to network but due to no centralized control attacks detection it becomes a difficult task. Numbers of intrusion detection systems were developed. In this paper, summary of different IDS which is developed till now is summarized.

References

1. Apurva Kulkarni, Mr. Prashant Rewagad, Mr. Mayur Agrawal; "Literature Survey on IDS of MANET". International Journal of scientific research and management (IJSRM), (2015).
2. Athira V Panicker, Jisha G; "Network Layer Attacks and Protection in MANETA Survey". International Journal of Computer Science and Information Technologies, Vol. 5 (3), (2014), 3437–3443.
3. Adnan Nadeem, Michael P Howarth, "A Survey of MANET Instrusion Detection & Prevention Approaches for Network Layer Attacks", Proc. IEEE Communications Surveys & Tutorials, Vol. 15, No. 4, Fourth Quarter (2013).
4. Ranjit j. Bhosale, Prof. R.K. Ambekar; "A Survey on Intrusion detection System for Mobile Ad-hoc Networks". International Journal of Computer Science and Information Technologies, Vol. 5 (6), (2014).
5. Anjum F, Talpade R, LiPaD: Lightweight Packet Drop Detection for Ad hoc Networks. In Proc of IEEE Veh Technol Conf (VTC) 2:1233–1237, (2004).
6. Elhadi, M. Shakshuki., Nan Kang, Tarek R. Sheltami, "Eaack—a secure intrusion-detection system for manets". IEEE transactions on industrial electronics, vol. 60, March (2013).
7. M. Mohanapriya, Ilango Krishnamurthy, —Modified DSR Protocol for Detection and Removal of Selective Black Hole Attack in MANET||, Computers and Electrical Engineering, Elsevier science, (2013).
8. R. H. Akbani, S. Patel and D. C. Jinwala, DoS attacks in mobile ad hoc networks: A survey, in Proc. 2nd Int. Meeting ACCT, (2012).
9. A. Nadeem and M. Howarth, "Adaptive intrusion detection and prevention of Denial of Service attacks in MANETs", Proceeding of ACM 5th International Wireless Communication and Mobile Computing Conference, (2008).
10. Nadeem, A., & Howarth, M., A generalized intrusion detection & prevention mechanism for securing MANETs. In Proceedings of IEEE international conference on ultra modern telecommunications & workshops, St. Petersburg, Russia, (2009).
11. A. Nadeem, M. Howarth, A. Nadeem, M. Howarth, Protection of MANETs from a range of attacks using an intrusion detection and prevention system, Telecommunications Systems Journal Springer, (2013).
12. A. Nadeem, M. Howarth, An intrusion detection & adaptive response mechanism for MANETs, Elsevier Journal of Ad Hoc Networks, 368–380, (2014).
13. Y. Zhang, W. Lee, and Y. Huang, "Intrusion Detection Techniques for Mobile Wireless Networks," ACM/Kluwer Wireless Networks Journal (ACM WINET), Vol. 9, No. 5, September (2003).
14. O. Kachirski and R. Guha, "Effective Intrusion Detection Using Multiple Sensors in Wireless Ad Hoc Networks,"Proceedings of the 36th Annual Hawaii International Conference on System Sciences (HICSS'03), p. 57.1, January (2003).
15. P. Brutch and C. Ko, "Challenges in Intrusion Detection for Wireless Ad-hoc Networks," Proceedings of 2003 Symposium on Applications and the Internet Workshop, pp. 368–373, January (2003).
16. Zhang Y, Lee W, Intrusion Detection in Wireless Ad Hoc Networks. In Proc of the 6th Int Conf on Mobil Comput and Netw (MobiCom): 275–283, (2000).
17. Huang Y, Fan W et al, Cross-Feature Analysis for Detecting Ad-Hoc Routing Anomalies. In Proc of 23rd IEEE Int Conf on Distrib Comput Syst (ICDCS):478–487, (2003).

Academic Dashboard—Descriptive Analytical Approach to Analyze Student Admission Using Education Data Mining

H.S. Sushma Rao, Aishwarya Suresh and Vinayak Hegde

Abstract Every academic year the institution welcome's its students from different location's and provides its valuable resources for every student to attain their successful graduation. At the present scenario, the institution maintains the details of students' manually. It becomes tedious task to analyze those records and fetching any information at short time. Data mining computational methodology helps to discover patterns in large data sets using artificial intelligence, machine learning, statistics, and database systems. Education Data Mining addresses these sensitive issues using a significant technique of data mining for analysis of admission. In this research paper, the analysis of admission is done with respect to location wise and comparison is done based on the year wise admission. The total admission rate for the current academic year and frequency of student admission across the state is calculated. The result of analyzed data is visualized and reported for the organizational decision making.

Keywords Educational data mining · Naive bayesian · Data mining
Admission · Analysis · Academic dash board

1 Introduction

One of the mottos of every institution is to achieve a higher admission rate and lower dropout rate. During the admission process, there is lots of paper work done and those papers are hard to maintain after afew years. By the method of manual

H.S. Sushma Rao (✉) · A. Suresh · V. Hegde
Deparment of Computer Science, Amrita Vishwa Vidyapeetham Mysuru Campus,
Amrita University, Mysuru, Karnataka, India
e-mail: sushmarao.1396@gmail.com

A. Suresh
e-mail: aishwarya3939@gmail.com

V. Hegde
e-mail: vinayakhegde92@gmail.com

D.K. Mishra et al. (eds.), *Information and Communication Technology for Sustainable Development*, Lecture Notes in Networks and Systems 10,
https://doi.org/10.1007/978-981-10-3920-1_43

entry and stored data in institutional, the institution might find it difficult to fetch any information regarding the admissions in short duration time span. The frequency of admission of students differs every year. Students from many places, for many courses join to the institution. Hence, the objective is to analyze admission from different locality who registers for various programs. The further analysis leads to the predication of the admission rate for the upcoming academic year. Data mining with a visualization technique using graph can be used to discover and extract knowledge. The EDM technique helps us to analyze and predict the future admissions of the institution by using the Naive Bayesian analysis methods.

In further, the paper is organized as follows: Sect. 2 describes the literature survey. Section 3 describes the methodology used to carry out the research. Section 4 describes the experimental results. Section 5 describes the implementation issues. Section 6 concludes the paper.

2 Related Work

Rakesh Kumar Arora and Dr. Dharmendra Badal (2012) have selected the top ten institutions which offer professional courses. The analysis of the admission rates is done on the basis of different locations for different years. The locations are divided into different zones such as North, South, East, West and the Central zone. Based on the actual values, tables are constructed for zone wise and year wise [1]. A percentage graph is plotted with the help of the tables constructed. The result observed as, the rate of admission is highest in the North zone and the least in the south zone and it is also observed that the North zone would have highest admission rate in the next consecutive years, while the least admissions would be in South zones.

Rakesh Kumar Arora and Dr. Dharmendra Badal (2013) focused on to identify the admission inquiries to actual admission. Analysis of 129 student is made and the parameters included which are taken as possible values [2]. Determining the reasons for the decline in admissions is taken as two different clusters giving results according to the data.

Kumar and Padmapriya (2014) have concentrated on efficient clustering technique for analysis on admission data. The model uses the K-means algorithm. The paper focus on the basis of the Davies–Bouldin's (DB) validity of the internal clustering validity index is measured using various clustering algorithms. K-Means performance is used to overcome both Fuzzy C-Means and SOM algorithms [3]. The accepted applicants' rates are shown in the form of 2D space. Three types of acceptance rate are calculated in low, average, and high acceptance rate for both males and females. The results shows 67% of the applicants are males with an acceptance rate 13.5% while they represent about 49% of the accepted applicants and percentages of accepted females are 51% with only 33% from the initial applications with acceptance rate 27.7% [3].

Kabakchieva, D (2012) gives the results of data mining research performed from Bulgarian universities. The main objective is to show the high rates of data mining applications in enrolment campaigns and in bringing more students [4]. It focuses on improving of data mining models for predictions of student performances based on certain characteristics.

Dorina Kabakchieva (2013) Academy, B., Sciences, O. F., & Technologies, I (2013) data mining methods are used for analyzing available datasets and extracting information for decision making [5]. It shows the initial results of the datasets implemented on the data mining research project. It is aimed to achieve high rates of use of, data mining applications in the institution.

Sundar, p. v. p (2013) the comparison of Bayesian network classifiers is done for prediction of student's academic performance and a model is generated based on that. The model helps in identifying the students who need extra attention towards their academics and allows their teacher in providing appropriate counseling's to the student. Prediction of student performance is also done which is useful in many different ways [6].

Saxena, R (2015) focuses on finding the most suitable technique in data mining, that is, between decision trees or clustering technique. The performance of both the algorithms is evaluated. Data is mined and the algorithms are applied to predict the results [7]. Ryan S. J .d. Baker (n.d.) focuses on the data mining methods which support the student modeling efforts and modeled with use of EDM that credit for details of student behavior and performance [8]. Dr. Rachel Rubin (2014) study provide high survey response rate of selective higher education institution (82%, n = 63) combined with process-oriented interviews [9].

3 Methodology

Machine learning provides a study of pattern recognition and computational learning theory in artificial intelligence. The learning mainly provides to analyze, extract information from existing student data. The main objective of this paper is to find students, which are likely to admit from different cluster of location and places. Increasing chances of admission can be done by maintaining historical data of admitted students. The place wise comparison can be visualized by any admitted years with the specified courses. The dashboard is created for viewing the total admission, prospective sold, fee payment, call recording, and visualization them according to a specific year's and courses. The parameters like sl. no branch, admission type, state, district, and city used for knowing the location clusters of admission. According to many results the Naïve Bayes algorithm provides better accuracy than any other model.

3.1 Data Preparation for Admission

Data plays a major role in data mining. Our research began with collecting required data using

(a) **Historical Admission Data (.csv files)**

This analysis mainly provides details of students who had already admitted to the institution. They mainly provide Serial No, branch, gender, admission type, district, occupation, caste, place, city, state, community, and year with their background details.

(b) **Manual Entered Data**

This analysis provides details of student who had already admitted to the institution. They mainly provide Serial No, branch, gender, admission type, district, occupation, caste, place, city, state, community, and year with their background details.

(c) **Call Analysis**

This analysis provides details of the caller. It mainly has their details like name, place, and program.

(d) **Walk-in**

It provides details of students with name, location, parent income, and program. Data are presently stored in hand written. These details are recorded manually. Hence cleaning and organizing these details correctly makes work faster to fetch and analyze. Spot-checking random collection of data is done to check discrepancies in the data groups while entering details.

3.2 *Data Selection and Transformation for and Admission*

Data transformation began after selecting required parameters by using data selection. The data set for transformation mainly targets on the admission rate of previous years to that of current rate, the program wise comparison between BCA, BCOM, BBM, MCOM, MCA with respect to the place wise admission. Hence, these details are interpreted to know the maximum, minimum, and moderate admission taken with different courses of given years.

Conversion rate (CR) is calculated by,

$$\text{CR} = \text{number of people enquired} - \text{number of people joined}$$

Algorithm:
The Naive Bayesian classifier is based on Bayes' theorem with independence assumptions between predictors. It is easy to build and without any complicated parameter for estimation in very large datasets. Bayes theorem provides a way of calculating the posterior probability, $P(c|x)$, from $P(c)$, $P(x)$, and $P(x|c)$. Naive Bayes classifier use the value of a predictor (x) on a given class (c) is independent of the values of other predictors.

$$P(c|x) = P(x|c)\,P(c)/P(x) \quad (1)$$

- P(c|x) is the posterior probability of class (target) given predictor (attribute).
- P(c) is the prior probability of class.
- P(x|c) is the likelihood which is the probability of predictor given class.
- P(x) is the prior probability of predictor.

3.3 Admission Process Diagrams

See Figs. 1 and 2.

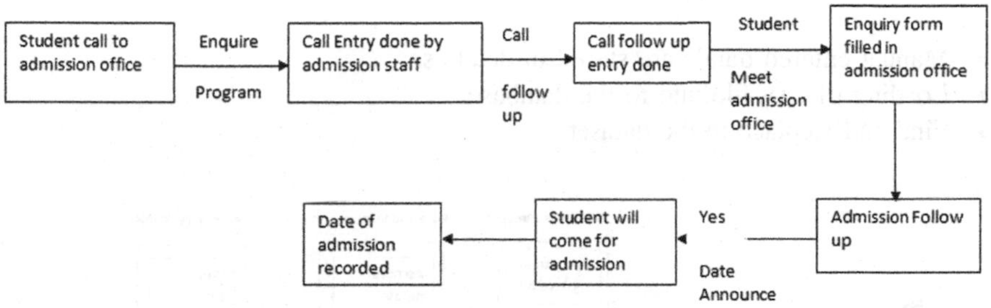

Fig. 1 Flow of student admission takes place. Student call to admission staff, then call follow up is entries. The enquiry form is filled and student details are recorded with date

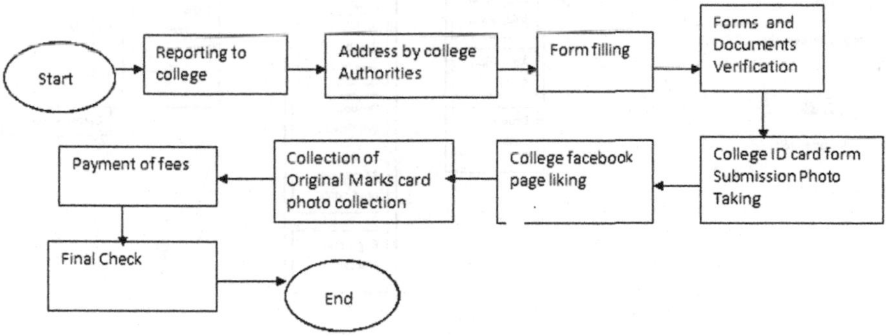

Fig. 2 Shows the flow of students after the admission takes place. Student report to the college, fill the form, get verify their documents, collect ID card, college page liking and collect their original certificate after the payment of fees final check

3.4 Module Design and Organization

See Fig. 3.

3.5 Work Methodology

After collecting of these data both manual and loading by (.csv) files into the database. These details are then preprocessed using the following steps for further analysis (Fig. 4).

3.6 Data Pre—Processing

- Manual entered data's are stored in database
- Loading of .csv file into to the database
- Find and Replace in the dataset

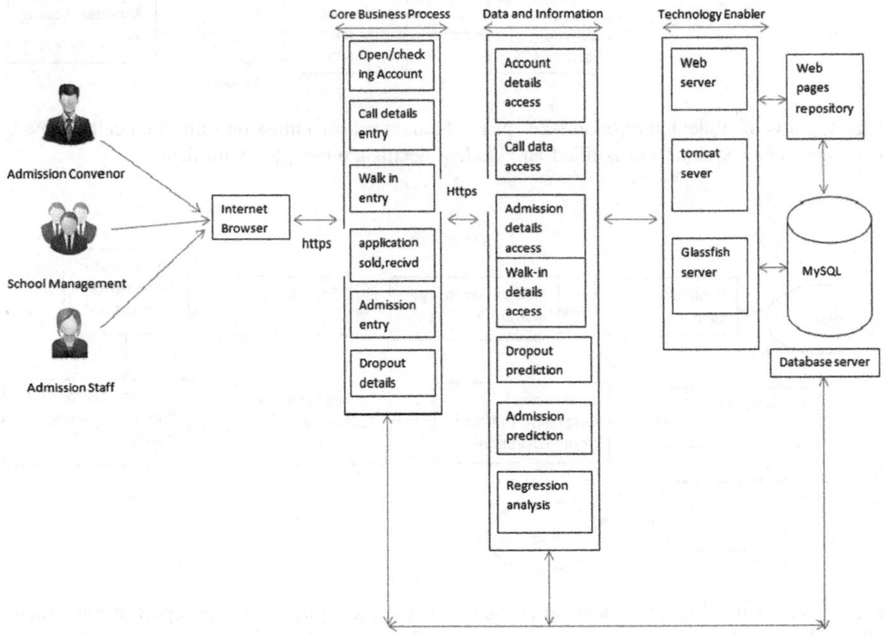

Fig. 3 Mainly provide the admission convener, school management and admission staff as users. These users are authorized to view the admission. The core business process mainly provides checking account, call details, walk-in, application downloaded and prediction of dropout. The data and information mainly provide the account details access, admission and walk-in details. The dropout and admission analysis are done using regression analysis and a database server

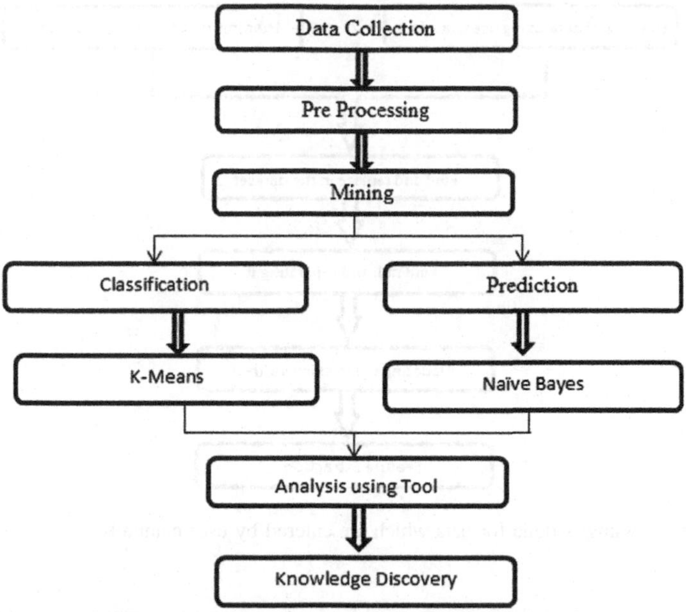

Fig. 4 In this figure, flow of work methodology is shown. The data gathering is done by manual entry and loading it from .csv files. These datasets are the preprocessed. The preprocessed data are classified in two mining techniques of classification and prediction. Based on the mining classifications, the algorithms and appropriate tools are used to get the output

- Finding null values and separating it
- Updating the missing values
- Feature extraction

loading method to reduce redundant and inaccurate data's the analysis of the district, state wise analysis based on the year and branch specified. The selected parameters values are displayed along with the frequency count and visualizing these details (Fig. 5).

4 Experimental Results

The admission of student with respect to the given year, branch displays the frequency count of the selected district, state. Plotting graph on these values provide easier way of increasing admission rates from different location at a short time interval (Figs. 6, 7 and 8).

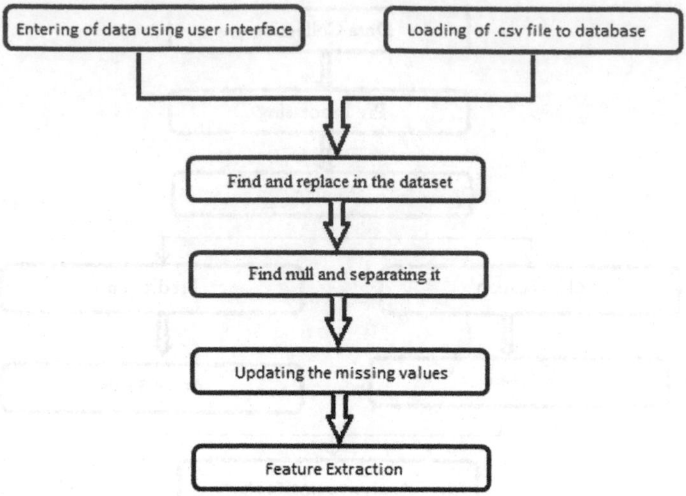

Fig. 5 Preprocessing is done for data which are entered by user manually

branch	gender	adtype	place	state	dist	comm	cast	occu	city	year
MCA	Male	Merit	Miao	Arunachal Pradesh	Changlang	Urban	Hindu	Officer	Dissing-Passo	2006
MCA	Male	Merit	Kharsang	Arunachal Pradesh	Changlang	Urban	Islam	Policeman	Pakke-Kessang	2004
MCA	Male	Merit	Diyun	Arunachal Pradesh	Changlang	Urban	Hindu	Business man	Pizirang(Veo)	2004
MCA	Female	Merit	Bordumsa	Arunachal Pradesh	Changlang	Urban	Hindu	Officer	Richukrong	2004
MCA	Female	Merit	Khimiyong	Arunachal Pradesh	Changlang	Urban	Islam	Teacher	Seijosa	2004
MCA	Male	Merit	Yatdam	Arunachal Pradesh	Changlang	Rural	Islam	Farmer	Dissing-Passo	2004
MCA	Male	Merit	Changlang	Arunachal Pradesh	Changlang	Urban	Islam	Policeman	Pakke-Kessang	2004
MCA	Male	Merit	Namtok	Arunachal Pradesh	Changlang	Urban	Hindu	Officer	Pizirang(Veo)	2004

Fig. 6 Part of dataset used for the analysis of admission for the state and district wise visualization based on the state, district and year

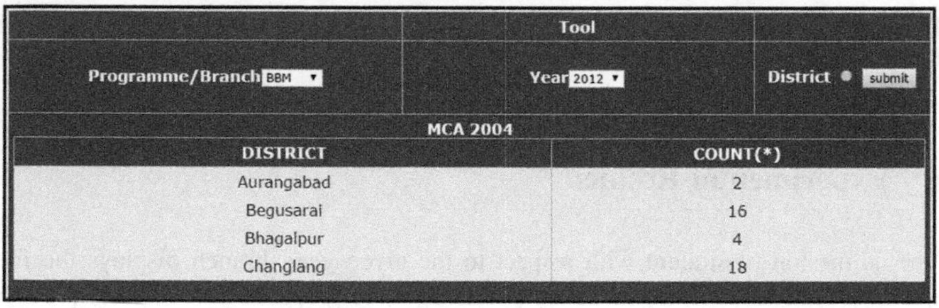
Tool
Programme/Branch BBM
Year 2012
District submit
MCA 2004
DISTRICT
COUNT(*)
Aurangabad 2
Begusarai 16
Bhagalpur 4
Changlang 18

Fig. 7 Frequency Count—district wise. Fetching the details of particular student details based on the selected branch as MCA and the year as 2004. These frequency counts help the organization to know better about the students who all came across the state. In future, it helps the management to take a decision on giving the advertisement for the future course

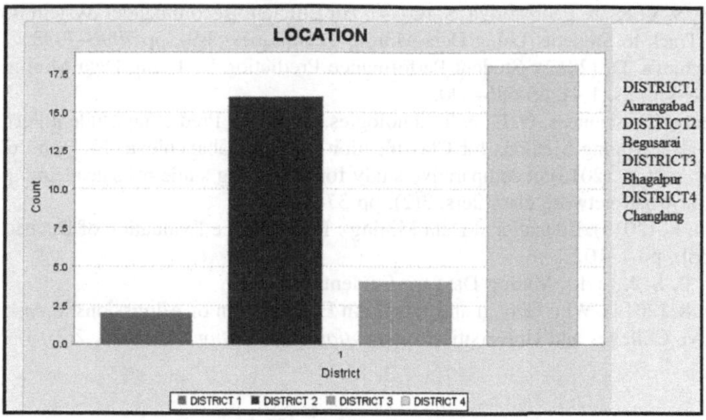

Fig. 8 Plotting graph—district wise based on branch and year visualizing the graph of required branch and year are selected of student

5 Implementation Issue

In our paper, we are mainly considering on analysis of admission but prediction on admission details are likely to be taken. Finding the prediction will be considered in next upcoming module.

6 Conclusion

The analysis is not only made to know the admission rate, but also creating a dashboard for the quick awareness of the admission details. The comparison or this analysis can make the growth in economic factors of the institution. Knowing the admission with respect to place wise comparison's, the comparison of program, admission rate with respect to year wise comparison can provide the visualized idea, to improve the admission rate for the successive year's. This paper serves as a platform for analyze the admission with respect to historical data. Organizations can know inflow and outflow of student better than past. Admission can be more for future marketing and improve organizational growth. The future, scope of the project is to predict the admission rate.

References

1. Arora, R. K. (2012). Location wise Student Admission Analysis, 2(6), pp 2010–2012.
2. Arora, R. K., & Badal, D. D. (2013). Admission Management through Data Mining using WEKA, 3(10), pp 674–678.

3. Kumar, S. V. K., & Padmapriya, S. (2014). An Efficient Recommender System for Predicting Study Track to Students Using Data Mining Techniques, 3(9), pp 7996–7998.
4. Kabakchieva, D. (2012). Student Performance Prediction by Using Data Mining Classification Algorithms, 1(4), pp 686–690.
5. Academy, B., Sciences, O. F., & Technologies, I. (2013). Predicting Student Performance by Using Data Mining Methods for Classification Dorina Kabakchieva, 13(1), pp 61–72.
6. Sundar, p. v. p. (2013). a comparative study for predicting stude nt' s academic performance using bayesian network classifiers, 3(2), pp 37–42.
7. Saxena, R. (2015). Educational Data Mining : Performance Evaluation of Decision Tree and, 14(April), pp 1–10.
8. Baker, R. S. J. (n.d.). Mining Data for Student Models.
9. Rubin, R. (2014). Who Gets In and Why? An Examination of Admissions to America's Most Selective Colleges and Universities. *International Education Research*, *2*(2), 1–18.

A New Approach for Suspect Detection in Video Surveillance

Manjeet Singh and Ravi Sahran

Abstract Face recognition is one of the most relevant applications of image analysis. Humans have very good face identification ability but not enough to deal with lots of faces. But computers have lots of memory and processing power to work with high speed. Our problem focused on detection of face from a video frame, extraction of the face, and to calculate the eigenface after normalizing the face image to match with the database of eigenfaces for the verification or identification propose. Here we are taking Vola johns algorithm into consideration for the face detection and eigenface algorithm for matching face. Face matching operation must be fast enough in video surveillance. We proposed these two methods in video surveillance for detection of suspect in video surveillance.

Keywords Video surveillance · Face detection · Face recognition

1 Introduction

Video surveillance is also known as closed-circuit television network. Video mean sequential frames or image, and the word surveillance comes from a French phrase for "watching over" ("sur" means "from above" and "veiller" means "to watch"). Video surveillance is a challenging research field in computer vision. It tries to detect, recognize, and track objects over a sequence of images. It also makes an attempt to understand and describe object behavior. Video surveillance came in research to replace the old traditional method of monitoring cameras by human operators [1].

M. Singh (✉) · R. Sahran
Central University of Rajasthan, Bandarsindri, NH8, Ajmer, Rajasthan, India
e-mail: manjeetsingh2805@gmail.com

R. Sahran
e-mail: ravisaharan@curaj.ac.in

D.K. Mishra et al. (eds.), *Information and Communication Technology for Sustainable Development*, Lecture Notes in Networks and Systems 10,
https://doi.org/10.1007/978-981-10-3920-1_44

1.1 Human Detection

Usually, CCTV cameras are used to monitoring human's behavior and humans activities. Usually, a prime concern in video surveillance is a human. When the lots of persons are there under the CCTV surveillance, then it is difficult for a person to monitoring all of them. There, an automatic human detection system in video surveillance is very useful. Detecting human in consecutive frames of video is really difficult task due to their changing appearance and different pose. There are number of application of human detection [2].

1.2 Face Detection

In video surveillance, face detection is an important stage of face recognition. It makes face recognition process somewhat easy. Face detection is defined as the method of extracting faces part from the given image or a video frame. A system is trained in such a way that it positively identities a certain part of image region as a face. Before detection of face component, the face part is first localized in the image; nowadays, most of the face recognition applications do not need the face detection separately, because it is predefined in the face recognition steps. A major example of this is a criminal data base, if there is a new subject and the police has his or her passport photograph, the records of people having a criminal report can be easily identified in Law enforcement agencies. Face detection is not mandatory all the time but at the same time face recognition play an important role [3].

1.3 Face Recognition

Face recognition has become a mostly used and famous research area in practical multimedia applications as well as computer vision, which includes the smart phones and video surveillance, face recognition has its non-intrusive nature because of it is mostly preferred among all the biometric techniques used and the wide range of availability of digital cameras and scanners. Although, the importance as well as performance of current face recognition systems can be greatly reduced, when taking the various components into the consideration which make face recognition as a challenging as well as difficult task. These challenging factors comprises of various changing face components as well as properties. Among them some of the commonly varying as well as difficult components include variations in illumination, facial expression, and pose variation the most varying component is the direction of face, etc. Various methods of face recognition includes Fisher-faces, Eigenfaces, Gabor wavelet transformation, and Laplacian have been proposed to and nonnegative matrix factorization, these methods are extensively used and adopted as elective and efficient feature extraction means for face recognition, because these method covering the problem of face recognition comparatively easy and make it much faster and efficient method to detect, recognize as well as classify the face [4].

2 Proposed Work

2.1 Human Detection

Human detection in video surveillance is a preliminary concern than face recognition. Human detection is an extension of object detection in video. When the idea of object classification came in practice human was one of the most crucial classes of object. Some famous techniques of human detection are here. Motion-based detection is the easy way to detect pedestrians in video frames, captured from a static camera. In this method static background is subtracted from upcoming frames. This is not a robust method, because this detects other moving object also [5]. HOG stands for 'Histograms of Oriented Gradients', for Human Detection by Navneet Dalal and Bill Triggs [6] for visual object recognition, adopting linear SVM-based human detection as a test case. After reviewing existing edge and gradient-based descriptors, they show experimentally that grids of Histograms of Oriented Gradient (HOG) descriptors significantly outperform existing feature sets for human detection. In this method image is divided in small cells and then calculate the histograms of block, which is collection of four cells. After that all these histograms of orientation gradients are passed in SVM (Support Vector Machine) classifier, which detects the presence of human in image. We also have to train SVM so that it can detect human and produce accurate result. Part-based detection Humans are modeled as collections of parts. In part-based detection features are designed to detect individual body parts of human. This method is helpful to detect human part in the presence of clutter and occlusion. Mikolajczyk et al. [7] followed this method in 'Human Detection Based on a Probabilistic Assembly of Robust Part Detectors'.

2.2 Face Detection

Face detection is defined as the method of extracting faces part from the given image or video scenes. Here system is trained in such a way that it positively identities a certain part of image region as a face. Before detection of face component, the face part is first localized in the image; nowadays most of the Face Recognition applications do not need the face detection separately, because it is predefined in the face recognition steps. A major example of this is a criminal data base, if there is new subject and the police has his or her passport photograph, the records of people having a criminal report can be easily identified in Law enforcement agencies. Face detection is not mandatory all the time but at the same time face recognition play an important role. Face detection deal with several well-known challenge [2, 8].

2.3 Face Recognition

The main concern of the face recognition is to either verify or identify a given image among the given set of available images in the data set. In other words face recognition is the mechanism of comparing an input image with database images. The recognition process consist of two types of images, training images and test images, training images are those images which are taken for the purpose of comparison among the different face images in the data has and the test images are those images which are used for the performing match or comparison. Many face recognition systems have a video sequence as the input. Those systems may require to be capable of not only detecting but tracking faces [9, 10].

Face recognition tasks face recognition perform two primary tasks

- **Verification**: It is also known as one-to-one matching, Here we have an image of an unknown individual along with its claim of identity, then need to ascertain that whether the specified individual is one who claimed to be or not. If it is verified with the claimed identity, then it is a one whose claimed is given otherwise it is someone else.
- **Identification**: It is also known as one-to-many matching, it is also similar to verification but difference is that here we need to compare the input image with a number of images available in the database. Given an image of an unknown individual, determining that person's identity by comparing (possibly after encoding) that image with a database of (possibly encoded) images of known individuals.

3 Proposed Method

There are three main stages in our proposed work.

1. Creation of face database
2. Detection of faces in a frame
3. Matching of detected faces to database.

3.1 Creation of Face Database

Creating eigenface space using N number of faces images of suspect. Let us have N face images each of size n × m [11].

1. Convert each image into a column matrix of [(n × m), 1].
2. Now we have a matrix of size [(n × m), N].
3. Calculate the average of all column and subtract each column matrix with average column matrix.
4. Now all the face images are normalized and we have a matrix 'A' of size [(n × m), N].
5. Calculate Covariance matrix $C = A'A$, and then calculate eigenvalues and eignvectors of Covariance matrix C.
6. Sort eignvectors and eliminate those whose eigenvalue is zero.
7. Save this as eigenface Space.

3.2 Detection of Faces in a Frame

For detecting faces in a frame, we are using Viola Jones algorithm. Face detection process mainly consists of four stages

- Haar Features Selection
- Creating Integral Image
- Cascaded Classifiers.

Haar Features Selection—Haar features are similar to convolution kernel, which is used to detect the feature in given image. Each feature results in a single value which is calculated by subtracting the sum of the pixel under white region from the sum of pixels under the black region. There are various types of Haar features, to design a new feature, depends on the user interest that which kind of kind of feature he is want to use or can use a general feature. Figure 1 show the kind of Haar features mostly used.

Fist feature is a convolution kernel which is of one row and two columns, where the right column is +1 and left column is −1. Similarly, each convolution vernal value is evaluated. The Haar features are much significant to detect the face and localize the face as well in a fast and accurate way. Vola johns use a 24 * 24 window. It calculates the feature all over the image by applying any Haar feature over the face image and gets some value when the Haar feature exactly matches the features. If we consider all the possible parameter of features like position, scale, and its type into the consideration it become a difficult and complex problem to calculated each and every type of Haar feature because in each face image there are more than 16000 similar features so become a toughest problem to detect face in real time. Properties that are similar for a human face are:

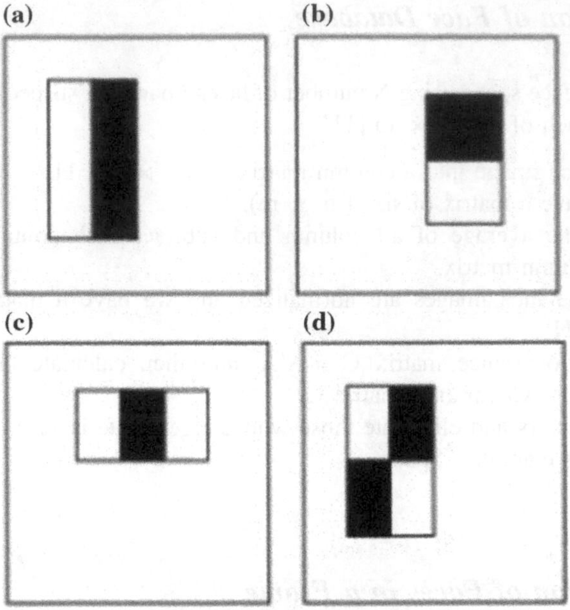

Fig. 1 Haar features [8]

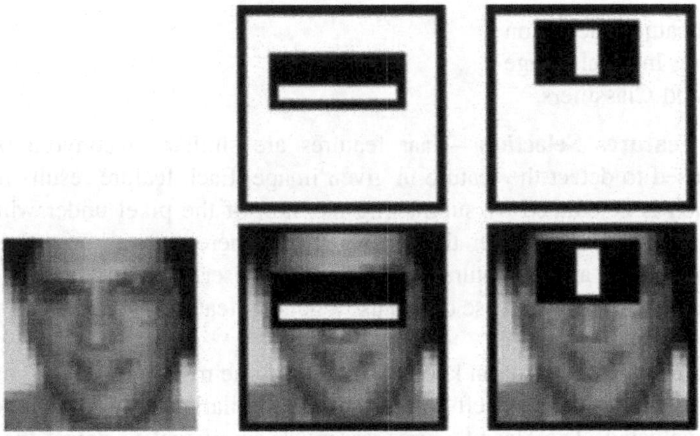

Fig. 2 Feature detection [8]

- Eyes region is darker than the upper cheeks.
- Nose bridge region is brighter than the eyes.
- Location size: eyes nose bridge region (Fig. 2).

Integral Image—The basic idea of integral image is to calculate the area of patch or feature under the consideration. Here corner value of the specified patch is

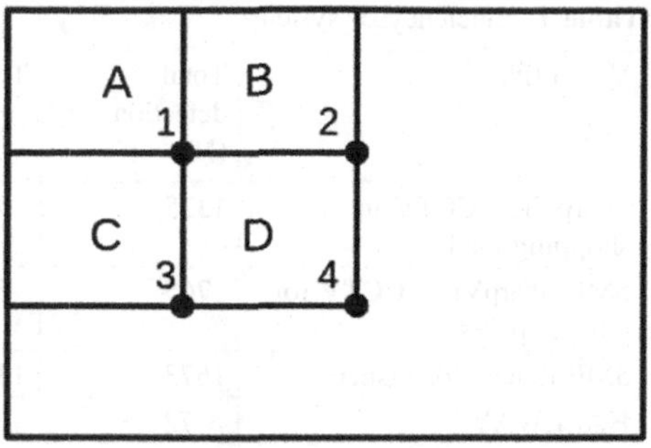

Fig. 3 Integral image [8]

needed, and does not need to sum the value of pixel. The sum of the pixels within rectangle D can be computed with four array references. The value of the integral image at location 1 is the sum of the pixels in rectangle A. The value at location 2 is A + B, The value at location 3 is A + C, and at location 4 is A + B + C + D. The sum within D can be computed as 4 + 1 (2 + 3) (Fig. 3).

3.3 Cascaded Classifiers

The basic principal behind the cascading is the time delay which rejects the non-face images or non-face window. Even if the face image contains one or more faces, it is obvious than an excessive large amount of evaluated sub-window would still be negative (non-face) (Fig. 4).

3.4 Matching of Detected Faces to Database

After detecting all faces in a frame, we extract those faces and try to identify them one by one. Take an input face image and normalized that converts the input face image into single column matrix. Then calculate the weights by multiplying input

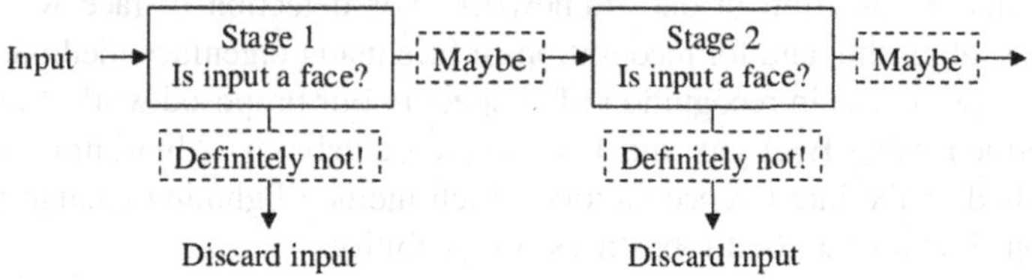

Fig. 4 Cascade classifier [8]

Table 1 Efficiency of system

Video title	Total detection (N)	True face detected		True face matched	
		FD	(FD/N) * 100 (%)	FM	(FM/FD) * 100 (%)
SharpView CCTV in shopping mall	1325	984	74.26	680	69.10
5MP SharpView CCTV for public spaces	765	528	69.01	345	65.34
5MP fathers for justice	1673	1321	78.95	978	74.03
HALLWAY	72	58	80.55	43	74.13
Average efficiency			**75.69**		**70.65**

face image to each eigenface. Store them into a vector and calculate the sum. If the sum of weights is above some threshold, then discard that input to being a face image, but if the sum is under that threshold then find the highest weight and its corresponding face image of suspect.

4 Experiments and Results

We are using MATLAB 2014a for performing these experiments. We used different quality videos as data sets. SharpView CCTV video of EyeLynx is used as data sets. Some snapshots captured during suspect detection process are shown in Table 1.

5 Conclusion

Suspect detection is a problem of face recognition in video surveillance. Face recognition is a challenging problem in the field of computer vision and image analysis that has received a great deal of attention over the last few years because of its huge applications in various domains. Here we have proposed both two face detection and recognition in one framework. For detection of face we are using Viola Jones algorithm and for recognition of face using eigenface method. We have achieved good results in recognition of suspect in our proposed work. Experiments are performed using EyeLynx and Local video databases. Algorithm validity has been verified in the interference factors which include lightning change and facial expression, it shows a good robustness and stability.

6 Future Works

In video surveillance, producing output stream should be at real time. Suspect detection system can be upgrade with another face recognition method which is fast enough to produce output video stream at real time.

References

1. Ojha, Shipra and Sakhare, Sachin, "Image processing techniques for object tracking in video surveillance-A survey", Pervasive Computing (ICPC), International Conference, 2015.
2. Menser, Bernd and Muller, F, "Face detection in color images using principal components analysis", In Seventh Int'l Conf. on Image Processing and Its Applications, 1999.
3. Gavrila, Dariu M, "The Visual Analysis of Human Movement: A Survey", Computer Vision and Image Understanding, Vol. 73, pp. 82–98, 1999.
4. Jafri, Rabia and Arabnia, Hamid R, "A Survey of Face Recognition Techniques", journal of Information Processing Systems, ACM computing Survey, Vol. 5, pp. 41–68, 2009.
5. Zhang, Ruolin and Ding, Jian, "Object tracking and detecting based on adaptive background subtraction", Procedia Engineering, Vol. 29, pp. 1351–1355, 2012.
6. Dalal, Navneet and Triggs, Bill, "Histograms of oriented gradients for human detection", Computer Vision and Pattern Recognition, IEEE Computer Society Conference on, Vol. 1, pp. 886–893, 2005.
7. Mikolajczyk, Krystian and Schmid, Cordelia and Zisserman, Andrew, "Human Detection Based on a Probabilistic Assembly of Robust Part Detectors", Computer Vision-ECCV 2004.
8. Viola, Paul and Jones, Michael J, "Robust real-time face detection", International journal of computer vision, Vol. 57, pp. 137–154, 2004.
9. Zhao, Wenyi and Chellappa, Rama and Phillips, P Jonathon and Rosenfeld, Azriel, "Face recognition: A literature survey", Acm Computing Surveys (CSUR), Vol. 35, pp. 399–458, 2003.
10. Wu, Jianxin and Hua Zhou, Zhi, "Face recognition with one training image per person", Pattern Recognition Letters, Vol. 23, pp. 1711–1719, 2002.
11. Turk, Matthew and Pentland, Alex P and others, "Face recognition using eigenfaces", Computer Vision and Pattern Recognition, pp. 586–591, 1991.
12. Brunelli, Roberto and Poggio, Tomaso, "Face recognition: Features versus templates", IEEE Transactions on Pattern Analysis & Machine Intelligence, 1993.
13. Cleber Zanchettin, "Face Recognition based on Global and Local Features", Proceedings of the 29th Annual ACM Symposium on Applied Computing, 2014.
14. Hennings-Yeomans, Pablo H and Baker, Simon and Kumar, BVKV, "Recognition of low-resolution faces using multiple still images and multiple cameras", 2nd IEEE International Conference, 2008.
15. Moghaddam, Baback and Jebara, Tony and Pentland, Alex, "Bayesian face recognition", Computer Vision and Pattern Recognition, Vol. 33, pp. 1771–1782, 2000.
16. Zhu, Xiangxin and Ramanan, Deva, "Face Detection, Pose Estimation, and Landmark Localization in the Wild", Computer Vision and Pattern Recognition (CVPR), IEEE Conference, 2012.
17. Ogale, Neeti A, "A survey of techniques for human detection from video", Survey, University of Maryland, 2006.
18. Chatrath, Jatin and Gupta, Puneet and Ahuja, Puneet and Goel, Ankush and Arora, Shaifali Madan, "Real time human face detection and tracking", Signal Processing and Integrated Networks, 2014.

Improved Segmentation Technique for Underwater Images Based on K-means and Local Adaptive Thresholding

Agrawal Avni Rajeev, Saroj Hiranwal and Vijay Kumar Sharma

Abstract In many cases, images are influenced by radiance and environmental turbulences owing to temperature variation, specifically in the case of underwater images. Due to lack of stableness in underwater circumstances, object identification in underwater is not easy in any aspect. As we know, the process of segmenting the image is a quite essential in automated object recognition systems. Subsequently, there is a necessity of segmenting the images. By means of segmenting the image, we split the image in meaningful fragments in a way to detect the concerned regions to annotate the data. We also need to process the image to eradicate the radiance effect. In this paper, we propose the improved technique to eradicate the effect of radiance and identify the object with more precision and accuracy. According to the proposed improved technique, the two segmentation techniques, k-means segmentation and local adaptive thresholding method, are merged. K-means deals with object detection whereas local adaptive thresholding eradicates the radiance effect. Lastly, the performance of improved technique is evaluated using objective assessment parameter namely, entropy, PSNR, and mutual information.

Keywords Underwater images · Illumination · Radiance · K-means
Local adaptive thresholding · Mutual information · Entropy

A.A. Rajeev (✉) · S. Hiranwal · V.K. Sharma
Computer science, Rajasthan Institute of Engineering and Technology,
Rajasthan Technical University, Jaipur, India
e-mail: vni.grawal@gmail.com

S. Hiranwal
e-mail: ersaroj.hiranwal@rediffmail.com

V.K. Sharma
e-mail: vijaymayankmudgal2008@gmail.com

D.K. Mishra et al. (eds.), *Information and Communication Technology for Sustainable Development*, Lecture Notes in Networks and Systems 10,
https://doi.org/10.1007/978-981-10-3920-1_45

1 Introduction

Nowadays, image processing is a necessity for human kind as the scientist, weather forecasters, medical images, and many more require the meaningful interpretation of the images. In order to identify and analyze the real problems and essential data in images and pictures, one needs to process the image obtained. One can use image segmentation to process image. Basically, image segmentation is the division of image into the region on basis of resemblance of features namely intensity. Segmentation is carried out in any of the two ways based on: (1) intensity similarity and (2) intensity dissimilarity. Intensity similarity segmentation is grouping the connected pixels with similar intensity values to form a region. Intensity dissimilarity segmentation is grouping of image on the basis of rapid changes in intensity [1]. Segmentation helps to recognize picture or image more proficiently. The segmentation of image is crucial in automated object recognition systems.

In several cases, temperature instabilities and atmospheric pressure cause illuminations circumstances, e.g., noise [2]. The underwater images are one such class of images which are directly influenced by illumination circumstances water medium, atmosphere, temperature, and pressure which in turn disturb the image. Consequently, it is challenging to analyze the image hence objects recognized may be erroneous and imprecise.

As noted earlier segmentation plays vital role in object recognition is not easy in such course of images. Due to instability of water ambiances and radiance situations, the contours and features of object in image fluctuate. Subsequently to evade such problems, we hereby propose a technique that uses segmentation which helps in improving the object recognition in more accurate and precise manner while it uses adaptive thresholding to remove illumination from underwater images.

2 Literature Review

The clustering analysis is the procedure which partitions the known set of data points into different cluster or regions. The main aim of clustering is to deliver beneficial information by forming meaningful clusters [3]. The data points inside clusters exhibits uniformity. Uniformity can be defined by a distance quantity. The data points in dissimilar clusters must possess distinct characteristics. Various clustering methods are designed, where efficient partitioning can be achieved by minimizing the objective function.

2.1 Fuzzy C-Means Segmentation

The fuzzy C-Means technique is soft clustering proposed by Dunn. This clustering technique was further improved by Bezdek. As stated, FCM is technique to identify groupings or clusters of data point from huge data set naturally [4]. Fuzzy c-means allows data point to belong to two or more clusters. However, membership of data point to one or more clusters varies to certain degree of their belongingness to the cluster. FCM algorithm is unsupervised as well as iterative based on minimizing objective function.

$$J = \sum_{j=1}^{c} \sum_{i=1}^{n} (u_{ij})^m \left\| X_i^j - C_j \right\|^2, \tag{1}$$

For minimizing the objective function, the algorithm considers the data set X of N data points such as X = (x_1, x_2,…x_n). It assumes C clusters such that $2 < C < N$. One needs to choose apt cluster fuzziness m where m є [1, ∞). u_{ij} is the membership degree of the x_i data point to c_j cluster. The cluster center can be obtained using Eq. (2). The u_{ij} membership degree of the data point can be obtained iteratively using Eq. (3)

$$c_j = \frac{\sum_{i=1}^{n} (u_{ij})^m X_i}{\sum_{i=1}^{n} (u_{ij})^m}, \tag{2}$$

$$u_{ij} = \left[\sum_{k=1}^{c} \left(\frac{d_{ij}}{d_{ik}} \right)^{\frac{2}{m-1}} \right]^{-1}, \tag{3}$$

The algorithm progresses until $\left\| \mu_{ij}^k - \mu_{ij}^{k+1} \right\| < \delta$, where $0 < \delta < 1$ is closure criterion. In other words, the algorithm converges when no change in membership is observed.

FCM is effective in many ways namely, it does not require any prior information. The data point can belong to more than one cluster as they are assigned as a degree of membership. Nonetheless this technique is subtle to noise and cluster center initialization.

2.2 Fuzzy C-means Thresholding

Fuzzy c-means thresholding method was proposed by Padmavathi, Muthukumar, and Thakur, which incorporates the conventional fuzzy c-means technique and global thresholding technique. Conventional FCM technique was used to reduce local minima. However, thresholding technique was used to extract the segmented part as segmented image from conventional FCM does not have clear significant boundaries. But simply extracting the segmented part was not sufficient [2]. They did not consider the effect of illumination on underwater images. Hence, in this paper, we propose a technique that deals with illumination effect on underwater images.

2.3 K-means Segmentation

K-means technique is most widely studied and implemented clustering technique. The algorithm k-means is nondeterministic and unsupervised solution that resolves the renowned clustering issue [5, 6]. K-means technique mainly focuses on minimizing the distortion and inter-cluster variance. K-means technique works primarily on selection of k-centroids for individual cluster. The centroids chosen must be placed in a sneaky way as diverse position yields different outcomes.

K-means technique splits *n* data points into k $(k \leq n)$ clusters in a way, where each data point belongs to the cluster with the closet mean. The algorithm considers the data set X of N data points such as $X = (x_1, x_2, \ldots x_n)$. The algorithm progresses iteratively until centroids of clusters do not alter any more [7]. K-means algorithm minimizes an objective function.

$$J = \sum_{j=1}^{k} \sum_{i=1}^{n} \left\| X_i^j - C_j^2 \right\|, \tag{4}$$

The closet mean can be obtained by using Euclidean distance between data point x_i and cluster center c_j. The new cluster center c_j can be computed using Eq. (5)

$$c_j = \frac{1}{n_j} \sum_{x_i \to c_j} x_i \tag{5}$$

The algorithm reaches termination when cluster centers remain intact. The main benefits of this algorithm are its ease and speed, which tenancies it to implement on huge datasets. Unlike FCM, it is hard partitioning and less subtle to noise. Hence here we prefer k-means technique to partition the image.

2.4 Thresholding

Thresholding technique is frequently used with grayscale images besides it is very easy to implement. In this technique, the decision is made on the basis of local point information of the image. An image is supposed to be separated into two fragments: objects and background using Eq. (6). The objects in the image are considered as a foreground and the remaining part is a background. The most crucial task in thresholding technique is to determine the threshold value T for precise results.

$$F(x,y) = \begin{cases} 0 & if\, I(x,y) \leq T(x,y) \\ 1 & otherwise \end{cases}, \tag{6}$$

where I(x, y) is the input image and F(x, y) is the output image Thresholding technique can be categorized as global thresholding approach and local adaptive thresholding. Global thresholding approaches such as one proposed by Otsu attempt to define a unique threshold value T for the entire image [8]. Local thresholding methods estimate a distinct threshold for each pixel on the basis of the grayscale data of the neighboring pixels. This technique is also referred as adaptive thresholding [9]. Global thresholding technique does not consider any relationships between the pixels, as a consequence, we may involve the undesired pixels in object and ignore the desired pixels. Furthermore, uneven illumination in image may result in improper objects. Thus to eliminate these problems, we can use local adaptive thresholding technique which determines the threshold locally. According to Singh et al. [10], efficient way of determining local threshold is using Eq. (7)

$$T(x,y) = m(x,y)\left[1 + k\left(\frac{\partial(x,y)}{1 - \partial(x,y)} - 1\right)\right], \tag{7}$$

Local adaptive algorithm can be implemented using local variance and local contrast methods. Local adaptive thresholding techniques produce improved result than global thresholding.

3 Proposed Technique

In this proposed approach, we have implemented k-means segmentation for effective clustering and to separate the object from the background. While adaptive thresholding method is applied to underwater images to eradicate the effects of illumination caused due to the instability of water and dispersion of light. This approach will give more suitable segmentation when compared to k-means and fuzzy c-means thresholding (FCMT).

The proposed improved segmentation technique takes color image as input and results in binary image. Initially, the proposed technique performs basic partitioning of grayscale image in object and background using k-means technique. The clusters can be computed using Eq. (5), where *k* must be chosen carefully for effective results.

Thereafter, the resultant image is processed using local adaptive thresholding. The image is divided into small parts known as windows. Now, we determine the value of threshold for each window considering their neighboring pixels. The threshold value is computed using Eq. (7) which is independent of window size. The smaller window is preferable as smaller the window less radiance effect. Thus eradication of radiance effect and identification of object is accurate and precise. Additionally, this adaptive thresholding technique eliminates the effect of noise as it partitions the image in small windows. Lastly, we compare our proposed technique with existing technique using objective assessment parameters.

3.1 Results and Evaluation

The proposed algorithm is evaluated using objective assessment parameters such as entropy, mutual information, and peak signal-to-noise ratio (PSNR).

Discrete Entropy: The degree of improbability of information content can be defined as discrete entropy. Discrete entropy can be computed using Eq. (8)

$$H(X) = \sum_{i=0}^{k} p(i) \log_2 \frac{1}{p(i)}, \tag{8}$$

Mutual Information: The measure of the mutual dependence between the two variables. More specifically, it quantifies the "amount of information" obtained about one random variable, through the other random variable. If H(X, Y) is the joint entropy of Image X and Y then MI can be defined as Eq. (9).

$$I(X{:}\,Y) = H(X) + H(Y) - H(X, Y), \tag{9}$$

Peak signal-to-noise ratio: PSNR is the ratio of the highest possible value of a signal and the power of fluctuating noise, where MSE is mean square error (Figs. 1 and 2, and Table 1).

$$PSNR = 10 \log_{10}\left(\frac{(255)^2}{MSE}\right), \tag{10}$$

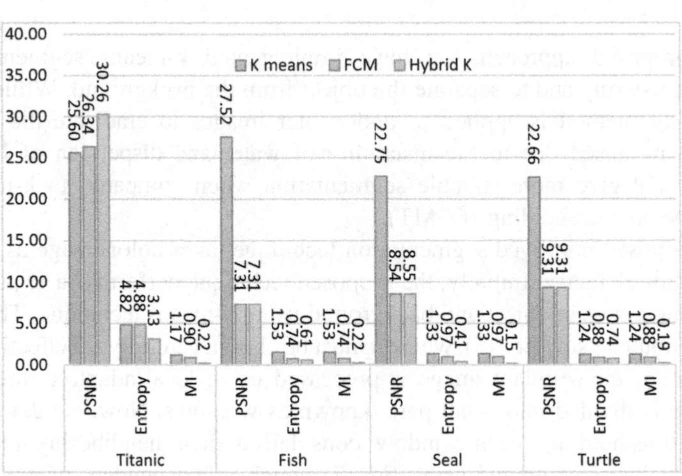

Fig. 1 Graphical performance evaluation of proposed technique and existing techniques

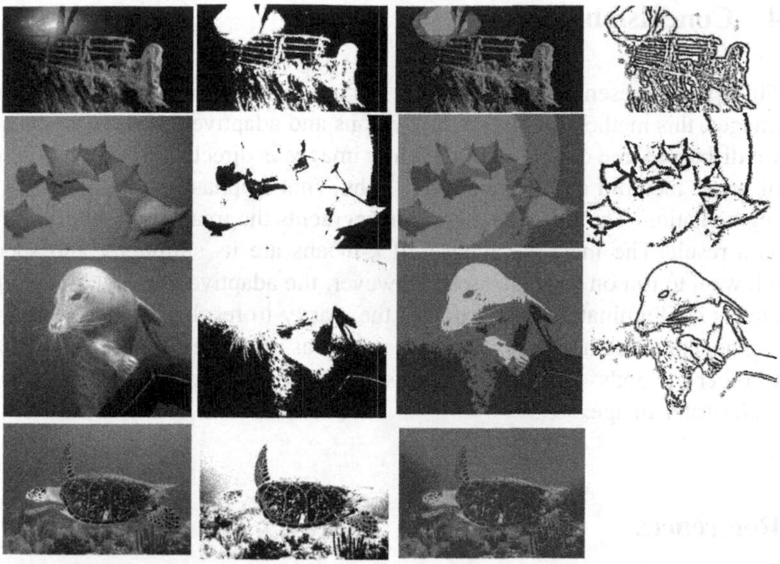

Fig. 2 Resultant segmented images after applying: *Column* 1 Original images, *Column* 2 FCMT, *Column* 3 K-means, *Column* 4 Proposed technique

Table 1 Performance evaluation of proposed technique and existing techniques

Input image	Objective assessment parameters	K-Means	Fuzzy C-means thresholding (FCMT)	Proposed technique
Titanic image	Entropy	4.8198	4.8775	3.1345
	PSNR	25.6002	26.335	30.256
	MI	1.1667	0.9047	0.2232
Fish image	Entropy	1.5252	0.7368	0.6070
	PSNR	27.5180	7.3096	7.3095
	MI	1.5252	0.7368	0.2248
Seal image	Entropy	1.3311	0.9694	0.4143
	PSNR	22.7476	8.5398	8.5481
	MI	1.3311	0.9694	0.1458
Turtle image	Entropy	1.2375	0.8804	0.7418
	PSNR	22.66	9.3114	9.3085
	MI	1.2375	0.8804	0.1883

4 Conclusion

This paper presents a new method for performing segmentation of underwater images, this method comprises of k-means and adaptive thresholding technique. As we discussed, the quality of underwater images is directly affected by the instability of water medium and dispersion of light. This emphasizes the necessity of image segmentation. The k-means algorithm segments the image by identifying the object as a result. The main advantages of k-means are its simplicity and speed, which allows it to run on large datasets. However, the adaptive thresholding eradicates the effects of illumination and provides the clarity to results. The proposed clustering segmentation method gives desirable results as illustrated in this paper, as compared to other methods. The proposed clustering segmentation method works well with underwater image.

References

1. Wen-Xiong Kang, Qing-Qiang Yang, Run-Peng Liang, "The Comparative Research on Image Segmentation Algorithms," First International Workshop on Education Technology and Computer Science IEEE-, Vol. 2, pp. 703–707, (2009).
2. Dr. G. Padmavathi, Mr. M. Muthukumar and Mr. Suresh Kumar Thakur, "Non linear Image segmentation using fuzzy c means clustering method with thresholding for underwater images", IJCSI, Vol. 7, Issue 3, (2010).
3. Gulhane, A., Paikrao, P.L. and Chaudhari, D.S., A review of image data clustering techniques. International Journal of Soft Computing and Engineering, Vol. 2, Issue 1, pp. 212–215, (2012).
4. Isa, N.A.M., Salamah, S.A. and Ngah, U.K., "Adaptive fuzzy moving K-means clustering algorithm for image segmentation", Consumer Electronics, IEEE Transactions on, Vol. 55, Issue 4, pp. 2145–2153, (2009).
5. J. Vijay, J. Subhashini, "An Efficient Brain Tumor Detection Methodology Using K-Means Clustering Algorithm", International conference on Communication and Signal Processing, IEEE, pp. 653–657, (2013).
6. Sugandhi Vij, Dr. Sandeep Sharma, Chetan Marwaha, "Performance Evaluation of Color Image segmentation using K Means Clustering and Watershed Technique", In Computing, Communications and Networking Technologies, Fourth ICCCNT, pp. 1–4, (2013).
7. Jay C. Acharya, Sohil A. Gadhiya, Kapil S Raviya, "Objective Assessment Of Different Segmentation Algorithm For Underwater Images", In Computing, Communications and Networking Technologies, Fourth ICCCNT, pp. 1–7 (2013).
8. Wu, Hongxia, and Mingxi Li. "A Method of Tomato Image Segmentation Based on Mutual Information and Threshold Iteration." Computer and Computing Technologies in Agriculture II, Springer US, Vol. 2, pp. 1097–1104, (2008).
9. Jichuan Shi, "Adaptive local threshold with shape information and its application to object segmentation", Robotics and Biomimetics (ROBIO), IEEE International Conference, pp. 1123–1128, (2009).
10. Singh TR, Roy S, Singh OI, Sinam T, Singh K., "A new local adaptive thresholding technique in binarization", IJCSI International Journal of Computer Science Issues, (2011).

HMM-Based Lightweight Speech Recognition System for Gujarati Language

Jinal H. Tailor and Dipti B. Shah

Abstract Speech recognition system (SRS) is growing research interest in the area of natural language processing (NLP). To develop speech recognition system for low resource language is difficult task. This paper defines a lightweight speech recognition system approach for Indian Gujarati language using hidden Markov model (HMM). The aim of this research is to design and implement SRS for routine Gujarati language which is difficult due to language barrier, complex language framework, and morphological variance. To train the HMM-based SRS we have manually created speech corpora that contained 650 routine Gujarati utterances which are recorded from total 40 speakers of South Gujarat region. Total numbers of speakers are selected on the basis of gender. We have achieved accuracy of 87.23% with average error rate 12.7% based on the word error rate (WER) computing.

Keywords Gujarati language • Hidden Markov model • Speech recognition
Word error rate (WER)

1 Introduction

Natural language processing(NLP) is an emerging technology that provides ease to analyze and understand human–computer interaction in natural languages. Development of NLP application is a challenging task due to language ambiguity, linguistic structure, and various regional dialects [1]. Speech recognition aims to provide accurate and efficient system that convert speech signal to text format [2].

J.H. Tailor (✉)
S.P. University, Vallabh Vidhyanagar, Gujarat, India
e-mail: jinal.tailorssa@gmail.com

D.B. Shah
G.H. Patel Post Graduate Department of Computer Science, S.P. University,
Vallabh Vidhyanagar, Gujarat, India
e-mail: dbshah66@yahoo.com

D.K. Mishra et al. (eds.), *Information and Communication Technology for Sustainable Development*, Lecture Notes in Networks and Systems 10,
https://doi.org/10.1007/978-981-10-3920-1_46

The most common applications of SRS are dictation in the domain of business, education, medical or legal notes, hands-free environment, and real-time embedded system as well as for the people with physical disabilities. SRS can be divided in different classes according to utterance of speech and speaker identification model. Classification of speech recognition includes single word and continuous word recognition. The system can be speaker dependent or speaker independent. SRS enhance its use if it is available in native language to the user [3].

1.1 Gujarati Language

SRS for Indian languages incorporate numerous technical and linguistic barriers and has the emerged area of growth. Gujarati language belongs to the Indo-Aryan Languages family which is the native language of Gujarat state in India [4]. Indian Gujarati language is one of the low resource languages due to the unavailability of language resources like corpora and Gujarati WordNet which makes the development of SRS for Gujarati very challenging.

1.2 HMM

HMM is a rich mathematical structure that is used for modeling data in multiple applications like speech recognition, artificial intelligence, data compression, and pattern recognition. HMM is used to model a nonstationary speech data. HMM model represents speech as sequence of probable observation and defines in different states [5]. In speech recognition system, HMM provide hidden state sequences which are visible to the observer through the observation sequence.

2 Related Work

Samudravijaya [6] studied performance of speaker-independent continuous speech for Hindi language using MLLR. This transform method improved accuracy of speaker adaption technique by 3%. They have concluded that using MLLR transform method for speaker adaption model average rate of error in ASR can be decrease by a factor of 0.19.

Samudravijaya et al. [7] presented speech recognition system for Hindi. The system included statistical pattern designed for different signal decoding using multiple knowledge sources. A semi-Markov model is formed to process frame level outputs to sequence segments with acoustic labels. Through dynamic programming lexical analysis is done for string matching. The database used for experiment included 200 words and commonly used in railway enquiry process.

Kumar and Aggarwal [8] have developed word recognizer for Hindi language. They have used HTK toolkit for the experiment. Total 30 Hindi words were collected from different eight speakers. They found average accuracy for the developed system 94.63%.

Kumar et al. [9] have presented two new techniques for speech recognition system in Hindi language. They have formed basic acoustic model by adding aligned trained data. The second method used to generate base form of phoneme of Hindi. They have included the approach to map base form with the phoneme by applying linguistic rules. Some of the major challenges involved with the approach like thorough knowledge of language rules and vagueness of words which are difficult to form in rules. They had proposed approach which combines rule-based and statistical methods.

Kumar et al. [10] presented a new approach for speech recognition system in Hindi language. They have used feature extraction method for processing speech. The outputs are combined using Rover technique. They have developed the system in Linux platform using tools like HTK Toolkit (V3.4) and audacity recording tool. They have selected 10 speakers to record 200 words using MFCC, PLP, and LPCC methods. They have measured 96% accuracy in quiet environment and 92% in noisy condition.

Gaurav et al. [11] presented a speech recognition system for Hindi language. The basic aim behind the system development was to use to teach Geometry in primary school. The system is designed to store 29 phonemes of Hindi. HTK 3.4 toolkit is used with MFCC feature extraction method. For the process of decoding they have used Julius recognizer. They have concluded that other feature extraction methods can be used to reduce speaker variability.

Thangarajan et al. [12] presented HMM-based continuous speech recognizer for Tamil language. They have developed two acoustic models word-based context Dependent with 371 words and triphone-based context Independent with 1700 words. They have concluded that word-based acoustic model can be use for small vocabulary. For medium and large vocabulary triphone acoustic model is well suited.

Das et al. [13] developed speech corpus for speaker-independent speech recognition in Bengali language. They have included two age groups that belong to 20–40 years and 60–80 years to collect speech corpora. They have created phone and triphone labeled speech corpora. HTK toolkit is used to process speech data and analyze the performance in phoneme and continuous word recognition.

Udhyakumar et al. [14] worked upon HMM-based multilingual speech recognition for Indian languages. They have designed system that integrates Indian accented English for the recognition of Hindi and Tamil language. They enclosed recognizer into information retrieval application and observed issues like real-time spontaneous telephony speech, identification of diverse languages.

Lakshmi et al. [15] defined a new approach for HMM-based speech recognition system for Indian languages. They have designed syllable model to process continuous speech in both training and testing. For the testing purpose, they have used Doordarshan database in which each session is of 20 min by both male and female.

They have achieved performance of language model as 80% while training and 65% in testing.

Dua et al. [16] presented the HMM-based automatic speech recognition system for isolated word in Punjabi language. Initially, they have collected 115 Punjabi words from eight speakers. For the testing purpose, they have selected six speakers in real-time environment. They have used JAVA platform with HTK Toolkit to process speech data. The average performance of system found between the ranges of 94 and 96% with the word error 4–6%.

Aggarwal and Dave [17] described the statistical approach to reduce number of Gaussian mixtures to achieve maximum accuracy in speech recognition system for Hindi language. They have implemented triphone HMM model for processing. Total 20 speakers, from which 10 male and 10 female were included for recording. Total 400 distinct Hindi words were recorded in database. MFCC and extended MFCC were used at front end. MLE and MPE methods were used in back end process.

Mishra et al. [18] presented HMM-based speaker-independent system for connected Hindi digits. They have performed and analyzed the system for both quite and noisy environments. They have prepared database collected from 40 speakers for training. In testing phase, they have included five persons from which two males and three females were there. They have performed comparison summary for different feature extraction methods like MF-PLP, PLP, and RPLP. The Performance of MF-PLP measured best compared to other two methods.

Kumar et al. [19] have presented the optimal design and development of speech corpora for Tamil, Telugu, and Marathi languages for speech recognition system. The system is developed to observe output for large vocabulary of languages. Total 560 speakers were selected for speech data collection process for Tamil, Telugu, and Marathi languages. They have also created speech recognizer to measure accuracy for the collected database speech using acoustic model created by Sphinx 2 speech toolkit.

3 Methodology

3.1 Feature Vectors

This phase describes a process to create feature vector from speech signal which is parametrically representation of speech content. Figure 1 presents main steps of information extraction from speech content that includes Preprocessing, frame blocking and windowing, feature extraction and postprocessing [20].

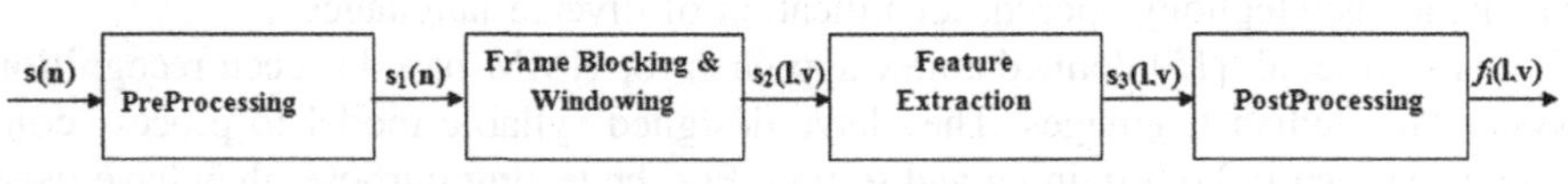

Fig. 1 Main steps in feature extraction

The notation used in Fig. 1, f_i (n, v), where v = 0, 1, 2,…,V−1 and n = 0, 1, 2,… N−1 means each vector V is size of N.

3.1.1 Preprocessing

It is initial step to create feature vector from speech signal. The main purpose of this step is to modify the raw speech content s (n) and convert into more generalized form which can be used for feature extraction analysis. Three basic operations for preprocessing are described in Fig 2.

Raw input data **s(n)** for speech recognition consistent of noisy **r(n)** and clean **c(n)** speech signal.

$$s(n) = r(n) + c(n)$$

In this research, spectral subtraction is used to reduce noisy speech signal from input. Voice activation detection for end point of utterance detection and reemphasis phases are also carried out to complete this process.

3.1.2 Frame Blocking and Windowing

More generalized form of speech signal that is produced by preprocessing which will use as input data in frame blocking. Input data $s_1(n)$ will divide into set of speech frame and apply a window on each frame. Figure 3 presents a process of blocking and windowing.

Produce a V vector of length L by applying frame blocking to $s_1(n)$. Output of frame blocking **s1(l, v)** will supply as input of windowing, here value of l = 0, 1, 2, L−1 and v = 0, 1, 2, V−1.

To increase signal continuity at end of the block window will apply on each frame. In this work, we have used Hamming window which can be calculate by Eq. (1)

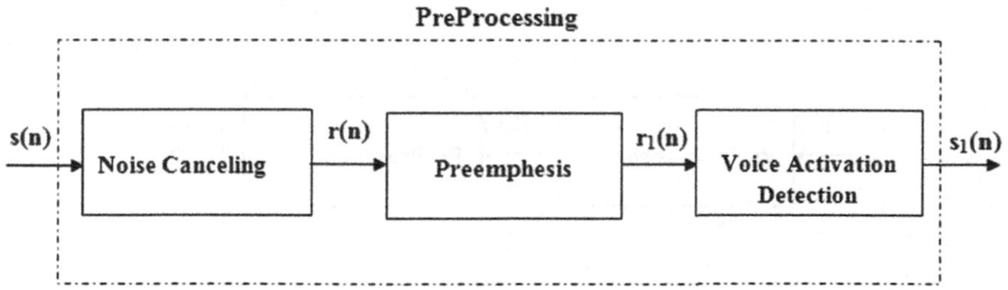

Fig. 2 Steps in preprocessing

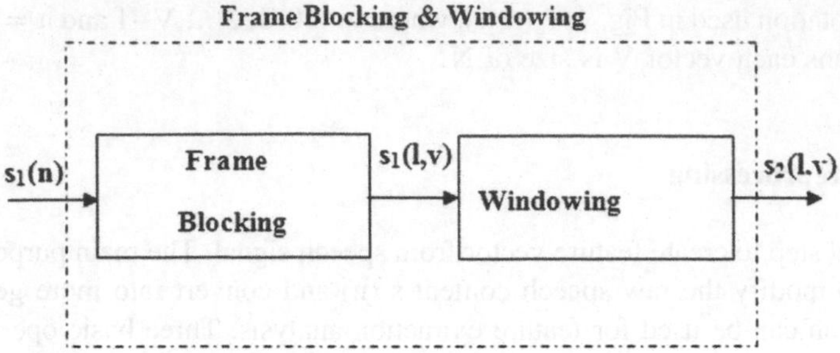

Fig. 3 Steps in frame blocking and windowing

$$w(l) = 0.54 - 0.46 \cos\left(\frac{2\pi l}{L-1}\right) \tag{1}$$

3.1.3 Feature Extraction

The main purpose of this step is to extract the most relevant information from speech block. In this work, we have used linear prediction approach for feature extraction because it works quite well in reorganization system [17].

3.1.4 Postprocessing

This is a final step to generate feature vector two operations weight function and normalization is done in this phase. This process is described in Fig. 4.

Here, weight function is used to determine influence of features which might be more or less important and normalization can be used to get zero mean for feature vector.

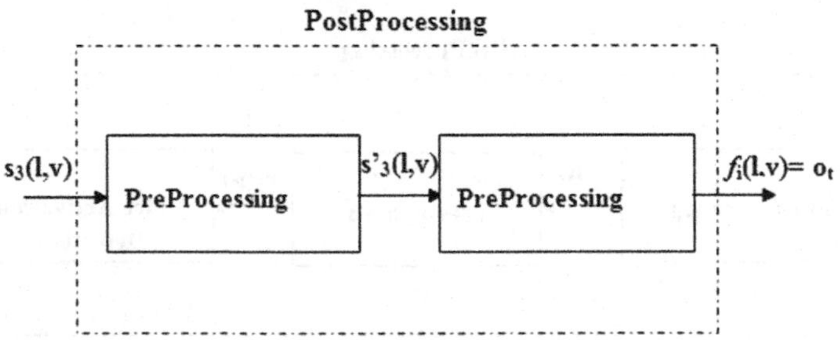

Fig. 4 Steps in postprocessing

3.2 *Decoding—Viterbi Algorithm*

Decoding method is used to find best match for the presented feature vector of utterance with the acoustic model. Decoding process uses the dynamic programming algorithm called as Viterbi Algorithm. Decoding phase includes acoustic modeling, language modeling, and pronunciation modeling [21]. In continuous speech recognition to find best path of word sequence W for the input signal X, system may find infinite sequences. The Viterbi algorithm finds most likely alternative for the input utterance. In decoding searching of the word W* can be define in Eq. (2) as

$$W^{*} = \text{argmax}_{w}(p(X \mid w)\ p(w)), \quad (2)$$

where p(w) calculated from language model

p(X|w) can be calculated from available sequence of phonemes of words available in Dictionary

$$p(X \mid w) = \text{argmax}_{s}\pi_{j}(p(x \mid s_{j})p(s_{j}))$$

3.2.1 Acoustic Modeling

Acoustic model of speech recognition system is a statistical representation of feature vector sequence and phoneme. One of the most common acoustic models used in speech recognition is HMM. Acoustic modeling also includes “pronunciation modeling” that defines combination of units of subwords to form large word or phrase in speech recognition. Acoustic model represent the statistical structure of speech for the input audio recordings.

3.2.2 Language Modeling

Language model is used to calculate the value of p(w) probability distribution that defines occurrences of the word W. For example, in language model representation for a word, we can have P(chhe) = 0.01 as after every hundred Gujarati sentences spoken by a speaker. Same for the other word sequence as P(khe) = 0 since it is a very strange word in Gujarati to speak for any speaker.

P(w) can be decomposed as

$$\begin{aligned} P(w) &= P(w1, w2, \ldots, wn) \\ &= P(w1)P(w2|w1)P(w3|w1, w2)\ldots P(wn|w1, w2, \ldots, wn-1) \\ &= \pi\ P(wi|w1, w2, \ldots wi-1), \end{aligned}$$

where p(wi|w1, w2,...wi−1) is the probability that wi will follow given that the word sequence w1, w2,..., wi−1 in above equation.

3.2.3 Pronunciation Modeling

In speech recognition system, decomposition process of a word into subwords are called pronunciation model. It is used for large number of words for the large vocabulary recognition system. Different variations in pronunciation include linguistic knowledge to generate possible alternatives pronunciation.

4 Experimental Setup

In preprocessing step the speech data are recorded at 16 kHz. The HMM-based SRS used collection of 650 routine Gujarati utterances from total 40 speakers. Speakers are selected with the gender criteria to compare accuracy level of the system. Total 40 speakers were included in the experiment from which 20 male and 20 female speakers were selected to analyze gender variability. To provide ease the individual speaker allowed speaking day to day routine Gujarati speech for recording. Recording tool audacity is used for the experiment. Total hours for recording are divided into 3 sessions according to speaker's comfort and availability. Each speaker used to utter minimum 25 words two times, one for training and second time for recording purpose. Total 1000 utterances were collected from 40 speakers that belong to the age group of 20–40 years. From which 350 utterances were deleted due to lower quality. Audio editing software tool was used to provide trimming process to each utterance. Acoustic model was used to process speech data with HMM five states model.

5 System Performance Evaluation

The word recognition rate (WR) can be measured as Eq. (3)

$$WR = (N - D - S / N)/100 \tag{3}$$

where

S Number of Substitutions
D Number of Deletion
I Number of Insertion
N Number of words in the reference.

Speech recognition system performance is adequately measured by the common measurement word error rate (WER). Basic measurement parameters for the speech are accuracy and velocity. WER represents the accuracy of the system using Eq. (4)

$$\mathrm{WER} = \frac{S + D + I}{N} \tag{4}$$

The real-time factor (RTF) is used to measure velocity parameter of speech recognition system. It is defined by the Eq. (5)

$$\mathrm{RTF} = P/I, \tag{5}$$

where

P Time to Process Input
I Duration.

6 Result and Discussion

Experimental results are:

Performance measurement of randomly selected eight sample data from dataset

Speakers	N	D	I	S	WR	WER
S1	25	2	0	1	88	12
S2	24	1	0	1	91.6	8.3
S3	22	3	1	0	86.3	18.1
S4	25	2	1	0	92	12
S5	26	1	1	2	88	15
S6	22	2	1	1	86	18
S7	20	0	1	1	95	10
S8	21	1	1	1	90	14

Experimental result showed average accuracy of SRS is 87.23% with average error rate 12.7%. Accuracy can be improved by overcoming factors that affect performance as form and mode of speech age, gender, and variations in dialects, language knowledge, and mental state of speaker. High-quality hardware can also reduce error rate for spoken word.

7 Conclusion and Future Work

This paper described a lightweight approach for Gujarati speech recognition system using HMM. This work is mainly focused on the routine for Gujarati language. Due to unavailability of resources in terms of dataset, literature, and tools for Gujarati language, we have manually created speech corpora that contained 650 routine Gujarati utterances which are recorded from 20 male speakers and 20 female speakers from south Gujarat region. By computing WER, we have achieved accuracy of 87.23% with average error rate 12.7%. The experimental results showed that the recognition rate is considerable. To improve the accuracy and efficiency of the system, we have applied Viterbi algorithm for pattern matching which will find best state path. For future work, dataset for training can be increase to achieve better results and HMM-based hybrid model can be apply to decrease the WER.

References

1. Weischedel, R., Carbonell, J., Grosz, B., Lehnert, W., Marcus, M., Perrault, R., & Wilensky, R. (1989, October). White paper on natural language processing. In *Proceedings of the workshop on Speech and Natural Language* (pp. 481–493). Association for Computational Linguistics.
2. Sandanalakshmi, R., Viji, P. A., Kiruthiga, M., Manjari, M., & Sharina, M. (2013). Speaker Independent Continuous Speech to Text Converter for Mobile Application. *arXiv preprint* arXiv:1307.5736.
3. Uchat, N. S. (2007). Hidden Markov Model and Speech Recognition. In *Seminar report, Department of Computer Science and Engineering Indian Institute of Technology, Mumbai.*
4. Jain, D., & Cardona, G. (2007). *The Indo-Aryan languages*. Routledge.
5. Gales, M., & Young, S. (2008). The application of hidden Markov models in speech recognition. *Foundations and trends in signal processing*, *1*(3), 195–304.
6. Samudravijaya, K. (1878). Computer recognition of spoken Hindi. *training*, *198*(56), 93. Samudravijaya, K., Computer Recognition of Spoken Hindi||. Proceeding of International Conference of Speech, Music and Allied Signal Processing, Triruvananthapuram, pages 8–13, 2000.
7. Samudravijaya, K., Ahuja, R., Bondale, N., Jose, T., Krishnan, S., Poddar, P., & Raveendran, R. (1998). A feature-based hierarchical speech recognition system for Hindi. *Sadhana*, *23*(4), 313–340.
8. Kuldeep Kumar and R.K. Aggarwal, "hındı speech recognition system using HTK", International Journal of Computing and Business Research, vol. 2, no. 2, 2011.
9. Kumar, M., Rajput, N., & Verma, A. (2004). A large-vocabulary continuous speech recognition system for Hindi. *IBM journal of research and development*, *48*(5.6), 703–715.
10. Kumar, M., Aggarwal, R. K., Leekha, G., & Kumar, Y. (2012). Ensemble feature extraction modules for improved Hindi speech recognition system. *Proc Int J Comput Sci*, (9), 3.
11. Gaurav, G., Deiv, D. S., Sharma, G. K., & Bhattacharya, M. (2012). Development of application specific continuous speech recognition system in Hindi.
12. Thangarajan, R., Natarajan, A. M., & Selvam, M. (2008). Word and triphone based approaches in continuous speech recognition for Tamil language. *WSEAS transactions on signal processing*, *4*(3), 76–86.

13. Das, B., Mandal, S., & Mitra, P. (2011, October). Bengali speech corpus for continuous automatic speech recognition system. In *Speech Database and Assessments (Oriental COCOSDA), 2011 International Conference on* (pp. 51–55). IEEE.
14. Udhyakumar, N., Swaminathan, R., & Ramakrishnan, S. K. (2004, May). Multilingual speech recognition for information retrieval in Indian context. In *Proceedings of the Student Research Workshop at HLT-NAACL 2004* (pp. 1–6). Association for Computational Linguistics.
15. Lakshmi, A., & Murthy, H. A. (2008). A new approach to continuous speech recognition in Indian languages. In *Proceedings national conference communication.*
16. Dua, M., Aggarwal, R. K., Kadyan, V., & Dua, S. (2012). Punjabi automatic speech recognition using HTK. *IJCSI International Journal of Computer Science Issues*, *9*(4), 1694–0814.
17. Aggarwal, R. K., & Dave, M. (2011). Using Gaussian mixtures for Hindi speech recognition system. *International Journal of Signal Processing, Image Processing and Pattern Recognition*, *4*(4), 157–170.
18. Mishra, A. N., Chandra, M., Biswas, A., & Sharan, S. N. (2011). Robust features for connected Hindi digits recognition. *International Journal of Signal Processing, Image Processing and Pattern Recognition*, *4*(2), 79–90.
19. Kumar, R., Kishore, S., Gopalakrishna, A., Chitturi, R., Joshi, S., Singh, S., & Sitaram, R. (2005). Development of Indian language speech databases for large vocabulary speech recognition systems. In *International Conference on Speech and Computer (SPECOM) Proceedings.*
20. Nilsson, M., & Ejnarsson, M. (2002). Speech recognition using hidden markov model.
21. Huang, X., & Deng, L. (2010). An Overview of Modern Speech Recognition.

Despeckling of SAR Image Based on Fuzzy Inference System

Debashree Bhattacharjee, Khwairakpam Amitab and Debdatta Kandar

Abstract Synthetic Aperture Radar (SAR) is a type of imaging radar system that is widely used for remote sensing of Earth. It is observed that the images obtained from the SAR systems are often corrupted with speckle noise which reduces the visibility of the image. Preprocessing such images is often essential to enhance the clarity of the image for acquiring the required information present in it. A nonlinear filtering technique is proposed in this work, based on fuzzy inference rule-based systems, which uses fuzzy sets and fuzzy rules that operates on the luminance difference between the central pixel and its neighbors in a 3 $\times$ 3 window to reduce the presence of speckle noise in the SAR images. A comparative evaluation is performed on the proposed filter with three other existing filtering methods namely Mean, Median and FIRE filter for mixed noise to evaluate its performance.

Keywords Synthetic Aperture Radar (SAR) · Speckle noise
Fuzzy inference system · Image processing · Image filtering

1 Introduction

The term **RADAR** stands as an acronym for **Radio Detection and Ranging**. Radar are electronic devices, equipped with electromagnetic sensors, for detecting the range and location of distant objects or targets by transmitting high-frequency radio waves [1–3]. There are various kinds of radar for different types of applications like nautical radars, aviation radars, weather sensing radars, etc.

D. Bhattacharjee (✉) · K. Amitab · D. Kandar
Department of Information Technology, School of Technology, North Eastern Hill University, Umshing Mawkynroh, Shillong 793022, India
e-mail: debashreeb21@gmail.com

K. Amitab
e-mail: khamitab@gmail.com

D. Kandar
e-mail: kdebdatta@gmail.com

D.K. Mishra et al. (eds.), *Information and Communication Technology for Sustainable Development*, Lecture Notes in Networks and Systems 10,
https://doi.org/10.1007/978-981-10-3920-1_47

The size of the antenna used by the radar determines the spatial resolution capability of the real aperture radar (RAR). To obtain better resolution, an antenna of a large size is required. However, it stands to be practically infeasible to have an extremely large antenna in order to obtain the required high-resolution image. This requirement has led to the development of SAR.

1.1 Synthetic Aperture Radar (SAR)

SAR is a class of high-resolution imaging radar system which may be either air borne or space borne. It is a side looking, day and night, weather independent active sensor device that uses microwave electromagnetic radiation [4, 5]. The antenna of SAR is mounted on a moving platform such as an aircraft or spacecraft which moves over a target area. The motion of SAR along the flight path synthesizes a large antenna and the length of the synthetic antenna is proportional to the distance that SAR travel.

The radar transmits an electromagnetic pulse at each position of the fight path and the returned echo from the target is stored at the receiver of the radar. The returned echo comprises of the amplitude of the reflected signal and its phase which creates a coherent target image at the receiver. The SAR image is formed by sampling the target pixels along the range and the azimuth direction.

Every pixel in the scene imaged by SAR can be considered to be formed from many scatterers reaching the antenna. There can be either destructive or constructive interferences of received signals from the scatterers depending on the difference of the relative phase between the signals due to which speckle noise arises in the image obtained. Pixels in the target image become bright whenever the interference is constructive and they become dark in destructive interference.

1.2 Speckle Noise

Speckle noise is defined as a multiplicative noise which is formed in the imaging systems that are coherent. Images corrupted with speckle noise $i[m,n]$ can be expressed as:

$$i[m,n] = p[m,n] + \eta[m,n]p[m,n], \tag{1}$$

where, $p[m,n]$ is the original pixel and $\eta[m,n]$ is the multiplicative component that is added to the original pixel [6]. The magnitude of the noise depends on the value of the original pixel, thereby, making the noise signal dependent. The speckle noise adds a black or a white grainy effect to the image which reduces the usability of the obtained SAR image.

1.3 Existing Filters for Reducing the Noise

The reduction of speckle noise has been studied widely over the years and various approaches have been proposed in the form of classical filters like the Mean filter which is a linear filter and Median filter, a nonlinear filter [6]. There are some nonlinear adaptive filters which exist for speckle noise reduction like Lee filter [7], Frost filter [8] and Kuan filter [9]. Based on fuzzy logic, there are a few speckle noise filters. One such filter [10] defines a fuzzy window and a membership function and the filtering is done based on the fuzzy-weighed MMSE (minimum mean square error). This filter falls in the class of Fuzzy-Classical filter. Another approach for speckle noise reduction using fuzzy logic involves developing several membership functions to express the existence of an edge between the central pixel, which is considered as the pixel of interest, of a 3 × 3 window and its neighboring pixels and the weights of the pixels in the window are computed. The normalized weight of the window is further obtained and the filter works iteratively to reduce the speckle noise that is present in the image [11].

1.4 Filtering Based on Fuzzy Inference Rules

Filters for noise reduction based on fuzzy inference rules are nonlinear filters which use fuzzy reasoning to reduce the noise present in an image. These filters use the IF-ElSE-IF structure to provide reduction of noise in a noisy image. A wide range of filters based on fuzzy inference rules have been implemented for impulse noise. These filters are called FIRE which is an acronym for Fuzzy Inference Ruled by Else action, originally proposed by Russo in 1992 [12]. A variety of modifications have been done on the FIRE filters [13] for impulse noise and Gaussian noise reduction.

2 Fuzzy Technique for Noise Reduction

2.1 Introduction on Fuzzy Sets

Fuzzy sets are generalized form of crisp sets that have ushered by introducing the degree of belongingness of a particular element in the set by making use of partial membership functions.

If U is considered to be the universal set, then a fuzzy set F can be defined on U as:

$$F = \{(u, \mu_F(u))|u \in U\}, \tag{2}$$

where, u is an element belonging to U and $\mu_F(u)$ is the membership function mapping U to [0, 1]. The membership function represents a value known as the degree of belongingness of u to the fuzzy set F [14]. The degree ranges between 0, which denotes that the element does not belong to F and 1 denoting that the element is present in F with the full value of membership. Membership functions are of various types namely triangular, trapezoidal, bell shaped, etc.

2.2 Fuzzy System

A fuzzy system falls in the class of nonlinear systems that comprise of fuzzy rules forming the knowledge base of the system. The result of the fuzzy system is generated by the numerical evaluation of these rules using an appropriate inference mechanism [14].

A fuzzy system maps n variables given as input to the system, namely $x_1, x_2, x_3, \ldots x_n$, to single variable p representing the output of the system, by using some m number of rules given by:-

$$IF\ (x_1, F_{11})\ AND \ldots AND (x_n, F_{n1})\ THEN\ (p, Q_1), \tag{3}$$

$$IF\ (x_1, F_{12})\ AND \ldots AND\ (x_n, F_{n2})\ THEN\ (p, Q_2), \tag{4}$$

$$IF\ (x_1, F_{1m})\ AND \ldots AND\ (x_n, F_{nm})\ THEN\ (p, Q_m). \tag{5}$$

where $F_{ij}\ (1 \leq i \leq n, 1 \leq j \leq m)$ represents the corresponding antecedent fuzzy set for the ith variable x taken as input in the jth fuzzy rule and Q_j represents the consequent fuzzy set for the output variable p in the jth rule [15].

The activations for each of the defined rules are then evaluated by the fuzzy inference mechanism.

If both F_{ij} and Q_j are described by triangular membership function defined as:

$$\mu_{F_{ij}}(x) = \begin{cases} 1 - \frac{|x - c_{F_{ij}}|}{w_{F_{ij}}}, & c_{F_{ij}} - w_{F_{ij}} \leq x \leq c_{F_{ij}} + w_{F_{ij}}\,. \\ 0, & otherwise \end{cases} \tag{6}$$

where $c_{F_{ij}}$ and $w_{F_{ij}}$ represent the center and the half-width of the fuzzy set F_{ij}. Then the activation degree of the jth rule, represented by λ_j, is described by the relation:

$$\lambda_j = \min\left\{\mu_{F_{ij}}(x_i); \mathrm{i} = 1, \ldots, \mathrm{n}\right\}. \tag{7}$$

Then each of the resulting activation value is combined to produce a single value of p, the output variable, for the entire rule base of the m rules. The obtained value of p is then defuzzified by using correlation-product inference [16]. As the consequent output fuzzy set Q_j is described by triangular membership function (6), the output expression [15] is obtained as:

$$p = \frac{\sum_{j=1}^{m} \lambda_j c_{Q_j} w_{Q_j}}{\sum_{j=1}^{m} \lambda_j w_{Q_j}}. \tag{8}$$

2.3 Working Principle of FIRE Filters

Let us consider, the window represented in Fig. 1 [14] on which the filter operates. The pixel at the center of the window, i.e., 0 is considered as the pixel of interest that is to be filtered. The FIRE filters operate by adopting directional fuzzy rules which deal with the luminance difference Δx, between the central pixel x and its neighbors, as its inputs. On evaluation of the rule base, if the rules are satisfied, then the central pixel is found noisy and a correction term Δp is added to the central pixel to reduce the noise which can have either a positive or a negative value depending on the luminance value of the central pixel [17]. As FIRE filter is recursive so every time a new value of the output variable p is added to x, i.e., $p = x + \Delta p$ when the processing for x gets over.

In case none of the rules in the rule base are satisfied then no changes are added to the value of x and it can be inferred that the central pixel is noise free.

2.4 Proposed Filtering Technique for Despeckling

In this proposed approach, fuzzy rules and two symmetrical fuzzy sets, namely medium positive (MP) and medium negative (MN) [14], represented graphically in Fig. 2, are used for noise reduction mechanism. The triangular membership functions μ_{MP} and μ_{MN}, are used to describe the fuzzy sets MP and MN. In Fig. 2, L represents the luminance range. As grayscale image is used, the L value ranges from −255 to +255 and $\pm c$ represents the centers of the fuzzy sets. The filter takes

Fig. 1 3 × 3 window

1	2	3
4	0	5
6	7	8

Fig. 2 Triangular fuzzy sets MN and MP

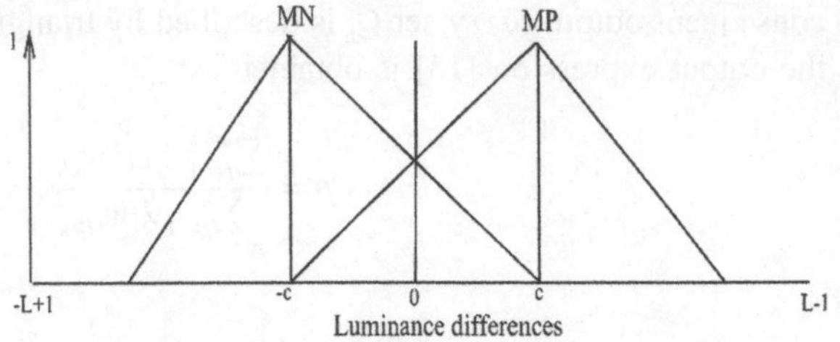

its inputs as $\Delta x = x_i - x$ [14] which represents the difference in the luminance value between the central pixel x and its neighbors x_i in a 3 × 3 window (Fig. 1).

It is seen that when an image is corrupted with speckle noise all the pixel values of the image tend to change within a specific range depending on the intensity of noise. So in this technique, eight patterns of pixels are considered as inputs to the fuzzy rule base from the 3 × 3 window in Fig. 1, considering the situation where groups of adjacent pixels are noisy. The patterns are $I_1 = \{4, 1, 2, 3\}$, $I_2 = \{2, 3, 5, 8\}$, $I_3 = \{5, 8, 7, 6\}$, $I_4 = \{7, 6, 4, 1\}$, $I_5 = \{2, 3, 5, 8, 7\}$, $I_6 = \{5, 8, 7, 6, 4\}$, $I_7 = \{7, 6, 4, 1, 2\}$, $I_8 = \{4, 1, 2, 3, 5\}$. Each pixel in the eight patterns contain the luminance difference between the central pixel 0 and itself.

Let us consider a pattern $I_1 = \{4, 1, 2, 3\}$ from the 3 × 3 window in Fig. 1. The two symmetrical fuzzy rules for I_1 are given as:

$$IF\,(\Delta x_4, MP)\,AND\,(\Delta x_1, MP)\,AND\,(\Delta x_2, MP)\,AND\,(\Delta x_3, MP)\,THEN\,\left(\Delta x_p, MP\right). \tag{9}$$

$$IF\,(\Delta x_4, MN)\,AND\,(\Delta x_1, MN)\,AND\,(\Delta x_2, MN)\,AND\,(\Delta x_3, MN)\,THEN\,\left(\Delta x_p, MN\right). \tag{10}$$

The rules (9) and (10) are designed to address noise pulses which may have positive or negative values with respect to the neighboring pixels for pattern I_1. If the value of Δx is negative for a particular pattern then it implies that the luminance value of the central pixel is higher that its neighboring pixels and so a positive correction term Δp is added to x to reduce its luminance value and this is addressed by rule (9). Similarly, rule (10) addresses that case when value of Δx is positive. As eight input patterns are considered here so a total of 16 such rules comprise the rule base of the proposed filter.

The activations of the rules (9) and (10) are [14]:

$$\beta = \max_j \left\{ \frac{1}{2}\left(\min_{i \in I_j}\{\mu_{MP}(\Delta x_i)\} \right) + \frac{1}{2}\left(\frac{1}{N_j} \sum_{i \in I_j} \mu_{MP}(\Delta x_i) \right) \right\}, \tag{11}$$

$$\beta^* = \max_j \left\{ \frac{1}{2}\left(\min_{i \in I_j}\{\mu_{MN}(\Delta x_i)\} \right) + \frac{1}{2}\left(\frac{1}{N_j} \sum_{i \in I_j} \mu_{MN}(\Delta x_i) \right) \right\}, \tag{12}$$

where, N_j represents the number of elements that are present in the pattern set $I_j(1 \leq j \leq 8)$.

The correction term Δp is then obtained by defuzzification using correlation-product inference method [16], the details of which can be inferred from [17]. So Δp is given by:

$$\Delta p = c(\beta - \beta^*). \tag{13}$$

The output p is finally obtained by adding the correction term Δp to the central pixel x, given by:-

$$p = x + \Delta p. \tag{14}$$

If the rules (9) and (10) are not satisfied then the central pixel remains unchanged.

The proposed filter is iterated when the value of speckle noise is high.

2.5 Experimental Results

The proposed method is tested on a SAR image (Fig. 3) [18] of size 100 × 150 corrupted with speckle noise variance in the range of 0.01–0.2. The proposed filter is compared with three other existing filters namely FIRE filter for mixed noise [14], median and mean filter. All the filters have been implemented using a 3 × 3 filter mask. For a quantitative evaluation of the proposed filter we use the peak signal-to-noise ratio (PSNR) value. The PSNR value is a ratio which is taken between the power of the maximum amplitude of the signal-to-noise present in the signal. Higher value of PSNR indicates that less value of speckle noise is present.

For variance values of speckle noise in the range of 0.01–0.05 a single iteration of the proposed filter is considered as it provides higher PSNR values than the FIRE filter for mixed noise, median and mean filter as it can be seen in Table 1.

For higher values of variance of speckle noise in the range of 0.06–0.2 the proposed filter is iterated twice.

From Table 2 it can be seen that the proposed filtering technique yields higher values of PSNR than the FIRE filter for mixed noise, median and mean filter. From

Table 1 PSNR values obtained from the four filters using 3 × 3 filter mask, namely Proposed Filter, FIRE filter for Mixed Noise, Median filter and Mean filter for Speckle Noise variance values in the range of 0.01–0.05

Noise variance	Noisy image	Proposed filter	Fire filter for mixed noise	Median	Mean
0.01	26.25	27.18	27.00	26.63	26.20
0.03	21.40	26.70	26.55	24.42	21.50
0.05	19.26	26.00	25.82	22.99	19.27

Table 2 For the second iteration, PSNR values obtained from the four filters using 3 × 3 filter mask, namely Proposed Filter, FIRE filter for Mixed Noise, Median filter and Mean filter for Speckle Noise variance values in the range of 0.06–0.2

Noise variance	Noisy image	Proposed filter	Fire filter for mixed noise	Median	Mean
0.06	18.39	25.53	25.21	22.46	18.45
0.1	16.19	24.74	22.00	20.87	16.22
0.16	14.18	21.66	17.72	18.97	14.26
0.2	13.26	19.61	16.4	18.18	13.31

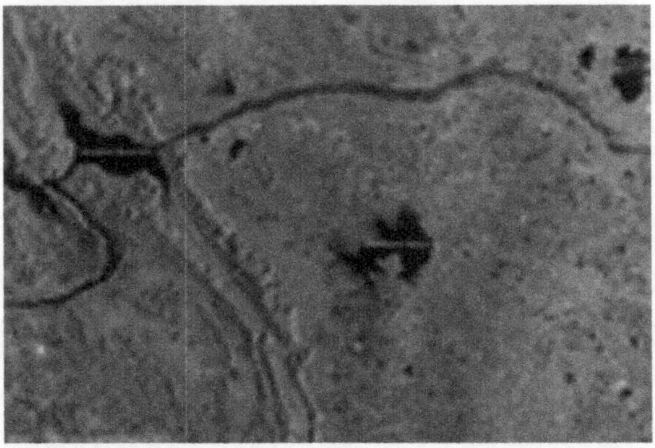

Fig. 3 SAR image in grayscale

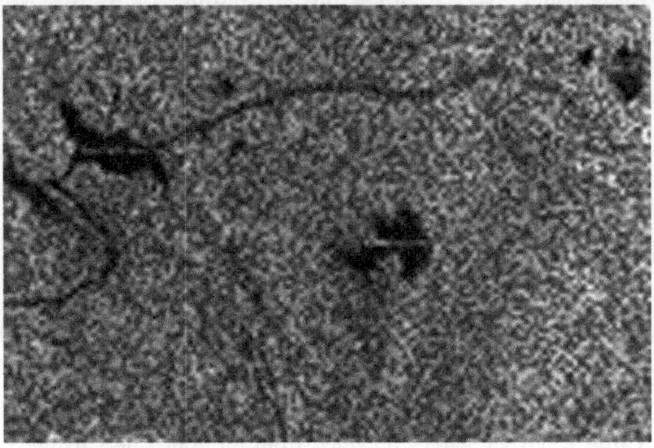

Fig. 4 SAR image corrupted with speckle noise having variance 0.06

the visual results, it can be seen that the filtered image from the proposed technique (Fig. 5) becomes blur although good amount of smoothing occurs and this effect continues as the value of speckle noise variance increases.

Now, graphical analysis of the results obtained for different values of speckle noise variance against the filtering techniques is provided in Fig. 9.

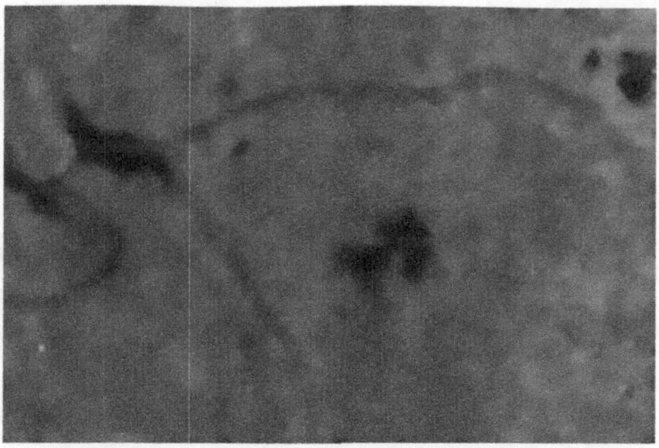

Fig. 5 Result obtained after filtering with the proposed technique

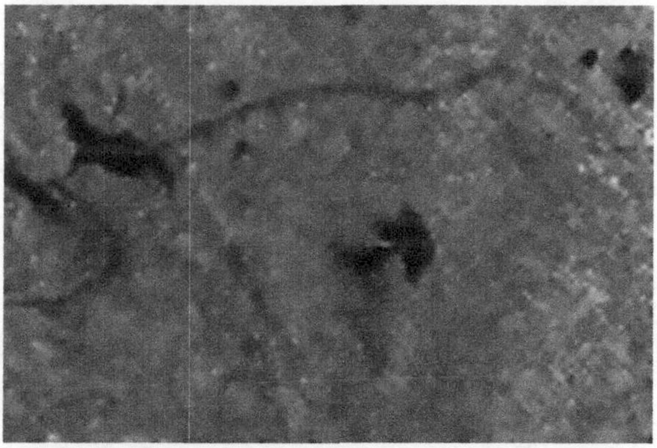

Fig. 6 Result obtained from 3 × 3 FIRE filter for mixed noise

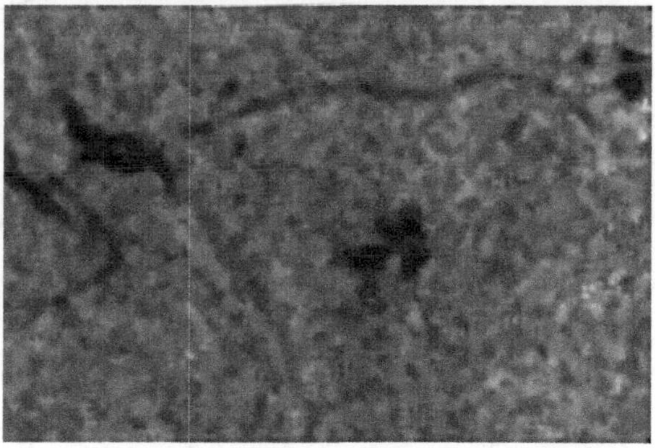

Fig. 7 Result obtained from 3 × 3 median filter

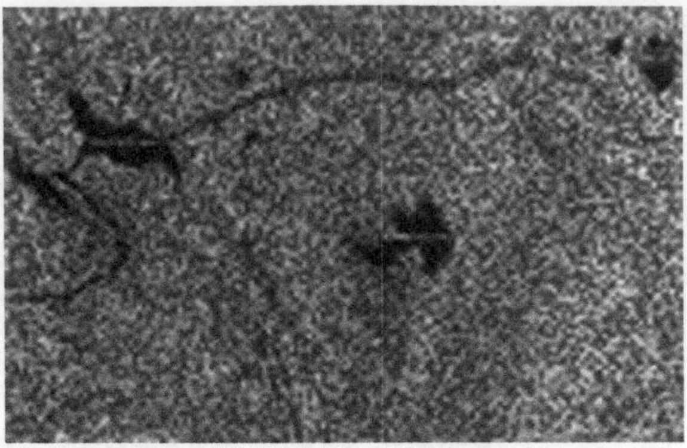

Fig. 8 Result obtained from 3 × 3 mean filter

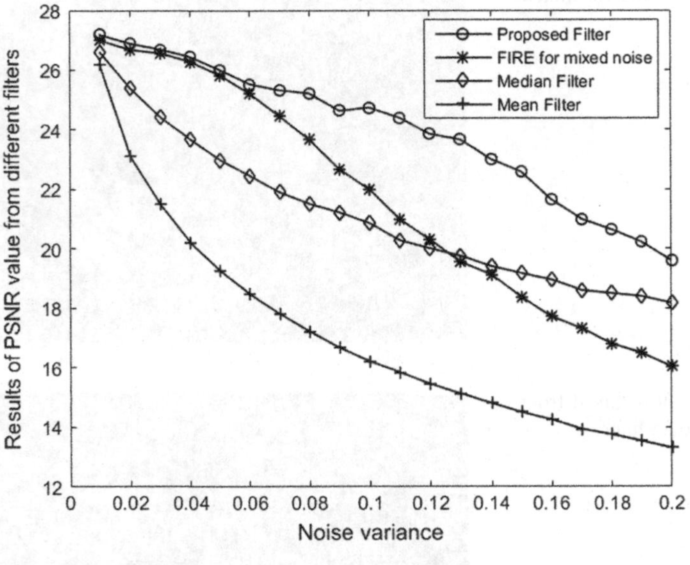

Fig. 9 PSNR values of the resulting images obtained from the different filters for filtering the SAR image corrupted with speckle noise

3 Conclusion

Synthetic Aperture Radar are widely used for providing high-resolution radar images. The coherent imaging process of SAR often causes speckle noise in SAR images which hinders the clarity of the image. In this work, a filter for de-noising speckle noise has been proposed which is based on fuzzy inference rule-based systems. The filter is used iteratively for higher values of speckle noise. The proposed method is compared with three existing methods of noise reduction namely

FIRE filter for mixed noise, mean and median filter. The proposed filter is found to give better results than the three other filters in terms of PSNR performance. Graphical analysis of performance of the proposed filter against the three other filtering techniques is provided for getting a clearer view on the results that have been obtained.

References

1. Skolnik, M. I.: Introduction to Radar Systems. McGraw-Hill, New York (1962)
2. Skolnik, M. I.: Radar Handbook. McGraw-Hill, New York (2008)
3. Radar basics, http://www.radartutorial.eu.
4. Moreira, A., Prats-Iraola, P., Younis, M., Krieger, G., Hajnsek, I., Papathanassiou, K.P.: A Tutorial on Synthetic Aperture Radar. In: IEEE Geoscience and Remote Sensing Magazine, vol. 1, pp. 6–43. IEEE (2013)
5. Chan, Y.K., Koo, V.C.: An Introduction to Synthetic Aperture Radar (SAR). In: Progress in Electromagnetics Research B, vol. 2, pp. 27–60 (2008)
6. Jain, A.N.: Fundamentals of Digital Image Processing. Prentice Hall, NJ, USA (1989)
7. Lee, J-S.: Speckle Suppression and Analysis for Synthetic Aperture Radar. In: Optical Engineering vol. 25(5) (1986)
8. Frost, V.S., Stiles, J.A., Shanmugan, K.S., Holtzman, J.C.: A model for radar images and its application to adaptive digital filtering of multiplicative noise. In: IEEE Transactions on Pattern Analysis and Machine Intelligence, vol. PAMI-4(2), pp. 157–166. IEEE (1982)
9. Kuan, D.T., Sawchuk, A.A., Strand, T.C., Chavel, P.: Adaptive Noise Smoothing Filter for Images with Signal-Dependent Noise. In: IEEE Transactions on Pattern Analysis and Machine Intelligence, vol. PAMI-7(2), pp. 165–177. IEEE (1985)
10. Chen, Y., Huang, F., Yang, J.: A Fuzzy Filter for SAR Image De-noising, 8th International Conference on Signal Processing, vol. 2. IEEE (2006)
11. Cheng, H., Tian, J.: Speckle Reduction of Synthetic Aperture Radar Images Based on Fuzzy Logic. In: First International Workshop on Education Technology and Computer Science, vol. 1, pp. 933–937. IEEE (2009)
12. Russo, F.: A user-friendly research tool for image processing with fuzzy rules. In: International Conference on Fuzzy System, pp. 561–568. IEEE (1992)
13. Nachtegael, M., Schulte, S., Van der Weken, D., De Witte, V., Kerre, E.E.: Fuzzy Filters for Noise Reduction: the Case of Gaussian Noise. In: The 14th IEEE International Conference on Fuzzy Systems, pp 201–206. IEEE (2005)
14. Mitra, S.K., Sicuranza, G.L.: Non Linear Image Processing, Elsevier, pp: 355– 374, (2001)
15. Russo, F.: Nonlinear Fuzzy Filters: An Overview. In: European Signal Processing Conference, EUSIPCO 1996, pp. 1–4. IEEE, Trieste, Italy (1996)
16. Kosko, B.: Neural Networks and fuzzy systems: a dynamical systems approach to machine intelligence. Prentice Hal, Inc. NJ, USA (1992)
17. Russo, F.: Fuzzy Systems in Instrumentation: Fuzzy Signal Processing. In: IEEE Transactions on Instrumentation and Measurement, vol. 4, pp. 735–740. IEEE, Waltham, MA, USA (1995)
18. National Remote Sensing Centre Indian Space Research Organisation, http://nrsc.gov.in/RISAT-1_Sample_Images?q=nellore

A Study of Representation Learning for Handwritten Numeral Recognition of Multilingual Data Set

Solley Thomas

Abstract Handwritten numeral recognition, a subset of handwritten character recognition is the ability to identify the numbers correctly by the machine from a given input image. Compared to the printed numeral recognition, handwritten numeral recognition is more complex due to variation in writing style and shape from person to person. The success in handwritten digit recognition can be attributed to advances in machine-learning techniques. In the field of machine learning, representation-based learning in deep learning context is gaining popularity in the recent years. Representative deep learning methods have successfully implemented in image classification, action recognition, object tracking, etc. The focus of this work is to study the use of representation learning for dimensionality reduction, in offline handwritten numeral recognition. An experimental study is carried out to compare the performance of the handwritten numerals recognition using SVM-based classifier on raw features as well as on learned features. Multilingual handwritten numeral data set of English and Devanagari numbers is used for the study. The representation learning method used in the experiment is restricted Boltzmann machine (RBM).

Keywords Handwritten recognition · Representation learning
SVM · RBM · Multilingual data set · Feature extraction

1 Introduction

Handwritten character recognition is one of the leading areas of research in the field of machine learning. Its application areas include processing bank cheque amounts, zip code identification for postal mail sorting, gadgets for disabled persons, numeric entries in form filled by hand etc. [1].

S. Thomas (✉)
Carmel College of Arts Science & Commerce for Women, Nuvem, Salcete
Goa, India
e-mail: sollybennet@gmail.com

D.K. Mishra et al. (eds.), *Information and Communication Technology for Sustainable Development*, Lecture Notes in Networks and Systems 10,
https://doi.org/10.1007/978-981-10-3920-1_48

Though a lot of research work is been carried out in the area of handwritten numeral recognition, there exists a potential for improvement. As Hindi and English are the most prominent languages used in India, and the constitution of India designates bilingual approach for all official language employing usage of Devnagari script as well as English, there is a need for the automatic handwritten recognition of this mixed data set. As such people in India commonly use Devnagari and English numerals as they fill up application forms, cheque, tax forms, etc. a mixture of Hindi and English numeral data set is chosen for the experimental study.

Handwritten character or numeral recognition can be categorized into two types: offline handwritten recognition and online handwriting recognition. In online method the recognition of the character is done at the time of writing. In the case of offline method, the recognition takes place from an image or the scan of the handwritten character/numeral. The major issue in handwritten numeral recognition is that the size, thickness, and orientation of the digits differ as the style of writing differs from person to person [2].

The various classifiers that can be used for handwritten numerals recognition include artificial neural network (ANN), Byes classifiers, K-nearest neighbor (KNN) classifier, Hidden Markov Models (HMM), and support vector machines (SVM). In the above-mentioned methods, support vector machines are more successfully implemented in recent years as it has great potential for classification task because of its strong discriminating abilities [3], and thus SVM is used for classification in this study.

Recent years, deep learning, a new way of learning, is being explored in machine learning. Deep learning is based on a set of algorithms that attempt to model high-level abstractions in data by using many layers of nonlinear information processing stages [4] and focus on learning representations of data. Representation-based learning in deep learning context is successfully implemented for various pattern recognition tasks [5].

This experiment is carried out to study the use of representation learning for dimensionality reduction in offline handwritten numeral recognition. Comparison of the performance of the handwritten numerals recognition using SVM-based classifier on raw features as well as on learned features is carried out to measure the effectiveness of representation-based method. Multilingual handwritten numeral data set of English and Devanagari numbers are used as the data set. Out of the various deep learning techniques which are discussed in Sect. 3 of this paper, RBM is used for the purpose of representational learning in the experimental study as RBMs have been successfully applied to problems involving high-dimensional data such as images and text [6]. The layout of the paper is as follows: Sect. 2—Literature survey, Sect. 3—Methodology, Sect. 4—Experimental study, Sect. 5—Result and Conclusion.

2 Literature Survey

Though Pattern recognition as a broad category and handwritten numeral recognition as a subset of it, is not a new field of study, it is still an active area of research. Over the years, researchers have worked on both offline handwritten recognition and online handwriting recognition and achieved promising results.

A general study of the earlier handwritten recognition methods [7], describes various online handwriting recognition methods before 1990.

In a unique work using SVM [8], 18 features were extracted from the image and obtained 99.48% recognition rate. The use of diffusion map for dimensionality reduction is experimented [9] to improve the handwritten digit recognition rate. Handwritten English numeral recognition using correlation method [10], focuses on segmentation for isolating the digits from multiple number images. An experiment on handwritten mixed numerals of south Indian scripts [11] reports an accuracy of 98.65% for Kannada numerals and 96.1% for Tamil numerals 96.5% for Malayalam, and 98.6% for Telugu numerals.

Recently, learning algorithms relying on models with deep architectures are implemented for solving harder learning problems [12].

Part of this work is based on the work done by Hinton and Salakhutdinov [6]. They explain method of dimensionality reduction using autoencoder.

3 Methodology

One of the promises of deep learning is replacing handcrafted features with efficient algorithms for unsupervised or semi-supervised feature learning and hierarchical feature extraction [13]. There are various deep learning architectures, such as deep neural networks (DNC), convolutional deep neural networks (CDNN), deep belief networks (DBN), Boltzmann machine (BM), RBM, and autoencoders [4].

Among these models, autoencoders directly learn a parametric feature mapping function by minimizing the reconstruction error between input and its encoding [14] and autoencoder is used in this experiment. A high-dimensional input data can be reduced to much smaller code space by using more number of layers in an autoencoder neural network [15].

The autoencoder in this experiment has multiple layers where the representation obtained in one layer is used as an input in the second layer and it is implemented using the RBM in such a way that the learned feature activations of one RBM are used as the input for training the next RBM in the stack. It consists of an encoder with layers of size 784-1000-500-250-30 and a symmetric decoder. Using, this method 784 raw features are reduced to 30 learned features. Classification is carried out on both the raw features and the learned features to compare the efficiency in terms of both accuracy and time complexity. Support vector machine is used for the classification. Support vector machines are supervised learning models based on the statistical learning techniques and associated learning algorithms that analyze data and recognize patterns, used as classification and regression prediction tool [16].

4 Experimental Study

The data used in this work consist of MNIST (Mixed National Institute of Standards and Technology) dataset for English numerals and ISI (Indian Statistical Institute) dataset for Devnagari.

All digit images have been size-normalized and centered in a fixed size image of 28 × 28 pixels. In the original dataset, each pixel of the image is represented by a value between 0 and 255, where 0 is black, 255 is white and anything in between is a different shade of gray. The ISI Data set is preprocessed to suit MNIST data set.

A mixed dataset is generated for this study, by evenly combining the MNIST dataset (English) and ISI Dataset (Devnagari) for the experiment. From the ISI data set we considered 3000 samples for testing and 19469 for training. From MNIST we considered 3000 for testing and 20531 for training. Thus, 40000 training samples and 6000 test samples formed the total data set. Some samples of both Devanagari and ISI data are given in Figs. 1 and 2 respectively.

4.1 Study on Raw Features

The data set is subjected to classification using the Support Vector Machine (SVM). The number of raw features is 784 (28 * 28 pixels) of each digit. The system is trained using the training data set of 40000 samples and 300 samples of each numeral of English and Devanagari is tested.

The experiment is carried out with different parameter values for Soft margin cost function (c) and Gaussian Kernel—gamma (g), and an accuracy of 96.74% is achieved, with parameter 'c' 1 and 'g' 0.07, on raw features. The LIBSVM [17] is used for training SVM (Gaussian Kernels).

4.2 Study on Learned Features

After completing the experiment on 784 (28 * 28) raw features the next step is to reduce the complexity by reducing the dimension of the data. By using the autoencoder of layer size 784-1000-500-250-30 and a symmetric decoder, 30 features were extracted. The learned feature activations of one RBM are used as the input for training the next RBM in the stack. Later, the deep autoencoder is created by unrolling the RBM. The data with 30 features were then subjected to classification using Support Vector Machine (SVM).

Fig. 1 Sample handwritten English numerals (*Source* MNIST)

The experiment is carried out with different parameters for Soft margin cost function (c) and Gaussian Kernel (g), and we achieved an accuracy of 96.7% with parameter 'c' 9 and 'g' 1.5.

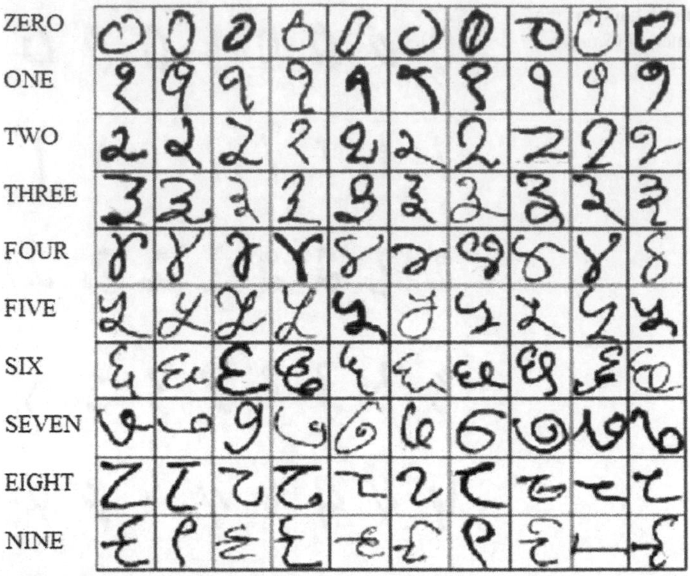

Fig. 2 Sample handwritten Devnagari script (*Source* ISI-Kolkota)

5 Result and Conclusion

Dimensionality reduction is of great importance for the efficiency of a classifier especially when large data base is concerned. In my experiment, it is observed that the time complexity is reduced by reducing the dimension of the features as shown in Tables 1 and 4. With the extracted features, the accuracy of numeral recognition is almost same as with raw features. By using representation learning method the number of features were reduced from 784 (raw feature) to 30 (learned features), thus the computational complexity is reduced to a great extent. One of major problems in classification task of large data set is the high-dimension feature space of the data set, and any step to reduce the feature space will add value to the classification task by way of reducing the complexity of the task (Tables 2 and 3).

Table 1 Classification result of SVM with Gaussian kernel on raw features

Method	Raw features	Parameter	Accuracy (%)	Time taken (in min)
SVM	784	c 1 g 0.01	95.1	19
		c 1 g 0.07	**96.74**	18

Table 2 Accuracy of English numeral recognition using SVM on raw features

English numerals	Correctly recognized	Accuracy (%)
0	299/300	99.6
1	294/300	98
2	291/300	97
3	290/300	96.6
4	290/300	96.6
5	289/300	96.3
6	285/300	95
7	277/300	92.3
8	286/300	95.3
9	273/300	91

Table 3 Accuracy of Devanagari numeral recognition using SVM on raw features of data

Devanagari numerals	Correctly recognized	Accuracy (%)
०	296/300	98.6
१	297/300	99
२	284/300	94.6
३	291/300	97
४	289/300	96.3
५	294/300	98
६	297/300	99
७	297/300	99
८	295/300	98.3
९	292/300	97.3

Table 4 Classification result of SVM with Gaussian kernel on extracted features of the data

Method	Features	Parameter	Accuracy	Time taken (in min)
SVM	30 extracted features	c 9 g 0.1	92	2.3
		c 9 g 0.5	93.1	2.4
		c 3 g 1.9	96.1	2.4
		c 4 g 1.9	96.1	2.5
		c 9 g 1.5	**96.7**	2.5

References

1. Ashwin S Ramteke, Milind E Rane: A Survey on Offline Recognition of Handwritten Devanagary Script. International Journal of Scientific and Engineering Research, May 2012.
2. Ahmad, Azad: Multi Script Handwritten Numeral Recognition using Neural Network Technique. Research Journal of Science, Engineering and Management, Vo. 1, April 2012.
3. Huang, B.Q.A: Hybrid HMM-SVM Method for Online Handwriting Symbol Recognition. IEEE- 6th International Conference on Intelligent Systems Design and Applications, Vol. 1, October 2006.
4. Deng, Li: Deep Learning: Methods and Applications. Foundations and Trends® in Signal Processing, Vol. 7, 2013.
5. Kongming Liang: Representation Learning with Smooth Auto encoder, Visual Information Processing and Learning, 2014.
6. Hinton, G. E & Salakhutdinov R.R: Reducing the Dimensionality of Data with Neural Networks. pp. 504–507, Science 2006.
7. Toru Wakahara Charles C. Tappert, Ching Y. Suen: The state of the art in on-line handwriting recognition, IEEE Transactions on Pattern Analysis and Machine Intelligence. Vol. 12(8), August 1990.
8. Shrivastava, Shailendra Kumar: Support Vector Machine for Handwritten Devanagari Numeral Recognition. International Journal of Computer Applications Vol. 7, October 2010.
9. Ritu Ashok Kambli, Yogesh Kailas Ankurkar and Ameya Vinay Mane: Handwritten Digit Recognition with Improved SVM. IJIST—International Journal of Innovative Science, Engineering & Technology, Vol. 1, Issue 4, June 2014.
10. Isha Vats, Shamandeep Singh: Offline Handwritten English Numeral Recognition using Correlation Method. IJERT—International Journal of Engineering Research & Technology, Vol 3, Issue 6, June 2014.
11. S.V Rajashekararadhya & P. Vanaja Rajan: Handwritten numeral/mixed numeral recognition of south Indian scripts: The zone based feature extraction method. Journal of Theoretical and applied Information Technology, 2005.
12. Larochelle, Hugo: An Empirical Evaluation of Deep Architecture on Problems with Many Factors Variation. International Conference on Machine Learning, 2007.
13. Schmidhuber, Juergen: Deep Learning in Neural Networks: An Overview, Vol 61, 2014, pp. 85–117.
14. Chet, Tan Chun: Auto encoder Neural Networks: A performance study based on Image recognition Reconstruction and Compression. Extended Abstract for masterwork Completion Seminar, 2008.
15. Hirwani, Amrita: International Journal of Advance Research in Computer Science and Management Studies, 2014, pp. 83–88.
16. Mansi Shah, Gordhan B Jethava: A Literature Review on Handwritten Character Recognition. Indian Streams Research Journal, Vol. 3, March 2013.
17. Chang, C.C. and C.J. Lin: LIBSVM: a library for support vector machines. http://www.csie.ntu.edu.tw/cjlin/libsvm.

Abandoned Object Detection and Tracking Using CCTV Camera

Parakh Agarwal, Sanaj Singh Kahlon, Nikhil Bisht, Pritam Dash, Sanjay Ahuja and Ayush Goyal

Abstract With the increase in crime and terror rate globally, automated video surveillance, is the need of the hour. Surveillance along with the detection and tracking has become extremely important. Human detection and tracking is ideal, but the random nature of human movement makes it extremely difficult to track and classify as suspicious activities. The primary objective of this is to detect the suspiciously abandoned object recorded by the closed-circuit television cameras (CCTV). The main aim of this project is to ease the load on the controller at the main CCTV station by generating and alarm, whenever there is a detection of an abandoned object. To solve the problem, we first proceeded by the background subtraction such that we obtain the foreground image. Further, we calculated the inter-pixel distance and used area-based thresholding so as to differentiate between the person and the object. The object will further be tracked for a previously set time, which will help the system to decide whether or not the object is abandoned or not. Such a system that can ease the load on single CCTV controller can be deployed in places which require high discipline and security and are more prone to suspicious activities like Airports, Metro station, Railway Stations, entrances and exits of buildings, ATMs, and similar public places.

P. Agarwal (✉) · S.S. Kahlon · N. Bisht · P. Dash · S. Ahuja · A. Goyal
Amity University Uttar Pradesh, Sector 125, Noida 201313, UP, India
e-mail: parakh1301@gmail.com

S.S. Kahlon
e-mail: sanajkahlon@gmail.com

N. Bisht
e-mail: nikhilbisht94@gmail.com

P. Dash
e-mail: dash.pritam@gmail.com

S. Ahuja
e-mail: sanjayahuja@india.com

A. Goyal
e-mail: agoyal1@amity.edu

D.K. Mishra et al. (eds.), *Information and Communication Technology for Sustainable Development*, Lecture Notes in Networks and Systems 10,
https://doi.org/10.1007/978-981-10-3920-1_49

Keywords Object detection · Video surveillance · Tracking · Image processing Algorithm · Security alarm system

1 Introduction

In spite of the upcoming technological developments to create automatic surveillance systems, there are a lot of different factors or variables that have affected the effectiveness of these systems, and thus diminishing their reliabilities. If the system could tell the user about the level of confidence in the alarms generated, then the user's trust on the system would be improved. The intensity of work for operators would decrease, and the efficiency of these automatic systems would be increased. The outcomes showed considerable reduction in workload when there is definite information. This project work not only anticipates the up gradation of automatic video surveillance but also the requirement for further research in this field."

1.1 General Introduction

CCTV have long been utilized as an effective tool for monitoring; CCTV is one of the finest machines for security purposes. The technology for video security and surveillance has developed faster in recent years than earlier. But automatic surveillance systems are modified such as to enhance the potency of CCTV cameras, although the effectiveness depends on different factors along with the technological advancements.

The automation process is used in complicated situations, where it reduces the workload on the human operator, and thereby helps to increase the productivity and improve the performances. An important factor is the design of automatic surveillance systems, which determines the level of reliability, and can be used for increasing the effectiveness of these systems. With the increase in the crime and terror activities, there is a need for an enormous number of CCTVs to keep a check on the surroundings, but monitoring each of the camera screens is a very difficult task.

Unlike robbery and assault cases, which can be solved through the captured footage of CCTV with the help of forensic analysis, there are cases where immediate responses to the conditions are required. Public transport systems such as the rail way stations are more vulnerable to severe events such as terrorist attacks, etc. which require continuous monitoring of video footage. Automatic surveillance systems are to assist operators in this matter.

1.2 Monitoring of CCTV Camera

CCTV has been extensively in use during the recent years over which they have changed from a simple recording system for general monitoring to automatic surveillance system for the purpose of suspect detection. A CCTV framework comprises of a few cameras, checking and recording frameworks and control room operations. The operator can monitor CCTV displays either reactively or proactively.

1.2.1 Reactive Checking

Reactive checking is uninvolved observing of CCTV cameras subsequent to being alarmed by sound. It is found that reactive checking is the most intensely utilized checking assignment which is typically activated by police and business radios.

1.2.2 Proactive Checking

Proactive observation alludes to online checking of the scenes to identify and foresee suspicious occasions and keep occurrences from happening. Notwithstanding the significance of proactive observation, the extent of proactive checking completed actually is practically nothing.

The automated surveillance concept has been mainly developed to monitor continuously and be in tandem with proactive monitoring. On the other hand, automated surveillance can also be applied in the reactive monitoring if the controller had the ability to switch to the automatic mode following the alert.

1.3 Previous Work and Scope for Future Work

There have been various studies and research in the field of automated surveillance and most of the work involved the machine learning so as to train the system to recognize and identify the different abnormal situations. Stereo cameras have been used to model the objects in 3-D to detect the target and further tracking has been done using the blob matching scheme along with the use of the Kalman filter. Subsequently the activity is classified into different previously defined suspicious activities by the semantics-based procedure.

But, we have developed an algorithm using background subtraction, so as to simplify the algorithm as it does not emphasizes on training of system. A simple and robust algorithm have been proposed by us in this project.

Furthermore, a system that allows the operator to identify the object manually and further let the system track the object or target and report its destination can be expected to be developed in the future. A further study in the human-automation interface can help us to specify more stricter guideline for the definition of suspicious activities and help improve the automated surveillance systems of the future.

2 Methodology

Detection and Tracking of Objects and then Classifying it into Abandoned or not, is a process, which will require low-level processing of image frame. Image Frame is extracted at a certain frame speed (in FPS) from a video file. Each and every image frame undergoes the following steps:

2.1 Background Model

It is not necessary that background remain same at all the time, there might some changes occur due to addition or subtraction of some objects or there might be some changes due to change in light intensity or illumination, so we have taken the average of some background images at regular time interval when no activity is going on, so that any new changes will get minimize. The resultant image is used as Background image for image Subtraction.

2.2 Low Level Processing

Before proceeding with object detection and tracking, we have segmented objects of interest and minimizes the noise. For that we have done Image Subtraction to obtain a foreground Image. Image Subtraction is done by subtracting Image with activity going on with the background model. There will be introduction of some noise due to intensity difference or some minor change in positioning of objects. To minimize the small noise, we have performed the Morphological open operation, which will remove the small noise to zero. Now, Using Otsu's Thresholding method, we have converted the Grayscale Image into Binary Image (Fig. 1).

2.3 Detection

In an image frame, there are both person and objects. So, to detect objects we have used two parameters: area-based thresholding and inter-pixel distance. In area-based thresholding, The size of suspicious object cannot be bigger than the size of a human, so on comparing the area of both, we came to a conclusion that,

$$\textbf{Area(object)} < \textbf{Area(Person)}$$

Fig. 1 System model

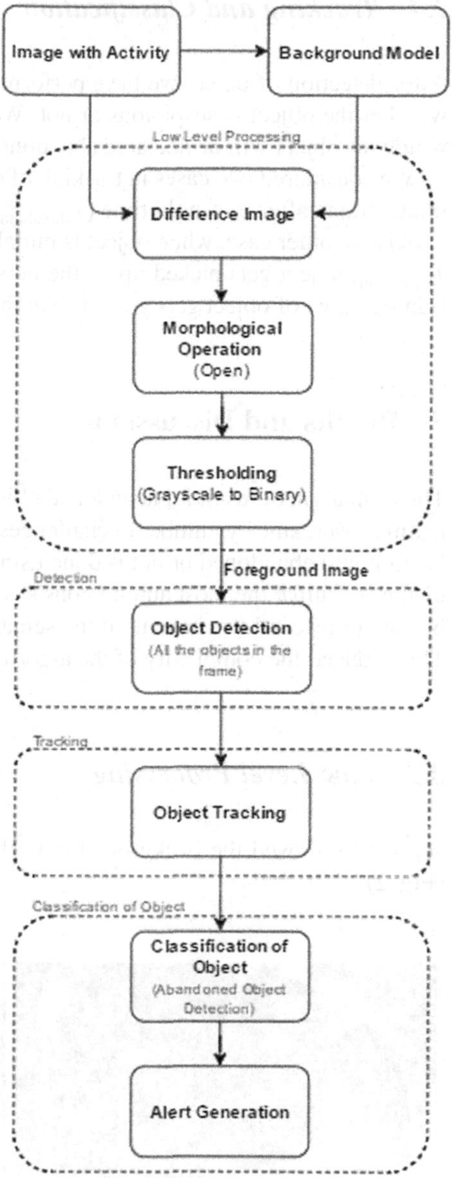

So, a threshold area ($A_{threshold}$) is defined which will distinguish object from person. In our case we have taken it as 300 pixel units. Also, Considering inter-pixel distance (IPD) as the parameter, the centroids of all the objects in the frame is taken, and the IPD is calculated between the object and the person nearby that particular object.

2.4 *Tracking and Classification*

After detection of object we have performed object tracking, in order to determine whether the object is suspicious or not. We have considered a threshold time, after which the object will be declared abandoned and suspicious, and alarm is generated.

We considered two cases in tracking of object, firstly when an object is still in the image frame after threshold time ($T_{threshold}$), and alarm is generated.

Also in other case, when object is initially alone, but after some time, less that of $T_{threshold}$, object gets picked up by the person, then the objects disappears from the frame as area of object gets greater than that of $A_{threshold}$.

3 Results and Discussions

The evaluation of result for abandoned object detection, that too using a simple algorithm is overcame by number of challenges. First to detect the object and then classifying it into abandoned or not is done using area-based thresholding and inter-pixel distance. Unlike the most autonomous systems there is the need to train the system, but in our research we have used the semantic approach of background subtraction. This reduced the complexity of the algorithm.

3.1 *Low-Level Processing*

We have removed the background only thresholded foreground image is obtained (Fig. 2).

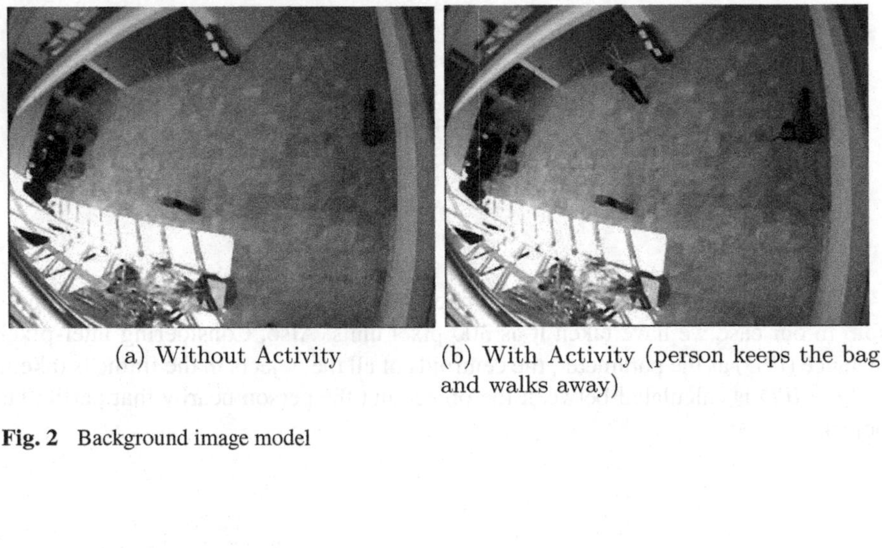

(a) Without Activity (b) With Activity (person keeps the bag and walks away)

Fig. 2 Background image model

(a) Subtracted and then noise suppressed image (b) Binary Image (Grayscale to Binary)

Fig. 3 Foreground and binary image

3.2 *Detection*

As in Fig. 3b, there is the presence of both person and object, so to distinguish between the two we have performed connectivity check to get the number of objects in the frame and to get the area of each and every boundaries, be it of person or object. Here, the number of objects are Two, with area of person as 575 pixel units and area of object as 222 pixel units. Since we have considered, $A_{threshold} = 300$ pixel units, so, all the boundaries with area above 300 pixel units will get removed from the frame (Figs. 4 and 5).

Now when we considered the another case, in which the guy comes and pick his bag back, Now, the number of objects turned to 1, as two boundaries merged into one, so the area on whole object increases, i.e., the addition of area of both person and object is the complete area, which is obviously greater than that of $A_{threshold}$. So, the object will vanish off from the frame (Figs. 6, 7 and 8).

```
CC =

    Connectivity: 8
       ImageSize: [288 384]
      NumObjects: 2
    PixelIdxList: {[575x1 double]  [222x1 double]}
```

Fig. 4 Connectivity check (Image information)

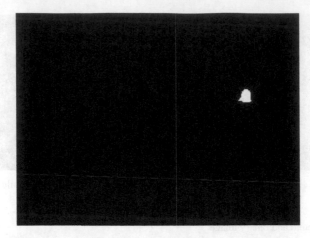

Fig. 5 Only object(s) in the frame image

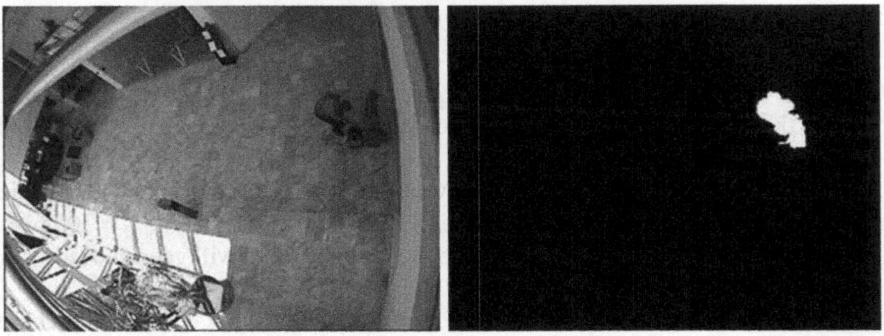

Fig. 6 Person pick his bag back (Both RGB and binary image)

Fig. 7 Connectivity check (Image information)

```
CC =

    Connectivity: 8
       ImageSize: [288 384]
      NumObjects: 1
    PixelIdxList: {[1205x1 double]}
```

3.3 *Tracking and Classification*

Since we have result of two cases, so now we have classified these two cases. For that we have tracked the object for a threshold time, $T_{threshold}$. When the object is there, a box will be marked outside the abandoned object, and alert will be generated on operators screen (Fig. 9).

Fig. 8 No objects appears in the frame as person picks his bag up

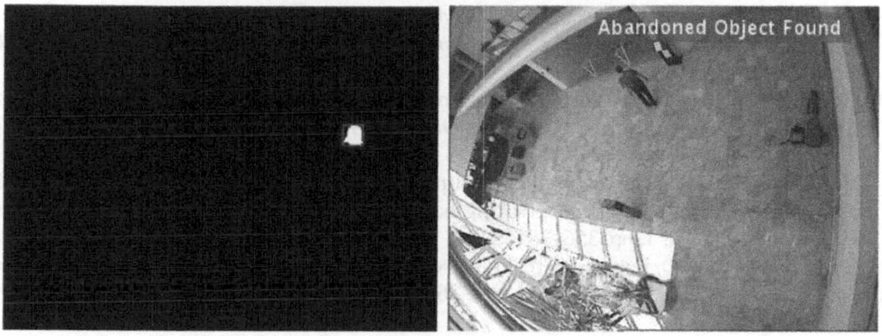

(a) Box is marked ouside the abandoned object

(b) Alert is generated for abandoned and suspecious object on CCTV Controller Screen (Grayscale to Binary)

Fig. 9 Results

4 Conclusion

Object detection via CCTV or for that matter any form of automated surveillance system is one for the near future, as it aims to make the world a lot safer. Object detection via CCTV cameras can be installed and applied to a whole lot of places where the need of security is high. The chances of improvement and further development in this area are enormous. With the increase in the terror and crime rates the use of this project will help to ease the load on human monitoring. Moreover, it will help us to stop the crime or act of terror as compared to using the CCTV recordings for post event analysis. The main process of detecting a human and an object uses the concept of background subtraction of the image with activity and an adaptive background model.The object detection is undertaken by the concept of area-based thresholding and inter-pixel distance, then the abandoned object tracking is done by setting a time threshold, once the object has been separated from the person for a time greater than the threshold time, it gets labeled as an abandoned object. Such a concept can be applied to other fields like locating lost baggage or lost cargo. There are many other possibilities for the further development of this algorithm based on the application and system installation point.

References

1. Sandesh Patil and Kiran Talele, "Suspicious Movement Detection and Tracking Based on Color Histogram", *International Conference on Communication, Information and Computing Technology, ICCICT-2015*, 15–17 January, 2015, pp. 1–6.
2. Mohannad Elhamod, and Martin D. Levine, Real-Time Semantics-Based Detection of Suspicious Activities in Public Spaces, *Proc. 9th Conf.* CRV, Toronto, ON, Canada, 2012, pp. 268–275.
3. H. Weiming, T. Tieniu, W. Liang and S. Maybank, "A survey on visual surveillance of object motion and behaviors", *IEEE Trans. Syst., Man, Cybern. C, Appl. Rev.*, vol. 34, no. 3, pp. 334–352, 2004.
4. Qian Zhang and King Ngi Ngan; Segmentation and Tracking Multiple Objects Under Occlusion From Multiview Video, *IEEE Transactions On Image Processing*, Vol. 20, No. 11, November 2011.
5. PETS-ECCV. (2004). [Online]. Available: http://wwwprima.imag.fr/PETS04/caviardata.html.
6. L. M. Fuentes and S. A. Velastin, "Tracking-based event detection for CCTV systems", *Pattern Anal. Appl.*, vol. 7, no. 4, pp. 356–364, 2004.
7. M. Elhamod and M. D. Levine, "A real time semantics-based detection of suspicious activities in public scenes", *Proc. 9th Conf.* CRV, pp. 268–275, 2012.

Accurate and Robust Iris Recognition Using Modified Classical Hough Transform

Megha Chhabra and Ayush Goyal

Abstract Circle Hough Transform (CHT) is a robust variant of the Hough Transform (HT) for the detection of circular features. Accurate iris recognition is one of the application areas of this technique. Robustness and accuracy are of utter significance but at the expense of high-computational time and space complexity as the method processes the entire image provided as an input. The present work formulates a computationally more efficient suggested solution as a modified Circle Hough Transform (MCHT) by fixating time–space complexity in terms of reducing the area of the image to cover (the number of pixels to process), and hence significantly decreasing computational time without compromising the accuracy of the method. The modified method is tested on a sample set of collected iris images. Each image is divided into three different sized skeleton grids of size 3×3, 5×5 and 7×7 (pixels). The center of each grid type applied on the image gives the Region of Interest (ROI) of the image sufficient to detect circular parameters as center and radius of the iris using CHT. The experiment shows the comparison of computational time required to detect the iris from CHT applied to the whole image versus the computational time required to detect the iris from just the ROI of the image using the grids. Additionally, the results of the comparison of the expected time and observed time of detection of the iris over a large number of images is presented. There is a substantial reduction in computational time complexity up to 89% using the 3×3 sized grids, and up to 96% using 5×5 sized grids and up to 98% in 7×7 sized grids with equally fair amount of reduction in space utilization. The experiment was performed to observe which grid size gave the most accurate center and radius values along with the most efficient performance. The results showed that the 3×3 and 5×5 sized grids provided better results as compared to the 7×7 sized grids, the results of which lacked accuracy for some images. From the results of the experiments with

M. Chhabra (✉) · A. Goyal
Amity University Uttar Pradesh, Sector 125, Noida 201313, UP, India
e-mail: megha.chhbr@gmail.com

A. Goyal
e-mail: agoyal1@amity.edu

D.K. Mishra et al. (eds.), *Information and Communication Technology for Sustainable Development*, Lecture Notes in Networks and Systems 10,
https://doi.org/10.1007/978-981-10-3920-1_50

varying grid sizes, the conclusion obtained is that the accuracy is compromised by grid sizes 7×7 and higher, and grid sizes of 3×3 or 5×5 provide the most accurate and efficient iris detection.

Keywords Hough transform · Iris · Circular hough transform

1 Introduction

Over last few years, Digital images are used in many application areas of wide range like remote sensing, mountaineering, x-rays, applications in multimedia, biometric security systems, inspection and detection of structural collapse of pipes, detecting atom structure, and endless list [1]. The features attracting these research fields and various other industrial applications are shape recognition and its extraction.

An efficient method for shape recognition is the need of present research work. Edges contain high quality data. Hence edge detection comprises of detecting curves like lines, circles, ellipses, corners, etc. Basically, this recognition is done using HT and its variants.

Much work on digital images is based on extraction of circular object(s). Detecting circles is now a well-exploited, well-known research area for computer vision systems. With increasing size of its research area, its applications are also increasing. This paper provides a modified HT algorithm for iris detection. As a significant part of the visual industry, object recognition has come out as a hot spot in biometric diagnostics field. Its a study of recognition of person on the basis of physiological or behavioral features. Among various identifications, one of the most promising is a human iris. This application of circle detection has gained too much attention of various researchers. Since the eye or iris is an unchanging feature of the face as opposed to other features of the face and since the iris is not dependent on other change prone biometric patterns like palm, fingerprints, face, etc., it is a suitable unique-to-individual biometric for our present research work to focus on. This is one such powerful area which rules important domains, such as criminal recognition, security agendas, biometric medical field works etc.

Area of research like time to detect and extract the iris from given iris image holds a good amount of space for improvement. Many researchers have studied iris recognition. Various methods are present which improve iris recognition. Hamming distance in combination with Gabor filters have previously been used by Sanchez-Avila et al. 2005 for iris detection [2]. Tsuyoshi Kawaguchi and Mohamed Rizon [3] give an algorithm which extorts intensity valleys from the face region. Utilizing the cost for each pair of iris potential matches the method detects an iris potential match analogous to each of the irises. Wojciech et al. [4] presents a method for reliable iris segmentation. Peihua Li et al. [5] contribute to accurate and robust iris detection in extremely noise-degraded images. It follows the notion of focusing on progressive detection of the iris in the eye, localizing the boundaries of the limbic and papillary type, and removing highlight of the specular type by locating the eyelids.

Wen et al. [6] presents the algorithm of the matching pursuit method. The iris is localized utilizing the difference in color in between regions of the eye. Extracted parameters are then translated into vectors representing features by the algorithm of the matching pursuit method. The work of Li ma et al. [7] shows iris detection based on the multichannel Gabor filtering. It shows filtering of iris images by Gabor filters. It results in a fixed length feature vector. Iris detection proposed in the study of Li ma et al. [8] works on a method utilizing circular symmetric filters. CHT is another variant of HT used for circular feature recognition [9]. It provides robust techniques for circular object detection using higher peak value analysis on the basis of a mechanism for determining circular parameters such as center and radius.

CHT has a major limitation in terms of time complexity. The research shows variants of CHT are more focused on reducing space utilization and thereby improving speed of detection. The fact that lesser the time utilization to detect iris with increased efficiency, better is the recognition process has opened much space for improvement in practice of CHT.

The study of Kimme et al. [10] proposed a method utilizing information about the direction of edges in the image. It uses the fact that the direction on the edge of a circular structure directs away from or toward its center. Hence, only an arc is required to be drawn perpendicular to the point at the edge to find the center. Xu et al. [11] suggests and proves that randomly chosen edge pixels and the circular geometry can be used methodologically to enhance speed. It further uses voting technique to find circular parameters. The study by Chen and Chung [12] chooses four random pixels and finds the parameters for circle detection. Also, another research presented in the study by Yip et al. [13] showcases that detection is also enhanced by reducing the dimensions of parameters by estimating the same parameters by local circular geometrical properties. The study by Ho and Chen [14] proposes another enhancement using global instead of a local geometrical property of circles. The study of Tsuji and Matsumoto uses the property of parallel type of circles to reduce the space of parameters [15].

2 Hough Transform for Circle Detection

The HT is a technique robustly suitable for the detection of line-like and curve-like structures in a digital image. It is an efficient method for confirming coordinates of a circle center point and the circle's radius. Certainly, with all the given invariants for extraction methods, HT and CHT are so far the most robust algorithms for line and circular feature extraction.

2.1 Hough Transform

When the elements of interest can be described in parametric form, one way to achieve the extraction is to use Hough Transform. HT transforms the pixels in the image space (in the x-y plane) into corresponding lines in parametric space. HT uses a convenient way to describe a set of lines parametrically:

$$r = x(\cos\theta) + y(\sin\theta) \tag{1}$$

Parametric space is now defined by θ along with r. Here r is the length of normal from the point of origin to the line and θ is the orientation of r with respect to the x-axis. For any point on this line as (x,y), r, and θ are constant values.

For the circle equation, HT uses a similar method to describe the equation in image space, i.e., the x-y plane, in parametric space. For a circle, the parametric equation is:

$$(x - h)^2 + (y - k)^2 = r^2 \tag{2}$$

The equation uses the coordinates of the center of the circle of radius r. For the same parametric equation, the equation is transformed with the same parameters as follows:

$$x = h(\cos\theta), y = k(\sin\theta) \tag{3}$$

In this case, the computational complexity of the algorithm increases in proportion to the increase in the parameters and a 3-D accumulator for voting is added up. So as the shapes get complex, the increasing complexity makes it difficult to work with HT, thereby making it more effective for simple curves only.

2.2 Circular Hough Transform

The main operation in feature extraction for any shape is locating interest spots in the input image at a certain point. CHT is a convenient way to achieve circle extraction from input image.

The definition of circle as Eq. 2. The method of the algorithm is to determine in the image, the parameters (h, k, r) of the circle. Post application of standard edge detection techniques, edge pixels are detected from the image. Each but all detected edge pixels contributes a circle of radius r and center (h, k) to an output accumulator space thereby requiring to make a 3-D array V(h, k, r) of the accumulator type.

Equation (2) after transformation from image space to HT space is given as follows:

$$(h - x)^2 + (k - y)^2 = r^2 \tag{4}$$

This transformation forces each edge pixel to map to a right circular cone centered at (x, y, 0). 3-D accumulator V(h, k, r) is now filled with votes for each edge pixel accumulated on the basis of Eq. 4. Now, the votes of these triplets as (h_i, k_i, r_i) are counted.

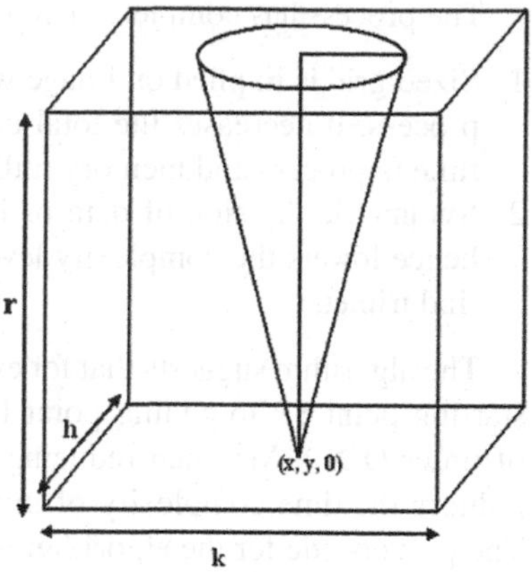

Fig. 1 Hough space for CHT

The basic operation of the procedure is to find value of triplet (h_o, k_o, r_o) corresponding to peak value of votes accumulated in accumulator array V. Figure 1 represents the parametric space of CHT after transformation.

3 Proposed Method

The aim of feature extraction is to locate interest spots in the input image at a certain precision. In this circle detection method as proposed, interest spot is basically ruled out and processed according to total seed pixels or boundary pixels processed which are more likely to give peak value for triplet $V(h_o, k_o, r_o)$ than rest of the data in the image. This area of interest spot is called region of interest (ROI).

In iris recognition, time to detect iris is a significant factor. The work proposed here aims at increasing the speed of extraction, at the same time keeping it accurate.

Next, ROI is nothing but the valid region of image where probability of finding the center is highest. Once this relevant region is found, that directly reduces the number of pixels to be processed by CHT in future.

Because, time of execution is concerned with pixels to process, thus this modification in algorithm allows lesser time for CHT to process image and find results.

The formal specification of modification suggested in CHT is the evaluation of the valid region. In general, if *f(x,y)* is a digital image matrix N × M in size. Given A defines the area of the digital image matrix *f(x,y)*, A = N × M. Let image is divided using varying size grid say of size P × P. Grids are nothing but symmetric division of the image. It is clear from the division that the most probable area for finding triplet $V(h_o, k_o, r_o)$ is the central area. Figure 5 shows the general structure of division. The central region is considered as valid region to be processed. This region is further accepted for CHT processing and rest of the data in image is discarded.

The process has completed two steps till now:

1. Sized grid is implied on image which helps in finding legitimate region(ROI) to process. It decreases the total count of pixels to be analyzed thereby reducing time to process and memory utilization.
2. Symmetric division of data of image helps in easy search of valid region and hence lowers the complexity level and decreases the execution time for CHT to find triplet.

The algorithm suggests that for every grid sized P, the ROI only takes region from starting point SP to ending point EP. Also, if applying CHT takes time complex of order O(N * M), then reducing the number of pixels to process to M/P × N/P reduces the time complexity of modified algorithm by order of O((M/P * N/P)). The pseudocode for the algorithm is given as:

Begin

1. Read image matrix *f(x,y)* of dimensions N × M.
2. Let P is the Grid Size then $S_p = (floor(P/2)/P)$ and $E_p = ((floor(P/2) + 1)/P)$.
3. For $i = S_p * M$ *to* $E_p * M$ repeat steps 4 to 5.
4. For $j = S_p * N$ *to* $E_p * N$repeat step 5.
5. Apply CHT and find triplet (h_o, k_o, r_o).

The experimental analysis is done on the basis of evaluation of the results of execution time of CHT when takes values; {3, 5, 7} for P x P sized grid. The motive behind using varying sized grids is to determine the future that in spite of using different areas resulting into different number of pixels to process, the robustness is still maintained or not. If not, then what is the threshold of size of grid?

Table 1 illustrates how taking grid size P reduces the complexity order from O($N * M$) to O(($N * M)/P^2$). For instance, 3 × 3 sized grid divides the image into 9 equal parts hence reduces the work area A of digital image matrix *f(x,y)* by a factor of 9 and hence the time complexity is reduced to O(($N * M)/9$). As shown in Fig. 2, the central area is the formal valid area to work with. This implies that now the image to be further processed by CHT is reduced to size M/3 × N/3. On similar grounds, 5 × 5 sized grid provides reserved processing image of size M/5 × N/5, thus reducing the area to process to A/25. In case of 7 × 7 size grid, as shown in Fig. 4, Image to be processed reduces to size A/49. It provides the valid area of size (Fig. 3).

4 Experimental Analysis

This section illustrates the analysis showing the calculated accuracy of the CHT method maintained with variation imposed on it. The experiments are performed on sequence of 100 images used for ophthalmology diagnosis. The images analyzed are the samples gathered from the captured eye images of patients gazing on various types of fixating devices. The work performed on original CHT is then compared

Table 1 Effect of various values of P on time and space complexity

34emFeature	Method				
	CHT	MCHT (Grid size)			
	No grid	3 × 3	5 × 5	7 × 7	P × P
Image size	M × N	M/3 × N/3	M/5 × N/5	M/7 × N/7	M/P × N/P
Area to process	A	A/9	A/25	A/49	A/p^2
Region coordinates	M-0 × N-0	2M/3-M/3 × 2N/3-N/3	3M/5-2M/5 × 3N/5-2N/5	4M/7-3M/7 × 4N/7-4N/7	(Floor(P/2)+1)M/P-(Floor(P/2))M/P × (Floor (P/2)+1)N/P-(Floor(P/2))N/P
Order of complexity	O(N × M)	O(M/3 × N/3)	O(M/5 × N/5)	O(M/7 × N/7)	O(M/P × N/P)

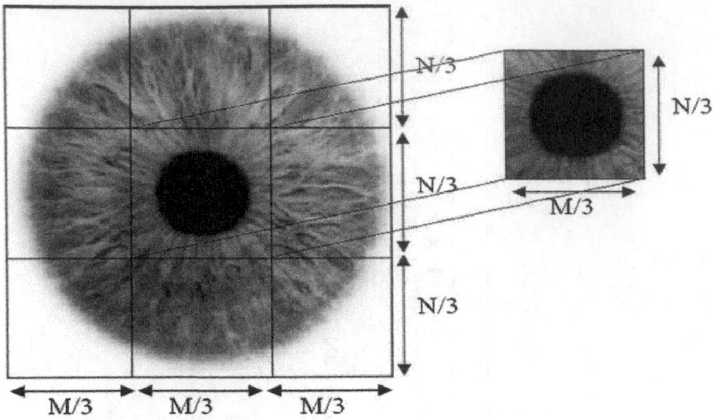

Fig. 2 Division of digital image matrix $f(x, y)$ into nine equal regions using grid size 3×3

Fig. 3 Division of digital image matrix $f(x, y)$ into 25 equal regions using grid size 5×5

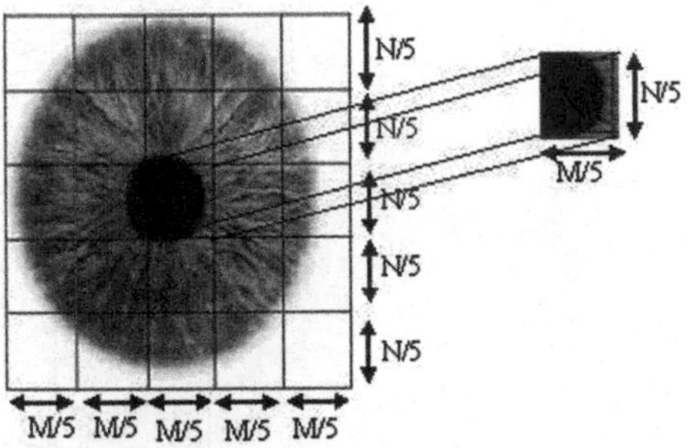

Fig. 4 Division of digital image matrix $f(x, y)$ into 49 equal regions using grid size 7×7

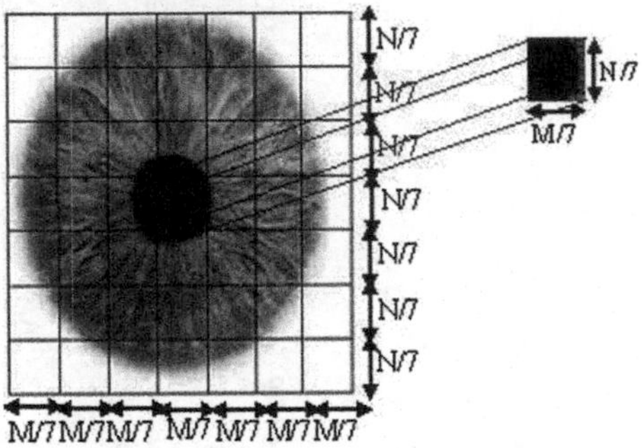

with output of proposed modified CHT. Time of execution of original CHT is compared and analyzed for accuracy with proposed CHT which uses preprocessing step of extracting valid region to be processed from image using varying sized grids. There is a significant increase in performance. Quality of results in Table 2 shows that grids of size 3×3 and 5×5 are more accurate in results in comparison to 7×7 sized grid results.

Table 2 Time analysis between CHT and MCHT

Name of the image	Number of edge pixels	Size of the image	Time analysis (In s)				Efficiency of proposed CHT using grid of size (in %)		
			Existing CHT	Proposed CHT using grid of size					
				3×3	5×5	7×7	3×3	5×5	7×7
1	6317	215×152	15.859	1.969	0.828	0.515	87.58433697	94.77898985	96.75263257
2	7532	215×144	17.844	2.14	0.906	0.563	88.00717328	94.92266308	96.84487783
3	4866	215×143	11.579	1.438	0.656	0.437	87.58096554	94.33457121	96.22592625
4	9551	215×156	24.406	2.859	1.172	0.687	88.28566746	95.19790216	97.18511841
5	8827	215×158	22.859	2.66	1.11	0.672	88.36344547	95.14414454	97.06023886
6	6209	215×215	21.954	2.578	1.094	0.656	88.25726519	95.01685342	97.01193404
7	17437	215×225	63.765	7.24	2.703	1.469	88.64580883	95.76099741	97.69622834
8	13825	215×137	30.922	3.532	1.406	0.813	88.57771166	95.45307548	97.37080396
9	6857	215×144	16.297	1.985	0.844	0.531	87.81984414	94.82113272	96.74173161
10	6138	215×146	14.813	1.797	0.797	0.5	87.86876392	94.6195909	96.62458651
11	6300	215×143	14.89	1.813	0.781	0.516	87.82404298	94.75486904	96.53458697
12	12943	215×214	45.078	5.125	1.969	1.079	88.63081769	95.63201562	97.60637118
13	11323	215×205	37.875	4.328	1.719	0.953	88.57293729	95.46138614	97.48382838
14	3285	215×162	8.907	1.172	0.562	0.391	86.84180981	93.6903559	95.61019423
15	12037	215×214	41.984	4.781	1.844	1.031	88.61232851	95.60785061	97.54430259
16	1784	215×143	4.406	0.671	0.391	0.313	84.77076714	91.12573763	92.89605084
17	1702	256×171	5.969	0.844	0.454	0.344	85.8602781	92.39403585	94.2368906
18	6060	256×171	20.329	2.438	1.016	0.625	22 00728024	95.00221359	96.9255743
19	12686	215×215	44.438	5.031	1.953	1.078	22 6786084	95.60511274	97.57414825
20	6632	215×144	15.766	1.922	0.828	0.515	3780920969	94.74819231	96.7334771

(continued)

Table 2 (continued)

Name of the image	Number of edge pixels	Size of the image	Time analysis (In s)				Efficiency of proposed CHT using grid of size (in %)		
			Existing CHT	Proposed CHT using grid of size					
				3×3	5×5	7×7	3×3	5×5	7×7
21	15840	256×224	68.687	7.688	2.844	1.578	22 80719787	95.8594785	97.70262204
22	10980	215×215	38.5	4.39	1.719	0.953	22 5974026	95.53506494	97.52467532
23	17958	256×178	61.875	6.984	2.593	1.438	88 71272727	95.80929293	97.6759596
24	19152	256×200	74.078	8.328	3.094	1.656	88.75779584	95.82332136	97.76451848
25	15695	256×208	63.218	7.14	2.641	1.438	88.70574836	95.82239236	97.72533139
26	15695	256×208	63.203	7.125	2.64	1.438	88.72680094	95.82298309	97.72479154
27	16793	256×255	82.813	9.375	3.453	1.828	88.67931363	95.8303648	97.7926171
28	12943	215×214	45.109	5.125	1.953	1.094	88.63863087	95.67048704	97.57476335
29	3969	256×192	15.016	1.859	0.796	0.515	87.61987214	94.69898775	96.57032499
30	4506	194×200	13.485	1.64	0.75	0.484	87.8383389	94.43826474	96.41082684
31	4254	203×175	11.672	1.469	0.672	0.454	87.41432488	94.24263194	96.11034955
32	6811	200×173	18.062	2.141	0.907	0.562	88.14638468	94.97840771	96.88849518
33	7745	247×164	23.953	2.812	1.125	0.688	88.26034317	95.3033023	97.12770843
34	7525	247×168	23.844	2.813	1.14	0.687	88.2024828	95.218923	97.11877202
35	6533	200×150	15.063	1.843	0.812	0.5	87.7647215	94.60930757	96.68060811
36	7028	247×185	24.515	2.86	1.172	0.703	88.33367326	95.21925352	97.13236794
37	5172	200×133	10.657	1.344	0.64	0.438	87.38857089	93.99455757	95.89002534
38	6890	200×134	14.204	1.718	0.766	0.5	87.90481554	94.60715291	96.47986483
39	7001	247×204	26.906	3.172	1.265	0.75	88.210808	95.29844644	97.21251765
40	5525	247×186	19.453	2.359	1.016	0.609	87.87333573	94.77715519	96.86937747

(continued)

Table 2 (continued)

Name of the image	Number of edge pixels	Size of the image	Time analysis (In s)				Efficiency of proposed CHT using grid of size (in %)		
			Existing CHT	Proposed CHT using grid of size					
				3×3	5×5	7×7	3×3	5×5	7×7
41	5512	250×170	19.5	2.141	0.937	0.562	89.02051282	95.19487179	97.11794872
42	8469	247×168	26.313	3.094	1.234	0.734	88.2415536	95.31030289	97.21050431
43	7381	215×161	19.547	2.297	0.984	0.61	88.24883614	94.96597943	96.87931652
44	6533	200×150	15.063	1.844	0.813	0.516	87.75808272	94.60266879	96.57438757
45	6074	200×150	14.016	1.719	0.781	0.5	87.73544521	94.4277968	96.4326484
46	5774	215×143	13.687	1.656	0.75	0.484	87.90092789	94.52034778	96.46379776
47	9900	247×165	30.672	3.578	1.422	0.813	88.33463745	95.36384977	97.34937402
48	5970	215×177	17.422	2.109	0.906	0.562	87.894616	94.79967857	96.77419355
49	4812	215×162	12.937	1.64	0.75	0.484	87.32318157	94.2026745	96.25879261
50	4449	215×146	10.813	1.39	0.703	0.422	87.14510312	93.49856654	96.0972903
51	7272	215×144	17.266	2.079	0.875	0.547	87.95899456	94.93223677	96.83192401

Table 3 Difference between expected R_E and observed R_E

34emImage name	34emImage size	Time analysis (s)				Expected R_E (s)			R_E(CHT) - R_E(MCHT) (s)		
		CHT	MCHT (Grid size)			MCHT (Grid size)					
		No grid	3×3	5×5	7×7	3×3	5×5	7×7	3×3	5×5	7×7
1	215×152	15.859	1.969	0.828	0.515	1.76211	0.6434	0.3236	0.0428	0.037496	0.036614
12	215×214	45.078	5.125	1.969	1.079	5.008667	1.80312	0.919959	0.013533	0.027516	0.025294
20	256×144	15.766	1.922	0.828	0.575	1.751778	0.63064	0.321755	0.028976	0.038957	0.064133
27	256×255	82.813	9.375	3.453	1.828	9.201444	3.3125	1.690031	0.030122	0.019735	0.019027
42	246×168	26.313	3.094	1.234	0.734	2.923667	1.0525	0.537	0.029013	0.032935	0.038809

Although the maximum time to detect iris taken by CHT is 82.813 s, if the proposed method is considered, it shows a maximum reduction of 89.02051282% in grid size 3×3. In case of grid size 5×5, maximum reduction is 95.859478% and in case 7×7, its 97.7926171%.

The test is repeated on different images and on an average, reduction in time in grids is 88.0339376%. In case of 5×5, its 94.90980612% and in case of 7×7, it is 96.85420798%.

The graph1 shows variation in the execution time of CHT from proposed CHT using grid size 3×3, 5×5, and 7×7.

Since only the absolute time take by images to detect the feature is shown in Table 2, so in order to see the efficiency in terms of reduced complexity factor R_E is calculated as follows:

R_E = Observed execution time of CHT/Observed execution time of MCHT

R_E is nothing else but relative measure of time of complexities of CHT and MCHT, hence RE is interpreted as time reduction factor which further measures the accuracy of the efficiency of exaction time of MCHT for each grid size.

Table 3 represents the expected RE and observed RE and its mean square error of each sample image taken. Table 4 represents the root mean square error of observed to expect execution time of each grid size considered.

Where Table 2 highlights the efficiency achieved in terms of substantial reduction in time to detect iris, Table 4 illustrates the root mean square error of each grid size stating that with increasing size of grid sizes or in other words minimizing ROI increases the rate of error (Figs. 5 and 6).

Table 4 Root mean square error for each grid size 3, 5, 7

Root mean square error (RMSE)	
RMSE(3×3)	0.190031
RMSE(5×5)	0.197885
RMSE(7×7)	0.214404

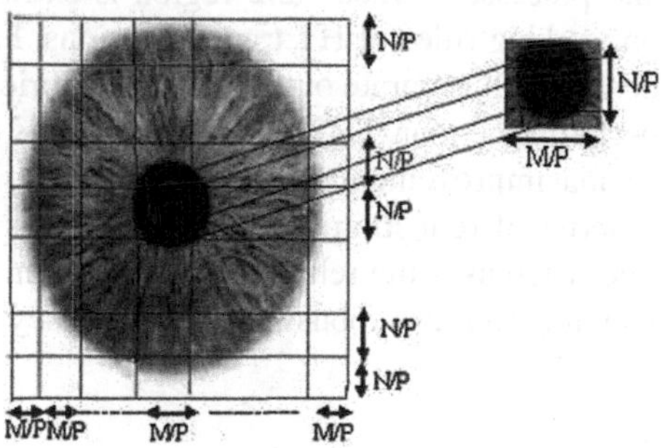

Fig. 5 Division of digital image matrix *f(x,y)* into P^2 equal regions using grid size P×P

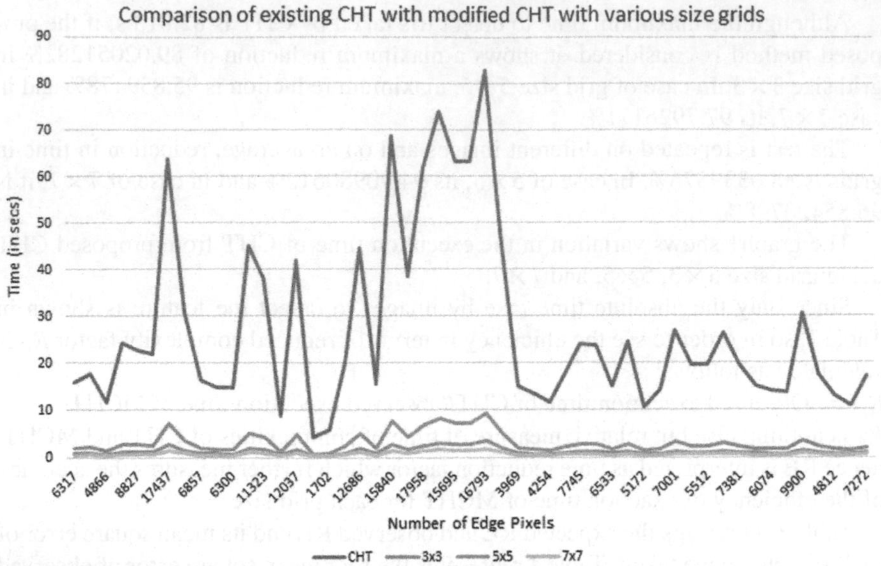

Fig. 6 Time comparison of the original CHT with the proposed modified CHT having a grid size of 3 × 3

5 Conclusion

CHT provides a robustly accurate method for detecting the iris, but is inefficient in terms of speed due to high-computational time and complexity. This study proposes and presents a modified CHT that significantly increases speed of detection while robustly retaining accuracy of iris detection. Dividing Hough space into varying size grids allows faster extraction of the valid region to process.

The valid region so obtained is the only region of which pixels are to be stored and processed. Once valid region is extracted, then execution time depends only on working rules of HT transformations. Fast detection using valid region strongly demands an accurate output as well. In brief, 3 × 3 and 5 × 5 sized grids show better performance than 7 × 7 sized grid in terms of robustness. Whereas 7 × 7 shows substantial improvement in detection time than other sizes used. The efficiency achieved in terms of reduction in time to detect iris is achieved with increasing size of grid sizes decreases the reliability of detection. Hence, time and space trade-off plays a vital role to grasp robustness and accuracy of detection.

References

1. H. Yazid et al., Circular discontinuities detection in welded joints using Circular Hough Transform, NDT & E International 40 (2007) 594–601.
2. C. Sanchez-Avila and R. Sanchez-Reillo, Two different approaches for iris recognition using Gabor fillters and multiscale zero-crossing representation, Pattern Recognition, vol. 38, issue 2, pp: 231–240, (2005).
3. R. Duda and P. Hart, Use of Hough transform to detect lines and curves in pictures, Communications of the ACM 15, pp: 11–15, (1975).
4. Wojciech et al., Reliable algorithm for iris segmentation in eye image, Image and Vision computing, vol. 28, issue 2, pp: 231–237, (2010).
5. Peihua Li et al., Robust and accurate iris segmentation in very noisy images, Image and vision computing, vol. 28, issue 2, pp: 246–253, (2010).
6. WL. Hwang, JL. Lin, CL. Huang and WG. Che, Iris Recognition Method, US Patent No. US 2008/0095411 A1.
7. L. Ma, Y. Wang and T. Tanm, Iris Recognition Based on Multichannel Gabor Filtering, Proceedings of ACCV, (2002).
8. L. Ma, Y. Wang and T. Tanm, Iris Recognition Using Circular Symmetric Filters, 16th International Conference on Pattern Recognition (ICPR'02) - Volume 2.
9. SJK Pedersen, Circular Hough Transform, Aalborg University, Novembre, (2007).
10. Kimme C., Ballard, D., and Sklansky, J., Finding circles by an array of accumulators, Proc. ACM 18, pp: 120–122, (1975).
11. Xu L., Oja, E., and Kultanan, P., A new curve detection method: randomized Hough transform (RHT), Pattern Recognition Letter, vol. 11, no. 5, pp: 331–338, (1990).
12. Chen T., and Chung, K., An efficient randomized algorithm, Computer Vision and Image Understanding, vol. 63, no. 83, pp: 172–191, (2001).
13. Yip R., Tam, P., and Leung, D., Modification of Hough transform for circles and ellipse detection using a 2-dimensional array, Pattern Recognition, vol. 25, no. 9, pp: 1007–1022, (1992).
14. Ho C., and Chen, L., A fast ellipse/circle detector using geometric symmetry, Pattern Recognition, vol. 28, no. 1, pp: 117 (1995).
15. Tsuji, S., and Matsumoto, F., Detection of ellipses by a modified Hough Transformation, IEEE Transactions on Computers, vol. C-27, no. 8, pp: 777–781, (1978).

Fault Detection and Mutual Coordination in Various Cyborgs

Amrita Parashar and Vivek Parashar

Abstract During multi-cyborgs area exploration fault may occur at any time. In case, if one cyborgs fails, another cyborgs can take over the task which is assigned to the failed cyborgs. In fact, fault tolerance and robustness to any cyborgs failures are major issues in favor of cyborgs systems, many studies have been devoted to this subject. If fault occurs, it is the duty of other cyborgss to take over his task and complete the process without any error in less time. So we have to design an approach for fault detection and tolerance which can work in any condition. Each cyborg updates its information in the global database and retrieves the information of other cyborgss from that database. Since cyborgs that are going through failures cannot update its information in the database periodically, they can be detected. Then any other cyborgs went there and complete the failed Cyborgs task then only the process stops. We show that down cyborgs are detected by functional cyborgss, and we assign their tasks to Working cyborgs using task allocation algorithm. Any one of the cyborgs, which is working takes over the failed cyborgs task and completes the process.

Keywords Multi-cyborgs · Area exploration · Fault tolerance
Task assignment · Coordination · Path planning

1 Introduction

If multi-cyborgs systems have to work properly in the real environment, they should be ready to deal with the sudden failures of other team members. A lot of amount of work in the area of detecting failures in the cyborgss and to recover from it [1, 2], and many multi-cyborgs planning systems gives various ways for reassignment of tasks

A. Parashar (✉) · V. Parashar
Amity University, Gwalior, Madhya Pradesh, India
e-mail: aparashar@gwa.amity.edu

V. Parashar
e-mail: vparashar@gwa.amity.edu

D.K. Mishra et al. (eds.), *Information and Communication Technology for Sustainable Development*, Lecture Notes in Networks and Systems 10,
https://doi.org/10.1007/978-981-10-3920-1_51

between other cyborgss after a particular cyborgs failure [3]. However, all of these methods are reactive only when failure occurs. Therefore, it is important to deal with these failures in real-world environments to be able to replan the system when any one or more cyborgs fails, that it should be useful to consider the probability of cyborgs failures when allotting initial task assignments.

1.1 Synchronization and Coordination

In this work, we are focussing on synchronization and coordination between cyborgss to obtain a simple and distributed approach to detect the faults in groups of autonomous cyborgss. We are applying concept of mutual coordination. By observing the presence of faults, the cyborgss can share their table entries and ensure that the process is continuing by functional cyborgss and performing the failed cyborgs's task also.

1.2 Area Exploration

Multi-cyborgs area exploration gets very much attention in the past decades. It is defined the process of effectively covering the whole area as soon as possible without any fault. The idea of multi-cyborgs has the advantage of more robustness and cooperation among the cyborgss. Single cyborgs is not efficient because of their less robustness and if fault occurs in the cyborgs our task remains incomplete so the concept of multi-cyborgs is introduced. Lots of work and studies have been given in the field of fault detection, that is, a cyborgs detects faults in itself, [4–8]. Some faults are complex to find among the cyborgss in which cyborgs it occurs. These faults include software failure that causes cyborgs program to hang and mechanical faults such as breaking of any part or an unstable connection to a power source. In this work, we present a dynamic and simple approach that allows all cyborgss to find the presence of faults in itself or in one another through local interactions, cyborgss are able to communicate with each other when they are in working situation and communication breaks, when they are having any fault in itself. Absence of messages shows other cyborgss that particular cyborgs having fault.

2 Related Work

Lynne E. Parker designs software named ALLIANCE [6]. It is a software architecture that gives the fault-tolerant control to groups of distinctive mobile cyborgss performing missions of approximately coupled subtasks that may have requesting conditions. ALLIANCE gives groups of cyborgss, each of which having a variety of large amount-works that it can perform during the mission, to separately select any movements all around the mission in light of the requirements of task, the exercises of other cyborgss, the current environment conditions, and the cyborgs's own interior states.

Goel et al. [8] they propose an approach to recognize faults in wheeled mobile cyborgss. The methodology behind the strategy is to utilize adaptive estimation to anticipate the result of s few faults, and to learn them by and large as a failure. Framework conduct displays under each sort of issue are coded in different Kalman channel (KF) estimators. Each KF is situated to some blame and predicts, utilizing its embedded model, the outcome values for the sensor readings.

Kakkar et al. [9] proposes a range techniques and algorithm for simultaneous area exploration, path planning and retrieval, obstacle avoidance, and object detection and retrieval by an autonomous multi-cyborgs system by the use of low-cost infrared sensors.

Canham et al. [4] describe an artificial immune system (AIS) that is used as an error detection system. The AIS learns normal behavior throughout a fault free learning time and then recognizes all error greater than preset error sensitivity. The AIS was implemented in programming but has the potential to be implemented in hardware.

Christensen et al. [5] they take inspiration from the occasional flashing behavior seen in a few types of fireflies. They determine a totally decentralized methodology to discover nonfunctional cyborgss in a cyborgs system. Every cyborgs flashes its light in onboard light-emitting diodes (Leds), and neighboring cyborgss are additionally flashes synchronously.

3 Problem Formulation

In the concept of fireflies flashing type system, all cyborgss have to be synchronized with each other and also to be present always in front of each other then only they are able to detect which cyborgs are flashing the LEDs. If they are not synchronized, then they cannot differentiate between the faulting and working cyborgs. Another disadvantages is if obstacles are present in the environment then they are unable to see each other because of obstacles and unable to synchronize and cannot see LED flashing. Even the range of LED flash system cannot penetrate the obstacles which are present in the environment.

So by LED flashing this will not give good performance. So, as per the research gap mentioned following objective can be made:

- Our approach should be dynamic and also for having obstacles in the environment.
- We use a global database scheme to maintain the information of all cyborgs. All the cyborgs maintain their information in that database and their tasks information which are provided to them and update their information that where they are currently.

- When any cyborgs fails it is unable its information in the database which leads to other we came to know that which one of the cyborgs fails.
- Then assign the task of failed cyborgs to one of the cyborgs who already finishes its task and nearer to that task of failed cyborgs.

4 Proposed Methodology

Each cyborgs maintains a global database that having all the information of other cyborgss and their tasks locations also. Each cyborgs updates 'I AM ALIVE' message with its ID number like 'I AM ALIVE1, 2..' in the database to show them that it is operational. When any one of the cyborgs stops updating this message then other cyborgss have the information that one of the cyborgs is dead. If any one of the cyborgs fails then the neighboring cyborgss identify that one cyborgs fails. Then they check their table entries and calculate the distance of that task which is assigned to the failed cyborgs. Then all cyborgss share the calculated path with each other and only cyborgs which are having the minimum distance to that task will proceed to their after finishing its own task. Each cyborgs maintains a global database which is get updated time to time having the information of all cyborgss and their tasks which are assigned to them. Time to time they retrieve the information from the database and perform their work accordingly (Table 1).

In the table shown above, ID represents the cyborgs number and count shows the message number. When Message number differs by 2 with all other cyborgs then we conclude that one of the cyborgss fails. When nay cyborgs will not able to update for 10 s, i.e., two timestamp then we conclude that one of the cyborgs fails. Each cyborgs calculates its minimum distance from all four tasks by distance from all four tasks by distance formula and finally moves forward their tasks to complete them. But in their way there are some obstacles also so after having obstacles also they choose minimum distance with their targets.

4.1 Proposed Path Planning Algorithm

Here, we discuss the algorithm for task assignment and detect the cyborgs failure and assigns its task to other cyborgs. Figure 1 shows four cyborgss and four tasks which they have to perform. All are on coordinates (x, y) which makes calculation of distance easy. Here, Cyborgs R1 is at (−12, −12) R2 is at (−12, 12) and R4 is at

Table 1 Task assignment with counter number

Cyborgs	Tasks	Assigned task	Counter
R1	T1, T2, T1, T3, T4		I am alive (R1, Count)
R2	T1, T2, T2, T3, T4		I am alive (R2, Count)
R3	T1, T2, T3, T3, T4		I am alive (R3, Count)
R4	T1, T2, T4, T3, T4		I am alive (R4, Count)

(12, −12). Tasks T1 is at (−3.5, −2.5) T2 is at (−6.5, 8) T3 is at (6, 11) and T4 is at (12, −4) (Table 2).

Figure 1 shows the task allocation to all cyborgs. So R1 will do task T1, R2 will do task T2 and so as R3 and R4.

Because, it is time efficient and all four tasks gets completed in 33.50 time units if there is no obstacles in the environment.

4.2 Fault-Tolerant Algorithm

Open log file.
Calculate its distance from all tasks which are given and find minimum distance task.
Write mytask and myloc in the log file.
If (timestamp=5)
Update its alive message with count numbers.
If (Ricount=Rjcount)
All cyborgs are alive
Else
If (Rjtimestamp>(currenttimestamp-10))
My new task distance=Rj task my task
if(my new task distance>other cyborgss task distance) $\sum$ then update my id and dead cyborgs id.
If (my loc=my task)
Find new task if any
If (Ri=my id)
Then goto new task
Else exit (1);
Else
Update alive message with count.

Fig. 1 Task assignment

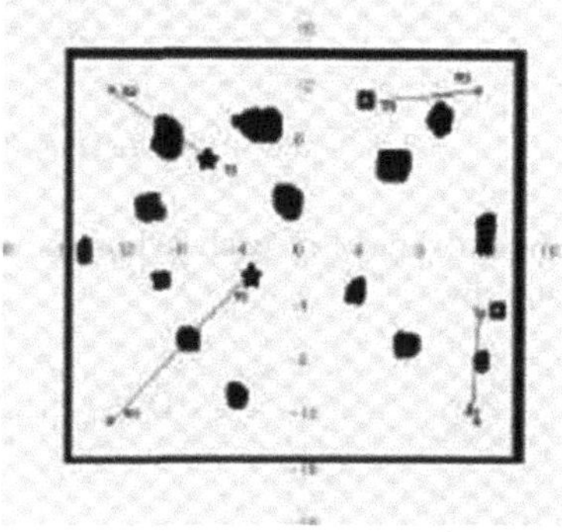

Table 2 Path planning and distance

Plan	Distance	Max. distance
R1T1	13.40	13.40
R2T2	6.80	13.40
R3T3	6.08	13.40
R4T4	8	

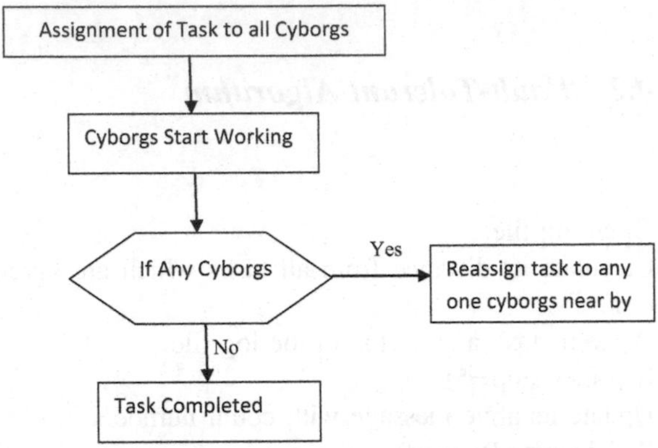

Fig. 2 Flow chart of task allocation

4.3 Block Diagram of Proposed Approach Is Shown Below

See Fig. 2.

5 Result and Simulation

All the four cyborgss complete their tasks in a particular time. The time T depends on three factors:

(1) Speed of cyborgs(S) in meter/sec.
(2) Distance of cyborgs from the task (D) in meters.
(3) Obstacle Delay (obs.-delay) in secs. Required time to complete the task is:

$$\mathbf{T = D/S + obs. - delay\ (in\ secs)}$$

Here, speed of each cybors is constant which is 0.4 m/s and distance is calculated by:

$$\mathbf{D = \sqrt{[(x2 - x1)2 + (y2 - y1)_2]}}$$

Now, the time required for each cybors to their tasks is shown in table given below (Table 3):

And extra delay is also added in the given time to avoid obstacles because it decreases the speed of the cyborgss and increases the distance. So the total time is 40.1 s in which 6.6 s is wasted in avoiding obstacles as shown in Fig. 3 (Fig. 4).

Table 3 Time and distance calculation

Cyborgs	Tasks	Distance	Time	Obs-Delay
R1	T1	13.04	33.50	6.6
R2	T2	6.80	17	7.5
R3	T3	6.08	15.2	4.2
R4	T4	8	20	

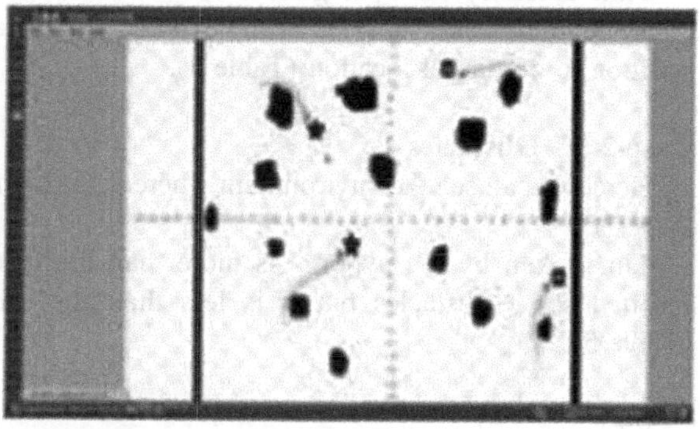

Fig. 3 Reached destination

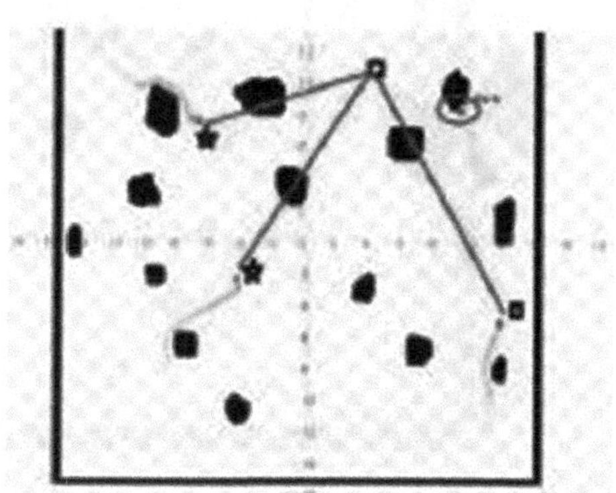

Fig. 4 When R3 fails

5.1 When 1 Cyborgs Fails

Now since R3 fails so one of the cyborgs have to perform its task which is at (6, 11) and distance from T2 (−6.5, 8) is 12.86 mm which is minimum in comparison on to all other cyborgs. So total time required for this task is 32.22 s. This time is when there are no obstacles in its way but as we see there are more obstacles in performing two tasks so overall delay is increased which is 25 s more than the original time. So total time required to complete the whole tasks is 57.30 s where 25 s are wasted to avoid obstacles shown in Fig. 5.

5.2 Performance in Different Environments

Case-1 0% Obstacles

Here, we can see the environment where no obstacles are present as shown in figure below (Fig. 6):

Time taken by the cyborgs is exactly the same which is estimated by our calculation on the basis of speed of the cyborgs and distance traveled from its original position to the target location (Table 4).

Case-2 5% Obstacles

Here, we can see the environment where it is having less percentage of obstacles (Fig. 7).

Time taken by the cyborgs is more than the calculate time because, here we inserted some obstacles but it is less than the time of an actual environment (Table 5).

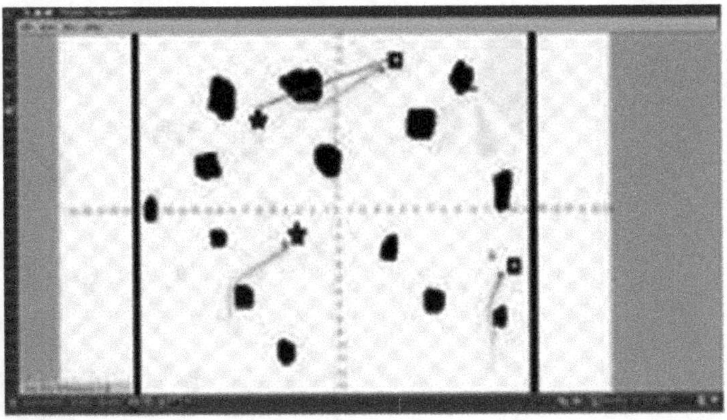

Fig. 5 R2 complete R3's tasks

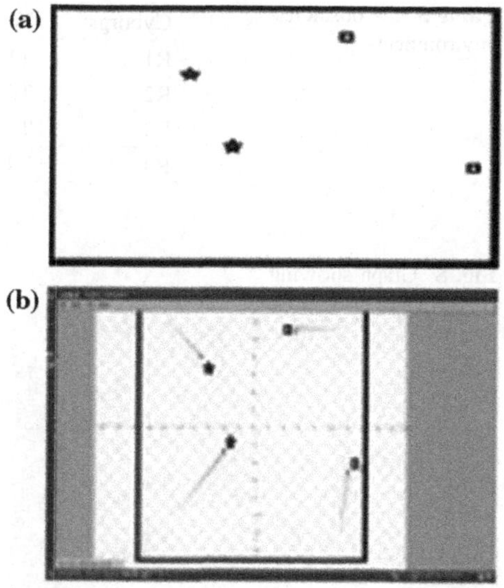

Fig. 6 **a** 0% obstacles. **b** Reached to targets

Table 4 0% obstacles environments

Cyborgs	Tasks	Distance	Time	Obs-Delay
R1	T1	13.40	33.50	0
R2	T2	6.80	17	0
R3	T3	6.08	15.2	0
R4	T4	8	20	0

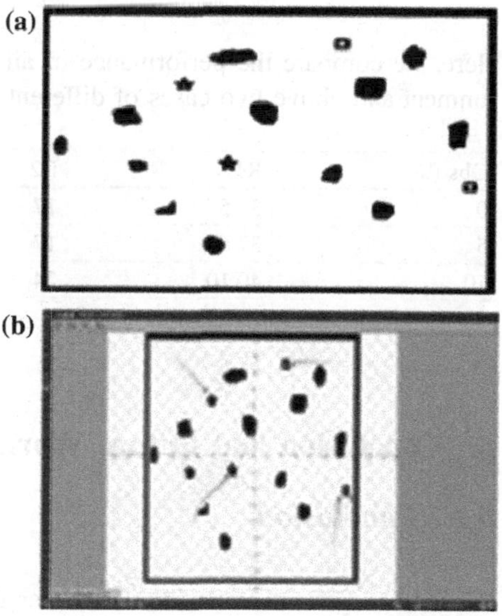

Fig. 7 **a** 5% obstacles. **b** Reached to targets

Table 5 5% obstacles environments

Cyborgs	Tasks	Distance	Time	Obs-Delay
R1	T1	13.40	33.50	4.10
R2	T2	6.80	17	0
R3	T3	6.08	15.2	5
R4	T4	8	20	0

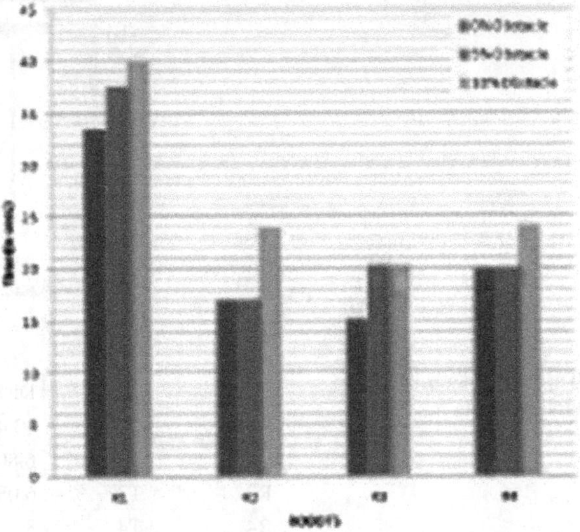

Fig. 8 Graph showing performance comparison

5.3 Comparison in All Environments

Here, we compare the performance of all three environment, i.e. our actual environment and above two cases of different obstacle percentage (Fig. 8).

Obs (%)	R1	R2	R3	R4
0	33.5	17	15.2	20
5	37.6	17	20.2	20
10	40.10	24	20.2	24.2

6 Conclusion and Future Work

6.1 Conclusion

When cyborgss interact with each other through global database then there is no need of synchronization between them or to look at each other. They maintain a global database which helps them to coordinate with each other. They can retrieve the information from the log file and fault detection becomes so easy. Log file contains all the information of all targets and all cyborgss current position also. After having the information of failed cyborgs and its task, any other cyborgs which is best suitable to complete that task will proceed to that task and completes it. Then any one Cyborgs takes over the failed cyborgs task and completes the process.

6.2 Future Work

We can try to implement radio communication in cyborgs so that they can send and receive by messages to each other and detect faults when any one of the cyborgs stop sending messages. Through radio communication they do not need any other log file. They can store the information in their own database and can communicate through TCP.

References

1. B. P. Gerkey and M. J. Mataric, "Sold!: Auction methods for multicyborgs coordination," Cyborgsics and Automation, IEEE Transactions on, vol. 18, no. 5, pp. 758–768, 2002.
2. G. A. Kaminka and M. Tambe, "Social comparison for failure detection and recovery," in Intelligent Agents IV Agent Theories, Architectures, and Languages, pp. 127–141, Springer, 1998.
3. M. Bernardine Dias, M. Zinck, R. Zlot, and A. Stentz, "Robust multicyborgs coordination in dynamic environments," in Cyborgsics and Automation, 2004. Proceedings. ICRA'04. 2004 IEEE International Conference on, vol. 4, pp. 3435–3442, IEEE, 2004.
4. R. Canham, A. H. Jackson, and A. Tyrrell, "Cyborgs error detection using an artificial immune system," in Evolvable Hardware, 2003. Proceedings. NASA/DoD Conference on, pp. 199–207, IEEE, 2003.
5. A. L. Christensen, R. O'Grady, and M. Dorigo, "From fireflies to fault tolerant swarms of cyborgss," Evolutionary Computation, IEEE Transactions on, vol. 13, no. 4, pp. 754–766, 2009.
6. M. B. Dias, R. Zlot, N. Kalra, and A. Stentz, "Market-based multicyborgs coordination: A survey and analysis," Proceedings of the IEEE, vol. 94, no. 7, pp. 1257–1270, 2006.
7. P. Goel, G. Dedeoglu, S. I. Roumeliotis, and G. S. Sukhatme, "Fault detection and identification in a mobile cyborgs using multiple model estimation and neural network," in Cyborgsics and Automation, 2000. Proceedings. ICRA'00. IEEE International Conference on, vol. 3, pp. 2302–2309, 2000.
8. L. E. Parker, "Alliance: An architecture for fault tolerant multicyborgs cooperation," Cyborgsics and Automation, IEEE Transactions on, vol. 14, no. 2, pp. 220–240, 1998.
9. A. Kakkar, S. Chandra, D. Sood, R. Tiwari, and A. Shukla, "Multi-cyborgs system using low-cost infrared sensors," IAES International Journal of Cyborgsics and Automation (IJRA), vol. 2, no. 3, pp. 117–128, 2013.
10. S. I. Roumeliotis, G. S. Sukhatme, and G. A. Bekey, "Sensor fault detection and identification in a mobile cyborgs," in Intelligent Cyborgss and Systems, 1998. Proceedings., 1998 IEEE/RSJ International Conference on, vol. 3, pp. 1383–1388, IEEE, 1998.

11. V. Verma and R. Simmons, "Scalable cyborgs fault detection and identification," Cyborgsics and Autonomous Systems, vol. 54, no. 2, pp. 184–191, 2006.
12. S. B. Stancliff, J. Dolan, and A. Trebi-Ollennu, "Planning to fail-reliability needs to be considered a priori in multicyborgs task allocation," in Systems, Man and Cybernetics, 2009. SMC 2009. IEEE International Conference on, pp. 2362– 2367, IEEE, 2009.
13. T. Kovács, A. Pásztor, and Z. Istenes, "A multi- cyborgs exploration algorithm based on a static bluetooth communication chain," Cyborgsics and Autonomous Systems, vol. 59, no. 7, pp. 530–542, 2013.
14. M. d. P. O. Cabrera, R. S. Trifonov, G. A. Castells, and K. Stoy, "Wireless communication and power transfer in modular cyborgss".
15. V. Verma, G. Gordon, R. Simmons, and S. Thrun, "Real- time fault diagnosis [cyborgs fault diagnosis]," Cyborgsics & Automation Magazine, IEEE, vol. 11, no. 2, pp. 56–66, 2004.
16. G. Lozenguez, A.-I. Mouaddib, A. Beynier, L. Adouane, and P. Martinet, "Simultaneous auctions for" rendezvous" coordination phases in multi-cyborgs multi-task mission," in Web Intelligence (WI) and Intelligent Agent Technologies (IAT), 2013 IEEE/WIC/ACM International Joint Conferences on, vol. 2, pp. 67–74, IEEE, 2013.
17. Y. Liu, G. Nejat, and J. Vilela, "Learning to cooperate together: A semiautonomous control architecture for multi- cyborgs teams in urban search and rescue," in Safety, Security, and Rescue Cyborgsics (SSRR), 2013 IEEE International Symposium on, pp. 1–6, IEEE, 2013.
18. L. Luo, N. Chakraborty, and K. Sycara, "Distributed algorithm design for multi-cyborgs task assignment with deadlines for tasks," in Cyborgsics and Automation (ICRA), 2013 IEEE International Conference on, pp. 3007–3013, IEEE.
19. A. Jevtic, A. Gutiérrez, D. Andina, and M. Jamshidi, "Distributed bees algorithm for task allocation in swarm of cyborgss," Systems Journal, IEEE, vol. 6, no. 2, pp. 296–304, 2012.
20. F. Arrichiello, A. Marino, and F. Pierri, "A decentralized fault detection and isolation strategy for networked cyborgss," Advanced Cyborgsics (ICAR), 2013 16th International conference on pp. 1–6 IEEE 2013.

Comparative Study of Inverse Power of IDW Interpolation Method in Inherent Error Analysis of Aspect Variable

Neeraj Bhargava, Ritu Bhargava, Prakash Singh Tanwar and Prafull Chandra Narooka

Abstract This paper deals with inherent error analysis of aspect variable using IDW interpolation method and its various power values. In the first section, it shows aspect analysis method and algorithm. In the second section, it creates a DEM model from various measured and erroneous elevated points which becomes input to calculate aspect of interpolated DEMs separately then error is calculated by calculating difference of the aspect for true and erroneous aspect. It explores error analysis of aspect with its practical implementation in ArcGIS. In the last section, it explains the comparison of errors on aspect for various power of IDW method. Result shows that aspect error decreases with the increment in the inverse power of distance in IDW method.

Keywords 3D GIS · Aspect · Error analysis · IDW · Interpolation

1 Introduction

In last decades, GIS technology is changing drastically over the world. It permits to query, interpret identify, and visualize spatial and nonspatial data in various ways that expose trends and patterns in the form of maps, graphs, and reports etc.

N. Bhargava (✉)
Department of Computer Science, School of Engineering & System Sciences, M.D.S.University Ajmer, Ajmer, India
e-mail: drneerajbhargava@yahoo.co.in

R. Bhargava
Department of Computer Science, Aryabhatt International College, Ajmer, India
e-mail: drritubhargava@yahoo.com

P.S. Tanwar · P.C. Narooka
Department of Computer Science, MJRP University, Jaipur, India
e-mail: pst.online@gmail.com

D.K. Mishra et al. (eds.), *Information and Communication Technology for Sustainable Development*, Lecture Notes in Networks and Systems 10,
https://doi.org/10.1007/978-981-10-3920-1_52

Some specific issues of 3D visualization with in 3D GIS are to explore for example, to visualize 3D spatial analysis result some tools are required to effortlessly explore and navigate real-time models [1–3].

Inverse distance weighting (IDW) is one of the most commonly used spatial interpolation method. This method is comparatively easy and fast to compute interpolation than other interpolation methods. This method assumes that attributes of unknown points are weighted average of known values within its neighboring points. Weights of the unknown points are inversely proportion to the power of distances between the unknown and known points [4].

There are many types of errors, which affects aspect, i.e., error in source data, data entry error, equipment error, error in algorithm used in interpolation, etc. Here, in this paper, we are considering the inherent errors in the data.

So what happens if the aspect is calculated from inherent erroneous data? This paper analyzes the comparison of inherent errors in aspect due to different powers of inverse distance weighting interpolation.

2 Purpose and Objectives

Zhou and Liu [5] compares various interpolation techniques with spatial data but they did not compare the effect of various inverse powers with aspect. Hickey [6] explained various errors noticeable errors due to errors in data, inherent in data structure, and created by algorithms in slope derived from grid DEM.

So what happens if the aspect is calculated from inherent erroneous elevation data? This paper analyzes the comparison of errors in aspect due to different powers of inverse distance weighting interpolation.

This study focuses on the IDW method and comparative study of various inverse powers in the interpolation of spatial data and its impacts on the aspect. The overall purpose is to better understand the IDW method and minimize the inherent error impact on the aspect.

3 Literature Review

Lu and Wong, worked on an adaptive IDW interpolation technique [4].

Hickey [6], described erosion modeling for calculating the cumulative downhill slope length. Along with it, methods for calculating slope and aspect were also defined.

Masaad and Moneim [7] explained suitable design of road pattern for Kosti town based on the TIN Analysis.

Bhargava et al. [3] creates the TIN Model from mass points given in GML form.

Azpura et al. [8] compared various spatial interpolation methods, i.e., spline, IDW and Krigging methods for estimation of average electromagnetic field magnitude.

Many researchers Like Chu, Tsai, O'Callagan, Flemming, and Hoffer etc. defined Mathematical model for slope and aspect on the basis of quantitative analysis, trend analysis, Vector-based analysis, and fast Fourier transform-based analysis.

Other researchers Like Hanjianga et al. [9], Vivoni et al. [2], Tucker et al. [10], Pajrola et al. [11] explained the 3D TIN models.

4 Methodology

4.1 Aspect

Aspect is direction angle of the TIN. Aspect is a circular measure, which starts with 0° at the north, moves clockwise, and ends with 360° also at the north. An aspect raster is typically classified into four or eight principal directions and an additional class for flat areas. Aspect is a measure of the rate of change of elevation at a surface location [6].

Aspect is the first horizontal derivative of the altitude and represents the direction of the slope. Classification of aspect directions according to its assigned color code is given in the following Table 1. The color combination according to classification of aspect in their direction is shown in Fig. 1.

Aspect (A) is defined as a function of gradients at X and Y (i.e., W–E and N–S) directions.

$$A = 270^\circ + \tan^{-1}\left(\sqrt{\frac{\left(\frac{dZ_y}{dy}\right)}{\left(\frac{dZ_x}{dx}\right)}}\right) - 90^\circ \frac{\left(\frac{dZ_x}{dx}\right)}{\left|\left(\frac{dZ_x}{dx}\right)\right|},$$

Table 1 Aspect classification

Direction	Value (in Angles)	Classes	Color
Flat	−1	D = infinitive	Gray
North	0–22.50	D = 0	Red
North–East	22.50–66.50	D = 45	Orange
East	66.50–112.50	D = 90	Yellow
South–East	112.50–157.50	D = 135	Green
South	157.50–202.50	D = 180	Cyan
South–West	202.50–247.50	D = 225	Sky blue
West	247.50–292.50	D = 270	Blue
North–West	292.50–337.50	D = 315	Pink
North	337.50–3600	D = 0	Red

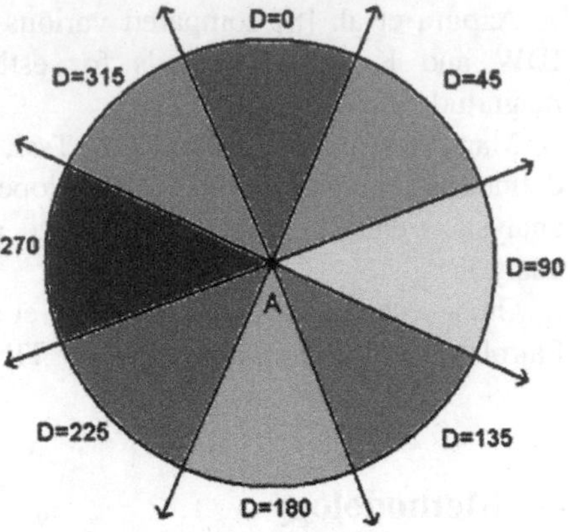

Fig. 1 Aspect color chart with their classification code

where,

$$dZ_x = Z_3 - Z_2, f_x = \frac{dZ_x}{dx}\ f_y = \frac{dZ_y}{dy},$$

where f_x and f_y are the gradients, at West–East direction (x-axis) and North–South direction (y-axis), respectively. From the above equations, it is clear that the key for slope and aspect computation is the calculation of f_x and f_y.

4.2 Interpolation

Interpolation is required to identify the heights for the unknown points. There are many interpolation methods which are used to interpolate unknown values from given points.

4.3 IDW Interpolation

IDW method is based on the supposition that the neighboring point contributes more in interpolation than the farther one. IDW method states that impact of known data point is inversely proportion to the distance from unknown to known data points [8].

Interpolated value in IDW method is defined as

$$\begin{aligned} w(x) &= \left(\frac{\sum_{i=1}^{n} \frac{w_i}{d(x,x_i)^p}}{\sum_{i=1}^{n} \frac{1}{d(x,x_i)^p}} \right) \quad \text{if } d(x, x_i) \neq 0, \\ w(x) &= w_i \quad \text{if } d(x, x_i) = 0 \end{aligned}$$

where,

$w(x)$	is the point to be interpolate
w_i	weight at point x_i
$d(x, x_i)$	distance from x to x_i
p	inverse distance power

IDW interpolation for height

$$h(x,y) = \left\{ \begin{array}{ll} \left(\dfrac{\sum_{i=1}^{n} \frac{h_i}{d((x,y),(x_i,y_i))^p}}{\sum_{i=1}^{n} \frac{1}{d((x,y),(x_i,y_i))^p}} \right) & \text{if } d((x,y),(xi,yi)) \neq 0 \\ h(x,y) = hi & \text{If } d((x,y),(xi,yi)) = 0 \end{array} \right\},$$

where

$h(x, y)$	is the height at x, y
hi	height at point (x_i, y_i)
$d((x, y), (x_i, y_i))$	distance from (x, y) to (x_i, y_i)
$d((x, y), (x_i, y_i))$	$\sqrt{((x-x_i)^2 + (y-y_i)^2)}$

Table 2 Aspect error analysis process due to IDW with inverse of power

S.No.	Process	Without Error	Errroneous
1	Elevation Data	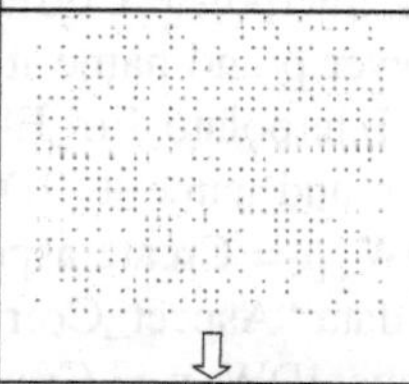	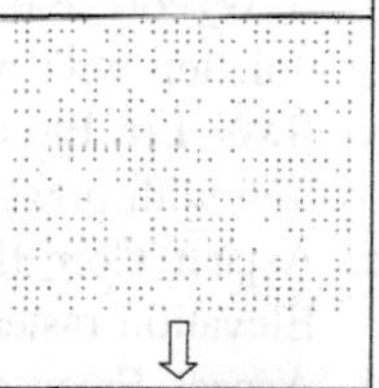
2	IDW Interpolation	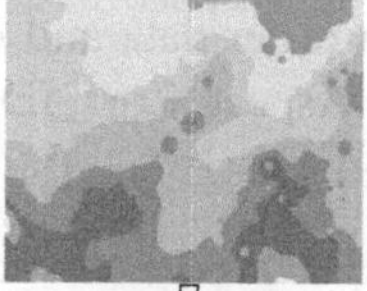	
3	Aspect	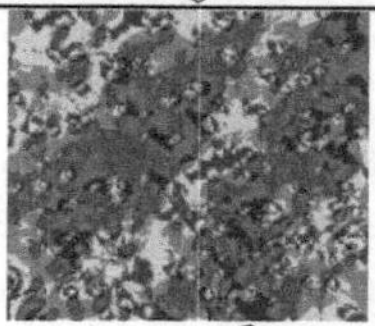	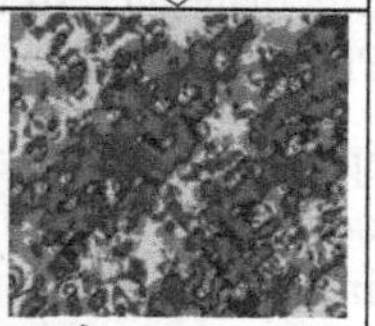
4	Difference (Correct Aspect - Errroneous Aspect)	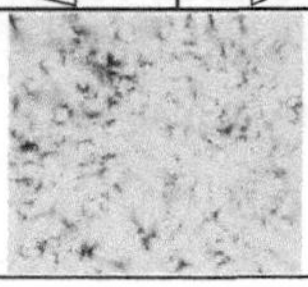	

$$h(x,y) = \left\{ \begin{array}{ll} \left(\frac{\sum_{i=1}^{n} \frac{h_i}{\left(\sqrt{((x-x_i)^2 + (y-y_i)^2)}\right)^p}}{\sum_{i=1}^{n} \frac{1}{\left(\sqrt{((x-x_i)^2 + (y-y_i)^2)}\right)^p}} \right) & \text{if } d((x,y),(xi,yi)) \neq 0 \\ h(x,y) = hi & \text{If } d((x,y),(xi,yi)) = 0 \end{array} \right\}.$$

4.4 *Error*

An "error" is a deviation from accuracy or correctness.

Error = actual data – experimental data

To make a fair comparison between erroneous data and true data, some error has been inserted into the true data so that correct results are obtained (Table 2).

5 Algorithm

Input: Measured Elevation data and call it Correct_Ele_data,

Step 1. Put some randomized error in the elevation data and call it "Err_Ele_data,"

Step 2. Repeat step 2(a) to 2(e) till p < n otherwise go to step 4

a. IDW_corr_Ele = Interpolate Correct_Ele_data Data With IDW interpolation with power p and name it "IDW_Corr_Ele_p" + p
b. IDW_Err_Ele = Interpolate Err_Ele_data Data With IDW interpolation with power p and name it "IDW_Err_Ele_p" + p
c. Aspect_Corr_IDW_p = Create aspect raster from interpolated Correct Elevation raster data "Aspect_Corr_IDW_p" + p
d. Aspect_Erroneous_IDW_p = Create aspect raster from interpolated Correct Elevation raster data and call it "Aspect_Error_IDW_p" + p
e. Diff_Aspect_p = Aspect_Corr_IDW_p - Aspect_Erroneous_IDW_p and call it "Diff_Aspect_p" + p

Step 3. Goto step 2

Step 4. Calculate the error

Step 5. Plot the graph

The step-by-step explanation of algorithm is shown with images in the Table 1. In this process, elevation data are used to create digital elevation model in raster format using IDW method for both erroneous data and error free data. In the next step, aspect analysis is carried out from the previous step's result and in the last step difference of each pixel is calculated which is shown in the figure of Table 1. This process is repeated for each inverse power values of the IDW method.

6 Results and Discussion

Various Inverse powers in IDW method for the Erroneous data and data without error are used to get the DEM of the surface, which is used to calculate the aspect.

Comparison of various aspect surfaces obtained from various inverse distance power of IDW method for erroneous and without error data are shown in Table 3. It is almost clear from the images of Table 3 that inverse power 2 gives a better aspect value than the higher inverse power values. As soon as the inverse power values increases, error in aspect analysis increases and the smooth curve changes in into rough zigzag aspect. It shows that it is better to use IDW with inverse power 2 than the higher inverse power values.

Error analysis graph for aspect with respect to various IDW's distance power 2, 3, 4, 5, and 6 are given in Fig. 2a, b, c, d, and e, respectively.

Table 3 Error Map with various inverse power in IDW methods

S.No	Inv. Distance Power	Error Map(Actual-Measured)	S.No	Inv. Distance Power	Error Map(Actual-Measured)
1	2	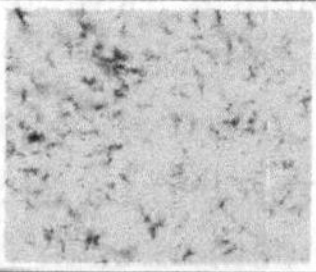	4	5	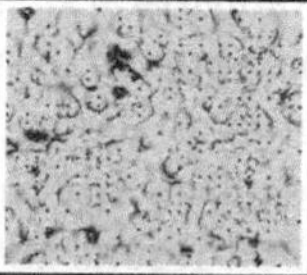
2	3	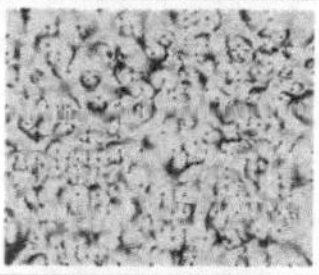	5	6	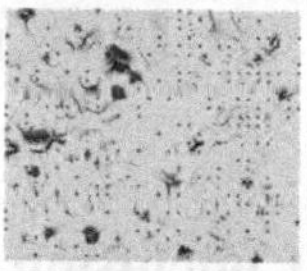
3	4	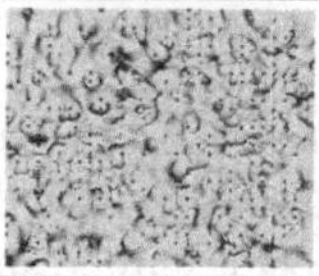	6	7	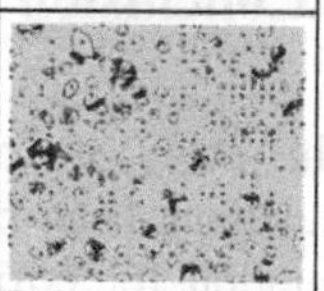

(a) (b) (c) (d) (e)

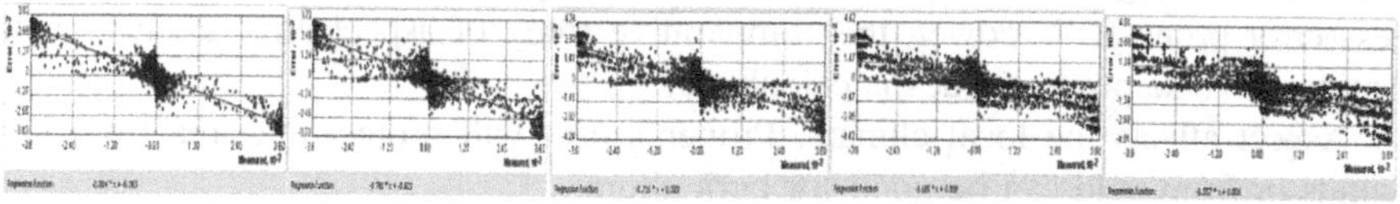

Fig. 2 Aspect error graph with IDW's power with **a** power 2, **b** power 3, **c** power 4, **d** power 5, **e** power 6

Table 4 Error range for various inverse distance power

S. No	Inverse distance power	Min error (10^{-2})	Max error (10^{-2})
1	2	−3.82	3.82
2	3	−3.73	3.73
3	4	−4.24	4.24
4	5	−4.42	4.42
5	6	−4.01	4.01
6	12	−3.61	3.61
7	20	−4.71	4.71

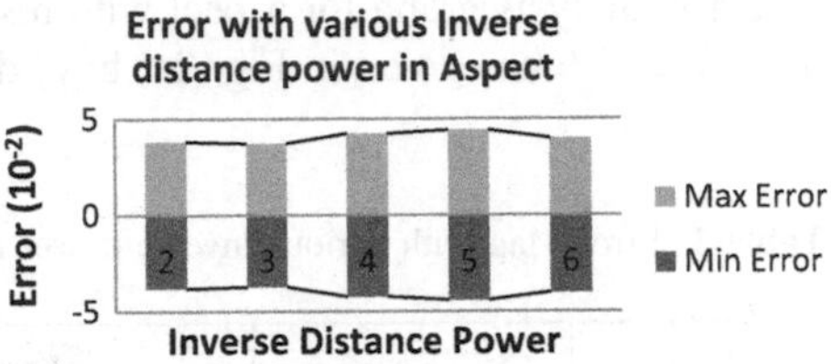

Fig. 3 Aspect error graph with IDW's power

Comparison of The IDW Interpolation method with different power parameters in the error analysis of aspect is specified in Table 4, and their graphical representation is shown in Fig. 3. From this experiment it is clear that as inverse power to the distance increases, range of error also increases. It is observed that for simple and best results IDW interpolation method with inverse power 2 is more suitable than other inverse powers.

7 Conclusion

In this study, we presented the comparison of various inverse powers for aspect calculations from raw elevation data of the site. The focus was on the algorithm generation for error analysis of aspect with respect to various inverse distance power values in the IDW method.

This paper discussed the possible practical solutions for creating error analysis of aspect, including theoretical aspects of error analysis with IDW method. To find less error prone IDW power, the comparative study of aspect error analysis with various inverse power values is calculated.

Aspect affects the local climate. Proper aspect will protect the area from hot, winds or from cold can be good for agriculture.

Error free or less error prone aspect analysis is required in almost every application areas like water flow analysis, vegetation analysis, terrain stability assessment, snow avalanche risk mapping.

References

1. A. A. Rahman, M. Pilouk, and S. Zlatanova, "The 3D GIS software development: global efforts from researchers and vendors," *Geoinformation Science Journal*, vol. 1, no. 13, 2001.
2. E. R. Vivoni, V. Y. Ivanov, R. L. Bras, and D. Entekhabi, "Generation of triangulated irregular networks based on hydrological similarity," *Journal of hydrologic engineering*, vol. 9, no. 4, pp. 288–302, 2004.
3. Neeraj Bhargava, Ritu Bhargava, and Prakash Singh Tanwar, "Triangulated Irregular Network Model from Mass Points," *International Journal of Advanced Computer Research*, vol. 3 No. 2, no. 10, pp. 172–176, June 2013.
4. George Y. Lu and David W. Wong, "An adaptive inverse distance weighting spatial interpolation technique," vol. 34, no. 9, pp. 1044–1055, September 2008.
5. Q. Zhou and X.Liu, "Error Analysis on Grid-Based Slope and Aspect Algorithms," *Photogrammetric Engineering & Remote Sensing*, vol. 70, no. 8, pp. 957–962, 2004.
6. R. Hickey, "Slope angle and slope length solutions for GIS," *Cartography*, vol. 29, no. 1, pp. 1–8, 2000.
7. E. M. Masaad and S. M. Moneim, "Suitable Design of Road Pattern for Kosti Town Based on TIN Analysis," *Khartoum University Engineering Journal*, vol. 2, no. 1, 2012.
8. M. A. Azpurua and K. D.Ramos, "A comparison of spatial interpolation methods for estimation of average electromagnetic field magnitude," *Progress In Electromagnetics Research M*, vol. 14, pp. 135–145, 2010.
9. X. Hanjianga, T. Limina, and S. Longa, "A Strategy To Build A Seamless Multi-Scale TIN-DEM Database," *The International Archives of the Photogrammetry, Remote Sensing and Spatial Information Sciences*, vol. XXXVII., no. B4, 2008.
10. G. E. Tucker, S. T. Lancaster, N. M. Gasparini, R. L. Bras, and S. M. Rybarczyk, "An object-oriented framework for distributed hydrologic and geomorphic modeling using triangulated irregular networks," *Computers & Geosciences*, vol. 27, no. 8, pp. 959–973, 2001.
11. R. Pajarola, M. Antonijuan, and R. Lario, "Quadtin: Quadtree based triangulated irregular networks," *Proceedings of the conference on Visualization'02*, pp. 395–402, 2002.

The Study of the Relationship Between Internet Addiction and Depression Amongst Students of University of Namibia

Poonam Dhaka, Atty Mwafufya, Hilma Mbandeka, Iani de Kock, Manfred Janik and Dharm Singh Jat

Abstract The aim of this preliminary study was to explore the impact of Internet usage on individual functioning by exploring the relationship between Internet usage and depression amongst the students of University of Namibia. An exploratory study was conducted amongst 36 conveniently selected males and females' students. This study investigated prevalence of internet addiction and its association with depression. In this study two tests were used, the Young Internet Addiction Test (YIAT) and the Patient Health Questionnaire (PHQ-9). The males scored an average of 36.6 on the YIAT whilst females scored an average of 33.9. On the PHQ-9, the males scored 14.7 and females scored 16 on average. The differences in hours spend per day for females were 4.5 h and males spend 6.4 on average. The result of this exploratory study reveals that there is a correlation between Internet addiction scores, depression scores and time spent online.

Keywords Internet addiction · Depression · Internet usage · Social networks

---

P. Dhaka (✉) · A. Mwafufya · H. Mbandeka · I. de Kock · M. Janik
University of Namibia, Windhoek, Namibia
e-mail: pdhaka@unam.na

A. Mwafufya
e-mail: attym8@gmail.com

H. Mbandeka
e-mail: ndapewambandeka@gmail.com

I. de Kock
e-mail: ianidk@gmail.com

M. Janik
e-mail: mjanik@unam.na

D.S. Jat
Namibia University of Science and Technology, Windhoek, Namibia
e-mail: dsingh@nust.na

D.K. Mishra et al. (eds.), *Information and Communication Technology for Sustainable Development*, Lecture Notes in Networks and Systems 10,
https://doi.org/10.1007/978-981-10-3920-1_53

1 Introduction

Internet use is a concept that can be dated back in history. It was considered to be a luxury and for only a few in the past [7]. Not everybody had access to it, as it was still spreading out to the rest of the nations. However, today a number of people worldwide use the Internet as a means of communication, research, buying flights as well as buying reading material amongst many other things [7]. As it is a relatively new technology that has become so prevalent in almost every household, not much is known about how it is impacting on its users in the long term. Available research suggests that excessive Internet usage may also impact our mood in various ways [6] and Internet addiction has increasingly been raised as a concern. Especially in countries like Namibia, the effects of Internet usage are still a relatively new area of study. Up to date, there has been minimal research done to assess the impact of Internet usage on individuals in a local context. In light of that, this study seeks to explore the impact of Internet usage on individual functioning by exploring the relationship between Internet addiction and depression amongst a small sample of the University of Namibia students.

Depression can be defined as "a common mental disorder that presents with depressed mood, loss of interest or pleasure, decreased energy, feelings of guilt or low self-worth, disturbed sleep or appetite, and poor concentration" [4]. This definition is inclusive of the criteria used for the diagnosis of major depressive disorder as listed in the DSM-5. These criteria are also included in the Patient Health Questionnaire-9 (PHQ-9), which is a commonly used screening tool for depression and, therefore, the tool that was selected for this study.

Alarmly, Internet addiction has found to be most prevalent amongst young people globally [6]. This phenomenon can be defined as a compulsive form where one constantly seeks to use information technology whilst abandoning other important aspects of their lives. This usage is also characterised by negative consequences on an individual's behavioural, psychological and cognitive well-being [6]. Many tools to measure Internet addiction have been measured, however for this study the Young's Internet Addiction Test (YIAT) has been selected as it is the most widely used Internet Addiction tool across the globe. It is against this background that this study seeks to measure the correlation between student scores on the YIAT, the amount of social media apps used by students, the amount of time students spend online as well as their scores on the PHQ-9.

1.1 Literature Study

Kindly increased accessibility to the Internet has improved our lives in various ways, especially in terms of serving as a platform for communication with a diverse group of people [7]. The Internet has also been said to be very effective during emergencies, especially in this global era. In the event that individuals may need to contact their loved ones, who may be thousands of miles away, the Internet has made it possible for mediums like Whatsapp and Facebook to exist in order to make this communication possible and easy. Other uses of the Internet extends to meeting new friends and interacting with friends and family who may be in different cities [2]. Some people may also use the Internet, through social media sites like Facebook, to network by channelling social links [8]. It is clear that the use of cyber—technology is being utilised for various needs that are useful and benefit us greatly.

Pempek as cited in [2] also found that social networking contributes to forming healthy social health and social development. Not only does the Internet and social networks aid development but they also provide a sense of belonging as argued by Sawyer and Chen as cited in [2]. These findings were also supported by other studies that found that more often, adolescence use Internet as a way of exploring themselves, linking themselves to society as well as seeking identity [8]. Various authors have differing views on whether social networks lead to healthy or unhealthy outcomes. A lot of them maintain that it can lead to either and is more dependent on the user of social network as opposed to the social network itself.

Consequently, in the midst of discussions in favour of the use of the Internet and social networks, there are also a vast amount of studies discussing the cons of the Internet and the use of social networks. One of the greatest discussions concerning the cons of the Internet centres on the Internet's great potential for addiction. Internet addiction is a universal concern. As the Internet and technology become increasingly available, the study of Internet addiction within our societies becomes a relevant focus of study [6]. Various studies have found the impact of Internet addiction to go beyond reduced productivity, but also have a profound implications on the psychological well-being of individuals and the stability of social units [6]. Although the amount of time individuals spend on the Internet has been found to have a strong positive correlation with Internet addiction in the past, it is advised that it not be used as the only indicator of determining problematic use of the Internet [6]. The study of Internet use and how Internet use becomes Internet addiction remains vital in order for healthier ways of using the Internet to be discovered, as the negative consequences of Internet use could lead to significant impairment as stated earlier in this paper.

One overly researched area pertaining to the use of the Internet is the effects of social media on relationships [8]. A lot of studies exploring the impact of social media on relationships found that social media brings people closer and enables convenient communication. In contrast, other studies found that social media impedes on relationships as physical contact is compromised. Findings also reveal that the use of Internet is linked to conflict amongst partners [2]. If people spend too much time on the Internet, it may create misunderstandings and people may feel neglected, thereby creating tension. Internet addiction was also found to correlate with a low self-esteem and a lack of social inhibition. As was mentioned earlier, time spent online, whilst being a contributing factor to Internet addiction, is not the only factor that studies have implicated. In addition, most findings argue that it causes isolation, which then leads to other undesirable effects such as loneliness and depression [7]. Anxiety was also found to be a negative effect of using social media

[2]. Posting pictures and updating one's status have been found to raise anxious feelings as one is constantly worrying about how this content will be received by other social media users. The opposite is also true, reading other people's status and viewing their photos also raises anxiety as the individual is constantly comparing the perceived happiness of their peers to that of their own [8].

According to research conducted with students, at the University of Venda in South Africa, 72% of the students admitted to staying away from social gatherings at school in preference of staying at home and using the Internet instead [2]. They further found that 63% of the students agreed that using Internet does lead to isolation from friends, and can actually also lead to limiting physical contact with other people. This was mediated by the use of computers and people usually spent more time on their computers than with other people [2]. Students would find the Internet interesting and focus on the various products that the computer may offer, which other people may not be able to yield. This was found to have a negative impact on social integration [7].

There could be a number of negative outcomes for social isolation and poor social integration. Individual's psychological health may be severely negatively impacted which could result in depression as discussed earlier. In support of this claim, students who spend more time on Facebook reported to be lonelier than those who didn't [8]. For university students specifically, Internet addiction has also been found to be positively associated with their propensity to engage in class digital distractions such as using digital devices in class to perform task in class which are totally unrelated to the course [6]. This could have a negative impact on their grades if they are distracted during the lectures. Generally, when someone is using the Internet they get seduced into the exciting picture of social media, and they may neglect their own needs. Internet is said to become a problem when it starts to lead to diminished impulse control, loneliness, depression, distraction and when individuals begin to use it as a tool for social comfort. Internet addiction is found to be most prevalent amongst young people globally [6], thus making the study of Internet addiction amongst university students a pertinent issue in need of research.

2 Methodology

As this is the first study conducted at the University on Internet addiction and depression, it was decided to conduct it as a pilot study to be used as a basis for and to guide future in-depth studies. A quasi-experimental quantitative format is used to allow for the statistical manipulation of data collected. Convenience sampling was used from a postgraduate writing class consisting of students that are representative of various fields and disciplines, as well as diverse ages. The purpose of the study was briefly explained, including that participation is voluntary.

2.1 Data Collection

All data was collected using a self-reporting questionnaire containing questions regarding biographical data (age; gender), Internet usage (average hours spent online per day; amounts of social media sites used), as well as the Patient Health Questionnaire-9 and the Young's Internet Addiction Test.

PHQ-9 is a screening tool that is based on DSM-5 criteria for MDD and has been validated in adult and adolescent populations [5]. The scale includes nine questions regarding the frequency of depression symptoms experienced in the last two weeks, such as depressed mood and hopelessness. Response categories include: *not at all (1), several days (2), nearly half the days (3), and nearly every day (4).* PHQ-9 scores range from 9 to 36 with a score of less than 14 suggests no depression, whilst a score of 15–36 or greater suggests mild to severe depression. However, as Major Depression is only prevalent in about 7% of the population, [1] (APA, 2013), we will be regarding depression on a continuum on the scale in correlation to the continuum of scores on the YIAT.

YIAT is the most widely used Internet Addiction tool across the globe with a Chronbach's Alpha of higher than 0.8 in more than 26 countries. This suggests a high international validity for usage with a diverse population [3]. It is also scored on a scale and response categories include: *Rarely (1), occasionally (2), frequently (3), Often (4), Always (5), does not apply (0).* The higher your score, the greater level of addiction is. A score between 20 and 49 points is indicative of an average online user that may surf the Web a bit too long at times, but has control over their usage. A score between 50 and 79 points could indicate occasional or frequent problems because of the Internet usage, whilst scores ranging between 80 and 100 points are indicative of problematic Internet usage.

3 Results and Analysis

36 completed questionnaires were collected, which is considered to be sufficient for a pilot study. The sex breakdown indicates the 44.4% of participants were female, whilst 55.6% were male. The mean age of participants was 27.1 years old, although the majority of participants (mode) were 23. The youngest participants were 23, whilst the oldest were 44. Of the 36 participants, two neglected to indicate their ages and were excluded from the calculations (Table 1).

Table 1 Age breakdowm (yrs)

Values	Age (yrs)
N	34
Missing	2
Mean	27.8
Mode	23
Range	23–44

Table 2 Data summary

Values	Sex	Mean	Range
Number of social used	Female	4.8	1–11
	Male	5.8	
	Average	5.3	
Hours spend online per day	Female	4.5	1–24
	Male	6.4	
	Average	5.5	
Young internet addiction test	Female	33.9	0–100
	Male	36.6	
	Average	35.3	
Patient health questionnaire 9	Female	16	9–36
	Male	14.7	
	Average	15.4	

As seen in Table 2, on average, students at UNAM use 5.2 social media applications, of which Facebook was by far the most prevalent with 35 out of 36 participants owning a Facebook account. Male students also use on average more social media sites (5.2) than female students (4.8).

The same pattern can be seen with male students spending more time on alone with an average of 6.4 h per day compared to 4.5 h for female students. Male students also had the highest average YIAT score of 36.6, which is indicative of average online use that is not generally problematic. Female students had an average score of 33.9 on the YIAT. Overall, only one student had a high enough score (80) to indicate problematic Internet usage. Female students recorded a higher PHQ-9 depression score of 16, which indicates mild depression. Male students scored 14.7 on average, which is also mildly depressed, albeit less so than their female counterparts. Pearson's correlation coefficients were used to calculate correlations between scores in the IAT and the PHQ-9, as well as scores on the IAT and time spent online, which are summaries in Table 3.

The correlation coefficient was the YIAT and the PHQ-9 was +0.185 considering a 2-tailed significance level. There is therefore a weak positive relationship between scores on the YIAT and the PHQ-9. However, due to the small sample size

Table 3 Correlations with the YIAT (PHQ-9 and Hrs online)

		YIAT	PHQ-9			Hrs Online/Day	YIAT
Young internet addiction test	Pearson correlation	1	0.226	Hrs online/day	Pearson correlation	1	0.182
	Sig. (2-tailed)		0.185		Sig. (2-tailed)		0.287
	N	36	36		N	36	36

(36 participants) to scores cannot be considered to be significant. There was also a weak positive relationship (+0.287) between scores on the YIAT and the hours spent online. Again, a larger sample size would be required for the relationship to be considered significant at a $p > 0.05$ level.

4 Discussion

The results indicated that male UNAM students, overall, use the Internet for longer and also use the most social media applications. In contrast, female students used the Internet for shorter periods and for fewer applications. As Internet addiction has been shown to relate the hours spent online, the YIAT test for male students was also higher for male students than female students, although both have average Internet usage rates that are not considered to be problematic.

Whilst both female and male students rated as mildly depressed, the female students rated higher on the PHQ-9 scale. This, however, is in correspondence with the DSM-5, which indicates that women are, overall, more prone to depression than males [1].

With regard to Pearson's r, weak positive correlations were found between the YIAT and the PHQ-9, as well as between the YIAT and hours spent online. However, for the relationship to be considered significant at the $p > 0.05$ level, a larger sample size would be required.

5 Conclusion and Recommendations

As global trends have indicated that Internet use is on the increase, research is required to understand what the impact on increased Internet use is on the overall mood and well-being of people and what the factors are that increase vulnerability to problematic Internet use or Internet addiction. As university students tend to especially spend a lot of time online, collecting data on university student Internet usage trends in the local Namibian context is critical if we are to understand and promote responsible Internet usage behaviours. As this exploratory quasi-experimental study has suggested, there does appear to be a correlation between Internet addiction scores, depression scores and time spent online. However, it is strongly suggested that this study be repeated with a larger, representative sample to conclusively claim a significant impact. It is also suggested that the covariance between depression and hours spent online is considered and how this covariance impacts on Internet addiction scores.

References

1. American Psychiatric Association: Diagnostic and Statistical Manual of Mental Disorders (5th ed). Arlington: American Psychiatric Publishing (2013).
2. Farhangpour, P., & Matendawafa, A.W.: The impact of social network sites on the social health of University of Venda students, South Africa. African Journal for Physical, Health Education, Recreation and Dance (AJPHERD), 412–425 (2014).
3. Laconi, S., Rodgers, R.F. & Chabrol, H.: The measurement of Internet addiction: A critical review of existing scales and their psychometric properties. Computers in Human Behavior, 41, 190–202 (2014).
4. Marcus, M., Yasamy, M.T., Ommerman, M., Chisholm, D., and Saxena, S.: Depression: A Global Public Health Concern. World Health Organisation (WHO): Department of Mental Health and Substance Abuse (2012).
5. Moreno, M.A., Christakis, D.A., Egan, K.G., Jelenchick, L.A., Cox, E., Young, H., Villiard, H. & Becke, T.: A Pilot Evaluation of Associations Between Displayed Depression References on Facebook and Self-reported Depression Using a Clinical Scale. Journal of Behavioral Health Services & Research, 39(3), 295–304 (2011).
6. Nath, R., Chen, L., Muyingi, H.N., & Lubega, J.T.: Internet Addiction in Africa: A Study of Namibian and Ugandan College Students. International Journal of Computing and ICT Research, 7(2), 9–22 (2013).
7. Weiser, E.B.: The Functions of Internet Use and Their Social and Psychological Consequences. Cyberpsychology and Behaviour 4(6), 723–743 (2001).
8. Young, C., & L, Strelitz. Exploring patterns of Facebook usage, social capital, loneliness, and well-being among a diverse South African student sample. Communicare, 33(1), 57–72 (2014).

Author Index

D.K. Mishra et al. (eds.), *Information and Communication Technology for Sustainable Development*, Lecture Notes in Networks and Systems 10,
https://doi,org/10.1007/978-981-10-3920-1

Printed in the United States
By Bookmasters